CORONAL AND STELLAR MASS EJECTIONS

IAU SYMPOSIUM No. 226

COVER ILLUSTRATION:

Inner image: SOHO/EIT 195Å image on October 28, 2003 at 1112 UT showing the flare accompanying the CME that erupted from near disk center. Outer image: Earth-directed 'halo' CME as seen by the SOHO/LASCO C2 coronagraph at 11:30 UT. The CME speed was 2500 $\mathrm{km\,s^{-1}}$ and its mass 1.7×10^{13} kg. The image was been filtered to enhance the fine structure of the ejecta.

The SOHO/EIT data are provided by the EIT consortium. The SOHO/LASCO data used here are produced by a consortium of the Naval Research Laboratory (USA), Max-Planck-Institut fuer Aeronomie (Germany)), Laboratoire d'Astronomie (France), and the University of Birmingham (UK). SOHO is a project of international cooperation between ESA and NASA.

INTERNATIONAL ASTRONOMICAL UNION

UNION ASTRONOMIQUE INTERNATIONALE

CORONAL AND STELLAR MASS EJECTIONS

PROCEEDINGS OF THE 226th SYMPOSIUM OF THE
INTERNATIONAL ASTRONOMICAL UNION
HELD IN BEIJING, CHINA
SEPTEMBER 13–17, 2004

Edited by

KENNETH DERE
George Mason University, Fairfax, VA, USA

JINGXIU WANG
National Astronomical Observatories, Chinese Academy of Sciences, Beijing, China

and

YIHUA YAN
National Astronomical Observatories, Chinese Academy of Sciences, Beijing, China

CAMBRIDGE
UNIVERSITY PRESS

CAMBRIDGE UNIVERSITY PRESS
The Edinburgh Building, Cambridge CB2 2RU, UnitedKingdom
40 West 20th Street, New York, NY 10011–4211, USA
477 Williamstown Road, Port Melbourne, VIC 3207, Australia
Ruiz de Alarcón 13, 28014 Madrid, Spain
Dock House, The Waterfront, Cape Town 8001, South Africa

First published 2005

Printed in the United Kingdom at the University Press, Cambridge

Typeset in System LaTeX 2_ε

A catalogue record for this book is available from the British Library

Library of Congress Cataloguing in Publication data

ISBN 0 521 85197 1 hardback
ISSN 1743-9213

Table of Contents

Topic 1. HISTORICAL INTRODUCTION

Topic 2. OBSERVATIONS OF CMES

Topic 3. CME SOURCE REGIONS

Topic 4. THEORETICAL MODELS OF CMEs

Topic 5. COMPARISONS OF CME MODELS AND OBSERVATIONS

Topic 6. CMES AND ENERGETIC PARTICLES

Topic 7. ICMEs IN THE HELIOSPHERE

Topic 8. CMES AND GEOMAGNETIC STORMS

Topic 9. STELLAR EJECTIONS

Preface

As global-scale ejections of magnetic flux and plasma from a normal star, our Sun, coronal mass ejections (CMEs) present a major challenge in astrophysics. The questions raised since their early detection 30 years ago on their physical nature and mechanism, such as the association of CMEs and other forms of stellar activity, CMEs' parent magnetic structures, the physics of CME initiation, CMEs as an element of the activity cycle, footprints of CMEs in the planetary system, CME's astrophysical analog and so on, continue as unsolved problems today. However, it has become more and more evident that Earth-direct CMEs are a primary agent in delivering the Sun's eruptive energy release to near-Earth space and the major source of disruptive space weather in the solar-terrestrial environment. Therefore, the study of coronal mass ejections has become of keen interest and focuses an interdisciplinary effort involving the fields of astronomy, space science and geophysics.

The success of the Solar and Heliospheric Observatory (SOHO) has greatly boosted CME studies. The 8 years of SOHO operations stand as a new era in CME astrophysics. For instance, the solar disk signatures of CMEs have been discovered as EUV and X-ray dimmings and have been systematically investigated. For many CMEs we have been able, for the first time, to identify their source region and activity on the solar disk and to scrutinize the magnetic evolution leading to the CME initiation. It has been timely to organize IAU symposium 226 in order to summarize and discuss the latest research on coronal and stellar coronal mass ejections. It is the only IAU meeting devoted solely to coronal and stellar mass ejections.

One of the goals of the meeting was to encourage the participation of as broad a segment of the international community as possible. The SOC of the symposium set a rule that colleagues *share the responsibility and honor.* This has enabled colleagues aside from the SOC members to have the opportunity to present invited reviews and a broader geographic distribution of SOC members and invited speakers. The spirit of the rule is a belief of Confucius that *when three walk together there must be one who can teach me.* In the same sprit, discussions on talks were encouraged and facilitated by careful time arrangement of the scientific program. There are 35 invited and roughly 70 contributed papers from colleagues from all continents in the proceeding. The proceedings of the symposium may serve as future guidance and new starting point for broad fields of studies.

In addition to the IAU, the symposium has been supported by the Chinese Academy of Sciences, the Ministry of Science and Technology of China, the National Natural Science Foundation of China, the National Astronomical Observatories of the Chinese Academy of Sciences, the National Science Foundation of the United States of American, the NASA Solar and Heliospheric Observatory program, the SOHO/LASCO program, and George Mason University. The SOC and LOC highly appreciate their kind support and thank all of the participants for their enthusiastic involvement in the scientific program.

Kenneth Dere and Jingxiu Wang, co-chairs SOC,
Yihua Yan, chair LOC
Beijing and Fairfax, December 9, 2004

THE ORGANIZING COMMITTEE

Scientific

K. Dere (co-chair, USA)
S. Antiochos (USA)
J. L. Bourgeret (France)
H. Cane (Australia)
C. Fang (China)
R. Harrison (UK)
H. Hudson (USA)
D. Reames (USA)
K. Shibata (Japan)
B. Tsurutani (USA)

J. Wang (co-chair, China)
V. Bothmer (Germany)
A. Cameron (UK)
I. Chertok (Russia)
T. Forbes (USA)
R. Howard (USA)
S. Kahler (USA)
R. Schwenn (Germany)
S. Solanki (Germany)
P. Venkatakrishnan (India)

Local

Y. H. Yan (chair)
J. X. Hao
A. Vourlidas

G. X. Dong
S. J Xue

Acknowledgements

The symposium is sponsored and supported by the IAU Division II (Sun and Heliosphere); and by the IAU Commissions No. 10 (Solar Activity), No. 12 (Solar Radiation and Structure), No. 49 (Interplanetary Plasma and Heliosphere).

The Local Organizing Committee operated under the auspices of the National Astronomical Observatories, Chinese Academy of Sciences.

Funding by the
International Astronomical Union,
Chinese Academy of Sciences,
The Ministry of Science and Technology (China),
National Natural Science Foundation of China,
National Astronomical Observatories (NAOC),
National Science Foundation (USA),
NASA SOHO project,
and
SOHO/LASCO project
is gratefully acknowledged.

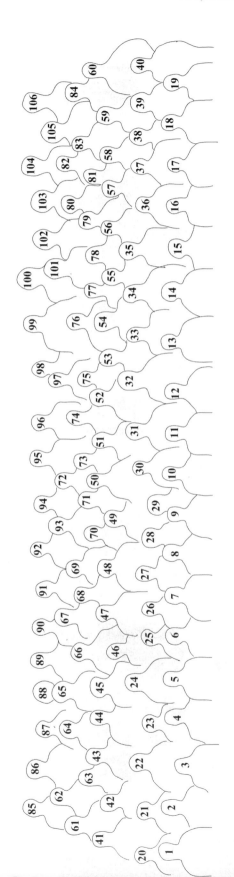

1. Raid Suleiman 2. Xuepu Zhao 3. Shahinaz Yousef 4. V. Grechnev
5. Vojtech Rusin 6. Kazunari Shibata 7. N. Gopalswamy 8. Jingxiu Wang
9. Cheng Fang 10. Kenneth Dere 11. Volker Bothmer 12. Rainer Schwenn
13. Stephen Kahler 14. Yihua Yan 15. Brigitte Schmieder 16. Terry G. Forbes
17. Shi Tsan Wu 18. Karin Muglach 19. Shujuan Wang 20. Yuying Liu
21. Zhuheng Li 22. Ruiguang Wang 23. Valeria Borovik 24. Marina Skender
25. Anqin Chen 26. Jianxia Cheng 27. Xiaoyan Xu 28. Hong Xie
29. Chinchun Wu 30. S. C. Kaushik 31. P. Raychaudhuri 32. G. S. Lakhina
33. Trevor Sanderson 34. Horst Kunow 35. T. Yokoyama 36. Meisheng Gao (Mrs Wu)
37. Alessandro Bemporad 38. Yanxin Zhu 39. Wenjuan Zhang 40. Hongxia Nie
41. Yiwei Li 42. Katya Georgieva 43. Andrei Zhukov 44. Vladimir Slemzin
45. Serge Koutchmy 46. Olga Sheiner 47. Boris Filippov 48. Eugene Romashets
49. Shinichi Watari 50. Marek Vandas 51. Qijun Fu 52. Jie Zhang
53. Pengfei Chen 54. Olga Malandraki 55. Kuniko Hori 56. Mrs Kunow
57. Chuan Li 58. Shuzhen Wang 59. Wei Wang 60. Pete Riley
61. Hongqi Zhang 62. Jun Lin 63. Kathy Reeves 64. Moira Jardine-Cameron
65. Catherine Dougados 66. David Ruffolo 67. Yuan Ma 68. Dalmiro Maia
69. Suli Ma 70. Emilia Huttunen 71. Yang Su 72. Baolin Tan
73. Huadong Chen 74. Robert Leamon 75. Alexander Nindos 76. Angelos Vourlidas
77. Hebe Cremades 78. Guadalupe Munoz 79. J-P. Delaboudiniere 80. Lidong Xia
81. Hui Li 82. Haisheng Ji 83. Sergio Dasso 84. Zhijun Chen
85. Prasad Subramanian 86. Rajaram Ramesh 87. Jianpeng Guo 88. Chengming Tan
89. Jianheng Guo 90. Shenghong Gu 91. Guiqing Zhang 92. Yingzhi Zhang
93. Chengwen Shao 94. Guiming Le 95. Zongjun Ning 96. Min Wang
97. Alysha Reinard 98. Youqiu Hu 99. Allan J. Tylka 100. Alphonse Sterling
101. Luciano Rodriguez 102. Shuhao Li 103. Fei Liu 104. Guangli Huang
105. Keping Qiu 106. Weiguo Zong

Participants

Guoxiang **Ai**, NAOC,Beijing , China — aigx@sun10.bao.ac.cn
Shudong **Bao**, NAOC, Beijing , China
Xingming **Bao**, NAOC, Beijing , China — baoxm@sun10.bao.ac.cn
Alessandro **Bemporad**, Dipart. Astron. Sci. Spa., Univ. Firenze, Italy — bemporad@arcetri.astro.it
Valeria **Borovik**, Main Astronomical (Pulkovo) Observatory, Sankt-Petersburg, Russia — borovik@saoran.spb.su
Volker **Bothmer**, MPI fur Sonnensystemforschung, Katlenburg Lindau, Germany — bothmer@linmpi.mpg.de
Anqin **Chen**, Department of Astronomy, Nanjing University, China — anqier911@sohu.com
Huadong **Chen**, NAOC/Yunnan Observatory, Kunming, China — 978210–czc@sohu.com
Jie **Chen**, NAOC, Beijing , China — chj@sun10.bao.ac.cn
Pengfei **Chen**, Department of Astronomy, Nanjing University, Nanjing, China — chenpf@nju.edu.cn
Zhijun **Chen**, NAOC, Beijing , China — zjchen@bao.ac.cn
Jianxia **Cheng**, Department of Astronomy, Nanjing University, Nanjing, China — chengjianxia@nju.org.cn
Hebe **Cremades**, MPI fur Sonnensystemforschung, Katlenburg-Lindau Germany — cremades@linmpi.mpg.de
Yanmei **Cui**, NAOC, Beijing , China
Yu **Dai**, Department of Astronomy, Nanjing University, Nanjing, China — njudai@sohu.com
Sergio **Dasso**, Instituto de Astronomia y Fisica del Espacio(IAFE), Argentine — dasso@df.uba.ar
J-P. **Delaboudiniere**, Institut d'Astrophysique Spatiale- Orsay, France — boudine@ias.fr
Yuanyong **Deng**, NAOC, Beijing , China
Kenneth **Dere**, George Mason University, VA, USA — kdere@gmu.edu
Mingde **Ding**, Department of Astronomy, Nanjing University, Nanjing, China — dmd@nju.edu.cn
Guoxuan **Dong**, National Natural Science Foundation of China, Beijing, China — donggx@rose.nsfc.gov.cn
Yunjiang **Dou**, NAOC, Beijing, China
Catherine **Dougados**, Laboratoire de Astrophysique Observatoire de Grenoble, France — dougados@obs.ujf-grenoble.fr
Cheng **Fang**, Department of Astronomy, Nanjing University, Nanjing, China — fangc@nju.edu.cn
Xueshang **Feng**, Center for Space Science and Applied Research, Beijing, China — fengx@spaceweather.ac.cn
Boris **Filippov**, IZMIRAN, Troitsk, Moscow Region, Russia — bfilip@izmiran.ru
Terry G. **Forbes**, University of New Hampshire, NH, USA — terry.forbes@unh.edu
Qijun **Fu**, NAOC, Beijing , China — fuqj@bao.ac.cn
Weiqun **Gan**, Purple Mountain Observatory, Nanjing, China — wqgan@pmo.ac.cn
Katya **Georgieva**, Sol.-Ter. Influ. Lab, Bulgarian Academy of Science, Sofia, Bulgaria — kgeorg@bas.bg
N. **Gopalswamy**, NASA/GSFC Greenbelt, MD, USA — gopals@fugee.gsfc.nasa.gov
V. **Grechnev**, ISTP, SD RAS, Irkutsk, Russia — grechnev@iszf.irk.ru
Shenghong **Gu**, NAOC/Yunnan Observatory, Kunming, China — shenggu@public.km.yn.cn
Jianheng **Guo**, NAOC/Yunnan Observatory, Kunming, China — guojh@ynao.ac.cn
Jianpeng **Guo**, NAOC, Beijing , China
Juan **Guo**, NAOC, Beijing , China — gj@sun10.bao.ac.cn
Jinxin **Hao**, Beijing, China — hjx@cashq.ac.cn
Han **He**, NAOC, Beijing , China
Kuniko **Hori**, National Institute of Information & Communications Technology, Ibaraki, Japan — hori@nict.go.jp
Youqiu **Hu**, University of Sciences and Technology of China, Hefei, China — huyq@ustc.edu.cn
Guangli **Huang**, Purple Mountain Observatory, Nanjing , China — glhuang@pmo.ac.cn
Emilia **Huttunen**, Department of physical sciences, University of Helsinki, Finland — emilia.huttunen@helsinki.fi
Moira **Jardine-Cameron**, School of Physics & Astronomy, University of St Andrews, Scotland — mmj@st-and.ac.uk
Haisheng **Ji**, Purple Mountain Observatory/(Big Bear Solar Observatory), Nanjing , China — jihs@bbso.njit.edu
Jie **Jiang**, NAOC, Beijing , China
Stephen **Kahler**, The Air Force Research Laboratory, MA, USA — stephen.kahler@hanscom.af.mil
Subhash C. **Kaushik**, Dept. physics, Government Autonomous PG College,India — subash_kaushik@rediffmail.com
Serge **Koutchmy**, Institut d'astrophysique de Paris- CNRS, Paris, France — koutchmy@iap.fr
Horst **Kunow**, IEAP, Universitaet Kiel, Germany — kunow@physik.uni-kiel.de
Kanya **Kusano**, Grad. Sch. Advanced Sciences of Matter, Hiroshima Univ., Japan — kusano@hiroshima-u.ac.jp
Kirill **Kuzanyan**, IZMIRAN, Troitsk, Moscow Region, Russia
G. Singh **Lakhina**, Indian Institute of Geomagnetism, New Panvel(W) Navi Mumbai, India — lakhina@iig.iigm.res.in
Guiming **Le**, Center for Space Science and Applied Research, Beijing , China — kjzhxsge@263.net
Robert **Leamon**, NASA Goddard Space Flight Center, MD, USA — leamon@grace.nascom.nasa.gov
Chuan **Li**, Department of Astronomy , Nanjing University, Nanjing , China
Gang **Li**, IGPP, Univ. of California at Riversider, CA, USA — gang.li@ucr.edu
Hui **Li**, Purple Mountain Observatory, Nanjing, China — lihui@mail.pmo.ac.cn
Li **Li**, NAOC, Beijing , China
Shuhao **Li**, NAOC, Beijing , China
Yiwei **Li**, NAOC, Beijing , China
Zhuheng **Li**, NAOC, Beijing , China
Jun **Lin**, Harvard-Smithsonian Center for Astrophysics, MA, USA — jlin@cfa.harvard.edu
Fei **Liu**, NAOC, Beijing, China
Jihong **Liu**, NAOC, Beijing, China — ljh@sun10.bao.ac.cn
Yuying **Liu**, NAOC, , Beijing, China — liuyy@bao.ac.cn
Zhenxing **Liu**, Center for Space Science and Applied Research, Beijing, China — liu@center.cssar.ac.cn
Zhong **Liu**, NAOC/Yunnan Observatory, Kunming, China
Suli **Ma**, NAOC/Yunnan Observatory, Kunming, China — masuli@163.com
Yuan **Ma**, NAOC/Yunnan Observatory, Kunming, China — mayuanf@public.km.yn.cn
Zhiwei **Ma**, Institute of Plasma Physics Hefei, China — weima511@yahoo.com
Dalmiro **Maia**, Faculdade de Ciencias da Universidade do Porto, Nova de Gaia, Portugal — dmaia@fc.up.pt
Olga **Malandraki**, Democritus University of Thrace, Space Research Lab., Xanthi, Greece — omaland@xan.duth.gr
Karin **Muglach**, NRL, Washington DC, USA — muglach@nrl.navy.mil
Guadalupe **Munoz**, MPI fur Sonnensystemforschung, Katlenburg-Lindau, Germany — munoz@linmpi.mpg.de
Hongxia **Nie**, NAOC, Beijing, China — nhx@bao.ac.cn
Alexander **Nindos**, Physics Department, University of Ioannina, Greece — anindos@cc.uoi.gr
Zongjun **Ning**, Department of Astronomy, Nanjing University, Nanjing, China — ningzongjun@sohu.com
Keping **Qiu**, Department of Astronomy, Nanjing University, Nanjing, China — kpqiu@sohu.com
Rajaram **Ramesh**, Indian Institute of Astrophysics, Koramangala, Bangalore, INDIA — ramesh@iiap.res.in
Probhas **Raychaudhuri**, Dept. Applied Mathematics, Calcutta University, INDIA — probhasprc@rediffmail.com
Kathy **Reeves**, University of New Hampshire, NH, USA — kreeves@unh.edu
Alysha **Reinard**, NRL/Artep, Washington DC, USA — reinard@nrl.navy.mil
Pete **Riley**, Science Applications International Corporation, San Diego, CA, USA — pete@peteriley.org
Luciano **Rodriguez**, MPI fur Sonnensystemforschung, Katlenburg-Lindau, Germany — rodriguez@linmpi.mpg.de
Eugene **Romashets**, IZMIRAN, Troitsk, Moscow Region, Russia — romash@izmiran.rssi.ru
Guiping **Ruan**, NAOC, Beijing, China — rgp@sun10.bao.ac.cn
David **Ruffolo**, Dept. of Physics, Faculty of Sciences, Mahidol Univ., Bangkok, Thailand — david_ruffolo@yahoo.com
Vojtech **Rusin**, Astronomical Institute, Slovak Academy of Sciences, Tatranska Lomnica, Slovakia — vrusin@ta3.sk
Trevor **Sanderson**, ESTEC, Noerdwijk, The Netherlands — trevor.sanderson@esa.int
Brigitte **Schmieder**, Observatoire de Paris, LESIA Meudon, France — brigitte.schmieder@obspm.fr
Rainer **Schwenn**, MPI fr Sonnensystemforschung, Katlenburg-Lindau, Germany — schwenn@linmpi.mpg.de
Chengwen **Shao**, NAOC/Yunnan Observatory, Kunming , China — cwshaoanhui@etang.com

Olga **Sheiner**, Radiophysical Research Institute, Nizhny, Novgorod, Russia   rfj@nirfi.sci-nnov.ru
Kazunari **Shibata**, Kwasan and Hida Observatories, Kyoto, Japan   shibata@kwasan.kyoto-u.ac.jp
Marina **Skender**, Rudjer Boskovic Institute Theorethical Physics Division Bijenicka, Croatia   marina@rudjer.irb.hr
Vladimir **Slemzin**, P.N. Lebedev Physical Institute, Moscow, Russia   slem@sci.lebedev.ru
Sami **Solanki**, MPI für Sonnensystemforschung, Katlenburg-Lindau Germany   solanki-office@linmpi.mpg.de
Wenbin **Song**, NAOC, Beijing, China   wenbin@ourstar.bao.ac.cn
Alphonse **Sterling**, NASA/MSFC/NSSTC/Space Science Department, Huntsville, AL,USA   alphonse.sterling@nasa.gov
Jiangtao **Su**, NAOC, Beijing , China   sjt@sun10.bao.ac.cn
Yang **Su**, Purple Mountain Observatory, Nanjing, China   say82@163.com
Prasad **Subramanian**, IUCAA, Pune, India   psubrama@iucaa.ernet.in
Raid **Suleiman**, BU/CfA, MA, USA   rsuleiman@cfa.harvard.edu
Yingzi **Sun**, NAOC, Beijing , China
Baolin **Tan**, Purple Mountain Observatory, Nanjing, China   baolintan@yahoo.com.cn
Chengming **Tan**, NAOC, Beijing , China   tanchm@bao.ac.cn
Chuanyi **Tu**, Peking University, Beijing, China   cytu@public3.bta.net.cn
Allan J. **Tylka**, Naval Research Laboratory, Washington DC, USA   allan.tylka@nrl.navy.mil
Marek **Vandas**, Astronomical Institute, Academy of Sciences, Praha, Czech Republic   vandas@ig.cas.cz
Angelos **Vourlidas**, George Mason University, VA, USA   vourlidas@nrl.navy.mil
Huaning **Wang**, NAOC, Beijing , China   hnwang@bao.ac.cn
Min **Wang**, NAOC/Yunnan Observatory, Kunming, China   wmynao@163.net
Ruiguang **Wang**, NAOC, Beijing , China   wangrg@ourstar.bao.ac.cn
Shujuan **Wang**, NAOC, Beijing , China   wsj888@263.net
Shuzhen **Wang**, NAOC, Beijing , China
Wei **Wang**, NAOC, Beijing , China
Xuyu **Wang**, Institute of Physics University of Bern , Switzerland   xuyu.wang@soho.unibe.ch
Yi **Wang**, NAOC, Beijing , China
Yuming **Wang**, School of Earth & Space Sci., Univ. of Sci & Tech of China, Heifei, China   ymwang@ustc.edu.cn
Jingxiu **Wang**, NAOC, Beijing , China   wjx@ourstar.bao.ac.cn
Shinichi **Watari**, National Institute of Information and Communications Technology, Tokyo,Japan   watari@nict.go.jp
Yayuan **Wen**, NAOC, Beijing , China   yayuanwen@ourstar.bao.ac.cn
Chinchun **Wu**, CSPAR/The University of Alabama in Huntsville, AL, USA   wuc@cspar.uah.edu
Shi Tsan **Wu**, CPSAR & Dept. Mech & Aerospcae Engineering, University Huntsville, AL, USA   wus@cspar.uah.edu
Lidong **Xia**, School of Earth & Space Sci., Univ. of Sci & Tech of China, Heifei , China   xld@ustc.edu.cn
Chijie **Xiao**, NAOC, Beijing , China   cjxiao@ourstar.bao.ac.cn
Hong **Xie**, NASA/GSFC, MD, USA   hong@aurora690.gsfc.nasa.gov
Lun **Xie**, Peking University, Beijing, China   xielun@pku.edu.cn
Haiqing **Xu**, NAOC, Beijing , China
Xiaoyan **Xu**, Department of Astronomy, Nanjing University, Nanjing, China   xyxu1122@126.com
Suijian **Xue**, NAOC, Beijing , China   xue@bao.ac.cn
Yihua **Yan**, NAOC, Beijing , China   yyh@bao.ac.cn
Xiulan **Yang**, NAOC, Beijing , China   yxl@bao.ac.cn
Yongtian **Yang**, Beijing, China
Zhiliang **Yang**, Normal University of Beijing, Beijing , China   zlyang@bnu.edu.cn
Chin-Teh **Yeh**, Department of Astronomy, Nanjing University, Nanjing, China   jdye@nju.edu.cn
T. **Yokoyama**, Dept.Earth & Planetary Sci., Univ. of Tokyo, Japan   yokoyama.t@eps.s.u-tokyo.ac.jp
Shahinaz **Yousef**, Astronomy & Meteorology Dept. Cairo University, Egypt   shahinazyousef@yahoo.com
Guiqing **Zhang**, NAOC, Beijing , China   zgq@bao.ac.cn
Hongqi **Zhang**, NAOC, Beijing , China   zhq@sun10.bao.ac.cn
Jie **Zhang**, George Mason University, VA, USA   jiez@scs.gmu.edu
Jun **Zhang**, NAOC, Beijing , China   zjun@ourstar.bao.ac.cn
Wenjuan **Zhang**, NAOC, Beijing , China
Yin **Zhang**, NAOC, Beijing , China   zhy@sun10.bao.ac.cn
Yingzhi **Zhang**, NAOC, Beijing , China   zhangyingzhi@ourstar.bao.ac.cn
Yuzong **Zhang**, NAOC, Beijing , China   yuzong@ourstar.bao.ac.cn
Hui **Zhao**, NAOC, Beijing , China
Liang **Zhao**, Peking University, Beijing, China
Meng **Zhao**, NAOC, Beijing , China
Xuepu **Zhao**, Stanford University, HEPL Annex, Stanford, California, USA   xpzhao@solar.stanford.edu
Guiping **Zhou**, NAOC, Beijing , China   zhougp@ourstar.bao.ac.cn
Aiping **Zhu**, NAOC, Beijing , China   apzhu@bao.ac.cn
Yanxin **Zhu**, NAOC, Beijing , China
Andrei **Zhukov**, Royal Observatory of Belgium, Brussels, Belgium   andrei.zhukov@oma.be
Weiguo **Zong**, Department of Astronomy, Nanjing University, China   zongweiguo_2003@hotmail.com

Session 1

Historical introduction

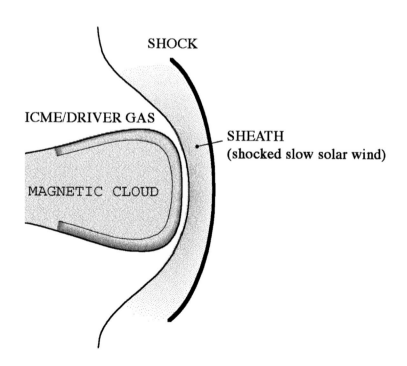

Coronal and Stellar Mass Ejections
Proceedings IAU Symposium No. 226, 2005
K. P. Dere, J. Wang & Y. Yan, eds.

Research on Historical Records of Geomagnetic Storms

G. S. Lakhina[1], S. Alex[1], B. T. Tsurutani[2], and W. D. Gonzalez[3]

[1]Indian Institute of Geomagnetism, Mumbai, India
email: lakhina@iigs.iigm.res.in,salex@iigs.iigm.res.in

[2]Jet Propulsion Laboratory, California Institute of Technology, Pasadena, CA, USA
email:Bruce.T.Tsurutani@jpl.nasa.gov

[3]Instituto Nacional Pesquisas Espaciais (INPE), Sao Jose dos Campos, Sao Paulo, Brazil
email:gonzalez@dge.inpe.br

Abstract.
  In recent times, there has been keen interest in understanding Sun-Earth connection events, such as solar flares, CMEs and concomitant magnetic storms. Magnetic storms are the most dramatic and perhaps important component of space weather effects on Earth. Super-intense magnetic storms (defined here as those with Dst < -500 nT, where Dst stands for the disturbance storm time index that measures the strength of the magnetic storm) although relatively rare, have the largest societal and technological relevance. Such storms can cause life-threatening power outages, satellite damage, communication failures and navigational problems. However, the data for such magnetic storms is rather scarce. For example, only one super-intense magnetic storm has been recorded (Dst=-640 nT, March 13, 1989) during the space-age (since 1958), although such storms may have occurred many times in the last 160 years or so when the regular observatory network came into existence. Thus, research on historical geomagnetic storms can help to create a good data base for intense and super-intense magnetic storms. From the application of knowledge of interplanetary and solar causes of storms gained from the spaceage observations applied to the super-intense storm of September 1-2, 1859, it has been possible to deduce that an exceptionally fast (and intense) magnetic cloud was the interplanetary cause of this geomagnetic storm with a Dst -1760 nT, nearly 3 times as large as that of March 13, 1989 super-intense storm. The talk will focus on super-intense storms of September 1-2, 1859, and also discuss the results in the context of some recent intense storms.

Keywords. Sun: solar-terrestrial relations

1. Introduction

  The history of geomagnetism is about 400 years old. The science of geomagnetism was born with the publication of De Magnete by William Gilbert in 1600 AD. The first map of magnetic field declination was made by Edmund Halley in the beginning of eighteenth century. We will go back about 200 years ago, specifically from May 1806 to June 1807 in Berlin, where Alexander von Humboldt and a colleague observed the local magnetic declination every half hour from midnight to morning. On December 21, 1806, for 6 consecutive hours, von Humboldt observed strong magnetic deflections and noted the presence of correlated northern lights (aurora) overhead. When the aurora disappeared at dawn, the magnetic perturbations disappeared as well. Von Humboldt concluded that the magnetic disturbances on the ground and the auroras in the polar sky were two manifestation of the same phenomenon (Schröder 1997; Tsurutani *et al.* 1997). He gave this phenomenon involving large scale magnetic disturbances (possibly already observed by George Graham) the name "Magnetische Ungewitter," or magnetic storms von Humboldt 1808. The world-wide network of magnetic observatories later confirmed that such "storms" were indeed world-wide phenomena.

An amateur German astronomer, S. Heinrich Schwabe, began observing the Sun and making counts of sunspots in 1826. In the year 1843, he reported a periodic behavior of 10 years in spot counts. A decennial period in the daily variation of magnetic declination was reported by Lamont from Munich in 1851, but he did not relate it to the sunspot cycle. From his extensive studies, Sabine (1852) discovered that geomagnetic activity paralleled the recently discovered sunspot cycle. However, it took nearly 100 years to gather sufficient statistics to make a convincing case for an association between large solar flares and severe storms (Hale 1931, Chapman & Bartels 1940, and Newton 1943).

2. Geomagnetic Storms

In recent times, there has been keen interest in understanding Sun-Earth connection events, such as solar flares, CMEs and concomitant magnetic storms. Magnetic storms are the most dramatic and perhaps important component of Space Weather effects on Earth.

A geomagnetic storm is characterized by a Main Phase during which the horizontal component of the Earth's low-latitude magnetic fields are significantly depressed over a time span of one to a few hours followed by its recovery which may extend over several days (Rostoker 1997). During intense magnetic storms, the auroral activity becomes intense and auroras are not confined to the Auroral Oval only, rather the Auroras could be seen at the sub-auroral to midlatitude stations. It is now believed that the major cause of solar wind energy transfer to the magnetosphere is magnetic reconnection between interplanetary magnetic fields and the Earth's magnetic field (Dungey 1961). Geomagnetic storms occur when solar wind-magnetosphere coupling becomes intensified during the arrival of fast moving ($\sim$700 km/s or more) solar ejecta, like CMEs, solar flares, fast streams from the coronal holes, etc. accompanied by long intervals of intense southward interplanetary magnetic field (IMF) (Gonzalez *et al.* 1994, Tsurutani & Gonzalez 1997) as in a "magnetic cloud" (Klein & Burlaga 1982). As a result, the magnetotail plasma gets injected into the nightside magnetosphere, with the energetic protons drifting to the west and electrons to the east, thus, forming a ring of current around the Earth. This current, called the "ring current", produces a diamagnetic decrease in the Earth's magnetic field measured at near-equatorial stations, and is the cause of the main phase of the magnetic storm. The decay of the ring current starts the recovery phase of the storm.

Super-intense magnetic storms (defined here as those with Dst < -500 nT) although relatively rare, have the largest societal and technological relevance. Such storms can cause life-threatening power outages, satellite damage, communication failures and navigational problems. The data for super-intense magnetic storms is rather scarce. For example, only one truly super-intense magnetic storm has been recorded (DST=-640 nT, March 13, 1989) during the spaceage since 1958 (Allan *et al.* 1989).

Last year, there was a great media-hype about the possible super magnetic storms in October-November, 2003. Though the solar flares on October 28 and 29 were of class X17 and X10, they failed to produce a super intense storm; they produced intense double storm of mere Dst -400 nT. A much weaker solar flare (and CME) of class M3.2/2N on 18 November resulted in a near super intense storm on November 20 with Dst -490 nT. This clearly shows that it is not only the energy of the solar flare and speed of the ejecta which control the strength of the geomagnetic storm, the solar magnetic field too play critical role!

Dessler & Parker (1959) and Sckopke (1966) have shown that the decrease in the equatorial magnetic field strength due to the ring current or Dst (disturbance storm time) index, is directly related to the total energy of the ring current particles, and thus is a good measure of the energetics of the magnetic storm. Though Dst index acts as a proxy for the strength of the ring current, other currents like magnetopause current can contribute to it as well. An empirical relationship between Dst and interplanetary parameters has been derived by Burton *et al.* (1975).

Although there is a record of only one or two super intense magnetic storms during the space age, many such storms may have occurred many times in the last 160 years or so when the regular observatory network came into existence. Thus, the research on historical geomagnetic storms can help to create a good data base for intense and super-intense magnetic storms. From the application of knowledge, of interplanetary and solar causes of storms gained from the spaceage observations, to this super-intense storm data set one can deduce their possible causes and construct a data base for solar ejecta, e.g., frequency of occurrence of extremely large solar flares, evolution of solar ejecta, etc.

An other important reason for undertaking such study is to answer some basic questions, namely, i) how many super-intense magnetic storms have occurred in the last 160 years and what were their probable solar and interplanetary causes? ii) the frequency of occurrence of super-intense storms and under what circumstances? iii) Is a prediction of a certain number of (say 3) most severe magnetic storm during a solar cycle possible? iv) Can the possible damaging effect of supper intense magnetic storms on the modern society be predicted in advance? and v) what is the energetics of eruptive phenomena on Sun and Stars, etc.

Table 1 gives a partial chronological list of some large magnetic storms which had occurred during the past 160 years or so. The list includes the "Remarkable Magnetic Storms" described in Moos (1910)and Chapman & Bartels (1940)(Tsurutani *et al.* 2003). One can see that some of the events fall under the category of super-intense magnetic storms. Analysis of these events can form a very useful data base for the super-intense storms.

3. Case History: Super-Intense Storm of September 1-2, 1859

We shall focus on the super storm of September 1-2, 1859 which was associated with the Carrigton flare that occurred on September 01, 1859. We use recently reduced ground magnetometer data of Colaba Observatory, Mumbai, India for the September 1-3, 1859, published papers (Carrington, 1859), auroral reports, based on newspapers (Kimball, 1960) and recently obtained (space-age) knowledge of interplanetary and solar causes of storms, to identify the probable causes of this super storm (Tsurutani *et al.* 2003). Similar methodology (with improved techniques) can be used to analyze other historical magnetic storms.

3.1. *Solar flare of September 1, 1859, magnetic storm and auroras*

The solar flare of September 1, 1859 was observed and reported by R. C. Carrington (Carrington, 1859) and Hodgson (1859) in the Monthly Notices of the Royal Astronomical Society and became the best known solar event of all times. Of particular note was the intensity of the event as quoted in the articles.

"For the brilliancy was fully equal to that of direct sunlight (Carrington, 1859)." "I was suddenly surprised at the appearance of a very brilliant star of light, much brighter than the sun's surface, most dazzling to the protected eye . . . . ." (Hodgson, 1859).

The solar flare was followed by a magnetic storm at the Earth. The time delay was 17 hrs and 40 min (stated in the Carrington paper). Although Carrington carefully noted this relationship, he was cautious in his appraisal: "and that towards four hours after midnight there commenced a great magnetic storm, which subsequent accounts established to have been as considerable in the southern as in the northern hemisphere". While the contemporary occurrence may deserve noting, he would not have it supposed that he even leans towards connecting them "one swallow does not make a summer" (Carrington 1859).

The auroras occurred globally and have been reported by many. Kimbal (1960) has provided the most complete indexing of auroral sightings. "Red glows were reported as visible from within 23° of the geomagnetic equator in both north and southern hemispheres during the display of September 1-2". This is perhaps the most equatorward sighting of aurora that can be confirmed for this or any other storm event in past history (Silverman, 2001). Loomis (1861) has reported that during this magnetic storm, many fires were set by arcing from currents induced in telegraph wires (in both the United States and Europe).

3.2. *Interplanetary Causes of major Geomagnetic storms*

There are several solar and interplanetary drivers which can give rise to magnetic storms. Solar ejecta (CMEs, solar flares etc) having high solar wind speeds and unusually intense magnetic fields seem to be the most important for causing intense geomagnetic storms (Gonzalez *et al.* 1994). Magnetic clouds within fast interplanetary coronal mass ejections (ICMEs) (Klein & Burlaga, 1982) are a source of intrinsically high magnetic field strengths. Gonzalez *et al.* (1998) and Dal Lago *et al.* (2001) have shown that there is an empirical relationship between extremely fast ICMEs and extremely high magnetic cloud field strengths. Figure 1 shows the structure of a typical ICMEs based on in situ observations at 1 AU.

Another important source is the strong sheath fields which could be produced by fast forward shock compression of slow stream magnetic fields (and plasmas), leading to larger (compressed) field strengths. If these sheath fields have strong southward components, they can cause major (Dst > -250 nT) magnetic storms (Gonzalez & Tsurutani, 1987; Tsurutani *et al.* 1988; Tsurutani *et al.* 1999). If both the sheath field and the cloud field (if present) have the proper orientation, a "double storm" (Kamide *et al.* 1998) will result. On the other hand, the compound streams, where one stream overtakes an upstream fast stream event (Burlaga *et al.* 1987) and the overtaking shock may compress the already compressed upstream sheath fields (Tsurutani *et al.* 1999) and magnetic cloud fields (Wei *et al.* 2003) may also lead to double storms. The triple and quadruple stream events, etc., produce even further magnetic compression and may lead to triple storms, etc.

3.3. *Magnetic Data of Colaba Observatory*

Magnetometers for measuring Declination and horizontal magnetic field component at Colaba Observatory during 1846-1867 were made by Thomas Grubb of Dublin and are described in Royal Society reports (1840; 1842). In the Declinometer, a scale and lens attachment to the magnet and the telescope set up made it possible to read the scale position manually based on the movement of the north end of the magnet. The absolute easterly declination (in minutes) was calculated from the relation: $d = 6'.841.(f - R).c$, where $6'.841$ is the adopted value of a unit of the declinometer scale, R is the true meridian reading, c the torsion co-efficient and f is the observed scale reading.

The Grubb Horizontal force magnetometer consisted of a rectangular bar magnet suspended horizontally, and carrying a collimator scale. The position of the magnet could

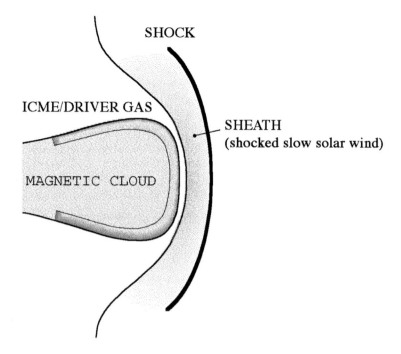

Figure 1. The configuration of a fast coronal mass ejection (CME) and its upstream sheath in the interplanetary medium, i.e., the so called ICME.

be determined by reading the scale with a properly placed telescope. The entries in the data book contained the scale reading of hourly observations taken at Gottingen mean time, which is almost one hour ahead of GMT. The computed hourly and fifteen minutes observations of the horizontal component from the scale readings were in units of grains and feet and the conversion factor used to compute the scale readings in to mm-mg-s was 0.46108. Measurements were taken at hourly intervals 24 hrs a day. When a magnetic storm (main phase) was occurring, measurements were made at 15 min. intervals. The final absolute values "H" plotted in Figure 2 are in nT (as converted from the c.g.s units.).

The magnetogram for the September 1-2, 1859 of the Colaba Observatory (Figure 2) shows that the magnitude of the storm sudden commencement (SSC) was about 120 nT. The maximum negative intensity recorded at Colaba was $\Delta H \approx$ -1600 nT, and the duration of the main phase of the storm (corresponding to the plasma injection) was $\sim$ 1-1/2 hour duration. The location of Colaba ($\sim$ 12 LT) was not ideal to detect the maximum magnetic response to the storm. However, based on observation from this one station, one can say that this is now the most intense magnetic storm on record. Magnetometers at high latitude, e.g. Kew and others, were either saturated or non-operational for this event.

3.4. *What caused Super storm of September 1-2, 1859?*

We will apply the recently gained knowledge about Sun-Earth connection and use other related information, and make these determinations by a process of elimination.

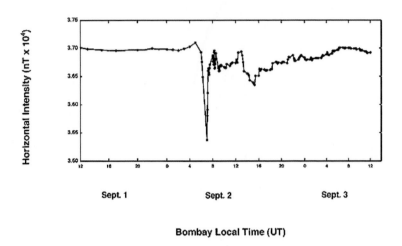

Figure 2. The Colaba (Bombay) magnetogram for the September 1-2, 1859 magnetic storm.

3.4.1. *Estimation of Magnetospheric Convection Electric Field*

The lowest latitudes of the auroras being 23° (Kimbal, 1960) was used to identify the Plasmapause location, which in turn was used to determine the magnetospheric convection electric fields, from the relation (Volland 1973; Stern 1975; Nishida 1978) for electric potential:

$$\Phi = -KR_E^2/r - A^*(r/R_E)^2 sin\Psi + \mu M/(qr^3), \tag{3.1}$$

where K=14.5 mV/m, R_E is the earth radius, r and Ψ are the radial distance and the azimuthal angle measured counter-clockwise from the solar direction, M is the Earth's dipole moment, q is the particle charge and μ is the particle transverse kinetic energy divided by the field magnitude, i.e., the first adiabatic constant. A^* is a coefficient given by Maynard & Chen (1975) and modified by Heppner (1977) and Wygant *et al.* (1998). The first and second terms on the left-hand side of equation (1) represent the corotation electric field and the shielded convection electric field, respectively. The third term represents the particle curvature and gradient B drifts.

A convection electric field, $E_C \sim 20$ mV/m, is needed for ring current at L=1.6 and plasmapause at L=1.3 (23° magnetic latitude).

These results are consistent with extrapolated magnetic latitude values for the auroral diameter given by Schulz (1997) as a function of Dst. Starting from a basic auroral boundary at about 65°, Schulz suggests that this boundary moves equatorwards 2° for each change of -100 nT in Dst.

3.4.2. *Estimation of the Interplanetary Electric Field*

From Carrington paper, the transit time of the ICME from the sun to Earth: ~ 17 hours and 40 min. This indicates an average shock transit speed of $V_{shock} \sim 2380$ km/s.

Cliver *et al.* (1990) found a relationship between the solar wind speed at 1 AU and the average shock transit speed of (limited to events below 1200 km s-1)

$$V_{sw} = 0.775 V_{shock} \tag{3.2}$$

Gonzalez *et al.* (1998) have found an empirical relationship between ejecta speeds at 1 AU and magnetic cloud magnetic field magnitudes given by:

$$B(\text{nT}) \approx 0.047 V_{sw}(\text{km s}^{-1}) \tag{3.3}$$

where V_{sw} is the peak solar wind speed of the ejecta at 1 AU. The expression was determined by a linear regression, where the correlation coefficient was 0.71. The data were limited to peak speeds less than ~ 750 km s^{-1} and peak magnetic fields less than ~ 35 nT.

Combining (2) and (3), the maximum possible electric field for extremely fast interplanetary events such as the September 1-2, 1859 event can be expressed as:

$$E_{IP} \approx 2.8 \times 10^{-5} V_{shock}^2 \text{mV/m} \tag{3.4}$$

Assuming $V_{shock} \approx 2380$ km s^{-1}, we get $E_{IP} \sim 160$ mV/m.

This estimates compares well with the convection electric field, $E_C \sim 20$ mV/m, derived above if a reasonable value of the penetration efficiency of $\sim 12\%$ of the interplanetary electric field is considered (Gonzalez *et al.* 1989).

3.4.3. *Estimation of Peak Storm Magnetic Intensity (Dst)*

Burton *et al.* (1975) gave an empirical relation for the evolution of ring current:

$$\frac{dDst}{dt} = Q - \frac{Dst}{\tau}, \tag{3.5}$$

where Dst is the disturbance storm time index which acts as a proxy for the energy of the ring current, Q is the energy input and τ is the decay constant. For energy balance of the ring current at the peak of the storm, we take

$$Dst = \tau Q \tag{3.6}$$

Further, for very intense storms, we can make use of the empirical relation derived by Burton *et al.* (1975) (neglecting the -0.5 mV/m constant value in Burton et al. due to the extremely large storm fields):

$$Q = \alpha V_{sw} B_S \tag{3.7}$$

where α is empirically $\sim 1.5 \times 10^{-3}$ nT s^{-1} (mV/m)$^{-1}$ and $V_{sw}B_S$ is in mV/m. Here B_S denotes the southward component of the interplanetary magnetic field. Considering $\tau = 1.5$ hrs (taken from Colaba magnetogram), we get from (6) and (7), $Dst \approx -1760$ nT, a value consistent with Colaba measurement of $\Delta H \approx -1600$ nT. This is also in fair agreement with the prediction of the theoretical model of Siscoe (1979).

The profile of the Dst index for this storm indicates that it was due to a simple plasma injection, and there is no evidence for the possibility of a complex storm. The most likely mechanism for this intense, short duration storm would be a magnetic cloud with intense B_S fields. Storm main phase "compound" events or "double storms" (Burlaga *et al.* 1987; Kamide *et al.* 1998) due first to sheath fields and then to cloud fields (Tsurutani *et al.* 1988) appear to be unlikely due to the (simple) storm profile The only other possibility that might be the cause of the storm is sheath fields. This can be ruled out because the compression factor of magnetic fields following fast shocks is only ~ 4 times (Kennel *et al.*

1985). Since typical quiet interplanetary fields ~ 3 to 10 nT, the compressed fields would be too low to generate the inferred interplanetary and magnetospheric electric fields for the storm. Thus by a process of elimination the interplanetary fields causing this storm have been determined to be part of a fast magnetic cloud.

3.4.4. *Solar Flare Energies*

How rare was the September 1-2, 1859 solar flare/solar ejecta event? Is it possible that an event of this intensity could happen again in the near future? To answer these question we note that in addition to "white light", solar flares radiate at a variety of other wavelengths as well. Using general scaling, Lin and Hudson (1976) have estimated total energy of August 1972 flare to be $\sim 10^{32}$ to 10^{33} ergs. Kane *et al.* (1995) has estimated the June 1, 1991 flare energy to be $\sim 10^{34}$ ergs.

The energy of the 1859 solar flare energy based on the white light portion as described in Carrington (1859) report, has been calculated by D. Neidig (private comm., 2001) to be $\sim 2 \times 10^{30}$ ergs. K. Harvey (private comm., 2001) has estimated the total energy of this event as $\sim 10^{32}$ ergs. The comparison shows that September 1, 1859 Carrington flare was not exceptional in term of total energy released.

Cliver *et al.* (1990) have pointed out that the 1972 event had the highest transit speed on record with a delay time of 14.6 hrs and the average ejecta speed ~ 2850 km s^{-1} (Vaisberg & Zastenker, 1976). The shock speed at 1 AU was > 1700 km s^{-1} (Zastenker *et al.* 1978). There was no measurement of the magnetic fields for the ejecta for the 1972 event at 1 AU. Using equations 2, 3 and 4, we get at 1 AU , $B \sim 103$ nT and a maximum interplanetary electric field $E_{IP} \sim 229$ mV/m.

If the August 1972 event had such high shock velocities why didn't the ejecta or sheath cause a great magnetic storm? To answer this, we note that Pioneer 10 measured $B \sim 15$ nT at 2.2 AU. Assuming an r^{-2} drop-off of the field intensity with radial distance and no super-radial expansion (due to high internal pressure), the extrapolated $B \sim 75$ nT at 1 AU.

The flux rope model (R. Lepping) indicate that Pioneer 10 passed through the edge of the cloud which was tilted at $84°$ relative to the ecliptic plane and cloud magnetic field orientation was northward (Tsurutani *et al.* 1992b). Extrapolating the data to the time of Earth passage, it was noted that during the interval when the magnetic cloud passed the Earth, the Dst index indicated a storm recovery phase, and AE and Kp were unusually low (<100 nT and 0+, respectively). This is consistent with the picture that the magnetosphere becomes extremely quiet during intense B_N events (Tsurutani *et al.* 1995; Borovsky & Funsten, 2002). Thus, the most probable reason for the failure of the August 1972 event to excite any major magnetic storm was due to the fact that the interplanetary magnetic field within the magnetic cloud was directed almost totally northward (rather than southward).

4. Summary and Conclusions

The September 1-2, 1859 magnetic storm is the most intense magnetic storm in recorded history. The auroral sightings were as low as $23°$ magnetic latitude (Hawaii and Santiago), and the estimated Dst ≈ -1760 nT. The Colaba station magnetic decrease of $\Delta H \sim$ - 1600 nT is consistent with this estimate.

The 1859 flare/CME ejecta was not unique. The August 1972 flare was definitely equally (or more) energetic, and the interplanetary ejecta speed faster. So, 1859 like super magnetic storms can occur again in the near future. How often can they occur? The one big flare per solar cycle (11 years) has the potential for creating a storm with a similar intensity. However in reality, we know that this was the largest storm in the last 143 years (13 solar cycles).

At this stage it is difficult to answer: "are even more intense events possible?, can one assign probabilities to the occurrence of a similar storm or to a greater intensity storm?"

The predictability of similar or greater intensity events requires knowledge of either full understanding of the physical processes involved in the phenomenon or a good empirical statistics of the tail of the energy distribution. For the former, if one knows the physical processes causing solar flares or magnetic storms, then the high energy tail (extreme event) distributions could be readily ascertained. Knowing the physical processes, of course means understanding mechanisms of saturation. The sun and the magnetosphere are of finite size, have finite magnetic field strengths, etc., and therefore will have cutoff energies.

Since we do not fully understand these specific saturation processes, it is therefore not known whether flares with energy $> 10^{34}$ ergs or magnetic storms with Dst < -1760 nT are possible or not. Then, the other possibility is to use statistics to infer the probabilities of flares with energies less than, but close to 10^{34} ergs and storms with Dst < -1760 nT? Unfortunately, the statistics for extreme solar flares with energies greater than 10^{32} ergs and extreme magnetic storms with Dst < -500 nT are poor. The shapes of these high energy tails are essentially unknown. One can therefore assign accurate probabilities to flares and storms for only the lower energies where the number of observed events are statistically significant.

There does not exit any strong relationship between the strengths of the flares and the speed and magnetic intensities of the ICMEs. Nevertheless, it is certainly noted that the most intense magnetic storms are indeed related to intense solar flares, i.e., the two phenomena have a common cause: magnetic reconnection at the sun. Recently it is found that the previously thought "upper limit" of 10^{32} ergs for the energy of a flare can be broken by a wide margin (Kane *et al.* 1995). It is quite possible that we may have not detected events at the saturation limit (either flares or magnetic storms) during the short span of only hundreds of years of observations. Most probably the sun cannot have flares at superflare energy ($10^{38} - 10^{39}$ ergs) levels (Lingenfelter & Hudson 1980), but perhaps 10^{35} ergs is feasible for our sun. If it were so, the effects of an accompanying super-intense magnetic storm might be catastrophic for the modern society!

References

Allen, J., Sauer, H., Frank,L. & Reiff,P. 1989, *EOS*, 70, 1479.

Borovsky, J. E. & Funsten, H. O., 2002, *J. Geophys. Res.*, .

Burlaga, L. F., Behannon, K. W., and Klein, L.W., 1987, *J. Geophys. Res.*, 92, 5725

Burton, R.K., McPherron,R.L., & Russell, C.T.,1975, A *J. Geophys. Res.*, 80, 4204

Carrington, R.C.,1859, *M. Not. Roy. Ast. Soc.*, XX, 13

Chapman, S. & J. Bartels, 1940, *Geomagnetism*, vol. I, pp 328-341, Oxford University Press, New york

Cliver, E.W., Feynman, J. and Garrett,H.B., 1990, *J. Geophys. Res.*, 95, 17103

Dessler, A.J. & Parker,E.N. 1959, *J. Geophys. Res.*, 64, 2239

Dal Lago, A., Gonzalez, W. D., Clua de Gonzalez, A. L. a& Vieira, L. E. A., 2001. *J. Atmos. Solar-Terres. Phys.*, 63, 451.

Dungey, J.W., 1961, *Phys. Res. Lett.*, 6, 47

Gonzalez, W.D. and Tsurutani, B. T.,1987, *Planet. Space Sci.*, 35, 1101

Gonzalez, W.D., B.T. Tsurutani, A.L.C. Gonzalez, E.J. Smith, F. Tang, and S.I. Akasofu, 1989, *J. Geophys. Res.*, 94, 8835

Gonzalez, W.D., Joselyn, J.A., Kamide, Y., Kroehl,H.W., Rostoker,G., Tsurutani,B.T. and Vasyliunas, V.M. 1994, *J.Geophys.Res.*, 99, 5771

Gonzalez, W.D., Gonzalez, A.L.C., Dal Lago, A. , Tsurutani,B.T., Arballo, J.K., Lakhina, G. S. , Buti,B., Ho,C.M.& Wu, S.T., 1998, *Geophys. Res. Lett.*, 25, 963

Hale, G.E., 1931, *Astrophys. J.*, 73, 379

Heppner, J.P., 1977, *J. Geophys. Res.*, 82, 1115

Hodgson, R., 1859, *M. Not. Roy. Ast. Soc.*, XX, 15

Kamide, Y., Yokoyama, N., Gonzalez, W., Tsurutani, B.T., Daglis,I.A., Brekke,A. & Masuda, S. 1998,*J. Geophys. Res.*, 103, 6917

Kane, S.R., Hurley, K., McTiernan, J.M., Sommer, M., Boer, M. , & Niel,M. , 1995, *Astrophys. J.*, L47, 446

Kennel, C.F., Edmiston, J.P. & T. Hada, 1985, in R..G. Stone and B.T. Tsurutani (eds.), *Collisionless shocks in the heliosphere: A tutorial review*, (Amer. Geophys. Union, Wash. D.C.), vol. 34, 1

Kimball, D.S., 1960, in: *Sci. Rpt. 6, UAG-R109*, University of Alaska

Klein, L.W. & Burlaga,L.F. *J. Geophys. Res.*, 87, 613

Lin, R.P. & Hudson, H.S., 1976, *Solar Phys.*, 50, 153

Lingenfelter, R. E., and Hudson, H. S., 1980, in: R. O. Pepin, J. A. Eddy, and R. B. Merrill (eds.), *The Ancient Sun*, (Pergamon Press), pp. 69-79

Loomis, E., 1861, *Amer. J. Sci.*, 82, 318

Maynard, N.C. & Chen,A.J. I *J. Geophys. Res.*, 80, 1009

Moos, N.A.F., 1910, *Magnetic observations made at the Government Observatory, Bombay 1846-1905, Pat II: The phenomenon and its discussion*, (Government Central Press, Bombay, India)

Newton, H.W., 1943, *M. Not. Roy. Ast. Soc.*, 103, 244

Nishida, A.,1978, *Geomagnetic diagnosis of the magnetosphere*, Springer-Verlag, N. Y.

Report of the Committee of Physics, including Meteorology on the objects of scientific inquiry in those sciences, Royal Society, R. and J.E. Taylor, London, (1840).

Revised instructions for the use of the magnetic meteorological observations and for magnetic surveys, prepared by The Committee of Physics and Meteorology of the Royal Society, London (1842).

Rostoker, G., 1997, in: B. T. Tsurutani, W. D. Gonzalez, Y. Kamide, and J. K. Arballo (EDS.), *Magnetic Storms*, Geophysical Monograph, (AGU, Washington D C), vol. 98, p. 149.

Sabine, E., 1852, *Phil. Trans. R. Soc. Lond.*, 142, 103

Schröder, W., 1997, *Planet. Space Sci.*, 45, 395.

Schulz, M., 1997,*J. Geophys. Res.*, 102, 14149

Sckopke, N.,1966, *J. Geophys. Res.*, 71, 3125

Siscoe, G.L., 1979, *Planet. Spa. Sci.*, 27, 285.

Stern, D.P., 1975, J. Geophys. Res., 80, 595

Tsurutani, B. T., Gonzalez, W. D., Tang, F., Akasofu, S.-I. & Smith E. J., 1988, *J. Geophys. Res.*, 93, 8519

Tsurutani, B.T., Gonzalez, W.D. Tang,F. & Lee, Y.T., 1992a, *Geophys Res. Lett.*, 19, 73

Tsurutani, B.T., W.D. Gonzalez, F. Tang, Y.T. Lee, M. Okada, and D. Park, 1992b, *Geophys. Res. Lett.*, 19, 1993

Tsurutani, B.T. and W.D. Gonzalez, 1995, *Geophys. Res. Lett.*, 22, 663

Tsurutani, B.T. & Gonzalez, W.D., 1997, in: B.T. Tsurutani, W.D. Gonzalez, Y., Kamide and J.K. Arballo (eds.), *Magnetic Storms*, (AGU Monograph, Amer. Geophys. Union, Wash. D.C.), vol. 98, 77

Tsurutani, B.T., W.D. Gonzalez, Y., Kamide and J.K. Arballo,1997, (eds.), *Magnetic Storms*, Amer. Geophys. Union, Wash. D.C., vol. 98.

Tsurutani, B.T., Y. Kamide, J.K. Arballo, W.D. Gonzalez and R.P. Lepping, 1999, *Phys. Chem. Earth*, 24, 101

Tsurutani, B.T., Gonzalez, W.D., Lakhina,G. S. & S. Alex, 2003, *J. Geophys. Res.*, 108, 1268, doi:10.1029/2002JA009504

Vaisberg, O.L. and G.N. Zastenker, 1976, *Space Science Rev.*, 19, 687

Volland, H., 1973, *J. Geophys. Res.*, 78, 171

von Humboldt, A., 1808, *Annales der Physik*, 29, 425.

Wei, F., R. Liu, X. Feng, D. Zhong and F. Yang, 2003, *Geophyys. Res. Lett.*

Wygant, J., Rowland,D., Singer, H.J., Temerin, M. , Mozer,F.S. & Hudson, M.K. 1998, *J. Geophys. Res.*, 103, 29527

Zastenker, G.N., Temny, V.V., d'Uston,C. & Bosqued, J.M. 1978, *J. Geophys Res.*, 83, 1035

Table 1. A partial chronological listing of large magnetic storms. The October-November 2003 storms have been added in the list.

Sr.No.	YEAR	MONTH	DAY	H Range[‡] (nT)	DST(nT)	Station	Geographic(Lat.,Long.)
1	1859	September "	1-2 "	1720 >700[†*]	- -	Bombay Kew	18.89° ; 72.82° 51.50°; 359.70°
2	1859	October	12	980	-	Bombay	18.89°; 72.82°
3	1872	February	4	1020	-	Bombay	18.89°; 72.82°
4	1882	November "	17 "	450 >1090[†*]	- -	Bombay Greenwich	18.89°; 72.82° 51.48°; 0.00°
5	1903	October "	31 "	820 >950[†*]	- -	Bombay Potsdam	18.89°; 72.82° 52.38°; 13.06°
6	1909	September	25	>1500[†*]	-	Potsdam	52.38°; 13.06°
7	1921	May "	13-16 "	>700[†*] 1060[†]	- -	Alibag Potsdam	18.63°; 72.87° 52.38°; 13.06°
8	1928	July	7	780	-	Alibag	18.63°; 72.87°
9	1938	April "	16 "	530 1900[†]	- -	Alibag Potsdam	18.63°; 72.87° 52.38°; 13.06°
10	1957	September	13	580	-427	Alibag	18.63°; 72.87°
11	1958	February	11	660	-426	Alibag	18.63°; 72.87°
12	1989	March	13	640	-589	Kakioka	36.23°; 140.18°
13	2003	October "	29 30	432 453	-370 -406	Alibag Alibag	18.63°; 72.87° 18.63°; 72.87°
14	2003	November	20	531	-491	Alibag	18.63°; 72.87°

Discussion

KAHLER: Does the size distributions of intense storms, measured by Dst, look like a power-law, similar to earthquakes and floods? If so, this suggests a self-organized system.

LAKHINA: Yes, the studies so far are consistent with power-laws, but only for weak magnetic storms. The data for super-intense storm is very scarce, I am not aware of self-organized criticality occurring in the Sun-Earth plasma system as far as magnetic storms (intense to super-intense) are concerned.

S. T. WU: Comments: According to the work of S. Kane in the 70's (If I remember correctly), the highest energy contents of a flare could be as high as 10^{40} ergs, which is very unusual, but it is possible.

[‡] H range is defined as the difference between the maximum and minimum value of H during the storm event.

[†] The values recorded at the mid-latitude stations could have an ionospheric component associated with the activity.

[†*] Saturation of the instrument. In addition, the value recorded at this station could have an ionospheric contribution.

LAKHINA: I am not aware of this work. Personally, I doubt that energies $\sim 10^{40}$ ergs for solar flares are possible.

GOPALSWARY: You mentioned that the Sep. 1-2 Storm was simple. But the plot shows one or two additional SSC's and a second dip suggesting a normal superstorm. Is it possible that the first big spike is an artefact? Also, what are the possibilities we are looking at a complex storm?

LAKHINA: The major main phase appears to be a clean single injection event. However there is a possibility of another pressure/shock wave during the recovery phase. So far we have not looked into this aspect, but will do that soon.

JUN LIN: The intensive flares occurring in October and November 2003 did not cause any significant geoeffectiveness. So according to your investigations, which kind of flares is most likely to cause magnetic storms?

LAKHINA: It is not only the energy of the flare and the ejecta speed at 1 AU, but the magnitude and southward direction of the IMF, which play important roles in geoeffectiveness. Although solar flares on October 28 and 29, 2003 had the energies to cause super-intense storm, the southward component of IMF in the magnetic cloud was not strong, therefore they could not produce a super storm. On the other hand, a weaker flare on Novermber 18, 2003, gave rise to a stronger magnetic storm than October 28/29 solar flares, as it has a stronger southward IMF lasting for several hours.

Session 2
Observations of coronal MASS ejections

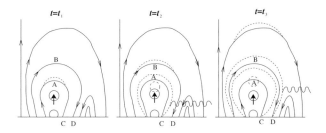

Coronal and Stellar Mass Ejections
Proceedings IAU Symposium No. 226, 2005
K. P. Dere, J. Wang & Y. Yan, eds.

What have we learned with SOHO?

Rainer Schwenn

Max-Planck-Institut für Sonnensystemforschung, 37191 Katlenburg-Lindau, Germany,
email: schwenn@linmpi.mpg.de

Abstract. The Solar and Heliospheric Observatory (SOHO), a space mission of international collaboration between ESA and NASA, has been operating almost continuosly since early 1996. The Sun and the heliosphere went through both: the minimum and maxumum of solar activity in 1996 and 2000, respectively. The perfectly working set of modern solar telescopes and in-situ instrumentation has been producing an unprecedented set of most valuable observational data that are almost immediately available to the public via the Internet. A wealth of new results has been published in innumerable papers. For CME research in particular, SOHO has started a new era. CME evolution can now be studied from their initiation up to the arrival of the ejecta clouds at 1 AU. For the first time, helioseismological observations reveal flow vortices underneath sunspots, i.e., activity centers that are involved in subsequebt eruptions. Combined EUV disk observations and coronagraph images allow to differentiate between CMEs pointed towards to or away from the Earth. Thus, space weather predictions have achieved a new quality. The occurrence of "EIT waves" at CME onset was discovered, the internal structure of CMEs (including "disconnection", magnetic topology and helicity, etc.) was made visible, statitics about CME properties and their change with solar activity were refined. Spectacular CME images and animations have been attracting the public to an unexpected extent, to the benefit of solar research in general.

Keywords. Sun: coronal mass ejections (CMEs)

Discussion

KOUTCHMY: From your talk we get the impression that there is a one to one correspondence between flares and CMEs. Is it always the case? What about the latest event producing a very large storm: why it should be related namely to the small flare you pointed out ?

SCHWENN: It is not always the case: there are many CMEs without associated flares and the other way round. For the big events, you usually have both flare and CME, but not in a unique cause-effect relation. My last example was just to illustrate that even minor flares may be associated with geo-effective CMEs. Be prepared for surprises.

RILEY: I like your idea about global CMEs being a coincidence. But I am a little bit surprised that you think that all "Sympathetic CMEs" are a coincidence. I would have thought that a large-scale eruption could provide a sufficient perturbation to initiate another eruption. Could you comment on that?

SCHWENN: I see no evidence for sympathetic events at large separations, except by mere coincidence. All events formally considered "global" could be traced back to single events samewhere close to Sun center.

ZHUKOV: You mentioned the result by C. St.-Cyr that the number of slow CMEs did not increase with the increased sensitivity of LASCO; you mentioned this in connection with the idea that slow wind might be composed of many small slow CMEs. Are the "blobs" described by Sheeley et al. included in St.-Cyr's statistics? Do you think these "blobs" are relevant to CMEs?

SCHWENN: Sheeley's "leaves in the wind" are not included in St.Cyr's study, to my knowledge. I do not consider them CMEs, although they seem to meet the definition. The leaves are definitely <u>not</u> sufficient to become the slow solar wind, nor are the other slow CMEs.

STERLING: a) you said that 85% of halo CMEs are geoeffective. Does that included only earth-directed events?
b) Can you see back side halos as efficiently as earth-side halos?

SCHWENN: a) Yes, we did not include back side events.
b) I do not know of any such study [somebody else responds: says that Andrews has studied this and found that you can see backside events as well as front side events]. Right.

TYLKA: I was somewhat surprised by your statement that "modern instrumentation has not increased the number of small faint CMEs". But one of the new insight on impulsive SEP events (i.e., ^{3}He/^{4}He>10%) is that at least half <u>are</u> accompanied by CMEs, that are small, faint, narrow, etc. Can you comment?

SCHWENN: It may deal with the definition of what we call a CME. In fact, there is no size specification in the classical CME definition. There is very much activity on the active Sun which does <u>not</u> produce CMEs and yet might energize particles.

Coronal and Stellar Mass Ejections
Proceedings IAU Symposium No. 226, 2005
K. P. Dere, J. Wang & Y. Yan, eds.

EUV observations of CME-associated eruptive phenomena with the CORONAS-F/SPIRIT telescope/spectroheliograph

V.A. Slemzin[1], V.V. Grechnev[2], I.A. Zhitnik[1], S.V. Kuzin[1], I.M. Chertok[3], S.A. Bogachev[1], A.P. Ignatiev[1], A.A. Pertsov[1] and D.V. Lisin[3]

[1]P.N. Lebedev Physical Institute, Moscow, Russia, email: slem@sci.lebedev.ru

[2]Institute of Solar Terrestrial Physics, Irkutsk, Russia, email: grechnev@iszf.irk.ru

[3]IZMIRAN, Troitsk, Moscow Region, Russia, email: ichertok.izmiran.ru

Abstract. A multi-channel SPIRIT telescope/spectroheliograph aboard the CORONAS-F satellite operating in soft X-ray and EUV ranges ($T \sim 0.05$–15 MK) is an effective instrument for complex studies of CME-associated phenomena such as eruptive filaments, dimmings, coronal waves, posteruptive arcades, etc. In particular, SRIRIT observations of high-temperature (T = 5–15 MK) plasma structures in the MgXII 8.42 Å line show specific pre-CME sigmoid magnetic field configurations. Eruptions of filaments (prominences) and dimmings in a CME process are seen with a high contrast in the coronal 175 Å band (FeIX–XI) and the transition-region 304 Å (HeII) images. Our results are illustrated by several powerful eruptive events of the current solar cycle. We compare SPIRIT data with observations at other spaceborne and ground-based instruments (SOHO/EIT, *Yohkoh*/SXT, and Hα images, etc.)

Keywords. Sun: corona, Sun: coronal mass ejections (CMEs), Sun: X-rays, gamma rays

1. Introduction

CMEs are complex phenomena associated with filament eruptions and solar flares. They show a variety of kinematic and thermal transformations in the solar plasmas (Zhang et al. 2001; Moon et al. 2002; Feynman & Ruzmaikin 2004). A pre-flare situation is often hinted by the presence of a sigmoidal loop system seen in hot ion lines of T ~ 2–8 MK (Gibson et al. 2002; DelZanna et al. 2002). To understand the origin and drivers of CMEs, important is to study the initial phase of the heating and acceleration of the erupting matter in the transition region and the corona at altitudes of several tenths $R_\odot$, where it can be observed better in the EUV spectral range. Disappearance of a filament visible in the chromospheric Hα images is often followed by its appearance in transition-region lines and, in some cases, in coronal lines later on. This stage of the CME process is not well studied so far, because necessary observations with sufficient time resolution simultaneously in several spectral lines usually are not available.

The SPIRIT experiment carried out aboard the CORONAS-F spacecraft since August 2001 (Zhitnik et al. 2002) provides synchronous observations of the whole Sun in several spectral bands (in particular, the transition-region band of 304 Å and coronal bands of 175 Å and 8.42 Å) covering the temperature range from 0.05 to 15 MK. Here we present some examples of SPIRIT observations of CME-related phenomena around the 23rd

solar maximum and compare it with observations at other spaceborne and ground-based instruments (SOHO/EIT, *Yohkoh*/SXT, Hα, etc.)

2. SPIRIT observations and their comparison with data from other instruments

The SPIRIT performs routine observations with an interval of 15 min during 45-min non-occulted period of the 93.5 min orbit. The observations are conducted simultaneously in two bands of the transition-region line 304 Å (HeII, $T \sim 0.05$–0.08 MK) and the coronal line 175 Å (FeIX–XI, $T \sim 0.9$–1.2 MK) with an exposure time of 2–9 s. In comparison with the SOHO/EIT FeXII 195 Å line, the 175 Å band contains cooler lines of FeIX–XI with lesser dynamic range from the quiet Sun to flares. EIT images produced in 195 Å line are often overexposured during flares and contaminated with strong scattered light, thus preventing observations of any features in the flaring region. By contrast, SPIRIT images remain clear.

The images in the MgXII 8.42 Å line ($T \sim 5$–15 MK) are produced, as a rule, with an exposure time of 37 s and with an interval of 3–20 min. These images revealed hot coronal features with a typical peak temperature of ~ 10 MK and $\log N_e \sim 9$–10 (Zhitnik et al. 2003). They do not always coincide with active regions and significantly differ from lower-temperature (1–2 MK) structures seen in EUV coronal lines, such as 175 or 195 Å. These structures are often localized at very high altitudes of 0.1–$0.3\,R_\odot$ above the solar limb and live from minutes to days. They show fast spatial variability and have a contrast of more than two orders of magnitude.

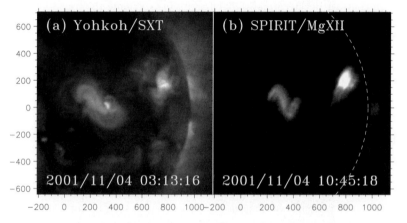

Figure 1. A 'sigmoid' in the flaring active region observed with *Yohkoh*/SXT and CORO-NAS-F/SPIRIT in MgXII line on November 4, 2001. White dashed line shows solar limb. Axes show arc seconds from solar disk center.

3. SPIRIT observations of CME-related phenomena

3.1. *Pre-event 'sigmoids' in MgXII images*

'Sigmoids' are believed to represent a highly sheared magnetic configuration overlying an extended filament channel (Glover et al. 2001) and to indicate a high probability of an eruption (Hudson et al. 1998). Sigmoids are clearly seen in soft X-rays, but poorly detectable in lower-temperature EUV emissions (Sterling et al. 2000). SPIRIT MgXII channel due to its high sensitivity to radiation of hot plasmas shows sigmoids with even better contrast than *Yokkoh*/SXT, indicating that the temperature of high pre-flare loops can exceed 5 MK. Fig. 1 shows a comparison of *Yokkoh*/SXT and SPIRIT/MgXII images for a powerful solar eruptive event of November 4, 2001 (Chertok et al. 2004). A sigmoid in AR 9684 preceded an X1.0 flare occurred at 16:03 UT and a fast (V~1800 km/s) halo CME at 16:35 UT. After the flare and CME, the loop structure transformed into a posteruptive arcade extending over 300 000 km for more than one day.

3.2. *Posteruptive dimmings in SPIRIT 175 Å and 304 Å lines*

Coronal ('EIT') waves and dimmings are regarded as the most reliable signatures of a CME (see, e.g., Zhukov & Auchère, 2004). Most often the coronal wave appears after a CME as a propagating bright front seen in coronal lines (e.g., EIT 195 Å line). Behind the wave, a dimming appears as temporary, localized depletion of EUV radiation: some of them are deep and compact, other are shallower, but more extended, being directed to neighboring active regions. Possibly, some of extended dimmings are associated with large transequatorial loops (Farnik et al. 2001; Glover et al. 2003).

SPIRIT observations show many examples of coronal waves and a variety of dimmings, in particular, observed during the period of solar extreme events of October–November 2003 (Slemzin et al. 2004). Coronal waves were observed on October 26 and November 18 by the SPIRIT in 175 Å and by the EIT in 195 Å. Dimmings observed for most powerful events of that period had very large scale to cover the hemisphere and showed conspicuous homology for many days, which suggests stability of the large-scale magnetic structure.

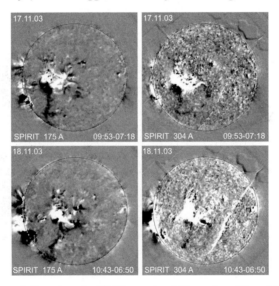

Figure 2. Structure of dimmings in CME events of November 17 and 18, 2003 in SPIRIT 175 Å and 304 Å channels

Fig. 2 shows a comparison of dimming structures observed on November 17 and 18, 2003 in the SPIRIT 175 and 304 Å bands. It is clearly seen that the main dimmings are located near the eruptive center, the most intense of them is directed to the boundary of the southern polar coronal hole. The dimmings in both 175 and 304 Å bands for two days are similar, but not fully coincide. A detailed analysis shows that they originated at the places of previous bright loop structures seen in the 175 Å band before the events. The dimmings in the transition-region HeII 304 Å line are similar in shape, but not coincide with those in 175 Å. As a rule, 304 Å band dimmings have lesser depth and sometimes appear with significant delay of several tens minutes with respect to coronal

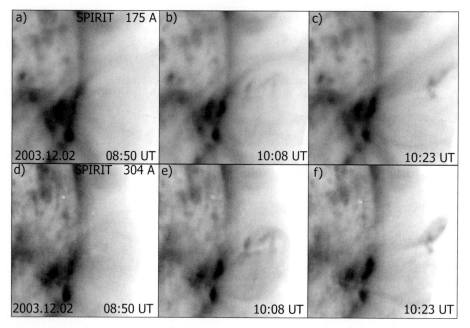

Figure 3. An eruptive event of December 2, 2003 in SPIRIT 175 Å (a-c) and 304 Å (d-f) bands (negative).

ones (Chertok et al. 2004). This probably suggests the development of the dimming from the higher coronal layers downward, up to the transition region.

3.3. *Eruptions of filaments and prominences*

Observations of filament (or prominence) eruptions are important to study the processes of CME formation and acceleration of eruptive mass. The temporal and spatial evolution of prominences is governed by dynamics of the magnetic field structure, but the trigger mechanism of eruption is still not well established (Schmieder et al. 2000, Marque et al. 2002). Simultaneous SPIRIT observations in 304 Å and 175 Å bands, due to their complementary temperature sensitivity ranges, are suitable to study plasmas in erupting structures at the initial stage of the CME launch.

We present two observations of eruptive events. In the first case, a fast ejection was observed after a prominence eruption on December 2, 2003 (fig. 3; Delaboudinière et.al 2005). A quiescent filament was observed during several days in Hα, 304 Å band, and 175 & 195 Å coronal lines before the eruption, which occurred on December 2 at 10:15 UT. It produced a large CME with a speed of ∼ 1300 km/s. The eruptive prominence was observed first with Pic du Midi coronagraph in Hα, then with EIT at 195 Å and SPIRIT at 304 & 175 Å (fig. 3). The prominence becomes bright in both 175 and 195 Å images, which implies its heating in the course of the eruption.

Another eruptive event was observed on June 8, 2004 during early Venus transit across the Sun (fig. 4). A large prominence slowly ascended (V∼ 55 km/s) above the eastern limb for 40 min of observations. It was much brighter in 304 Å line, than in the coronal 175 Å line; hence, it was rather cold. It left the field of view of the SPIRIT at 05:41 UT at a height of 0.27 R$_\odot$, and the frontal structure of the corresponding CME appeared in LASCO/C2 image at 06:26 UT at 1.86 R$_\odot$. LASCO has not registered this event, because there were no subsequent C2/C3 observations during several hours.

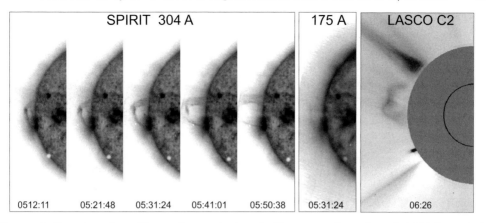

Figure 4. Eruption of a cold prominence during the Venus transit on June 8, 2004 (negative).

4. Conclusions

A multi-channel SPIRIT telescope/spectroheliograph operating aboard the CORONAS-F satellite allows studying various CME-associated phenomena in wide temperature range of 0.05–15 MK. The SPIRIT is able to record images in several spectral bands simultaneously with high cadence that is important to study dynamics of eruptive processes.

Acknowledgements

We are grateful to Prof. I.I. Sobelman and Prof. V.D. Kuznetsov, scientific leaders of the CORONAS-F project, for their interest and support of this study; A. Zhukov for fruitful discussions; and the CORONAS-F team in IZMIRAN for the assistance in the telemetry data supply. In this study we used SOHO data. SOHO is a project of international cooperation between ESA and NASA. This work is supported by the Russian Foundation for Basic Research (grants 02-02-17272, 03-02-16049, and 03-02-16591), the Federal Ministry of Industry and Science (grants NSh 477.2003.2 and 1445.2003.2), and the Program No. 16 of the General Physics Department of RAS.

References

Chertok, I.M. et al. 2004, *Astronomy Rep.* 48, 407.
Delaboudinière, J.-P., et al. 2005, *this volume.*
DelZanna, G., et al. 2002, *Adv. Space Res.* 30(3), 551.
Farnik, F. et al. 2001, *Sol. Phys.* 202, 81.
Feynman, J. & A. Ruzmaikin 2004, *Sol. Phys.* 219, 301.
Gibson, S., et al. 2002, *ApJ* 574, 1021..
Glover, A., et al. 2001, *A&A* 378, 239.
Glover, A. et al. 2003, *A&A* 400, 759.
Hudson, H.S. et al. 1998, *Geoph. Res. Lett.* 25(14), 2481.
Marque, C. et al. 2002, *A&A* 387, 317.
Moon, Y.-J., et al. 2002, *ApJ* 581, 694.
Sterling, A.C. et al. 2000, *ApJ* 532, 628.
Schmieder, B. et al. 2000, *A&A* 358, 728.
Slemzin, V.A. et al. 2004, *Proc. of IAUS 223* (in press).
Tripathi, D. et al. 2004, *A&A* 422, 337.
Zhang, J., et al. 2001, *ApJ* 559, 452.
Zhitnik, I.A. et al. 2002, *ESA SP-506* 915.
Zhukov, A. & Auchère, F. 2004, *A&A* 427, 705.

Discussion

KAHLER: What is the time cadence of the SPIRIT images?

SLEMZIN: we have synoptic observations in several wavelengths at least once per 1.5h orbit for 15 orbits a day. For some periods we take simultaneous images in 175 Å/304 Å bands with a 15 min cadence during 45 min interval interrupted by 47-48 min occultations. We also participate in the "EIT Shutter-less" or "High cadence" program, when observations are done with a cadence less than 1 min.

P.F. CHEN: Are the data available online?

SLEMZIN: We can provide on request. Some data can be found in MEDOC. Solar data center (France), http://www. medoc.ias.u-psud.fr.

Coronal and Stellar Mass Ejections
Proceedings IAU Symposium No. 226, 2005
K. P. Dere, J. Wang & Y. Yan, eds.

X-Ray and EUV Observations of CME Eruption Onset

Alphonse C. Sterling[1]

[1]NASA/MSFC/NSSTC, SD50/Space Science Depart., Huntsville, AL, USA
email: alphonse.sterling@nasa.gov

Abstract.
Why CMEs erupt is a major outstanding puzzle of solar physics. Signatures observable at the earliest stages of eruption onset may hold precious clues about the onset mechanism. We summarize and discuss observations from SOHO/EIT in EUV and from Yohkoh/SXT in soft X-rays of the pre-eruption and eruption phases of three CME expulsions, along with the eruptions' magnetic setting inferred from SOHO/MDI magnetograms. Our events involve clearly-observable filament eruptions and multiple neutral lines, and we use the magnetic settings and motions of the filaments to help infer the geometry and behavior of the associated erupting magnetic fields. Pre-eruption and early-eruption signatures include a relatively slow filament rise prior to eruption, and intensity dimmings and brightenings, both in the immediate neighborhood of the "core" (location of greatest magnetic shear) of the erupting fields and at locations remote from the core. These signatures and their relative timings place observational constraints on eruption mechanisms; our recent work has focused on implications for the so-called "tether cutting" and "breakout" models, but the same observational constraints are applicable to any model.

Keywords. Sun: coronal mass ejections (CMEs), Sun: filaments, Sun: flares, Sun: magnetic fields, Sun: UV radiation, Sun: X-rays, gamma rays

1. Introduction

Over the past decade it has become clear that a Coronal Mass Ejection (CME) is just one aspect of a general solar magnetic eruption process, which also involves the release of energy in the form of solar flares, and often also involves expulsion of a solar filament. Understanding the solar eruption process is key to eventually being able to predict when eruptions will occur; such understanding will also give us insight into basic solar, stellar, and astrophysical phenomena, many of which are based upon magnetic activity.

Here we report on some of our recent work trying to understand what drives solar eruptions. We have been examining satellite and ground-based data of the onset phase of solar eruptions in an effort to try to understand the eruption trigger mechanism. We primarily consider data from the Soft X-ray Telescope (SXT) on the *Yohkoh* satellite and from the EUV Imaging Telescope (EIT) on *SOHO*. We also use magnetograms from *SOHO*'s Michelson Doppler Imager (MDI).

Several models suggest that interactions among coronal magnetic fields are responsible for eruption onset. In order to test these ideas it would be best to directly observe the coronal magnetic field at the start of eruptions, but of course this is not possible. Instead, we have been selecting eruption events which involve observable erupting filaments; such filaments are sometimes visible in absorption in EUV images. We take the motions of the filaments early in the eruption to be a proxy for the pre-eruption and eruption-time evolution of the coronal magnetic fields. In addition to the filaments in the EUV images, images in both EUV and soft X-rays can show prominent intensity brightenings and intensity dimmings, and both of these intensity changes can give us further information on the eruption process. In the following, we will present background for two specific models we have been examining, the "tether cutting" and the "breakout" models. We will also discuss "intensity dimmings" and their use as a diagnostic of eruptions. We will then present three examples of eruptions, and consider the implications for these two eruption theories based on observed relative timings of intensity brightenings and dimmings and observed filament motions in the eruption examples.

2. Two Eruption Theories

There are several ideas for the cause of solar eruptions; for reviews see, e.g., Forbes (2000), Klimchuk (2001), and Lin et al. (2003). In our recent work we have been using observations to test two specific ideas: the tether cutting model, as developed by Moore & LaBonte (1980), Sturrock (1989), and Moore et al. (2001); and the "breakout model," put forth by Antiochos (1998) and Antiochos et al. (1999).

Tether cutting holds that the key energy release mechanism for eruptions involves a single highly-sheared magnetic bipole. Reconnection among highly-sheared magnetic fields below a filament (or low in a filament channel) in the core of the bipole initiates and releases the eruption; the reconnection "cuts" (rearranges) magnetic field lines ("tethers") that tie down the core magnetic field, coronal material, and filament (if present), unleashing the sheared core field to erupt.

In contrast to tether cutting, breakout requires a multi-bipolar magnetic field and has the initial reconnection at a neutral point far-removed from the core field that explodes in the eruption. An example is a quadrupole configuration where a highly-sheared "inner" bipole is initially trapped beneath field of an enveloping "outer" bipole. If flux emergence or some other process causes the inner bipole to push upward, then reconnection ("external reconnection") between the inner bipole and outer bipole fields can result in creation of new "side lobe" coronal loops. If the early reconnection between the inner and outer bipoles is slow enough, large stress can build up at the boundary as the slow reconnection progresses. In the model, eventually the reconnection rate increases, and the pent-up inner fields explosively "breakout" through the field of the outer bipole and escape into the heliosphere as a CME. As the inner-bipole field is escaping, tether-cutting-like internal reconnection will occur among its outstretched fields reaching back to the surface, resulting in a standard solar flare.

We can inspect observations for signatures consistent with these models. In particular, breakout predicts activity far from the core during the pre-eruption phase. In contrast, tether cutting requires early activity only in the core region, although this does not preclude concurrent activity far from the core as a byproduct of the core activity.

3. Intensity Dimmings

Intensity dimmings are commonly observed in regions associated with eruptions, and they are one tool that we use to assist in checking for eruption signatures consistent with the tether cutting or breakout model. In on-disk coronal images, they are most easily visible in EUV movies, such as those from EIT, and they also can appear in soft X-ray images, such as those from SXT. Some early examples of dimmings are Manoharan et al. (1996), Sterling & Hudson (1997), Thompson et al. (1998), Thompson et al. (2000), and Gopalswamy & Hanaoka (1998). Two factors which could be responsible for the dimmings are temperature changes occurring over the duration of the eruption, or mass loss. Howard & Harrison (2004) and Sterling & Moore (2004a) present examples of dimmings due to temperature changes, but we believe that most of the primary eruption-associated dimmings are due to mass loss rather than temperature change. An argument against temperature-change-induced primary dimmings is that the dimmings frequently occur contemporaneously and cospatially in EIT images (showing 1.5 MK or 2.0 MK plasmas) and SXT images (showing plasmas hotter than 2.0 MK), which argues against the possibility that the plasma is being heated from cooler to hotter coronal temperatures (since this would lead to a brightening in SXT images coinciding with the dimming in EIT). Harra & Sterling (2001) present more direct evidence for dimmings resulting from mass loss, using spectral data combined with imaging data from the Coronal Diagnostics Spectrometer (CDS) instrument on *SOHO*.

Frequently, dimmings and brightenings show up best in "difference images" (where an earlier image is subtracted from a later image), although care must be exercised to avoid spurious intensity changes when the difference in time between the two images being subtracted is large, as discussed in Sterling & Moore (2004a).

4. Eruption Examples

Here we present three examples of eruptions. We will summarize our observations of each event, and discuss the possibility that they are initiated by the processes described by the tether cutting or the breakout model. Full details for each case appear in the referenced publications for each respective event.

Broadly speaking, we used two criteria for selecting events for detailed study. First, the events had to include an erupting filament; from the filament motions in the early stages of eruption, we infer properties of the dynamical evolution of the coronal magnetic fields in which the filaments were embedded. Second, we restricted our data sets to events which evolved slowly enough to be resolved by EIT, which has a time cadence of about 12-minutes. Consequently, all of our events are slowly-evolving quiet-region eruptions involving large-scale filaments.

4.1. *Event 1*

This event involves a quiet-region filament eruption of 1999 April 18; Sterling et al. (2001) and Sterling & Moore (2004a) discuss the event in detail. As we will see in our other examples also, this filament showed a two-stage eruption profile, rising slowly at first and then rapidly at the time of eruption (Fig. 1); in this case the slow-rise phase lasted for about six hours. Post-flare loops and other eruption-related phenomena occurred approximately concurrent with the onset of the fast eruption.

If breakout is responsible for this eruption, then we might expect to see signatures of high-altitude breakout reconnection prior to the onset of the fast-rise phase. Sterling & Moore (2004a) do indeed observe phenomena consistent with such breakout reconnection occurring during the slow-rise phase. For example, they find subtle brightenings and dimmings in EUV, and brightenings in soft X-rays during the slow rise that are consistent with formation of new side lobe loops in a quadrupolar magnetic geometry, as predicted by breakout. These observations therefore are consistent with early breakout reconnection occurring, although they are not conclusive proof of such reconnection (see Sterling & Moore (2004a)).

Tether cutting-type of reconnection does occur in this event, in association with a soft X-ray flare and the formation of post-flare loops. The question is: is that reconnection the main agent responsible for triggering the onset of the fast eruption? If so, then we would expect this reconnection to have started prior to the onset of the fast rise. Although the EUV data alone suggest that this may be the case (Sterling et al. (2001)),

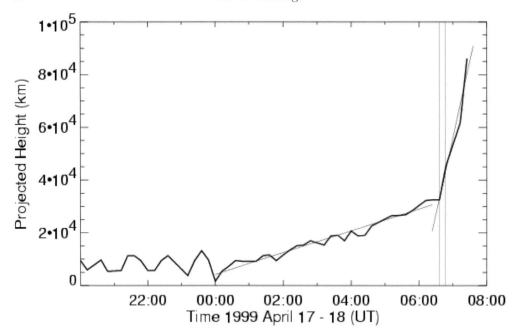

Figure 1. Rise trajectory of an erupting filament as a function of time, where the height is measured projected against the solar disk in EIT 195 Å images for an event of 1999 April 17—18 (Event 1). A nearly linear slow-phase "pre-eruption" rise between 0 UT and about 6:30 UT is followed by a rapid fast-phase rise; we define the eruption onset as occurring between the times of the two vertical lines. Overplotted on the trajectory are two linear fits, giving line-of-sight velocities of ~ 1 km s^{-1} and ~ 15 km s^{-1} for the slow and fast phases, respectively.

Sterling & Moore (2004a) point out that the soft X-ray emission begins prior to the EUV emission. They were not, however, able to say whether the soft X-ray emission began before or just after the start of the fast rise.

In conclusion, this eruption shows characteristics of both breakout and tether cutting. Supplemental studies, e.g. combining our observations with theoretical predictions, should be able to tell us which, if either, mechanism was actually responsible for the onset of the fast eruption.

4.2. *Event 2*

Our second eruption example occurred on 1999 February 8—9, and involved a large-scale prominence that erupted from the north polar crown region. It was well observed in the EIT 284 Å (Fe xxv) filter, and by SXT. Sterling & Moore (2003) give full details of this event.

As in Event 1, this eruption also shows characteristics of a two-phase eruption (Fig. 2), but here the slow-rise phase is not as close to linear as the Event 1 case. This case provides a nice example of concurrent dimmings visible in soft X-rays and EUV, lending further evidence that the dimmings result from mass loss rather than from heating of plasma.

This was a particularly slow eruption, and the associated magnetic fields weak, making it difficult to search for expected pre-eruption breakout signatures. Strong soft X-ray emission did not begin until well after the start of the fast rise of the prominence, and this seems to be inconsistent with the idea that tether cutting reconnection is responsible for the onset of the fast eruption. Sterling & Moore (2003) show, however, that this is not necessarily the case; tether cutting could have occurred prior to onset of the fast

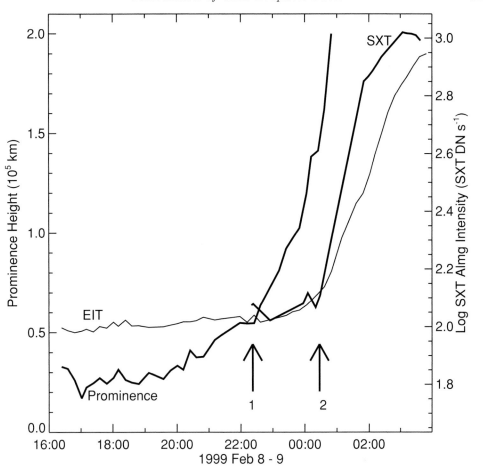

Figure 2. Curve labeled 'prominence' shows the height of the top of the prominence as a function of time for the eruption of 1999 Feb 8—9 (Event 2). Curves labeled 'SXT' and 'EIT' respectively show averaged integrated SXT and EIT intensity lightcurves over a spatially-localized region where flare brightenings first occurred (Sterling & Moore 2003 show the precise location of the region); the EIT lightcurve is plotted with an arbitrary vertical scale. Arrow 1 indicates the time where the prominence trajectory undergoes acceleration, i.e. the start of the transition from the slow-phase to the fast-phase of the eruption. Arrow 2 indicates the time where SXT intensity shows a sharp increase above background level.

eruption, but at such a weak level that soft X-ray emission was lost in the background coronal emission. Therefore, this example does not give us direct evidence for or against either eruption model.

4.3. *Event 3*

This event involved the eruption of a large filament near the solar limb, and was observed in EIT 195 Å images, SXT, and other instruments, as discussed in Sterling & Moore (2004b). Once again, it showed a two-stage evolution for the filament rise (Fig. 3). In addition to the filament itself, we also follow two features in the corona above the filament, one is a "suspended feature" consisting of a short filament-like patch of cool material, and the other is a bright coronal loop; both moved in conjunction with the rising filament. Sterling & Moore (2004b) argue that the filament and the suspended feature belonged to the same coronal magnetic cavity, and that the coronal loop was either part of the

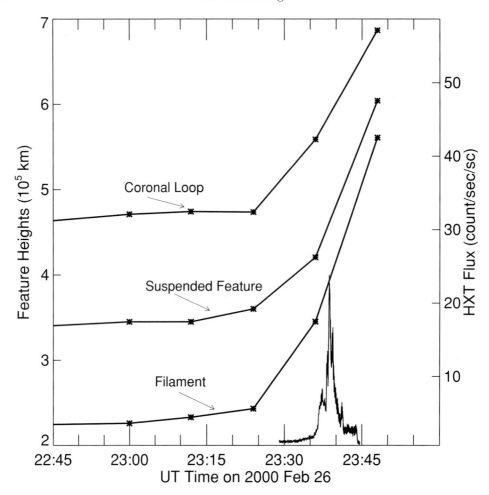

Figure 3. Trajectories of the filament, suspended feature, and coronal loop as functions of time for the eruption of 2000 February 26 (Event 3), measured projected against the disk in EIT images; flux from the Lo channel of *Yohkoh*'s hard X-ray telescope (HXT) is overplotted from 23:29 UT. Slow pre-eruption motion starts near 23:00 UT and fast eruption starts near 23:25 UT.

cavity or arched over the cavity. This example, along with a second, similar example in Sterling & Moore (2004b), suggest that entire magnetic systems, in this case consisting of the filament, suspended feature, and the larger magnetic cavity in which they resided, underwent a two-stage rise process during eruption.

As with Event 1, this event occurred in a quadrupolar magnetic geometry, and showed evolution consistent with breakout during the pre-eruption slow-rise phase. Also, brightenings occurred in the core region consistent with tether-cutting reconnection. Once again, we need additional information (e.g., combinations of our data with numerical simulations) to tell us whether either of the two models was responsible for triggering the onset of the eruption.

5. Discussion

Each of our three filament eruption examples show a slow-rise phase followed by a fast-rise (eruptive) phase, although the change in rise rate is less obvious in the Event 2 case. This change in rise rate has been observed previously, e.g. Kahler et al. (1988), and similar rise trajectories have been seen in other phenomena (e.g., Ohyama & Shibata (1997)). For our Events 1 and 3, we find evidence that breakout-type coronal reconnection may be occurring in the respective magnetic systems during the slow-rise phases. For all three cases, tether-cutting-type reconnection occurs, resulting in soft X-ray flares and associated phenomena. With these observations alone we are not able to say conclusively which, if either, of these proposed mechanisms is responsible for triggering the onset of these eruptions. We encourage modelers to combine our observations with detailed numerical simulations mimicking the setup of these specific events; such studies would lend support for or against these (or other) theories. Further detailed morphological and quantitative studies of similar events promise to yield further insights into the eruption-onset process.

Acknowledgements

This work was supported by funding from NASA's Office of Space Science through the Solar Physics Supporting Research and Technology Program and the Sun-Earth Connection Guest Investigator Program. *Yohkoh* is a mission of the Institute of Space and Astronautical Sciences (Japan), with participation from the US and UK, and *SOHO* is a project of international cooperation between ESA and NASA. The author thanks the IAU Symposium committee for inviting him to speak at this conference.

References

Antiochos, S. K. 1998, *apj* 502, L181
Antiochos, S. K., DeVore, C. R., & Klimchuk, J. A. 1999, *apj* 510, 485
Forbes, T. G. 2000, *JGR* 105, 23,153
Gopalswamy, N., & Hanaoka, Y. 1998, *ApJ* 498, L179
Harra, L. K. & Sterling, A. C. 2001, *ApJ* 561, 215
Howard, T. A., & Harrison, R. A. 2004, *Solar Phys.* 219, 315
Hudson, H. S. et al. 1998, *Geophys. Res. Lett.* 25, 2481
Kahler, S. W. et al. 1998, *Solar Phys.* 328, 824
Klimchuk, J. A. 2001, in: Space Weather, Geophysical Monograph, 125, 143
Lin, J., Soon, W., & Baliunas, S. L. 2003, *New Astron. Reviews* 47, 53
Manoharan, P. K. et al. 1996, *ApJ* 468, L73
Moore, R. L., & LaBonte, B. 1980, in: Proc. Symp. on Solar and Interplanetary Dynamics, Reidel, Boston, 207
Moore, R. L., Sterling, A. C., Hudson, H. S., & Lemen, J. R. 2001, *ApJ* 552, 833
Ohyama, M., & Shibata, K. 1997, *PASJ* 49, 249
Sterling, A. C., & Hudson, H. S. 1997, *ApJ* 491, L55
Sterling, A. C., Moore, R. L., & Thompson, B. J. 2001, *ApJ* 561, L219
Sterling, A. C., & Moore, R. L. 2003, *ApJ* 599, 1418
Sterling, A. C., & Moore, R. L. 2004a, *ApJ* 602, 1024
Sterling, A. C., & Moore, R. L. 2004b, *ApJ* 613, 1221
Sturrock, P. A. 1989, *Solar Phys.*, 121, 387
Thompson, B. J. et al. 1998, *Geophys. Res. Lett.* 25, 2465
Thompson, B. J. et al. 2000, *Geophys. Res. Lett.* 27, 1431

Discussion

SCHMIEDER: 1. In the Moore cartoon, there are loops on both sides of the core of the flare which expand. Did you see them in your observations?
2. In your example "events", you say that there is a quadrupolar reconnection. Did you see the brightening of the ribbons corresponding to such a reconnection?

STERLING: 1. Fig. 1 of Moore et al. (2001) shows a cartoon where loops bulge out from both ends of the core; we sometimes refer to these bulging loops as "elbows" to the core field. Prior to eruption we have most often seen these elbow loops in hotter coronal (soft X-ray) images, as in the examples of Moore et al. (2001). But our three events here all occur in quiet regions, where the magnetic fields are not strong enough to result in significant pre-eruption soft X-ray emission. Therefore we do not see these loops before eruption for these cases. For at least the first event, however, we can infer that such loops existed prior to eruption based on double dimming patterns in EUV difference images of that event (see Sterling and Moore 2004a, Fig. 2).
2. Not really, the images are too noisy. We see some evidence of both dimming and brightening at the expected outer ribbon locations, but these are too close to the noise level to be sure; recall that this event is very weak, so such ribbons might be present, but very weak. The inner ribbons correspond to the main flare ribbons, which we do see.

JIE ZHANG: 1. Are EIT waves and dimming different phenomena, or the same phenomena but only different in intensity?
2. The breakout model involves a multipolar region and reconnection at the top. But in terms of the main energy release, do they both involve reconnection in the deep core field? Do they both have the same main energy source?

STERLING: 1. There seems to be at least two phases of wave propagation: (1) A phase where the waves have dimming behind its front, and (2) A phase where the wave continues on after the dimming stops propagating outward. For phase (1), the waves may be the same as the spread of the dimming, but I am not sure about this. For (2), the waves and dimming are almost certainly different. This two-phase wave idea is consistent with the model of Chen & Shibata (2000) as discussed in Harra & Sterling (2003).
2. I agree that in breakout and in tether cutting the main energy release is in the core, and you are right in saying that this is an important point. But another important question is "what are the conditions necessary for violent eruption?" The breakout proponents argue that early, slow reconnection far away from the core is essential for violent eruption to start. Tether cutting proponents say that reconnection in the core alone is sufficient for violent eruption.

DELABOUDINIERE: What is the role of the filament in the eruption ? – Disappearance of filament coincident with acceleration and change of ionization state of cold matter which is heated at coronal temperature. This is "explosive" and may create the EIT shock(?) waves and dimming. Is a pressure wave pushing the field open after energy has been deposited suddenly from contact between the filament and corona at about 0.5 solar radius altitude?

STERLING: These are good questions that are worthy of future consideration. So far we have only considered the filament to be a passive marker of the coronal magnetic field.

SHIBATA: 1. You mentioned that the tether-cutting model is fundamentally bipolar, and the breakout model is a multi-polar model. But I think this is not a good classification, because the tether-cutting process can occur in a multi-polar geometry. In other words, an observed multi-polar geometry does not necessarily support the breakout model.
2. In addition to this, your cartoon of event 3 is not the same as the breakout model, although the magnetic field configuration is multi-polar.
3. Your finding that the events tend to show an initial slow rise of a filament followed by a fast rise is very interesting and important. If you normalize the time scale by Alfven time, what will you obtain?

STERLING: 1. Yes, you are correct that multi-polar geometry does not necessarily support breakout; I try to emphasize this point in my papers. Our events 1 and 3 are consistent with breakout, in terms of multi-polar geometry, brightening of side lobes, etc. This does not prove breakout is occurring. Indeed, it could be that some other mechanism is responsible for triggering the eruption in a quadrupole geometry. In that case breakout-like effects would result as a byproduct of a more fundamental non-breakout mechanism.
2. That event's geometry is different from that of the standard breakout picture, but we argue that event 3 is basically similar to breakout if reconnection at the elevated null point has to occur in order to for the explosive eruption to be triggered. Although we see evidence for reconnection occurring at the elevated null point, we cannot say whether that reconnection is essential for the eruption.
3. Thank you, we are working on doing such a normalization.

Coronal and Stellar Mass Ejections
Proceedings IAU Symposium No. 226, 2005
K. P. Dere, J. Wang & Y. Yan, eds.

Flare-induced coronal disturbances observed with Norikura "NOGIS" coronagraph

K. Hori[1], K. Ichimoto[2], T. Sakurai[2], I. Sano[2], and Y. Nishino[2]

[1]Hiraiso Solar Observatory, National Institute of Information and Communications Technology, 3601 Isozaki, Hitachinaka, Ibaraki 311-1202, Japan
email: hori@nict.go.jp

[2]National Astronomical Observatory, National Institutes of Natural Sciences, 2-21-1 Ohsawa, Mitaka, Tokyo 181-8588, Japan

Abstract. A 2-dimensional Doppler coronagraph "NOGIS" (NOrikura Green-line Imaging System) at the Norikura Solar Observatory, NAOJ, is a unique imaging system that can provide both intensity and Doppler velocity of 2 MK plasma from the green coronal line emission $\lambda5303$ Å of Fe XIV. We present the first detection of a CME onset by NOGIS. The event was originally induced by a C9.1 confined flare that occurred on 2003 June 1 at an active region NOAA #10365 near the limb. This flare triggered a filament eruption in AR 10365, which later evolved into a partial halo CME as well as an M6.5 flare at the same AR 10365 on 2003 June 2. The CME originated in a complex of two neighboring magnetic flux systems across the solar equator: AR 10365 and a bundle of face-on tall coronal loops. NOGIS observed i) a density enhancement in between the two flux systems in the early phase, ii) a blue-shifted bubble and jet that later appeared as (a part of) the CME, and iii) a red-shifted wave that triggered a periodic fluctuations in Doppler shifts in the face-on loops. These features are crucial to understand unsolved problems on a CME initiation (e.g., mass supply, magnetic configuration, and trigger mechanism) and on coronal loop oscillations (e.g., trigger and damping mechanisms). We stress a possibility that interaction between separatrices of the two flux systems played a key role on our event.

Keywords. Sun: corona, flares, coronal mass ejections (CMEs), oscillations

1. Introduction

The green coronal line $\lambda5303$ Å of Fe XIV (2 MK formation temperature) is important to diagnose various coronal disturbances such as coronal waves or periodic oscillations. Spectroscopic observation is useful to get Doppler information with a high time resolution, although spatial information is limited to 1-dimensional along the slit (e.g., Koutchmy et al. 1983; Tsubaki 1988). The Fe XIV line is also used for 2-dimensional imaging observations such as the LASCO C1 coronagraph on the SOHO spacecraft (1.1–3 Rs with a pixel resolution of 5.6″), the Mirror Coronagraph for Argentina (MICA, 1.05–2 Rs with 3.7″/pix), and the Solar Eclipse Corona Imaging System (SECIS, 4.07″/pix) for the total eclipse. These instruments can monitor global coronal disturbances but Doppler information has not been available so far.

The intensity and Doppler imaging observation with the 2-dimensional Doppler coronagraph, NOGIS started in 1997 at the Norikura Solar Observatory, NAOJ (Ichimoto et al. 1999). The Doppler images (Dopplergrams) are constructed by subtracting a $\lambda - 0.45$ Å image from a $\lambda + 0.45$ Å image, which can provide the line-of-sight velocity up to ± 25 km s^{-1} with an accuracy of 0.6 km s^{-1}. Hence, the target phenomena suitable for NOGIS are coronal waves and flows, rather than fast ejections. NOGIS has a field of view of 2000×2000 pixels in a full frame mode and a spatial resolution of 1.84″ in a partial frame mode. Time resolution is reduced to 40 sec to increase S/N.

On 2003 June 1–2, NOGIS continuously observed a birth place of a CME that originated in a complex of two neighboring magnetic flux systems across the solar equator; a flare-productive active region, NOAA #10365, and a bundle of face-on coronal loops overarching a quiescent filament. By combining optical, EUV, and radio data, Hori et al.(2005) reported this event in detail with a scenario that can explain the whole observed evolution. Here we present an outline of Hori et al. (2005). We briefly discuss a propagating wave that was detected in Doppler shifts but not in intensity variations.

2. Event Overview

The huge coronal disturbances were observed above the west limb, in association with two limb flares that successively occurred at AR 10365. In Figure 1 (d), the GOES soft X-ray light curves of the two flares are shown with a solid line (1.0–8.0 Å) and a dashed line (0.5–4.0 Å). The two vertical dashed lines in Figure 1 indicate the start time of each flare; a C9.1 flare from 23:23 UT on June 1 and an M6.5 flare from 00:07 UT on June 2. The first flare was less eruptive while the second flare was associated with metric Type-II and -IV bursts, and a partial halo CME. According to TRACE 195Å (1.6 MK) images, the C9.1 flare activated a filament in AR 10365. Although the filament started to erupt with a mean plane-of-sky speed of 64 km s^{-1}, the eruption seems to have stagnated at 0.04 Rs above the limb. The height-time profile of the filament is given with asterisks in panel (d). After 15 min TRACE data gap (indicated by an arrow in panel (d)), the filament again started to erupt at 23:46 UT with a velocity of ∼67 km s^{-1} during the decay phase of the C9.1 flare. This eruption (or an associated disturbance) appeared in the NOGIS field-of-view from 00:05 UT as an expanding dark bubble (or a less dense region at 2 MK), whose height-time profile is shown with triangles in panel (d).

In Figure 1, the first and second panels show time slices of NOGIS Dopplergrams (a) and intensity maps (b) at a height of 0.15 Rs from the west limb. The vertical axes show the heliographic latitude in degree and the horizontal axes show the time in UT for the same period as in panels (c) and (d). White means red-shifts (a) or intensity (density) enhancement (b), while black means blue-shifts (a) or intensity decrease (b). In diagram (a), the color is normalized to the line-of-sight velocity of ±5.5 km s^{-1}. These diagrams show coronal disturbances propagating in north-south direction within or above two neighboring magnetic flux systems; a bundle of face-on coronal loops (N20–S05) and AR 10365 (S05–S29). Before the start of the two flares, NOGIS observed a formation of dense, 2 MK region in the space bounded between AR 10365 and the southern legs of the face-on loop system. The bright horizontal bands in panel (b) correspond to a slice of the dense region. According to NOGIS radial slices (not shown), this dense region slowly moved upward, apparently tracing EUV elongated structures standing within ±10° in latitude. In panel (c), from the top to the bottom, the plots in different colors show time evolutions of the maximum value in NOGIS intensity slices at heights of 0.08, 0.1, 0.13, 0.15, and 0.18 Rs from the west limb. As clearly seen in the lowest slice (black line), the coronal intensity above the limb peaked and then turned to decrease a few minutes prior to the start time of each flare.

As the filament erupted from AR 10365, the height of the outermost part of AR 10365 grew upward in NOGIS intensity images. The dark bubble mentioned above started to expand from the boundary of the growing AR 10365 and the overlying dense region when the two regions came into contact (00:05 UT). In Figure 1 panel (b), the cone-shaped weak dimming appearing from the latitudinal range of S7.5–S12.5 corresponds to the region swept by the bubble. From NOGIS radial slices (not shown), we confirmed that the bubble expanded both inward and outward with a projected speed of ∼140 km s^{-1}.

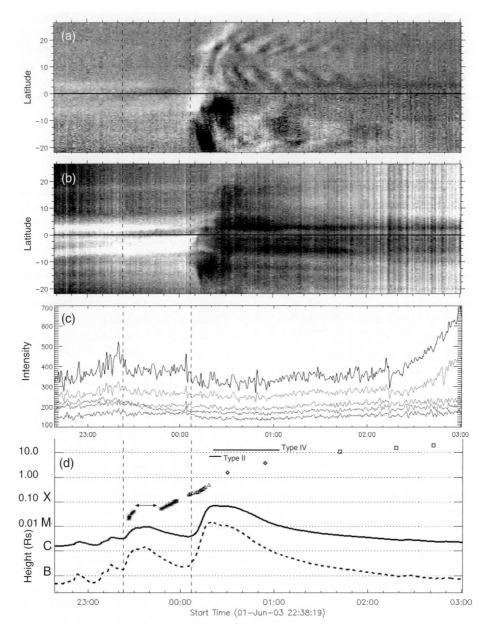

Figure 1. Time slices of NOGIS Dopplergrams (a) and intensity maps (b) at a height of 0.15 Rs from the west limb. The vertical axes show the heliographic latitude in degree. (c) From the top to the bottom, time profiles of the maximum intensity in the NOGIS time slices at heights of 0.08, 0.1, 0.13, 0.15, and 0.18 Rs from the west limb (in arbitrary scale). (d) Top: the height-time plots for the filament observed by TRACE (asterisks, the arrow indicates the period of a data gap), the expanding bubble observed by NOGIS (triangles), and the CME leading front observed by LASCO C2 (diamonds) and C3 (squares). Bottom: Time profiles of GOES 10 X-ray flux in 1.0–8.0 Å (thick solid line) and 0.5–4.0 Å (thick dashed line). All horizontal axes show the time in UT for the same period. The vertical dotted lines indicate the start time of two GOES flares; C9.1 (23:23–23:37–23:48 UT on June 1) and M6.5 (00:07–00:22–00:43 UT on June 2). The two horizontal bars indicate the periods of the metric Type-II (00:19–00:26) and -IV (00:20–01:03) from Culgoora.

When the downward edge of the bubble arrived at the lower part of AR 10365 (00:10 UT), a blue-shifted jet was ejected upward with a velocity of $\sim$400 km s^{-1} from the interface of the two neighboring magnetic flux systems (S05). The jet soon overtook the preceding upward front of the bubble, pushed it from its behind, and accelerated it to $\sim$360 km s^{-1}. The impulsive phase of the second flare (M6.5) also started at 00:10 UT in AR 10365 (S08), which was defined by 17 GHz radiation from Nobeyama radioheliograph.

The collision of the two dense regions also produced a red-shifted velocity disturbance. In Figure 1 panel (a), the white arc appearing around S05 at 00:05 UT together with the cone-shaped dimming in panel (b) suggests that the disturbance propagated northward with a velocity of > 1000 km s^{-1}, which is much faster than the projected expansion speed of the bubble. In the northern hemisphere, the disturbance pushed the face-on loop system anti-earthward and triggered damping oscillations in Doppler shifts among the adjacent loops within the system (the zebra pattern in panel (a)). The oscillations continued over 100 minutes with amplitudes of $< \pm5$ km s^{-1} and periods in the range of 8–16 min.

The upward front of the bubble blew off the overlying dense region northward and extended toward the face-on loop system in the northern hemisphere (see the expansion of the dimming region in Figure 1 panel (b)). After the passage of the bubble in the NOGIS field-of-view, a partial halo CME appeared above the west limb that had a mean plane-of-sky speed of $\sim$1500 km s^{-1} (or an acceleration of ~ 30 m s^{-2}). In Figure 1 panel (d), the height of the CME leading edge estimated from LASCO C2 and C3 images are plotted with diamonds and squares, respectively. The CME had an angular extent covering the latitudinal range of the two neighboring magnetic flux systems. The face-on loop system apparently remained at the same place, forming a cusp on its top.

3. Discussion

Using NOGIS, we observed huge coronal disturbances that were produced by a combined activity of two neighboring magnetic flux systems. The observed features (e.g., an expanding bubble, a propagating wave, and a jet) suggest that the two flux systems interacted (or reconnected) each other at the intersection of their magnetic separatrices (Beveridge, Priest, and Brown 2002). In between the two flux systems, 2 MK plasmas originated in the low corona had been accumulated since early phase. Through this dense region, the interaction was presumably triggered by a filament eruption that was induced by a C9.1 flare. The filament eruption resulted in a partial halo CME, an M6.5 flare, and coronal loop oscillations. The accumulated plasmas were blown off by the expanding bubble and thus (partially) contributed to the CME mass.

In our event, the propagating wave, or a red-shifted fast velocity disturbance, triggered damping oscillations in Doppler shifts among face-on coronal loops. Note that the wave was produced in the proximity of separatrices, which is a key to excite coronal loop oscillations as pointed by Schrijver and Brown (2000). On the basis of TRACE oscillation events (e.g., Aschwanden et al. 1999), Nakariakov (2003) described a flare-generated blast wave (fast-mode magnetoacoustic wave) as a possible excitation mechanism of kink oscillations of coronal loops (Figure 2, top). Hudson & Warmuth (2004) supports this idea considering a strong association of TRACE oscillation events with type II bursts and their temporal relationship. In our event, however, the wave was generated a few minutes earlier than the start time of the M6.5 flare (as well as the flare associated metric Type II burst). The M6.5 flare was not a generator of the wave. Instead, the wave might have induced the flare (Hori et al. 2005). Therefore, we consider another scenario that includes the role of magnetic separatrices (Figure 2 bottom). As the disturbance did not appear in

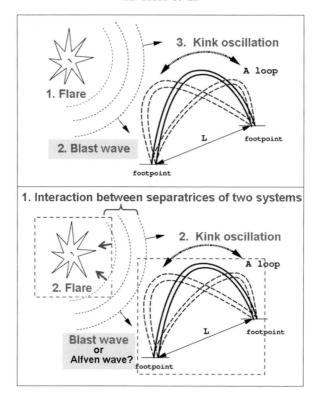

Figure 2. A possible mechanism for excitation of kink oscillations of coronal loops. Top: A flare-excited (#1) coronal blast wave (#2) may excite oscillations (#3) in a nearby flux system. After Nakariakov (2003) Bottom: An interaction between separatrices (#1) may excite flare (#2) and oscillations (#2) in each system.

intensity images (compare panels (a) and (b) in Figure 1), it might be an incompressible Alfvén wave, rather than a blast wave. It is, however, still possible that the oscillation in intensity (density) variations was too weak to be detected by NOGIS.

From the NOGIS observations, we learn the importance of ground-based imaging spectroscopy using visible lines for diagnosis of global coronal disturbances.

References

Aschwanden, M.J., Fletcher, L., Schrijver, C.J., Alexander, D. 1999, *ApJ* 520, 880

Beveridge, C., Priest, E. R., Brown, D. S. 2002, *Sol.Phys.* 209, 333

Hori, K., Ichimoto, K., Sakurai, T., Sano, I., Nishino, Y. 2005, *ApJ* 618, 1001.

Hudson, H.S., Warmuth, A. 2004, *ApJ* (Letters) 614, 85

Ichimoto, K., Noguchi, M., Tanaka, N., Kumagai, K., Shinoda, K., Nishino, T., Fukuda, T., Sakurai, T. 1999, *pasj* 51, 383

Koutchmy, S., Zhugzhda, Ia, D., Locans, V. 1983, *Sol.Phys.* 120, 185

Nakariakov, T 2003, in: B.N. Dwivedi (ed), *Dynamic Sun* (Cambridge University Press), pp.314-334

Schrijver, C.J., Brown, D.S. 2000, *ApJ* (Letters) 537, 69

Tsubaki, T 1988, in:Solar and stellar coronal structure and dynamics, Proceedings of the 9th Sacramento Peak Summer Symposium, pp.140-149

Discussion

KOUTCHMY: Before the flare of June 1-2 you observed a density enhancement and after the flare you observed a dimming. Would you suggest this is due to a mass loss or can you explain this observation in term of temperature changes? (assuming that the mass of the CME is provided by the filament.)

HORI: That's a good question. We actually discussed it very seriously. In our event the CME occurred in between the two flux systems and both regions showed the dimming; not only the flare region but also a relatively cold region where the kink oscillation was observed. If the dimming was due to a temperature effect, we expect that only the hot region might become dark. Thus we concluded that this was due to the mass loss, i.e, the CME.

JIE ZHANG: Could you say something about the instrument, the quality, say, how many CME are observed per year?

HORI: The quality of the observation depends on the weather at the top of the Norikura Mountain. The event which I showed you was the clearest example. I do not know the exact number of the CMEs observed so far.

GOPALSWARY : Comment: You mentioned that this is the first detection of CME onset in coronal green line. However, a large number of coronal green line transients were observed and studied in the seventies. (De Mastus et al. 1971).

HORI: Yes. But there was no 2D Doppler observations in coronal green line before NOGIS (Dr. Koutchmy answered).

Coronal and Stellar Mass Ejections
Proceedings IAU Symposium No. 226, 2005
K. P. Dere, J. Wang & Y. Yan, eds.

Determination of geometrical and kinematical properties of frontside halo coronal mass ejections (CMEs)

XuePu Zhao

W. W. Hansen Experimental Physics Laboratory, Stanford University, Stanford, CA
94305-4085, USA
email: xpzhao@solar.stanford.edu

Abstract. Recent studies show that the cone model with a circular cross section can be used to determine the geometrical and kinematical properties only for a class of halo CMEs with the semi-minor axis of the elliptic halo threading the solar disk center. This work shows how to use an improved cone model with an elliptic cross section to determine the geometrical and kinematical properties for another class of halo CMEs with the semi-major axis threading the solar disk center.

Keywords. coronal mass ejections (CMEs)

1. Introduction

Determination of the geometrical and kinematical properties of halo CMEs is necessary for understanding the cause of halo CMEs. It is also necessary for predicting the geoeffectiveness of frontside halo CMEs. For broadside CMEs, i.e., the CMEs with the latitudinal span of their bright feature in the plane of the sky being less than $120°$, the latitudinal span and its bisector are the measure of the angular width and central position angle of the broadside CMEs, respectively. Such geometrical properties for halo CMEs, i.e., the CMEs with the latitudinal span greater than $120°$, can not be measured directly from white-light images. The central position of halo CMEs is often assumed to be located near the associated surface activities such as solar flares, though it has been reported that CME-associated flares or active regionsare often located near one leg of CMEs.

The speed profiles of CMEs are often derived by choosing a specific feature in a time-lapse movie and tracking its position outward with time. For broadside CMEs the measured velocity and acceleration are nearly radial directed; they are, however, projected against the plane of the sky for halo CMEs.

We have developed a cone model with a circular cross section (Zhao, *et al.* (2002)) for determining the geometrical and kinematical properties of halo CMEs from coronal images. The present work shows how to use an improved cone model with an elliptic cross section to determine the geometrical and kinematical properties for another class of halo CMEs that can not be modeled using the original cone model.

2. Halo CMEs with minor axis threading solar disk center

Halo CMEs are interpreted as the result of the Thompson scattering of Sun's white light by a broad shell or bubble of dense plasma of CMEs along the line-of-sight (Howard, *et al.* (1982)). Specifically, the broad shell of dense plasma may be a conical shell with a circular cross section, as shown by Figure 4 of Howard, *et al.* (1982). It has been shown recently that the projection of the circular cross section against the plane of the sky has elliptic shape with the semi-minor axis threading the solar disk center (Xie, *et al.* (2004) and Zhao (2004)). Figure 1 shows the images for the 6 January 1997 halo CME observed at various time, t, and the ellipse calculated at corresponding heliocentric distance $r(t)$. The determined angular width and central position of the 6 January 1997 halo CME are shown at the top of the first panel.

3. Halo CMEs with major axis threading solar disk center

There are halo CMEs with their semi-major, instead of semi-minor, axis threading the solar disk center, suggesting that the corresponding cross section must be elliptic if the halo CMEs are formed by Thompson scattering of a conical shell of dense plasma. Figure 2 shows the images for the 7 April 1997 halo CME observed at various time, t, and the ellipse calculated at corresponding $r(t)$. The determined angular width and central position of the 7 April 1997 halo CME are shown at the top of the first panel.

Figure 3 shows the hight-time scatter-point plot based on the fitting result in Figure 2 and the velocity and acceleration inferred for the 7 April 1997 halo CMEs.

4. Summary and Disscussion

Halo CMEs exhibit various shapes. Some show ellipse (including circle), and others show ellipse-like but with ragged structure. Many elliptic halo CMEs may be interpreted as the Thompson scattering of Sun's white light by a conical shell of dense plasma along the line-of-sight (Howard, *et al.* (1982)). The halo CMEs with ragged structure is interpreted as the deflection of preexisting coronal features by super-Alfvenic CMEs (St. Cyr & Hundhausen (1998)). The ragged structure may be used to distinguish the deflection-formed halo CME from the scattering-formed halo CME (Sheeley *et al.* (2000)).

Recent studies have shown that the cone model with a circular cross section can be used to determine geometrical and kinematical properties only for a specific class of scattering-formed elliptic halo CMEs of which the minor axis threading the solar disk center. We show that the scattering-formed elliptic halo CMEs with their major axis threading the solar disk center may be modeled using the cone model with an elliptic cross section, and their geometrical and kinematical properties may also be determined.

There are evidence that the plasma structure for some halo CMEs may be approximated using the spherical shell model (Howard, *et al.* (1982)) or the ice-cream cone model (Fisher & Munro (1984)). It is interesting to develop a model for determining the geometrical and kinematical properties for those halo CMEs.

Acknowledgements

References

Fisher, R.R. & Munro, R.H. 1984, *Astrophys. J.* 280, 428

Howard, R.A., Michels, D.J., Sheeley Jr., N.R., & Koomen, M.J. 1982, *Astrophys. J.* 263, L101

Sheeley, Jr., N.R., Hakala, W.N. & Wang, Y.-M. 2000, *J. Geophys. Res.* 105, A3, 5081

St. Cyr, O.C. & Hundhausen, A.J. 1998, in: V.P. Pizzo, T. Holzer, & D.G. Sime (ed.), *Proceedings of the 6th International Solar Wind Conference* (Boulder), p. 235

Xie, H., Ofman, L. & Lawrence, G. 2004, *J. Geophys. Res* 109, A8

Zhao, X.P., Plunkett, S.P., & Liu, W. 2002, *J. Geophys. Res.* 107, A8

Zhao, X.P. 2004, in: A.V. Stepanov, E.E. Benevolenskaya & A.G. Kosovichev (eds.), *Multi-Wavelength Investigations of Solar Activity*, Proceedings of IAU 223, in press

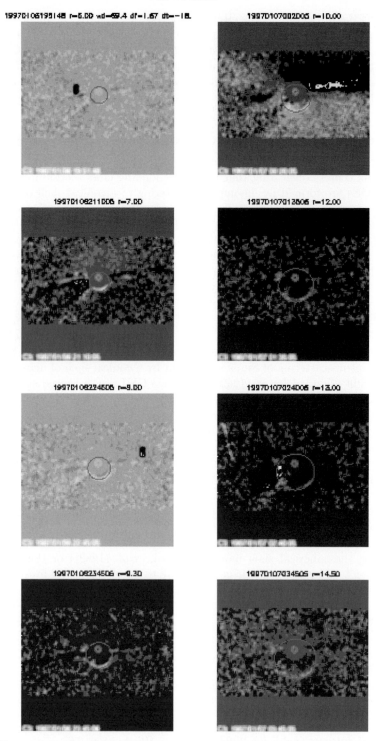

Figure 1. Comparison of the plane-of-sky projection of circular cross sections calculated using the cone model at various heliocentric distances $r(t)$ with the halos observed in difference images at various times, t.

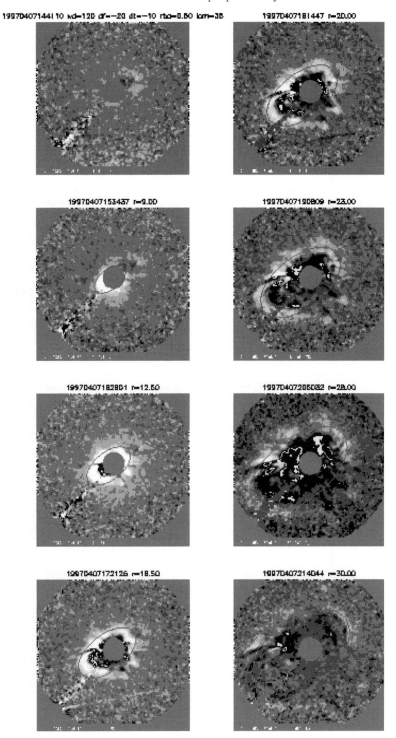

Figure 2. Comparison of the plane-of-sky projection of elliptic cross sections calculated using the cone model at various heliocentric distances $r(t)$ with the halos observed in difference images at various times, t.

X. P. Zhao

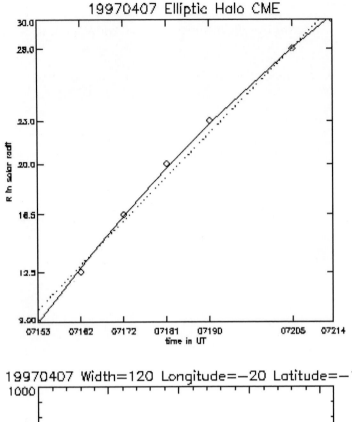

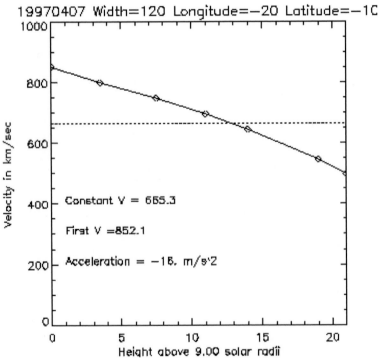

Figure 3. The hight-time scatter-point plot and fitting curve for the 7 April 1997 halo CMEs (the top panel) and the inferred variation of the velocity with hight (the bottom panel).

Discussion

P.F. CHEN: Have you statistically investigated the CME properties after the correction by your model ?

ZHAO: No, I just examined a few events. The number is not big enough to do any statistical analysis.

Coronal and Stellar Mass Ejections
Proceedings IAU Symposium No. 226, 2005
K. P. Dere, J. Wang & Y. Yan, eds.

Geometrical Properties of Coronal Mass Ejections

Hebe Cremades[1] and Volker Bothmer[1]

[1]Max-Planck-Institut für Sonnensystemforschung, Katlenburg-Lindau, D 37191, Germany
email: cremades@mps.mpg.de

Abstract. Based on the SOHO/LASCO dataset, a collection of "structured" coronal mass ejections (CMEs) has been compiled within the period 1996-2002, in order to analyze their three-dimensional configuration. These CME events exhibit white-light fine structures, likely indicative of their possible 3D topology. From a detailed investigation of the associated low coronal and photospheric source regions, a generic scheme has been deduced, which considers the white-light topology of a CME projected in the plane of the sky as being primarily dependent on the orientation and position of the source region's neutral line on the solar disk. The obtained results imply that structured CMEs are essentially organized along a symmetry axis, in a cylindrical manner. The measured dimensions of the cylinder's base and length yield a ratio of 1.6. These CMEs seem to be better approximated by elliptic cones, rather than by the classical ice cream cone, characterized by a circular cross section.

Keywords. Sun: corona, Sun: coronal mass ejections (CMEs), Sun: prominences

1. Introduction

Coronal Mass Ejections (CMEs) constitute magnificent explosive events observed in the solar corona (e.g. Munro et al. 1979; Hundhausen et al. 1984). They were discovered by space-borne coronagraphs. Coronagraphs record the density of free electrons in the coronal plasma, integrated along the line of sight. The two-dimensional nature of coronagraphic images as well as the current restriction to single viewpoint observations constrain our view and knowledge regarding the three-dimensional configuration and geometry of CMEs. In the earliest studies it had been proposed that CMEs may be planar looplike structures (e.g. Trottet & MacQueen 1980). On the contrary, other studies supported rather a shell-like configuration (e.g. Webb 1998). CMEs are presently known as three-dimensional structures seen in projection on the plane of the sky. However, it is still a subject of debate whether the 3D structure resembles more an arcade or a spherical bubble.

Cremades & Bothmer (2004) studied a set of 276 structured CMEs selected from the full set of LASCO (Large Angle Spectroscopic Coronagraph) C2 images during the period 1996-2002, in order to gain information about the three-dimensional configuration of CMEs. Their associated source regions in the low corona and photosphere could be uniquely identified in 124 of the cases. These were located essentially in limb and near-limb regions, as expected from the clarity of the observed (projected) CME profile and direction of propagation. From the analysis of the structured CMEs and their associated source regions, based on the comparison of magnetic structures in the low corona as recorded by the Extreme UV Imaging Telescope (EIT) with the profile of the structured white light CME as observed by the LASCO coronagraphs, a generic scheme for CME configurations was deduced. It basically considers the structured CMEs as cylindrical entities (as supported by the later study of Moran & Davila 2004), whose projected

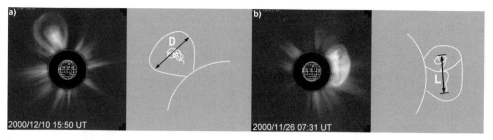

Figure 1. Examples of two CMEs seen in extreme projection: a) main axis primarily oriented along the line of sight, i.e. CME seen in cross-section with diameter D; b) main axis L oriented perpendicular to the line of sight.

profiles depend on the neutral line locations and orientations, consistent with the axes of the approximately cylindrically-symmetric CMEs.

In the next section we present width measurements for the subset of structured CMEs that exhibited extreme projections: seen approximately along their axis, or perpendicular to it. These two groups of projection yielded average widths rather distinct, revealing an asymmetry between the axial direction and the perpendicular one. In Sections 3 and 4 we attempt to find the same discrepancy in halo CMEs, via an elliptical cone model based on the circular cone model introduced by Zhao et al. (2002).

2. Typical Dimensions of cylindrically shaped CMEs

The high order of cylindrical symmetry evident from the analyses of the 124 structured CMEs with associated source regions listed in Cremades & Bothmer (2004) raises the question of whether the angular widths (AWs) of CMEs seen along their symmetry axis differ from those seen perpendicularly to it (see Fig. 1). In order to investigate the degree of similarity in the AWs of these two views, the structured CMEs exhibiting the two purest cases of projection -seen along their axis or perpendicular to it- were identified from the set of 124 structured CMEs. Examples of the two cases of projection are depicted in Fig. 1. Their AWs were measured in LASCO/C2 and, if the (leading edge was visible) in EIT 195 Å images, in EIT as well. Altogether we identified 16 events seen along their axis, out of which 9 were also measurable in EIT 195 Å; and 17 events seen perpendicular to their axis, out of which 9 were measurable in EIT 195 Å. The angular width "D" of the axial cases indicates the diameter of the imaginary cylinder's cross-section (Fig. 1a), while the width "L" of the events observed perpendicular to their axis denotes the length of the imaginary cylinder's axis (Fig. 1b), most likely aligned with the extended prominence material.

The AWs derived from the analyzed events were put together in distribution histograms for D and L as shown in Fig. 2. Each histogram shows in different colors the values measured in the LASCO/C2 and EIT 195 Å fields of view. Although limb CMEs seem to keep their AWs in the field of view of LASCO/C2 and C3, there is a marked difference between the widths measured in EIT and in LASCO/C2, supporting the assumption that the CME's greatest expansion takes place already between 1 and 2.2 solar radii, i.e. into the region of transition from open to closed coronal fields. The averages of D in EIT and in LASCO/C2 were 14° and 37°, measured at average heights of 1.3 and 4 solar radii, respectively. The larger averages of L in EIT and in LASCO/C2 yielded 22° and 58° respectively, measured at identical heights. The ratio L/D, considered with the purpose of comparing the average D and L of the imaginary cylinder, yielded the value of 1.6 for both EIT and LASCO/C2 data. That the ratio does not change from the EIT to the

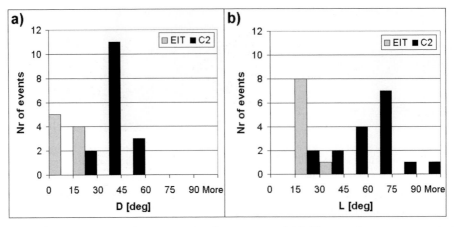

Figure 2. AW distribution histogram for the structured CMEs seen in extreme projection. Measurements carried out in EIT 195 Å are shown in light grey, and those in LASCO/C2 in dark grey. a) Main axis primarily oriented along the line of sight, b) main axis oriented perpendicular to the line of sight.

LASCO/C2 field of view for this particular set of events can be interpreted as a sign of self-similarity expansion, as addressed by some theoretical models (e.g. Gibson & Low 2000).

The arcade-like rather than bubble-like structure derived from the basic CME scheme implies an essential difference in the profiles projected in the plane of the sky, depending on the arcade's orientation with respect to the observer. If the arcade is approximated by a cylinder, the measured widths of its cross-section D and length L can be substantially different. It is expected that halo CMEs will present cases of asymmetry as well, as a consequence of the imprints of the source region characteristics. Halos entirely seen from top, i.e. originating from exactly the center of the solar disk, should not look circular, but rather elliptical. On the other hand, the presence of elliptical full halos in LASCO/C2/C3 observations could merely be a result of projection effects, i.e. of intrinsically circular halo CMEs traveling in other directions than exactly the line of sight.

The circular cone model introduced by Zhao et al. (2002) approximates CMEs as cones with circular base and allows the user to observe the base of this cone as projected on the plane of the sky for events propagating in different directions, apart from the line of sight. The result of projecting the base of a circular cone on the plane of the sky is typically an ellipse. Though after the implementation of the circular cone model from Zhao et al. (2002) still a substantial number of events could not be fitted, suggesting the existence of real elliptical halos, observed as such not only as an effect of projection. In order to prove the existence of asymmetric halos, an elliptical cone model was developed, in an effort to reproduce fairly well a set of halo CMEs observed by LASCO.

3. The elliptical cone model

The elliptical cone model was developed on the basis of the circular cone model presented by Zhao et al. (2002), assuming as well radial propagation and constant AWs. To generate an elliptical cone, not only one half-AW of the cone is needed, but two: one that grows in the direction of the semi minor axis and another that grows in the direction of the semi major axis, called ω_a and ω_b respectively (see Fig. 3). From these two half-AWs arise the two slant heights s_a and s_b, which are related by the condition $s_a \cos(\omega_a) = s_b \cos(\omega_b)$. The fact of producing an elliptical base involved the insertion of

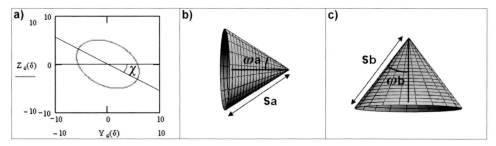

Figure 3. The three views of the elliptic cone. a) Plane Z_c-Y_c: the base of the cone. b) Plane Z_c-X_c: minor width of the cone. c) Plane Y_c-X_c: major width of the cone.

a new variable χ that defines the tilt of the semimajor axis with respect to the horizontal. Then the equations that define the cone in its own coordinate system are given by:

$$X_c(\delta) = s_a \cos(\omega_a) \tag{3.1}$$

$$Y_c(\delta) = s_b \sin(\omega_b) \cos(\delta) \cos(\chi) + s_a \sin(\omega_a) \sin(\delta) \sin(\chi) \tag{3.2}$$

$$Z_c(\delta) = s_a \sin(\omega_a) \sin(\delta) \cos(\chi) - s_b \sin(\omega_b) \cos(\delta) \sin(\chi) \tag{3.3}$$

Where, using the same nomenclature of the circular cone model when applicable, X_c is directed along the central axis of the cone, Y_c and Z_c are oriented in the directions of the semi major and semi minor axis respectively (whenever tilt $\chi = 0$), and δ is the parameter that varies from $0°$ to $360°$. The three different views of the cone are shown in Fig. 3.

To project the base's rim of the elliptical cone against the plane of the sky, the same transformation as applied by Zhao et al. (2002) was employed, by means of the heliographic latitude λ and longitude ϕ in which the central axis of the cone is oriented.

A system of six equations with six unknowns arises after matching the parametric equation of the ellipse measured in the LASCO/C2/C3 images and that of the ellipse synthetically generated by the model (representing the rim of the cone's base projected on the plane of the sky); making it possible to find analytical solutions for all the variables ω_a, ω_b, s_a, s_b, λ, ϕ and χ. Unfortunately, due to the fact of dealing with transcendent equations, the solution is not unique though constrained to four equivalent ranges of non-complex values, depending on the location of the start point of the ellipse. The central solution of the value range has been established as the most probable one, as a tradeoff solution for all cases.

4. Application to halo CMEs

In an effort to test whether the ratio L/D also compares to measurements of halo CMEs, the elliptical cone model was applied to a set of halo CMEs, selected from the halo database at ftp://ares.nrl.navy.mil/pub/lasco/halo. The 32 selected events were characterized for being full halos, front sided and very bright. The latter attribute was preferred because of the simplicity while choosing the boundary that defined the ellipse to be approximated by the model. To construct the ellipse that best approximated each full halo CME, the coordinates of five points that defined the ellipse were determined in the image of the sequence that showed the sharpest leading edge (see an example in Fig. 4). The parametric equations of the so obtained ellipse were matched with those of the projected cross section of an elliptical cone, obtaining in this way its characteristic geometric parameters. Although these geometric parameters are not unique, they have been calculated for all cases under the same conditions, which allows for comparisons between them.

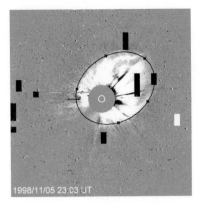

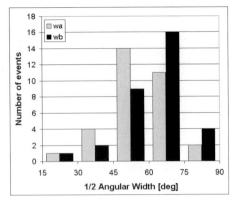

Figure 4. Running-difference image of a bright full halo recorded on November 5th 1998 by the LASCO/C3 coronagraph. The thick black line marks the fitted ellipse and the crosses the five points from which the ellipse was derived.

Figure 5. Half-AW distribution histogram for the analyzed halo CMEs. Light grey columns correspond to half-AWs in the direction of minor width (ω_a), while dark grey columns refer to half-AWs in the direction of major width (ω_b).

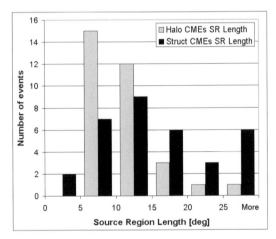

Figure 6. Distribution histogram for the source region lengths of both sets of analyzed CMEs.

The distribution histogram of the AWs ω_a and ω_b as measured in the semi minor and semi major directions (see Fig. 5) indicates that the asymmetry does also exist in halo CMEs, due to the shift visible in the two distributions for this set of 32 halo CMEs. The average ratio $< \omega_b/\omega_a >$ =1.11 is though not so remarkable when compared with the value $< L > / < D > $ = 1.6 calculated from the subset of structured CMEs having extreme projections.

The reason for the discrepancy in the symmetry ratio for the compared groups of CMEs (structured and halo) may lie in the different features that characterize both groups of events. The findings from Cremades & Bothmer (2004) suggest that the degree of cylindrical symmetry depends on the length of the neutral line that is associated with the CME origins. Fig. 6 displays the lengths of the associated source regions for both sets of CMEs. The distribution for the analyzed halo CMEs is narrower, with the average located at 11°; while that of the structured CMEs is broader, reaching the 40°, with the average located at 16°. The dissimilitude in the length distributions likely explains the discrepancies found in the two sets of CMEs.

5. Summary and conclusions

We have investigated a subset of structured CMEs analyzed in a previous study by Cremades & Bothmer (2004), characterized by the alignment of the main axis of symmetry either primarily along the line of sight or perpendicular to it. The average width of structured CMEs with the axis oriented along the line of sight (D) yielded 14° in EIT and 37° in LASCO/C2 images. Likewise, average widths of structured CMEs with the axis perpendicular to the line of sight (L) yielded greater values: 22° in EIT and 58° in LASCO/C2. The ratio L/D was identical as measured in the field of view of both instruments, yielding a value of 1.6.

An elliptical cone model was developed in a later step, on the basis of the circular cone model by Zhao et al. (2002). It was applied to a set of full, bright, and front-sided halos, looking forward to investigate a relationship similar to L/D in the fitted halos. The values representative for D and L, the half-widths ω_a and ω_b, yielded an average ratio $<\omega_b/\omega_a>$=1.11, substantially different from the expected value of 1.6.

If indeed the source region lengths play a prime role for the configuration of CMEs, as suggested by Cremades & Bothmer (2004), then it is expected that the source region characteristics for the set of analyzed halos should differ from those of the analyzed subset of structured CMEs. It is also important to know that the group of selected halos is most likely not representative of all CMEs, because the applied requirement of enhanced brightness implied attached selection effects, which might explain why most of these halo CMEs originated from compact, active regions. Finally, Cremades & Bothmer (2004) found for the structured CMEs that the lengths of their associated source regions systematically increased with latitude. These extended source regions at higher latitudes do not seem to be typical sources of very bright full halos. Out of the ecliptic missions like the ESA Solar Orbiter or the NASA Solar Probe would allow for the first time to prove these associations.

Acknowledgements

This work is part of the scientific investigations of the project Stereo/Corona, in context of the International Max Planck Research School, supported by the German "Bundesministerium für Bildung und Forschung" through the "Deutsche Zentrum für Luft- und Raumfahrt e.V." (DLR, German Space Agency) under project number 50 OC 0005. Stereo/Corona is a science and hardware contribution to the optical imaging package SECCHI currently being developed for the NASA STEREO mission to be launched in 2005. LASCO images used in this study are courtesy of the SOHO/LASCO consortium. SOHO is a project of international cooperation between ESA and NASA. The authors are thankful to Dr. Bernd Inhester and Rajat Thomas for their disinterested help.

References

Cremades, H., & Bothmer, V. 2004, A&A, 422, 307

Gibson, S. E., & Low, B. C. 2000, J. Geophys. Res., 105, 18187

Hundhausen, A. J., Sawyer, C. B., House, L., Illing, R. M. E., & Wagner, W. J. 1984, J. Geophys. Res. 89, 2639

Moran, T. G., & Davila, J. M. 2004, Science, 305, 66

Munro, R. H., Gosling, J. T., Hildner, E., et al. 1979, Solar Phys., 61, 201

Trottet, G., & MacQueen, R. M. 1980, Solar Phys., 68, 177

Webb, D. F. 1988, ApJ, 93, 1749

Zhao, X. P., Plunkett, S. P., & Liu, W. 2002, J. Geophys. Res., 107, 10.1029/2001JA009143

Discussion

GOPALSWAMY: I can understand that the filaments will not be seen well in white light when they are parallel to the limb. However, in the example you showed (SW limb), the filament (prominence) is also not seen well in EUV images, why is this?

CREMADES: Well, that is the case for other events as well. Surprisingly, even though we don't see the filament in EIT, the configuration of the CME observed in C2 is still consistent with the of the orientation of the neutral line. In any case, it is generally more difficult to identify filaments (prominences) in EIT 195 Å than in Hα.

KOUTCHMY : Regarding the source region of the CME, did you pay attention to the filament length and strength as well, or did you pay attention only to the magnetic configuration?

CREMADES: I also paid attention to the filament length (measured from Meudon and /or BBSO images). Neutral lines might be very large, but for the purpose of the measurement of the source region length, just the length of the erupting filament was the one taken into account. In the case of an active region in which no filament was observed, the length of the post-eruptive arcade of loops was considered as the source region length.

JIE ZHANG: The CME structure is related to the source region neutral line length and orientation, as you suggested. The question is, in many cases, the CME has a circular shape, but the source region is elongated. How can these be reconciled?

CREMADES: Not only do source region length and orientation play a role, but also other factors such as location, ambient coronal configuration, etc. A long source region perpendicular to the limb, would yield a CME seen along its axis, with circular shape and overlying loops. But as this source region becomes more parallel to the line of the limb, the elongation of the filament should become more apparent. Concerning halos, the high degree of circularity (although there were ones with eccentricity^{-1} = 1.3) is for the moment attributed to the relatively short length of the source regions.

Coronal and Stellar Mass Ejections
Proceedings IAU Symposium No. 226, 2005
K. P. Dere, J. Wang & Y. Yan, eds.

EIT waves – A signature of global magnetic restructuring in CMEs

P. F. Chen and C. Fang

Department of Astronomy, Nanjing University, Nanjing 210093, China
email: chenpf@nju.edu.cn

Abstract. The discovery of "EIT waves" after the launch of SOHO spacecraft sparked wide interest among the coronal mass ejection (CME) community since they may be crucial to the understanding of CMEs. However, the nature of this phenomenon is still being hotly debated between fast-mode wave explanation and non-wave explanation. Accumulating observations have shown various features of the "EIT waves". For example, they tend to be devoid of magnetic neutral lines and coronal holes; they may stop near the magnetic separatrix between the source region and a nearby active region; they may experience an acceleration from the vicinity of the source active region to the quiet region, and so on. This paper is aimed to review all these features, discuss how these observations may provide constraints for the theoretical models, and point out their implication to the understanding of CMEs.

Keywords. Sun: CMEs, waves, Sun: UV radiation, Sun: activity, magnetic fields

---

After the launch in 1995 December, *Solar and Heliospheric Observatory* (*SOHO*) space-craft (Domingo, Fleck, & Poland 1995) provided unprecedented views of the structures and activities of the Sun, which greatly enhanced our understanding of the solar interior, solar atmosphere, solar wind, coronal mass ejections (CMEs), and so on. One of the most striking discoveries is the ubiquity of waves propagating in the low corona, conventionally referred to as "EIT waves", concurrent with the launch of front-disk CMEs as revealed by the Fe XII 195Å running difference images of EUV Imaging Telescope (EIT; Delaboudinière *et al.* 1995) on *SOHO*. They not only manifest the low coronal signature of these more or less earth-directed CMEs, thus are very important for the space weather forecast, but also sparked hot debate of the relation between them and the Moreton waves discovered more than 40 years ago. This paper is aimed to piece the main properties of "EIT waves" together to see how they can be understood, and how the understanding can shed light on the nature of CMEs. § 1 gives a general description of the "EIT waves", § 2 reviews their properties which led to a fast-mode wave explanation, § 3 is devoted to their properties which led to a non-wave explanation, and § 4 summarizes how the diversity of their features can fit into a self-consistent picture, followed by a discussion of the implication of the understanding of "EIT waves" to the nature of CMEs in § 5.

1. Introduction

"EIT waves" are often observed as almost circular diffuse emission enhancements (ranging from 25% to less than 14%) propagating across the whole solar disk immediately followed by an expanding dimming region when the magnetic structure on the Sun is simple with only one active region on the disk (Thompson *et al.* 1998), as shown by figure 1. While, when the global magnetic structure gets complicated, they propagate rather inhomogeneously, avoiding strong magnetic features and neutral lines, as pointed

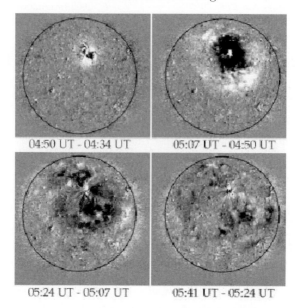

Figure 1. An example of an "EIT wave" event showing almost circular fronts when the magnetic structure on the Sun is simple (Thompson *et al.* 1998).

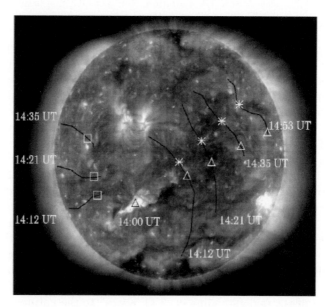

Figure 2. An example of an "EIT wave" event where fronts are propagating inhomogeneously when there are several active regions on the Sun (Thompson *et al.* 1999).

out by Thompson *et al.* (1999), who also found that "EIT waves" generally stop near coronal holes, as depicted in figure 2.

A careful examination by Biesecker *et al.* (2002) revealed that there is always a one-to-one correspondence from "EIT waves" to CMEs, though the contrary is not true. Their correlation with solar flares is significantly weaker, and they frequently are not accompanied by type II radio bursts (note that at least 90% type II bursts are associated with "EIT waves", as pointed out by Klassen *et al.* 2000). Therefore, "EIT waves" are a phenomenon intrinsically connected with CMEs. The speeds of these "EIT waves"-associated CMEs range from $\sim$200-1800 km s^{-1}, and there is no correlation between the occurrence of an "EIT wave" and the linearly fitted speed of the CME (Kay *et al.* 2003).

The statistical study by Klassen *et al.* (2000) indicates that the typical velocities of "EIT waves" range from 170 to 350 km s^{-1}, with a mean velocity of 271 km s^{-1}. Velocity as low as 80 km s^{-1} has also been reported (Dere *et al.* 1997). Roughly these values are three or more times smaller than the speeds of chromospheric Moreton waves (Moreton & Ramsey 1960). The start times of "EIT waves" seem to be earlier than the type II radio bursts when they are associated, and be roughly concurrent with the associated type III bursts, which can serve as the flare onset reference.

It is noted that some telescopes other than *SOHO*/EIT can also observe this wave phenomenon. For example, *TRACE* spacecraft captured several "EIT wave" events in both 171 and 195Å bands with better cadence but narrower field of view than *SOHO*/EIT (e.g., Wills-Davey & Thompson 1999). Besides, He I waves cospatial with "EIT waves" are observed by CHIP telescope at Mauna Loa via He I 10830Å line, the formation of which is strongly affected by the thermal parameters in the low corona (Gilbert *et al.* 2004).

2. Properties leading to the wave explanation and relevant modeling

More than 40 years ago, Hα off-band observations of some big flare events showed that a kind of chromospheric disturbances propagate from the flare site to distances on order of 5×0^5 km with a velocity ranging from 500 to 2000 km s^{-1}, which were later called Moreton waves (Moreton & Ramsey 1960). Since they could not be any wave of chromospheric origin, Uchida (1968, 1974) then proposed that the skirt of the wave front surface of a coronal fast-mode wave or even shock wave sweeps the chromosphere, and produces the Moreton waves. The shocked coronal wave is also the source region of the associated type II radio bursts. According to this self-consistent model, there should exist the coronal counterparts of the chromospheric Moreton waves, which could be observed in soft X-ray or EUV wavelengths.

EUV emission of the Sun comes mainly from the transition region and low corona up to heights of $0.15R_\odot$. Since "EIT waves", originating from the source active region, propagate across the quiet regions where the magnetic field lines are more or less radial, with a mean velocity significantly above the sound speed in the low corona (i.e., $\sim$179 km s^{-1}), they are widely believed to be fast-mode magnetoacoustic waves, i.e., the coronal counterparts of Hα Moreton waves (e.g., Klassen *et al.* 2000). The belief was strengthened by the fact that in some "EIT wave" events there is a sharp wave front which is cospatial with almost simultaneous Moreton wave front (Thompson *et al.* 2000b) as seen in figure 3, though the "EIT wave" speeds are statistically three or more times smaller than Moreton wave speeds. Regarding to this incompatibility, Warmuth *et al.* (2001, 2004) noticed that Moreton waves are visible only near the flare sites, while "EIT waves" are mostly observed at larger distances. Therefore, they postulated that there might be a deceleration during the wave propagation.

The wave hypothesis is tested by the model simulation of Wang (2000) and 3D numerical simulations of Wu *et al.* (2001) and Li, Zheng, & Wang (2002). All of them found that the propagation of fast-mode wave can match the observed "EIT wave" fronts, keeping in mind that all of them used the large-scale magnetograms of the Wilcox Solar Observatory to extrapolate the coronal potential field. As for the velocity discrepancy between EIT and Moreton waves, Wang (2000) postulated that the former could be ordinary waves, while the latter be super-Alfvén wave shock waves.

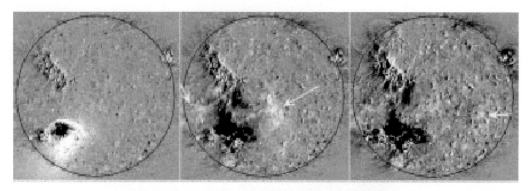

Figure 3. An example of an "EIT wave" event where one sharp bright EUV front is cospatial with the Hα Moreton wave (Thompson *et al.* 2000b).

3. Properties leading to a non-wave explanation and relevant modeling

Straightforward evidence to support non-fast mode wave explanations is the velocity discrepancy. Different from the proposal of Warmuth *et al.* (2001, 2004), Foley *et al.* (2003) found in an event that the "EIT waves" experience an acceleration rather than deceleration from the vicinity of the source active region to a large distance in the quiet region. Moreover, it is inferred from a winking filament by Eto *et al.* (2002) that the Moreton waves propagate much ahead of the "EIT waves" without deceleration at a large distance.

In the apparent appearances, it is noted by Thompson *et al.* (1999) that the relatively weak amplitudes and the diffuse fronts of the "EIT waves" indicate that they are not always be shocklike in nature. Another important point is that "EIT waves" often have circular shapes, while Moreton waves rarely span an angle of more than 160°. In the relation with type II radio bursts, it was well established that the speeds of the radio bursts exceed, but are typically proportional to, the associated Moreton wave velocities (Pinter 1977). In contrast, the recent statistical study by Klassen *et al.* (2000) pointed out that the radio burst speeds are not correlated with the "EIT wave" speeds, which also indicates that the "EIT waves" are not the coronal counterparts of the Hα Moreton waves. More serious doubt comes from the discovery of a stationary "EIT wave" front in some events by Delannée & Aulanier (1999) and Delannée (2000) (as shown by figure 4), who found that the "EIT waves" propagate initially, and then stop just near the footpoints of the magnetic separatrix, a characteristic far different from that of a magnetohydrodynamic wave. They suggested that the "EIT waves" should be related to the rearrangement of the magnetic structure in CMEs, where the bright fronts are due to the compression during the opening process of the magnetic structure.

There is some other indirect evidence to support that the "EIT waves" are not the coronal Moreton waves. For example, assuming that the shocked coronal Moreton waves are the acceleration site of energetic particles associated with CMEs, Bothmer *et al.* (1997) and Krucker, Larson, & Lin (1999) compared the "EIT wave" positions and the footpoint of the Parker spiral connecting the Sun to Earth along which energetic particles are transported, and found that the diffuse "EIT waves" are two times slower to propagate from the flare site to the particle acceleration site (or the coronal Moreton wave front).

By performing numerical simulation of the eruption of CMEs, Chen *et al.* (2002) found that two types of wave structures are associated with CMEs. One is the piston-driven shock straddling over the erupting flux rope and sweeping the solar surface with a

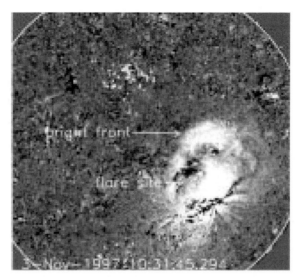

Figure 4. An example of an "EIT wave" event where the propagating "wave" stops near the footpoint of magnetic separatrix to form a stationary front (Delannée & Aulanier 1999).

super-Alfvén speed, which they proposed corresponds to the coronal counterparts of Hα Moreton waves; the other wavelike structure appears behind the coronal Moreton wave as propagating density enhancements, followed by an expanding dimming region, with a speed about three times smaller, which they proposed corresponds to the observed "EIT waves", as shown by figure 5. They further put forward a theoretical model for the phenomenon, which is sketched in figure 6, where "EIT waves" are thought to be formed by successive opening of magnetic field lines covering the erupting flux rope, and therefore, they are not real waves. The model self-consistently explained the relationship between "EIT waves" and EIT dimming, and solved the velocity discrepancy between "EIT waves" and Moreton waves. A further consideration of the model indicated that the thickness of "EIT wave" fronts increases as they propagate, and it would be comparable to the size of the dimming region behind it (Chen & Shibata 2002). This model was strongly supported by the delicate observational analysis by Harra & Sterling (2003), who found that two waves emanate from the flare site with different propagation velocities, and the Doppler motions are significant only behind the slower wave.

4. A full view of "EIT waves"

As pointed out by Biesecker & Thompson (2002), there is confusion in the literature as to what an "EIT wave" is, and multi-wavelength observations indicate that two types of waves are associated with CMEs, which correspond to the classical Moreton wave and to what we call the "EIT wave". Several factors are responsible for the long-standing confusion. The first is that the time cadence of EIT instrument is ∼15 min, which makes it difficult for the fast moving coronal Moreton wave to be detected in at least two consecutive images. Though *TRACE* spacecraft has a much higher cadence, its field of view at 195Å is a little small. The second factor is that there are far less Hα Moreton wave observations than "EIT waves". Besides, the velocities of "EIT waves" and Moreton waves overlap around 300-400 km s^{-1}, hence, it is hard to tell whether some events are high-speed "EIT waves" or low-speed coronal Moreton waves in EIT images. Even though, the hints of the existence of two types of EUV waves, i.e., coronal Moreton wave

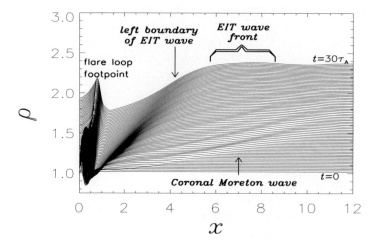

Figure 5. Numerical results showing a wavelike structure propagating behind the coronal Moreton wave with a velocity about three times smaller than coronal Moreton wave (Chen *et al.* 2002).

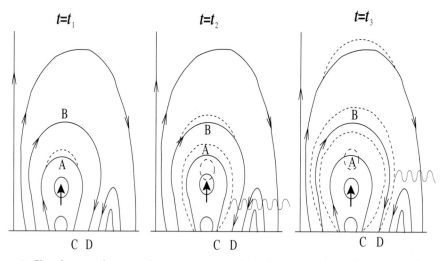

Figure 6. Sketch map showing the propagation of the EIT wave front from point C to point D, where the rising flux rope is the original source for the successive opening of all field lines. Solid lines represent the initial configuration, dashed lines correspond to the new configuration at each new time, while wave lines represent the fast-mode waves emitted from the EIT wave fronts. Moreton waves, much ahead of the EIT wave fronts, are not shown. (Chen *et al.* 2002).

and "EIT wave", distribute widely in the literature. Recently, Chen, Fang, & Shibata (2004) summarized the different observational characteristics of the two types of waves detectable in EUV images, and discussed how these properties can fit the model proposed by Chen *et al.* (2002):

(1) The coronal Moreton wave, like the first EUV wave front cospatial with Hα More-ton wave in Thompson *et al.* (2000b), appears as a sharp, extremely bright feature; In contrast, the "EIT wave" fronts are rather diffuse and faint. This is because the coronal Moreton wave is a shock wave in nature, while the "EIT wave" corresponds to propagat-ing large-amplitude perturbations formed by successive opening of closed field lines;

(2) Several fronts of the "EIT wave" may be captured in each event, while at most only one front of the coronal Moreton wave, which may have also been seen as the soft X-ray wave (Khan & Aurass 2002; Hudson *et al.* 2003), can be detected by *SOHO*/EIT in most "EIT wave" events owing to the low cadence. Only when the coronal Moreton wave has a speed near its lower limit, i.e., ~300-400 km s^{-1}, several EUV fronts can be found to be cospatial with Hα Moreton waves in a single event. Some of this kind of events were probably analyzed by Gilbert *et al.* (2004), Okamoto *et al.* (2004, the second event in their Table 1), and Vršnak *et al.* (2002);

(3) The coronal Moreton wave tends to propagate in a narrow direction where the magnetic field is weak owing to wave refraction. In contrast, a lot of "EIT waves" propagate almost circularly as all the field lines covering the flux rope are pulled up to open. They are devoid of magnetic neutral lines since no field lines are rooted there; they stop near magnetic separatrix, either between two active regions or near the boundary of coronal holes, to form a stationary front since the field lines outside the separatrix do not cover the erupting flux rope, and therefore do not open;

(4) The velocities of Moreton waves are almost proportional to those of type II radio bursts, since these two phenomena have the same driving agent, i.e., the piston-driven shock straddling over the CME. In contrast, there is a lack of correlation between the velocities of "EIT waves" and those of type II bursts since the former are determined by both the magnetic field strength and the magnetic geometry;

(5) Hα Moreton waves experience little deceleration during their propagation, while "EIT waves" may experience an acceleration from the vicinity of the source active region to quiet region far away (e.g., Foley *et al.* 2003). The smaller speed of the "EIT waves" near the active region is strongly against the wave explanation, and is accounted for by the stretched magnetic configuration in our model;

(6) Moreton wave fronts and their wake are not associated with strong Doppler motions, while the "EIT waves" are followed by an expanding dimming region where substantial Doppler motions are observed (Harra & Sterling 2001, 2003). This is because "EIT waves" are not oscillating waves. They are formed by successive opening of field lines covering the erupting flux rope, therefore, correspond to the propagation of large-amplitude perturbations with substantial Doppler motions.

5. Implication to CMEs and a remark

One of the intriguing questions about the onset of CMEs is how to connect relatively small-scale activity in the lower corona (e.g., solar flares) to the large-scale , often global, CME structures that are observed in the outer corona. As pointed out by Plunkett *et al.* (2002), "EIT waves" may be the means by which this cross-scale connection is established. After the onset of a CME, "EIT waves" propagate outward from the flare site to a large distance, immediately followed by an expanding dimming region. It was found that the EIT dimming region map the footprint of the CME (Thompson *et al.* 2000b; Harrison *et al.* 2003). Considering the spatial roughness and the nature of the two phenomena, we propose that it is the "EIT waves" that map the footprint of the CME leading edge, and the dimming region maps the the bottom of the CME cavity. Therefore, we extend the model sketch for the typical CMEs of Forbes (2000) to include "EIT waves" and EIT dimming as shown in figure 7. Along this line of thought, it is tentatively suggested that CME leading edges and cavities have the same formation mechanism as "EIT waves" and EIT dimmings, i.e., they are formed by successive opening of field lines covering the erupting flux rope, which was proposed by Chen *et al.* (2002). In a word, "EIT wave" fronts correspond to the lower legs of CME leading edges near the solar surface.

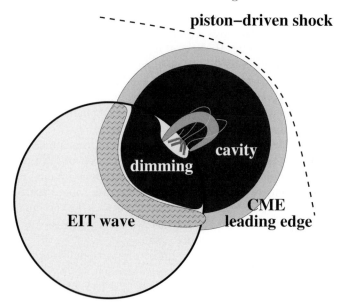

Figure 7. A sketch of our model indicating the relationship between CMEs and EIT waves/dimming.

Finally, a remark is made regarding the occurrence of "EIT waves". It has long been noticed that EIT dimmings occur much more frequently than "EIT waves". Similarly, Biesecker *et al.* (2002) pointed out that "EIT waves" are always associated with CMEs, which is consistent with our model that "EIT waves" are generated by successive opening of field lines in CMEs, while many CMEs have no associated "EIT waves", though large-scale magnetic rearrangement always occurs in CMEs. Chen, Fang, & Shibata (2004) suggested that the magnetic geometry could contribute to it since the compression outside the opening field lines is significant when the field lines are more or less perpendicular to the solar surface. If the field lines are strongly divergent outward as the initial condition in Chen & Shibata (2000), the compression outside the opening field lines is substantially weak, therefore, the resulting "EIT wave" fronts would be too faint to be detected. Besides the geometry effect, we emphasize here the role of temperature variation, since it is often assumed implicitly that "EIT waves" are purely due to the density enhancement. In fact, however, a 10% increase of the plasma density would be associated with an $\sim$ 7% increase of the plasma temperature in an adiabatic process. It is noted that the line emission contribution function for the Fe XII 195Å is sensitive to the temperature around 1.4×10^6 K. The $\sim$ 7% increase of the plasma temperature would result in an $\sim$ 17% decrease of the emissivity, which may explain the absence of "EIT waves" in many CME events. Or, the occurrence of the "EIT waves" favors the low corona environment with a temperature a little lower than 1.4×10^6 K, where the line emission contribution function increases with an increasing temperature.

Acknowledgements

The research is supported by FANEDD (200226), NSFC (10221001, 10333040 and 10403003) and NKBRSF (G20000784).

References

Biesecker, D. A., Myers, D. C., Thompson, B. J. *et al.* 2002, *ApJ* 569, 1009

Biesecker, D. A. & Thompson, B. J. 2002, *BAAS*, 34, 695

Bothmer, V. *et al.* 1997, in: A. Wilson (ed.), *Proc. 31st ESLAB Symposium* (Noordwijk: ESA), vol. (SP-415), p. 207

Chen, P. F., Fang, C., & Shibata, K. 2004, *ApJ*, submitted

Chen, P. F. & Shibata, K. 2000, *ApJ* 545, 524

Chen, P. F. & Shibata, K. 2002, in: S. Ikeuchi, J. Hearnshaw, & T. Hanawa (eds.), *Proc. of the IAU 8th Asian-Pacific Regional Meeting*, vol. II, p. 421

Chen, P. F., Wu, S. T., Shibata, K., & Fang, C. 2002, *ApJ*(Letters) 572, L99

Delaboudinière, J.-P. *et al.* 1995, *Solar Phys.* 162, 291

Delannée, C. 2000, *ApJ* 545, 512

Delannée, C. & Aulanier, G. 1999, *Solar Phys.* 190, 107

Dere, K. P., Brueckner, G. E., Howard, R. A. *et al.* 1997, *Solar Phys.* 175, 601

Domingo, V., Fleck, B., & Poland, A. I. 1995, *Solar Phys.* 162, 1

Eto, S., Isobe, H., Narukage, N. *et al.* 2002, *PASJ* 54, 481

Foley, C. R., Harra, L. K., Matthews, S. A., Culhane, J. L., Kitai, R. 2003, *A&A* 399, 749

Forbes, T. G. 2000, *J. Geophys. Res.* 105, 23153

Gilbert, H. R., Holzer, T. E., Thompson, B. J., & Burkepile, J. T. 2004, *ApJ* 607, 540

Harra, L. K. & Sterling, A. C. 2001, *ApJ*(Letters) 561, L215

Harra, L. K. & Sterling, A. C. 2003, *ApJ* 587, 429

Harrison, R. A., Bryans, P., Simnett, S. M., & Lyons, M. 2003, *A&A* 400, 1071

Hudson, H. S., Khan, J. I., Lemen, J. R., Nitta, N. V., & Uchida, Y. 2003, *Solar Phys.* 212, 121

Kay, H. R. M., Harra, L. K., Matthews, S. A. *et al.* 2003, *A&A* 400, 779

Khan, J. I. & Aurass, H. 2002, *A&A* 383, 1018

Klassen, A., Aurass, H., Mann, G., & Thompson, B. J. 2000, *A&A Suppl.* 141, 357

Krucker, S., Larson, D. E., & Lin, R. P. 1999, *ApJ* 519, 864

Li, B., Zheng, H.N. & Wang, S. 2002, *Chinese Phys. Lett.* 19, 1639

Moreton, G. E. & Ramsey, H. E. 1960, *PASP* 72, 357

Okamoto, T. J., Nakai, H., Keiyama, A. *et al.* 2004, *ApJ* 608, 1124

Plunkett, S. P., Michels, D. J., Howard, R. A. *et al.* 2002, *Adv. Space Res.*, 29, 1473

Thompson, B. J., Cliver, E. W., Nitta, N. *et al.* 2000a, *GRL* 27, 1431

Thompson, B. J., Gurman, J. B., Neupert, W. M. *et al.* 1999, *ApJ*(Letters) 517, L151

Thompson, B. J., Plunkett, S. P., Gurman, J. B. *et al.* 1998, *Geophys. Res. Lett.* 25, 2465

Thompson, B. J., Reynolds, B., Aurass, H. *et al.* 2000b, *Solar Phys.* 193, 161

Uchida, Y. 1968, *Solar Phys.* 4, 30

Uchida, Y. 1974, *Solar Phys.* 39, 431

Vršnak, B., Warmuth, A., Brajša, R., & Hanslmeier, A. 2002, *A&A* 394, 299

Wang, Y. -M. 2000, *ApJ*(Letters) 543, L89

Warmuth, A., Vrsnak, B., Aurass, H., & Hanslmeier, A. 2001, *ApJ*(Letters) 560, L105

Warmuth, A., Vršnak, B., Magdalenić, J., Hanslmeier, A., & Otruba, W. 2004, *A&A* 418, 1101

Wills-Davey M. J. & Thompson, B. J. 1999, *Solar Phys.* 190, 467

Wu, S. T., Zheng, H. N., Wang, S. *et al.* 2001, *J. Geophys. Res.* 106, 25089

Discussion

KAHLER: Do you expect to see a correlation between EIT wave speeds and associated CME speeds in your model?

CHEN: Not for the leading edge speeds of CMEs. There may be a correlation between EIT wave speeds and the expansion speeds of the CME footpoints, which in fact map into the EIT wave front.

SLEMZIN: In your model of coronal wave, do the waves exist in the transition region or not?

CHEN: They should exist there, while the strength of the wave front will rapidly decrease as the formation height of the emission line decreases.

KOUTCHMY: I am wondering how important is gravity in your simulations? I see a good reason to have a faster speed (velocity): they propagate horizontally without work produced against gravity, which is not the case of EIT (coronal) waves!

CHEN: We did not include gravity for the current simulations. The important thing regarding the velocity discrepancy might be their different mechanisms for the two kinds of waves. The coronal counterpart of Moreton waves are fast-mode (shock) waves in nature, while EIT waves are formed by successive opening the closed field lines covering the erupting flux rope. The role of gravity is not so important, I suppose.

ZHUKOV: I have two comments: one to support your model and the second is against it. First, any interpretation of EIT waves as fast-mode wave - regardless of deceleration (as suggested by Warmeth et al., or presence of a shock - has the plasma beta in the quiet Sun more than 1. This assumption is doubtful. Second, in your model EIT wave propagates together with a dimming as a result of successive opening of field lines. However, in some events (e.g. May 12, 1997) the EIT wave propagates to longer distances than the dimming.

CHEN: Yes, the plasma beta in a major part of the lower corona might be much smaller than unity. As for your second comment, I should say that the EIT wave front is quite diffuse, with large width, which was explained by Chen & Shibata (2002).

JUN LIN: How fast is the CME associated with the EIT wave in your simulation?

CHEN: A little larger than the Moreton wave speed, i.e., about 800 Km/s, which is about 3 times faster than EIT waves.

IAU Symposium No. 226 on Coronal and Stellar Mass Ejection
Proceedings IAU Symposium No. 226, 2005
A.C. Editor, B.D. Editor & C.E. Editor, eds.
© 2005 International Astronomical Union
doi:10.1017/S1743921305000153

A Study on the Acceleration of Coronal Mass Ejections

Jie Zhang[1]

[1] School of Computational Sciences, George Mason University,
4400 University Drive, MSN 5C3, Fairfax, VA 22030, USA
email: jiez@scs.gmu.edu

Abstract.
We present a statistical study on the acceleration of CMEs. This study is based on 23 CME events best observed by SOHO LASCO/C1 coronagraph, which observes the inner corona from 1.1 to 3.0 R_S. The kinematic evolution of a CME has a distinct acceleration phase that mainly takes place in the inner corona. We find that the acceleration duration distribution ranges from 10 to 1100 min with a median (average) value at 54 min (169 min). The acceleration magnitude distribution ranges from 6 m s^{-2} to 947 m s^{-2} with a median (average) value at 209 m s^{-2} (280 m s^{-2}). We also find a good correlation between CME acceleration magnitude A (in unit of m s^{-2}) and acceleration duration T (in unit of min), which can be simply described as A=10000/T.

Keywords. Sun: corona, Sun: coronal mass ejection (CMEs)

1. Introduction

CME is a phenomenon observed as a systematic motion of a large scale coronal structure. It has been found that a CME's full kinematic evolution may undergo three distinct phases: an initiation phase of slow rising, a major acceleration phase of fast velocity increasing, and finally a propagation phase with minor velocity changing (Zhang *et al.* 2001). The first two phases mainly occur in the inner corona (e.g., $< 2.0\ R_S$), while the third phase is largely observed in the outer corona (e.g., $> 2.0\ R_S$) by traditional white light coronagraphs.

Based on thousands of CMEs observed by Solwind (Howard *et al.* 1985), SMM (Hundhausen *et al.* 1994) and SOHO/LASCO (Moon *et al.*2002, Yashiro *et al.* 2004), it has been consistently found that CME velocity distribution in the outer corona ranges from about 50 km s^{-1} to 2600 km s^{-1}; the average velocity in the distribution is about 400 km s^{-1} with a median velocity at 350 km s^{-1} and a peak velocity at 300 km s^{-1}.

On the other hand, the statistical knowledge about CME acceleration in the inner corona remains relatively poor. St.Cyr et al (1999) made a statistical study on CME acceleration using 46 CME features observed by ground-based Mauna Loa K-coronameter in the inner corona combined with SMM observations in the outer corona. They found that acceleration distribution ranged from -218 m s^{-2} to 3270 m s^{-2} with an average (median) value at 264 m s^{-2} (44 m s^{-2}). This might be the only statistical study on CME acceleration so far. However, they did not distinguish acceleration phase from subsequent propagation phase in their calculation. In this paper, we present a study on CME acceleration based on advanced LASCO observations.

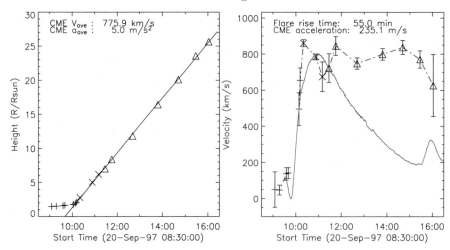

Figure 1. Kinematic plots for the CME on 1997 Sep. 20. The left panel shows the CME height-time measurements from LASCO C1(plus), C2(cross) and C3(triangle) coronagraphs. The straight line is a linear fit to the H-T in the outer corona. The right panel shows the CME velocity-time evolution (indicated by dotted-line-connected symbols with error bars) . The velocity is obtained by applying piece-wise numeric derivative to the H-T measurement. The smooth solid curve denotes the intensity-time profile of the associated soft X-ray flare.

2. Observation and method

We make uses of SOHO LASCO C1, C2 and C3 coronagraph observations, which have fields of view from 1.1 to 3.0, 2.2 to 6.0 and 4.0 to 30.0 R_S, respectively (Brueckner et al. 1995). The combined field of view of LASCO enables one to investigate the full kinematic evolution of CMEs. The C1 coronagraph provides key observations of CME acceleration in the inner corona. While C2 and C3 are typical white-light coronagraphs that observe Thomson-scattered photospheric light from free electrons in the corona, C1 observes the coronal emission lines (forbidden optical lines) from highly ionized ions. LASCO C1 has successful observations for a period of two and half year from 1996 Jan. to 1998 June; it unfortunately became out of function after the SOHO interruption in 1998. We have explored all C1 images obtained during that period, in total about 100,000 image. We are able to find 101 C1 CMEs. A C1 CME is that at least one CME leading edge is seen in C1. As a comparison, there are about 860 C2 CMEs during the same period. The much smaller number of C1 CMEs is simply due to the fact that the stray light level is much higher in C1 because of its closeness to the bright solar disk. As a consequence, most C1 CMEs are strong CMEs originated close to the limb. Please refer to Zhang et al. (2004) for a more detailed discussion about C1 observations. Because a full CME kinematic evolution may undergo three distinct phases: initiation phase, acceleration phase and propagation phase, a realistic measurement of CME acceleration should be limited to the time range of acceleration phase only.

We illustrate the method of calculating CME acceleration in Figure 1. The CME occurred on 1997 Sep. 20, observed by C1, C2 and C3 with 8, 3 and 7 snapshot images, respectively. The left panel shows the height-time (H-T) plot of the CME from direct measurement of CME leading edge in the images. The right panel shows the CME velocity-time (V-T), derived from first-order numerical derivative of the H-T measurement. Numerical derivative is essentially a piece-wise fitting to obtain local velocity at each time based on a small set of neighboring H-T points. This is in contrast to the nominal functional fitting, which is essentially a global method to obtain the average velocity (first order polynomial fitting) and average acceleration (second order polynomial fitting) of CME over the entire time range, without differentiating different kinematic phases.

CME acceleration is obtained by examining the V-T plot (see Figure 1, right panel), Based on the plot, the acceleration started at about 9:40 UT, at which time the CME already had a velocity of about 120 km s^{-1}. Note that this CME started to rise at about 09:05 UT, but velocity kept low till 09:40 UT; this phase is so called initiation phase. The CME velocity peaked at about 10:20 UT, or about 40 minute after the onset of fast acceleration; this 40 minute is the acceleration duration of the CME acceleration phase. Because of the limitation of C1 observing cadence, which is irregular from a few minute to 20 minute, the acceleration duration thus obtained may subject to an uncertainty up to 20 minute. Another way to obtain the acceleration duration is to use the flare rise time as proxy. It has been found that CME major acceleration occurs largely during the rise phase of associated soft X-ray flare (Zhang *et al.* 2001, Zhang *et al.* 2004). While there may be time difference between CME and flare timing, the difference largely falls into the uncertainty in CME measurement. We calculate the CME acceleration magnitude to be 235 m s^{-2}, by simply adopting the flare rise time 55 minut as the CME acceleration duration and adopting the CME average velocity 776 km s^{-1} in the C2/C3 field of view as the CME velocity increasing during the acceleration phase. Note that thus calculated CME acceleration magnitude has an uncertainty, which may range from 10% for well observed events to up to 50% for relatively poor observed events.

To carry out the statistical study on CME acceleration, we choose 23 best observed C1 CME events among the 101 C1 CMEs in our list. Each of these 23 events has at least 3 CME leading edges seen in C1, as well as good observations in C2 and C3. We use the method illustrated above to calculate CME acceleration magnitude, which is based on the obtained CME acceleration duration and CME velocity increasing during this duration. This relatively large set of 23 events further demonstrate that there is a temporal coincidence rule between CME acceleration and flare flux rise. Therefore, we use flare rise time as CME acceleration duration when a CME is assocaited with a flare. When a CME is not associated with a flare (e.g., gradual event), or when a CME is originated from backside of the Sun with the possible accompanied flare not observable, we directly use V-T plot to estimate CME acceleration time.

3. Results

The CME acceleration distribution for the 23 events is shown in Figure 2 (cross symbols), along with average acceleration in the outer corona (plus symbols). Apparently, CME acceleration has a wide distribution from event to event. For the 23 events, the maximum value is 947 m s^{-2} and the minimum value is 6 m s^{-2}, with a medium value at 209 m s^{-2} and an average value at 280 m s^{-2}. These acceleration parameters should be treated as the CME true acceleration, which is measured during the acceleration phase only. These numbers are much different from that of CME acceleration measured in the outer corona. The acceleration distribution in the outer corona is within a much narrower range, with a minimum value at -13 m s^{-2} (or deceleration), a maximum value at 40 m s^{-2}, the medium value is 3 m s^{-2}, and the average value at 4 m s^{-2}. CME has a much smaller acceleration magnitude in the outer corona, in other words a more or less constant speed; this is so called propagation phase.

Another interesting result is shown in Figure 3. It seems that there is a linear anti-correlation between CME acceleration magnitude and CME acceleration duration (in logarithm scale). Both CME acceleration magnitude and duration vary by more than 2 orders of magnitude. The acceleration duration ranges from 10 to 1100 minute with

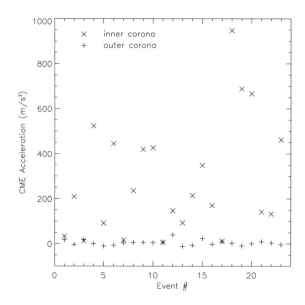

Figure 2. The comparison of CME acceleration in the acceleration phase (mainly in the inner corona, cross symbols) and in the outer corona (or in the propagation phase, plus symbols), for the 23 events with the best LASCO/C1 observations. The events are displayed in time order from left to right along the X-axis.

a median (average) value at 54 minute (170 minute), while the acceleration magnitude distribution ranges from 6 m s^{-2} to 947 m s^{-2}. The linear correlation is prominent across this broad range of distribution. Nevertheless, there is large scattering at local level, e.g, acceleration magnitude may vary by a factor of 5 for those CMEs with acceleration duration of tens of minutes. In a general sense, the weaker the CME acceleration magnitude, the longer the CME acceleration duration. For the cases of gradual CMEs with acceleration magnitude of less than 10 m s^{-2}, the acceleration duration can be as long as 1000 minute. On the other hand, the very impulsive events with acceleration duration less than 10 minute, the acceleration magnitude can be as high as 900 m s^{-2}. The linear fitting to the correlation yields a relation of acceleration magnitude A (in unit of m s^{-2}) and acceleration duration T (in unit of minute), simply described as A=10000/T.

4. Discussion

A CME's kinematic evolution has a distinct acceleration phase. The acceleration magnitude during this phase can vary significantly from event to event. For the 23 best observed C1 CMEs events from 1996 Jan. to 1998 June, the acceleration distribution is found to be from 6 m s^{-2} to 947 m s^{-2}. If considering all other events, e.g., extremely impulsive events, the acceleration magnitude distribution may extend to even higher values. The CME on 1997 Nov. 6 (associated with an X9.4 flare) was found to have an acceleration about 7300 m s^{-2} (Zhang *et al.* 2001). Alexander, Metcalf & Nitta (2002) reported that a CME-associated X-ray ejecta had a peak acceleration of about 4800 m s^{-2} based on YOHKOH SXT observations. Gallagher,Lawrence & Dennis (2003) found a CME-associated EUV ejection feature reaching a peak acceleration of about 1500 m s^{-2} based on TRACE observations. Therefore, it seems that there exists a continuous distribution of CME acceleration from extremely impulsive events down to extremely gradual events. However, it is still an open question whether there exists a single peak or multiple

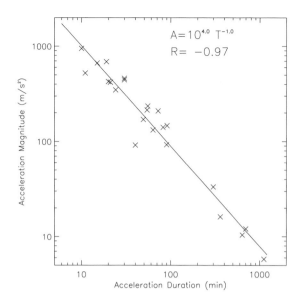

Figure 3. The scattering plot of CME acceleration magnitude versus CME acceleration duration measured in CME's acceleration phase, for the 23 events with the best LASCO/C1 observations. Note that there is a good linear correlation (correlation coefficient -0.97) between the two parameters

peaks on the acceleration distribution, which is related to the issue of CME classifications. To answer this question, a statistical study on a much larger number of events is needed.

Acknowledgements

J. Zhang is supported by NASA grants NAG5-11874 and 200730, and NSF SHINE grant ATM-0203226.

References

Alexander, D., Metcalf, T.R., Nitta, N.V. 2002 *Geophys. Res. Lett.* 29, No.10, 41
Brueckner, G.E., et al. 1995 *Sol. Phys.* 162, 357
Howard, R.A., Sheeley, N.R., Jr., Michels, D.J., Koomen, M.J. 1985 *J. Geophys. Res.* 90, 8173
Hundhausen, A.J., Burkepile, J.T., St. Cyr, O.C 1994. *J. Geophys. Res.* 99, 6543
Gallagher, P.T., Lawrence, G.R., Dennis, B.R. 2003 *ApJ* 588, L53
Moon, Y.-J., Choe, G.S., Wang, H., Park, Y.D., Gopalswamy, N., Yang, G., Yashiro, S. 2002 *ApJ* 581, 694
St. Cyr, O.C., Burkepile, J.T., Hundhausen, A.J., and Lecinski, A. R. 1999 *J. Geophys. Res.* 104, No. A6, 12493
Yashiro, S., Gopalswamy, N., Michalek, G., St. Cyr, O.C., Plunkett, S.P., Rich, N.B., Howard, R.A. 2004, *J. Geophys. Res.* 109, A7, 7105
Zhang, J., Dere, K.P., Howard, R.A., Kundu, M.R., and White, S.M. 2001, *ApJ* 559, 452
Zhang, J., Dere, K.P., Howard, R.A., Vourlidas, A. 2004, *ApJ* 604, 420

Discussion

MUNOZ: How were the events used in the analysis selected?

ZHANG: The two dozen events used in the analysis are those events having very good observations in LASCO C1. Specifically, at least three CME snapshot image are observed by C1. This is necessary for further analysis of velocity evolution in the inner corona.

GOPALSWARY: Comment: You talked about "true" acceleration in the inner corona ($< 2R_s$) and "residual" acceleration in the outer corona. This is some what misleading because the propelling and retarding forces act at all heliocentric distance. Propelling, retarding (gravity and drag) may have different magnitudes at different heliocentric distance, but the resultant is always "true".

ZHANG: The "true" acceleration I used refers to the phase of rapid velocity increase till the peak velocity. But I agree with your comment. The point is that the acceleration, in the inner corona, is much different from that in the outer corona.

Coronal and Stellar Mass Ejections
Proceedings IAU Symposium No. 226, 2005
K. P. Dere, J. Wang & Y. Yan, eds.

UVCS Observations of a Helical CME Structure

R. M. Suleiman[1,2], N. U. Crooker[2], J. C. Raymond[1] and A. van Ballegooijen[1]

[1]Harvard-Smithsonian Center for Astrophysics, 60 Garden Street, Cambridge, MA 02138, USA
[2]Center for Space Physics, Boston University, 725 Commonwealth Avenue, Boston, MA 02215, USA

Abstract.
A helical structure in the coronal mass ejection (CME) of 12 September 2000 was observed by the Ultraviolet Coronagraph Spectrometer (UVCS) aboard the Solar and Heliospheric Observatory (SOHO) at heliocentric distances of 3.5 and 6 $R_\odot$. A difference of 300 km sec^{-1} in line-of-sight velocities for two segments of the helix obtained from Doppler measurements implies expansion and allows one to distinguish which segment was closest to the observer. The tilt of the segment then determines the handedness. Observed Lyα and C III line emissions indicate that the helix was threaded with filament plasma of varying density. While the helix constituted the bright core of filament plasma, the helix itself was most likely not the pre-existing filament structure.

Keywords. Sun: corona, Sun: coronal mass ejections (CMEs), Sun: filaments, Sun: UV radiation

1. Introduction

In this paper we use the observations by instruments aboard SOHO to study a helical structure in a CME that occured on 12 September 2000. In particular, we use the Doppler shift of UVCS (Kohl *et al.* 1995) to derive its handedness. In addition, we combine the white-light data of LASCO C2 and C3 (Brueckner *et al.* 1995) and the UVCS spectroscopic observations to address questions about the morphology of filament plasma.

2. Analysis

2.1. Spectroscopic Observations

The Ultraviolet Coronagraph Spectrometer (UVCS) on the SOHO spacecraft was well-positioned to observe a complicated CME on 12 September 2000. Observations were made from the O VI spectrometer channel, which is optimized to observe the O VI $\lambda\lambda$1032, 1037 doublet and the H I λ1026 (Lyβ) line and covers the spectral range 945-1123 Å (473-561 Å in second order). The O VI includes a redundant Lyα path which has a convex mirror between the grating and the O VI detector that focuses the H I 1216 Å (first order) onto the O VI channel.

In the 12 September observations, the Si XII λ1041.26 and Mg X λ1219.7 lines are not present which implies that the temperature of the plasma is significantly less than 10^6 K.

The density of this cold plasma (e.g., plasma emitting at C III λ977; Akmal *et al.* 2001) was calculated using a commonly used estimate of the emission measure:

$$\int n_e\, n_H\, dh = I_\lambda\, /\, \langle P_\lambda \rangle\ cm^{-5} \tag{2.1}$$

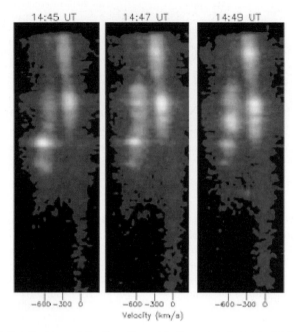

Figure 1. Spectral distribution of the Lyα line at heliocentric height of 6 R$_\odot$, which shows the Doppler shift. Each panel respents 2 minutes of observation. A Doppler shifts of 300 km sec^{-1} and 600 km sec^{-1} can be easily seen.

where I$_\lambda$ is the intensity of the spectral line, $\langle P_\lambda \rangle$ is the radiative loss for a given line, h is distance along the line-of-sight, and n_e and n_H are the electron and hydrogen densities. The radiative losses for the lines used in determining the emission measure are obtained from an updated version of Raymond & Smith (1977) code. The densities in the C III emitting plasma exceed 3×10^5 cm^{-3}, and the ionization temperature ranges from below 25,000 K to above 150,000 K. This is the ionization temperature in ionization equilibrium. Although there is no good reason to believe that equilibrium was reached in this case, but based on the excitation rates the temperature cannot be < 25,000 K and based on the observed line widths it cannot be > 150,000 K.

Of particular interest were the Doppler shifts for the Lyα and C III lines. Doppler shifts give line-of-sight velocities of the erupting plasma. The velocity Doppler shift is governed by the usual formula $\Delta\lambda / \lambda = v/c$, for collisionally excited lines. For lines excited by radiation scattering, the wavelength shift can be smaller because the profile is weighted by the emission line profile from the solar disk (Noci & Maccari 1999). Due to the high velocities of the CME material, the radiation scattering components are severely Doppler dimmed (Withbroe *et al.* 1982), and the lines are mostly collisionally excited.

Examples of UVCS Doppler shift measurements from the 12 September CME are given in Figure 1. They show spectral images of the Lyα line at three different times. The vertical axis represents position along the UVCS slit and the horizontal axis is the wavelength, in this case converted to velocity. A Doppler shifts of 300 km sec^{-1} and 600 km sec^{-1} can be seen in all three panels.

In contrast to Figure 1, the two right panels in Figure 2 give spatial rather than wavelength information. They show Lyα and C III λ977 intensity maps as a function of time at 6 R$_\odot$. The UVCS slit was positioned directly south of the Sun (at position angle (PA) 180 degrees South). The vertical axis shows time increasing from the bottom, which converts to radial distance under the assumption of constant flow speed. The horizontal

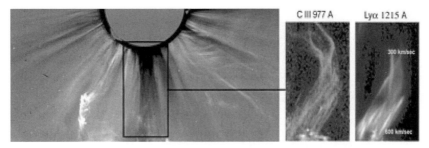

Figure 2. Lyα and C III line intensity maps along the UVCS slit (horizontal axis) as a function of time (vertical axis) at heliocentric height of 6 R$_\odot$ (right panel). The left panel shows LASCO C2 image with UVCS slit superimposed

axis is the spatial coordinate in arcmin along the entrance slit. The panels cover three hours of observations, from 14:45 UT to 17:45 UT. At 14:45 UT the Doppler shift is 600 km sec^{-1} while later 16:45–17:45 UT the Doppler shift is 300 km sec^{-1}. The shift of plasma from east to west and then to the east along the UVCS slit implies helical structure for ∼ 3 hours in both lines. Combined with Doppler shifts, we deduce that the portion of the helix tilted upward east to west is the left leg of the flux rope.

The two values of Doppler shift imply that the helix is expanding. Further, the faster Doppler shift of 600 km sec^{-1} must apply to the near side of the helix tilting to the West, while the slower shift of 300 km sec^{-1} must apply to the far side of the helix, tilting to the East. This is counter-clockwise for an observer looking radially out from the Sun, implying a right-handed helix. Thus Doppler shifts can uniquely determine the handedness of a helical structure.

To gain an understanding of the context of the observed plasma, we combine the spectroscopic observations with LASCO C2 white light images. In this way we can identify the location of the plasma with respect to the 3-part CME structure. The left panel of Figure 2 is the LASCO C2 image at 14:45 UT with the region corresponding to the UVCS observations superimposed. The vertical side of the box represents the distance that the plasma traveled in three hours, where the horizontal side represents the field of view of the UVCS slit. The plasma traveled at a speed of ∼ 390 km sec^{-1}. If the plasma structure observed in the UVCS panels were advecting with the flow and undergoing no dynamic evolution, the structure in the rectangle on the LASCO panel would be the same as that observed by UVCS. Since the helical structure in the UVCS panel is not apparent in the LASCO panel, we conclude that dynamic evolution over the 3-hour interval obscured an exact match. Over the shorter interval at 3.6 R$_\odot$ in Figure 3, however, an excellent match is obtained.

Figure 3 corresponds to 32 minutes of observations, starting at 13:28 UT and ending at 14:00 UT. The structure in the UVCS data, especially that of the C III line, is a remarkable match to the structure seen by LASCO C2. The helical form is more pronounced in the Lyα compared to the C III line. On the other hand, compared to Figure 2 at 6 R$_\odot$, the helical structure in the C III line at 3.6 R$_\odot$ is less pronounced. This is due to the fact that the observation at 3.6 R$_\odot$ is significantly shorter than that at 6 R$_\odot$, as pointed out above.

Also as pointed out above, the C III emission indicates cool material (< 150,000 K) typical of a filament. However, the location of some of this cool plasma, as clearly seen by Figure 3, is at the edge of what appears to be a tightly wound flux rope, beyond most of the more loosely wound helical structure apparent in the UVCS data in Figure 3. The tightly wound flux rope is the type thought to be generated by reconnection of

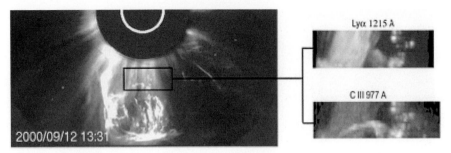

Figure 3. Lyα and C III line intensity maps along the UVCS slit (horizontal axis) as a function of time (vertical axis) at heliocentric height of 3.6 R$_\odot$ (right panel). The left panel shows LASCO C2 image with UVCS slit superimposed.

coronal arcade loops and thus should contain hot plasma. Evidence of this reconnection is a concurrent, long lasting arcade event in EIT data (e.g., Vršnak *et al.* 2003).

3. Discussion

We have presented an analysis of a helical structure observed by UVCS/SOHO in the 12 September 2000 CME. The UVCS Doppler shift observations can distinguish field lines in the near side of the helix from those on the far side and consequently the skew of these fields and the handedness of what appeared to be a portion of a flux rope in the CME. We conclude that the helix is right-handed.

The analysis suggests that the cool plasma of the filament appeared in the newly-formed helix of the CME which is usually associated with the dark cavity of 3-part CME structure rather than the bright core where current understanding of filaments would suggest. Basic reconnection through the current sheet (Lin & Forbes 2000), which presumably creates the helix, is generally thought to allow only ambient very hot ($>$ million K) coronal plasma to travel upward into the helix. A possible explanation of the apparent contradiction is that the cooler filament plasma somehow became caught up in this reconnection and gained access to the helix by losing its magnetic coherence. A more interesting possibility is that the field lines carrying filament material remained coherent but moved from the axis of the flux rope to the edge by an internal kink instability. In future work we will explore these possibilities in detail.

Acknowledgements

R. M. S. thanks the Scientific Organizing Committee for travel support to the IAU 226 symposium. This work was supported by NASA grant NNG04GE84G to the Smithsonian Astrophysical Observatory and NSF grant ATM-0119700 to Boston University. The authors would like to thank A. Vourlidas for the LASCO data and L. Strachan for useful discussions.

References

Akmal, A., Raymond, J. C., Vourlidas, A., Thompson, B., Ciaravella, A., Ko, Y.-K., Uzzo, M., & Wu, R. 2001, *ApJ* 553, 922
Brueckner *et al.* 1995, *Sol. Phys.* 162, 357
Kohl, J. L., *et al.* 1995, *Sol. Phys.* 162, 313
Lin, J., & Forber, T.G. 2000, *J. Geophys. Res.* 105, 2375
Noci, G., & Maccari, L. 1999, *Astron. Astrophys.* 341, 275-285
Raymond, J.C., & Smith B.W. 1977, *ApJ* 35, 419

Withbroe, G.L., Kohl, J.L., Weiser, H., & Munro, R.H. 1982, *Space Sci. Rev.* 33, 17
Vršnak, B., Warmuth, A., Maricic, Otruba, W., & Ruzdjak, V. 2003, *Sol. Phys.* 217, 187

Discussion

KOUTCHMY : UVCS reconstructed images taken in Lyα and C III look quite different from images seen on W-L, Do you have any comment on that?

SULEIMAN: This is true only in the UVCS observations at 6.0 $R_\odot$, The UVCS observation at 6.0 $R_\odot$ was for over 3 hrs. The dynamics of the CME changed and thus the helical structure was not evident in LASCO/C2 data but it shows the location of the leg of the flux rope. In the case of UVCS observations at 3.6 $R_\odot$, there was a very remarkable match between UVCS and LASCO/C2 data. As a matter of fact UVCS was able to provide temperature and density information about the CME. In this case UVCS showed that the filament plasma resided on the edge of the flux rope rather than at the axis of the flux rope. This is a very important piece of observation since CME models predict that filament material should be at the axis of the flux rope. The remarkable match between UVCS and LASCO/C2 at 3.6 $R_\odot$ is very clear since the time period of the observation is 32 min, comparing to over 3 hrs at 6 $R_\odot$ where the dynamics of the CME have changed.

BOTHMER: 1. Handedness of flux rope overlying sinistral filament?
2. The analyzed magnetic structure is not the flux rope but expanded prominence (prominence is base of overlying flux rope).

SULEIMAN: 1. The handedness of the overlying arcade has an inverse relationship with the chirality of the filament. Since the filament is sinistral then the overlying arcade is right skewed.
2. The flux rope did not exist prior to the eruption but was formed later on as part of the CME liftoff process. If the flux rope did exist prior to eruption then the filament channel would have disappeared, after the eruption, which Hα observations do not show. In addition the EIT observations show the appearance of the overlying arcade as the filament disappears in EIT images.

Coronal and Stellar Mass Ejections
Proceedings IAU Symposium No. 226, 2005
K. P. Dere, J. Wang & Y. Yan, eds.

Error Estimates in the Measurements of Mass and Energy in White Light CMEs

Angelos Vourlidas

Naval Research Laboratory, Washington DC, USA
email: angelos.vourlidas@nrl.navy.mil

Abstract. Due to the optically thin nature of the white light emission, all measurements of the energetics and dynamics of a CME are based on sky-plane projected quantities. The extent and distribution of the CME material along the line of sight is unknown. Thus, CME measurements have an inherent degree of uncertainty. In this paper, I identify the various (possible) sources of errors associated with measurements of CME mass and energy (e.g., instrumental, random, projections effects, etc) and give an error budget for the final measurements. I apply these errors to the statistics of mass and energy for several thousand CMEs observed with LASCO in 1996-2003.

Keywords. Sun: coronal mass ejections (CMEs)

Discussion

KAHLER: Are your error estimates independent of any uncertainties in the F-corona brightness?

VOURLIDAS: In this talk, I refer only to excess brightness (mass) measurement (and errors). In this case, the F-corona is subtracted automatically since we take $I_{CME} - I_{PREEVENT}$ and the F-corona does not affect the measurements.

KOUTCHMY: Looking along a time sequence, did you check the mass conservation equation or did you see decrease or increase of the overall mass of the CME(s)?

VOURLIDAS: We find that for the majority of the events we have studied ($\sim 2500/4500$) the mass reached a value above 10 R$_\odot$ (e.g. Vourlidas *et al.* 2002, in Proc. of the 10th Europe. Sol. Phys. Mtg, Prague, Czech Rep., Wilson, A. (ed), ESA SP-506, Dec 2002, p. 91).

Coronal and Stellar Mass Ejections
Proceedings IAU Symposium No. 226, 2005
K. P. Dere, J. Wang & Y. Yan, eds.

© 2005 International Astronomical Union
doi:10.1017/S1743921305000189

Post–CME events: cool jets and current sheet evolution

A. Bemporad[1], G. Poletto[2] and S. T. Suess[3]

[1]Astronomy Dept., University of Florence, L.go E. Fermi 2, 50125 Florence, Italy
email: bemporad@arcetri.astro.it

[2]INAF – Arcetri Astrophysical Observatory, L.go E. Fermi 5, 50125 Florence, Italy

[3]NASA Marshall Space Flight Center, Mail Stop SD50i, Huntsville, AL 35812, USA

Abstract. In this work we focus on UVCS data acquired during the November 2002 SOHO–Ulysses quadrature, at an altitude of 1.7 $R_\odot$ over a range of latitudes centered around 27°N in the western quadrant. A couple of hours before our observations started, a CME event (November 26, 17:00 UT) originating at about 27°N, disrupted the coronal configuration of the region. In the $\sim$ 2.3 days following the event UVCS detected emission in the neutral H $Ly\beta$ and $Ly\gamma$ lines as well as in lines from both high and low ionization ions such as C III, O VI, Si VIII, IX and XII, Fe X and XVIII. Enhanced emission from the hot Fe XVIII ion ($\log T_{max} = 6.7$), lasting nearly to the end of our observations and originating in a region between 10°N and 30°N, has been identified with a post–CME current sheet. Our interpretation is supported by EIT Fe XII images which show a system of loops at increasingly higher altitudes after the event. Northward of the CME, UVCS observed repeated, sudden and short lived emission peaks in the "cool" $Ly\beta$, $Ly\gamma$, C III and O VI lines. These events seem to be the extension at higher altitudes of the chromospheric plasma jets observed in the EIT He II images. Electron temperatures of both the hot and cool region will be presented here and their time evolution will also be illustrated.

Keywords. Sun: coronal mass ejections (CMEs), UV radiation.

---

1. Introduction

Many CMEs have been observed by SOHO instrumentation: the combination of white light (LASCO coronagraphs) and EUV (EIT telescope) images with UV spectral data (CDS, SUMER and UVCS instruments) have provided a comprehensive view of Coronal Mass Ejections (CMEs) structure and evolution. However, mainly because of the uncertainties in the determination of their three–dimensional structure (as they are always seen projected onto the plane of the sky) and in the coronal magnetic field measurements, this phenomenon is far from being completely understood. Recent theoretical works (see e.g. Lin *et al.* 2004) suggest that magnetic reconnection can play a key role, in particular in the CME development and the plasma heating during and after the eruption. CME models predict that open magnetic fields relax, via magnetic reconnection, into a closed configuration, with the formation, between the chromosphere and the ejected bubble, of a current sheet (CS); moreover reconnection heats the plasma converting magnetic energy into kinetic and thermal energies. In the last few years, observations made by the UVCS instrument (see, e.g., Akmal *et al.* 2001; Ciaravella *et al.* 2002; Ko *et al.* 2003) described CSs as long lasting ($\sim$ 2 days) structures with high plasma temperature ($\log T \sim 6.6 - 6.8$ K), as inferred from the observed Fe XVIII emission. Such temperatures have been also observed *in situ* by ACE as high Fe charge state associated with ICMEs (Lepri *et al.* 2001; Lepri & Zurbuchen 2004). However, the CS evolution and the explanation for its very long lifetime are still under discussion.

Figure 1. The coronal morphology in the North-West quadrant as seen by LASCO/C2 coronagraph from the beginning (left panel) to the end (right panel) of UVCS observations; in the figure we show also the position of the UVCS slit.

In this work we focus on CME data acquired over ∼ 2.3 days by UVCS instrument during the SOHO–Sun–Ulysses quadrature (Suess *et al.* 2000; Poletto *et al.* 2002; Bemporad *et al.* 2003a; Suess *et al.* 2004) of November 2002. In section 2 we describe the post–CME evolution as seen in LASCO and EIT images, while in section 3 we describe the UVCS observations and data analysis. The origin for the high temperature plasma observed by UVCS is discussed in section 4, while in section 5 we focus on the comparison with Ulysses *in situ* data. A short discussion of our results concludes the paper.

2. LASCO and EIT observations

As shown in figure 1, on November 26, 2002 LASCO/C2 (*Large Angle Spectroscopic COronagraph*, see Brueckner *et al.* 1995) images show in the NW quadrant two coronal streamers centered at about 10°N and 50°N (hereafter, respectively, "streamer 1" and "streamer 2"). Starting from about 17:00 UT a CME occurs on the northward side of streamer 1: a plasma bubble is ejected at a latitude of about 27°N with an initial speed (as we estimate from LASCO images) of ∼ 120 km/s. The CME partially disrupts streamer 1, which appears displaced southward by about 7° after the event; on the contrary, streamer 2 seems to be unaffected by the CME.

EIT (*Extreme ultraviolet Imaging Telescope*, see Delaboudiniere *et al.* 1995) Fe XII images show, in the NW quadrant, two Active Regions (ARs USAF/NOAA 10197 and 10199) respectively at 25°N and 28°N, close to the position of the southward edge of streamer 2. These ARs approximately cross the solar limb on November 26-27: if the CME material originates from these ARs, the plasma is likely to propagate mainly on and around the plane of the sky. Moreover, before and after the CME, EIT He II images show recursive ejections of chromospheric plasma from these ARs. Radial speeds of these short lived (30 – 60 min) jets, evaluated from the variation with time of their altitude, are strongly variable with time (∼ 20–200 km/s) and the motion of each jet has peculiar kinematical properties. On the plane of the sky the plasma is ejected at an angle of about 20°N with respect to the radial from the Sun, hence towards latitudes higher than 25 – 28°N.

3. UVCS observations

3.1. *UVCS instrumental settings*

The UVCS (*Ultra Violet Coronagraph Spectrometer*, see Kohl *et al.* 1995) observations start on November 26, 18:39 UT (left panel in figure 1) and last through November 29, 02:56 UT (right panel). This time interval (∼ 2.3 days) is covered with a time resolution

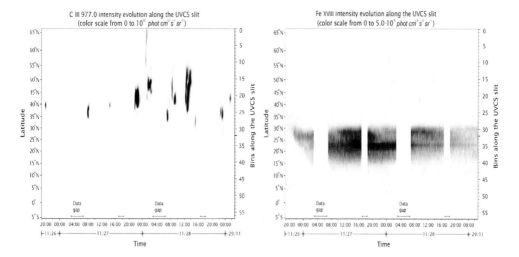

Figure 2. Intensity evolution along the UVCS slit for the "cool" C III ($\log T_{max} = 4.9$, left panel) and the "hot" Fe XVIII spectral lines ($\log T_{max} = 6.7$, right panel).

of 120 s and some gaps in the observations cut the total observing time to ~ 1.8 days (see also Bemporad *et al.* 2003b for a preliminary analysis of the initial ~ 7 hours of observations). The spectrometer slit is at an altitude of 1.7 R$_\odot$ and is centered at a latitude of 27°N (North – West quadrant), as shown in figure 1; the slit width was 100 μm. In this position UVCS observes the solar corona between the latitudes of 5.8°S and 67.3°N with a spatial binning of 42 *arcsec/bin*. Because at the beginning of UVCS observations the CME flux rope is, in LASCO images, between ~ 2.4 and 2.6 R$_\odot$, hence above the UVCS slit height, in this work we focus on the post CME evolution. The UVCS grating position has been choosen in order to observe line emission from ions with both high and low temperature of maximum formation T_{max}, such as the O VI 1031.91 Å – 1037.61 Å ($\log T_{max} = 5.5$), the Si XII 520.67 Å – 499.37 Å ($\log T_{max} = 6.3$) and the Si VIII 949.35 Å – 944.47 Å ($\log T_{max} = 5.9$) doublets, the Si IX 950.15 Å ($\log T_{max} = 6.0$), the H $Ly\beta$ 1025.67 Å and $Ly\gamma$ 972.54 Å, the Fe X 1028.06 Å($\log T_{max} = 6.0$), Fe XV 481 Å ($\log T_{max} = 6.3$), Fe XVIII 974.86 Å ($\log T_{max} = 6.7$), C III 977.02 Å ($\log T_{max} = 4.9$) and Ca XIV 943.61 Å ($\log T_{max} = 6.5$) lines; these lines are observed with a spectral binning of 0.1986 Å/*bin*. The line intensities are derived summing over the line profile and subtracting an average background evaluated over a spectral interval near the line.

3.2. *UVCS data analysis & interpretation*

As shown in figure 2 (left panel), many peaks in the C III spectral line intensity are detected by UVCS in the northward half of the UVCS slit. Because of the very low T_{max} value of the C III ion, this spectral line is usually absent in coronal spectra. Taking into account a), the position and time when these "cold" C III peaks appear in the UVCS slit and b), the analogous evolution of other "cold" spectral lines like $Ly\beta$ and $Ly\gamma$, we conclude that we are sampling the low temperature ($T \approx 8 \times 10^4$ K) plasma jets observed in the EIT He II images. Doppler line shifts up to 250 km/s indicate that these jets have also a strong inclination away from the plane of the sky.

As shown in figure 2 (right panel), between the latitudes of $\sim 15 - 30$°N, UVCS data reveal intense emission from the Fe XVIII ion, whose T_{max} is unusually high for coronal plasma even above ARs (see e.g. Ko *et al.* 2002). At the latitude where the main Fe XVIII

974.4 Å emission is concentrated (2.5°N) the line intensity increases continuously up to a value of about $4.2 \cdot 10^9\, phot\, cm^{-2}s^{-1}sr^{-1}$ on November 27, ~15:30 UT. Such intensity increase is unusual, because the energy released by reconnection is expected to decrease with time. Over the following hours, the Fe XVIII emission decreases down to $\simeq 1.0 \cdot 10^9\, phot\, cm^{-2}s^{-1}sr^{-1}$ at the end of UVCS observations (November 29, 02:56 UT). At the same latitude the Fe XV line intensity, initially undetectable, increases progressively up to a value of $4.6 \cdot 10^8\, phot\, cm^{-2}s^{-1}sr^{-1}$ at the end of UVCS observations. To understand this peculiar behaviour of different spectral lines from the same element, we have to consider that in coronal conditions, spectral lines form mainly by electron collisional excitation followed by spontaneous emission. Hence in general, the line intensity I_{line} is:

$$I_{line} = \frac{1}{4\pi} \int_{LOS} \epsilon_{line} n_e^2\, dl \;\; ; \;\; \epsilon_{line} = h\nu_{line} \frac{n_{el}}{n_H} \frac{n_{ion}}{n_{el}} (T_e)\, B_{line} q_{line}(T_e)$$

where n_{el}/n_H is the elemental abundance relative to H, n_{ion}/n_{el} is the ionic fraction, q_{line} is the collisional excitation rate (both function of the electron temperature T_e), B_{line} is the branching ratio for the line transition, n_e is the electron density and ϵ_{line} is the line emissivity. From this formula we can see that, in first approximation, the ratio between the observed intensities of spectral lines from different ions of the same element is equal to the ratio between their emissivities, hence depends only on T_e: this implies that from the observed ratio we can estimate T_e. Using the Fe XV/Fe XVIII line intensity ratio observed at 22.5°N and the theoretical line emissivities from the CHIANTI spectral code (v. 4.01, computed with the ionization equilibria of Mazzotta *et al.* 1998), we find $\log T_e > 6.94$ at the beginning of our observations, followed by a continuous, slow decrease over the following ~ 2.3 days down to $\log T_e \simeq 6.52$. The Fe XVIII intensity increase and the following decrease can be explained with the continuous decrease in T_e we derived, because the $\epsilon_{FeXVIII}$ peaks at $\log T = 6.7$. For the same reason, because the ϵ_{FeXV} peaks at $\log T = 6.3$, the plasma cooling gives rise to a continuous increase in the Fe XV intensity. In the next section we discuss the origin of this high temperature plasma.

4. The post–CME high temperature plasma

Because the position of the Fe XVIII 974.4 Å main emission and the CME ejection latitude are about the same, we identify this high temperature plasma as the CS, associated with the post CME magnetic field reconfiguration. However, the slow T_e decrease we derived might as well be ascribed either to plasma cooling inside the CS or to plasma cooling in the newly closed loops. To distinguish between these possibilities we need to know whether, during our observations, the height of the rising neutral point (approximately equal, at each time, to the height of the last newly formed loop) was below or above the height of the UVCS slit. The speed of the rising neutral point depends on the local Alfvèn speed, hence is in general unknown. However, EIT images show a system of newly formed loops after the CME above the ARs: the loops rising speed is, on average, nearly constant and equal to ~1.8 km/s: with this speed, it takes 3.4 days for the rising neutral point to reach the UVCS slit (0.7 $R_\odot$ above the solar limb). We conclude that, during our observations, the neutral point could not have reached the height of the UVCS slit: hence, we are looking at the CS plasma. Moreover, order of magnitude estimates of the conductive and radiative cooling time for a semi–circular loop of height of 0.7 $R_\odot$ shows that the loops cool mainly by conduction over times on the order of 1h, hence much faster than the slow cooling we may infer from the temperature vs. time profile derived from the line ratio technique: again, we conclude that the high temperature plasma we observe with UVCS was not in closed loops, but inside the CS.

5. Ulysses *in situ* observations

During the November 2002 quadrature the Ulysses spacecraft was at a distance of about 4.3 AU. The November 26, 2002 CME has been identified with the *in situ* plasma (see Poletto *et al.* 2004) observed at Ulysses on 14–15 December, 2002 (DoY 348–349). During this time interval the Ulysses/SWICS (*Solar Wind Ion Composition Spectrometer*) instrument observed an unusual abundance of the high charge ion (Fe^{16+}), in agreement with remote observations of high temperature plasma in the aftermath of the CME event, implying that the source temperature was ~ 6–$7.5 \cdot 10^6$K. We refer the reader to Poletto *et al.* 2004 for further details on the association between coronal and *in situ* plasma.

6. Conclusions

In this brief paper we have provided evidence for the presence of a CS in the aftermath of a CME event, as inferred from UVCS observations of very hot plasma at the position where closed loops form at increasingly high altitudes. While this is not the first time that CS have been observed by UVCS, this event is unique in that the behavior of the CS has been followed over a two day time interval and evidence for the same hot plasma sampled in the corona has been found *in situ* at about 4 AU. We have given here a qualitative view of the electron temperature profile with time at the current sheet position, showing how temperature slowly decreases with time. A more thorough description of the physical properties of the post-CME ambient is in progress (Bemporad *et al.* 2005). Jets of "cold" plasma have been detected over the same time interval when the hot CS plasma was observed, but there is apparently no relationship between the cold jets and the behavior of the reconnecting magnetic field in the CME region.

Acknowledgements

The work of GP has been partially supported by the Italian Space Agency (ASI) and by the Italian Ministry of University and Scientific Research (MIUR). SOHO and Ulysses are missions of international cooperation between ESA and NASA.

References

Akmal, A., *et al.* 2001, *ApJ* 553, 934
Bemporad, A., *et al.* 2003, *ApJ* 593, 1163
Bemporad, A., *et al.* 2003, *ESA SP*–535, 567
Bemporad, A., *et al.* 2005, *ApJ* submitted
Brueckner, J. E., *et al.* 1995, *SP*, 162, 357
Ciaravella, A., *et al.* 2002, *ApJ* 575, 1116
Delaboudiniere, J. -P., 1995, *SP*, 162, 291
Ko, Y.K., *et al.* 2002, *ApJ* 578, 979
Ko, Y.K., *et al.* 2003, *ApJ* 594, 1068
Kohl, J., *et al.* 1995, *SP* 162, 313
Lepri, S.T., *et al.* 2001, *JGR* 106, 29231
Lepri, S.T., & Zurbuchen, T.H. 2004, *JGR* 109(A1), A01112
Lin, J., *et al.* 2004, *ApJ* 602, 422
Mazzotta, P., *et al.* 1998, *A&AS* 133, 403
Poletto, G., *et al.* 2002, *JGR* 107, A10, SSH9-1
Poletto, G., *et al.* 2004, *ApJL* 613, 173
Suess, S.T., *et al.* 2000, *JGR* 105, 25033
Suess, S.T., *et al.* 2004, *GRL* 31(5), 801

Discussion

BOTHMER : Why are the conclusions unambiguous? - i.e, why is the hot coronal plasma really the one seen at Ulysses?

BEMPORAD: The quadrature geometry allows us to identify the plasma sampled at Ulysses with the CME plasma remotely observed by SOHO by extrapolating backwards from the time of Ulysses observations to the time when the plasma was ejected from the Sun. The extrapolation has been made using, as a first approximation, the average solar wind speed; then this first extrapolation has been refined taking into account typical CME signatures such as counterstreaming electrons, magnetic clouds, low plasma β and high α particles flux.

Coronal and Stellar Mass Ejections
Proceedings IAU Symposium No. 226, 2005
K. P. Dere, J. Wang & Y. Yan, eds.

© 2005 International Astronomical Union
doi:10.1017/S1743921305000190

Low Frequency (30-110 MHz) Radio Imaging Observations Of Solar Coronal Mass Ejections

R. Ramesh

Centre for Research and Education in Science and Technology, Indian Institute of
Astrophysics, Hosakote 562 114, Bangalore, INDIA
email: ramesh@iiap.res.in

Abstract. Ground based radio imaging observations play an useful role in the study of
mass ejections from the solar corona since they do not have the limitation of an occulter and
both the disk/limb events can be detected early in their development, particularly via the thermal bremmstrahlung emission from the frontal loop of the CME. I present here some of the
recent results on the above topic using data obtained with the Gauribidanur radioheliograph,
near Bangalore in India.

Keywords. Sun: general, Sun: corona, Sun: coronal mass ejections (CMEs), Sun: radio
radiation

1. Introduction

Studies of the influence of solar activity on our terrestrial environment has taken on
increasing importance in recent years, as the realization of just how damaging space
influences can be. The state of near-Earth space environment is significantly controlled
by CMEs, the most-geoeffective manifestation of solar activity (Gosling *et al.* 1991).
CMEs are large-scale magneto-plasma structures that erupt from the Sun and propagate
through the interplanetary medium with speeds ranging from only a few km s^{-1} to nearly
3000 km s^{-1}. They carry typically 10^{15} gm of coronal material. Observations from instruments such as the Large Angle and Spectroscopic Coronagraph (LASCO: Brueckner
et al. 1995) onboard the Solar and Heliospheric Observatory (SOHO: Fleck *et al.* 1995)
have now revolutionized our perception and understanding of the solar eruptive events.
But, by their very nature, the coronagraphs have an occulting disk to block the direct
photospheric light and so the early life/kinematics of a CME in the low corona cannot be
studied using them. One needs non-coronagraphic data to obtain information on the early
evolution of CMEs, in particular for those directed along the Sun-Earth axis which occur
far from the plane of the sky. The latter originate on the visible hemisphere of the Sun
and appear as a 'halo' of expanding, circular brightening that completely surrounds or
spans a large angle outside the occulting disk of a coronagraph (Howard *et al.* 1982). The
Earthward-moving events are geophysically important, in the context of space-weather
related phenomena such as geomagnetic storms. Measurements of CME properties in the
lower corona are significant for several reasons. Foremost among them is the general assumption that CMEs have a constant speed behind the occulting disk of a coronagraph.
This has often caused controversial results while comparing the CME onset with other solar activity signatures (see Ramesh & Ebenezer 2001b; Ramesh & ShanmughaSundaram
2001c for possible radio signatures in the solar corona prior to as well as during the lift-off
time of CMEs). The kinematic evolution of a CME can be described in a three-phase scenario: the initiation phase, impulsive acceleration phase, and propagation phase. Among

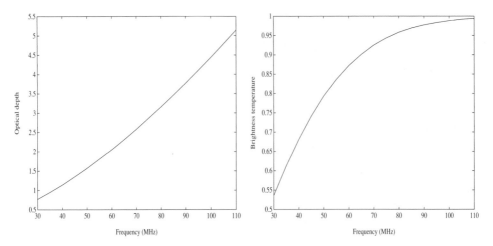

Figure 1. Variation of optical depth and observed brightness temperature in the solar corona as a function of plasma frequency. The electron density model of Newkirk (1961) and a coronal electron temperature (T_e) of $\sim 10^6$ K was used for the calculations. The frequency range in the above Figure corresponds to a height range of $\sim 1.2-2.0$ R$_\odot$ (R$_\odot$ = solar radius = 6.96×10^5 km), from the center of the Sun.

these, the first 2 phases take place primarily in the low corona. These suggest that a more complete description of the motion of a CME, close to the Sun, is crucial for a better prediction of its characteristics at higher altitudes. Again, what one measures usually from a time-lapse movie of the images obtained with a coronagraph is the speed at which a CME spreads in the plane of the sky. This will at best be only a lower limit to the true value (i.e. the speed in the three-dimensional space) especially in the case of an Earth-directed 'halo' CME as it will lie away from the plane of the sky. In this respect, imaging observations particularly at frequencies $\leqslant 110$ MHz play an important role since the radio instruments do not have the limitation of an occulting disk and a CME can be detected early in its development via the thermal bremmstrahlung radiation from its frontal loop (Sheridan et al. 1978; Gopalswamy & Kundu 1992; Kathiravan, Ramesh & Subramanian 2002). The frontal structure of a CME has a large optical depth in the above frequency range, and can be readily observed (Bastian & Gary 1997; Gopalswamy 1999; Ramesh, Kathiravan & Sastry 2003). Note that the optical depth of the background corona is not so large in the range $30-110$ MHz (left hand side of Figure 1). Due to this, the observed brightness temperature is generally less than the electron temperature of the corona (right hand side of Figure 1). Therefore the presence of a weak density enhancement like a CME in the solar atmosphere can be noticed with comparitively better contrast. Also one can observe activity at any longitude similar to X-ray and EUV wavelengths (Ramesh 2000a). In addition to direct imaging, ray tracing analysis of the thermal radio counterpart of a CME also play a vital role since one can localize the position of the associated density enhancement in a three-dimensional space (Ramesh & Sastry 2000b; Kathiravan, Ramesh & Subramanian 2002; Kathiravan & Ramesh 2004). One can also use imaging observations of the apparent angular broadening of a distant cosmic source when the radio waves from it undergo scattering while happen to pass through a CME, to infer characteristics of the latter particulary at large distances from the Sun (Ramesh, Kathiravan & Sastry 2001a).

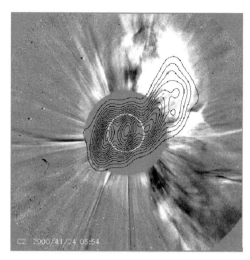
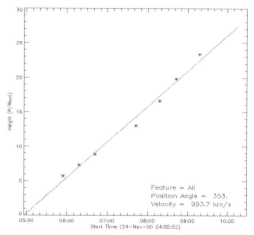

Figure 2. LEFT - A composite of the LASCO C2 difference image (05:54-05:06 UT) and the Gauribidanur radioheliogram (05:25 UT) of the CME event observed on November 24, 2000. The inner circle indicates the solar limb, and outer circle is the occulting disk of the coronagraph. It extends approximately upto 2.2 $R_\odot$ from the center of the Sun. Solar north is straight up and east is to the left. The CME can be clearly noticed as a bright structure above the north-west quadrant of the occulting disk in the coronagraph image and as a well-defined bulging of the contours at the corresponding location in the radioheliogram. **RIGHT -** Height-time plot of the whitelight CME obtained using LASCO C2/C3 measurements. Note that the first data point for the plot was available only at a height of about 5.71 $R_\odot$ from the center of the Sun.

2. Case studies

2.1. *The event of November 24, 2000*

The radio data reported were obtained at 109 MHz with the Gauribidanur radioheliograph (GRH) operating near Bangalore in India (Ramesh *et al.* 1998a; Ramesh 1998b; Ramesh, Subramanian & Sastry 1999). Note that after the closing down of the Culgoora and Clark Lake radioheliographs, the GRH is the only instrument that is presently in operation for radio observations of the solar corona in the frequency range 30-110 MHz. The latter corresponds to an altitude range of $\sim 1.2 - 2$ $R_\odot$ (from the center of the Sun) in the solar atmosphere. According to the CME list for the year 2000 (http://cdaw.gsfc.nasa.gov/CME_list), the LASCO C2 coronagraph onboard SOHO observed a full 'halo' CME on November 24, 2000 around 05:30 UT, the time at which it was first noticed in its field of view. The left half of Figure 2 shows a composite of the LASCO C2 difference image of the event at 05:54 UT obtained by subtracting a pre-event image at 05:06 UT, and the radioheliogram obtained with the GRH at 05:25 UT. The CME can be noticed as a bright feature above the north-west quadrant of the occulting disk of the coronagraph. The contours in the radioheliogram also show a well-defined bulging at the corresponding location. The extrapolated lift-off time of the CME was 04:55:52 UT. Its estimated linear speed in the plane of the sky was 994 km/s (right half of Figure 2). A quadratic fit to the height-time data gave an acceleration of 72 m/s^2 for the event. Figure 3 shows the radioheliogram obtained with the GRH at 04:55, 05:05, 05:15 & 05:25 UT on that day. In addition to the discrete sources on the disk, one can clearly observe enhanced radio emission in close spatial correspondence with the whitelight CME described above, at all the above 4 epochs. Its estimated peak brightness temperature (T_b) was found to be $\sim 10^5$ K. A comparison of the LASCO and GRH images clearly indicates that the radio enhancement moved in the same direction as the whitelight CME. Also their

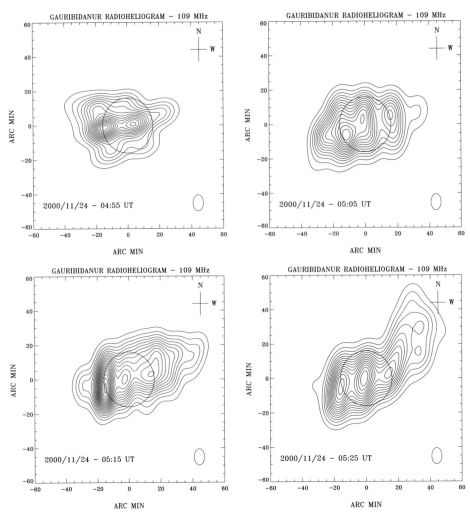

Figure 3. Radioheliogram obtained with the GRH on November 24, 2000 at different epochs. The open circle at the center is the solar limb. The instrument beam is shown near the bottom right corner. **04:55 UT** - The peak T_b is $\approx 4.18 \times 10^6$ K, and the contour interval is 0.27×10^6 K. **05:05 UT** - The peak T_b is $\approx 6.08 \times 10^6$ K, and the contour interval is 0.34×10^6 K. **05:15 UT** - The peak T_b is $\approx 6.1 \times 10^6$ K, and the contour interval is 0.39×10^6 K. **05:25 UT** - The peak T_b is $\approx 5.6 \times 10^6$ K, and the contour interval is 0.29×10^6 K.

appearances are closely similar. We therefore conclude that the former is the radio counterpart of the latter. We estimated the velocity of the 'radio CME' from the displacement of its centroid, and the values are 1087 ± 73 km s^{-1} and 1185 ± 73 km s^{-1} during the interval 05:05-05:15 UT and 05:15-05:25 UT, respectively. This gives it an effective acceleration of $\sim 157 \pm 121$ m s^{-2}. Note that this is about a factor of 2 more than the acceleration of the whitelight CME obtained using LASCO C2/C3 measurements whose first point was available only at a height of about 5.71 R$_\odot$ from the center of the Sun.

2.1.1. *Analysis and Results*

Single frequency observations of CME associated discrete radio sources travelling outwards to large heights ($\sim 2 - 3$ R$_\odot$) in the solar corona are generally attributed to non-thermal continuum emission from moving type IV radio bursts [see Dulk (1980) and

Table 1. Characteristics of 'radio CME'

Time	Co-ordinates of the centroid	Position angle	Brightness temperature (°K)	Electron density (cm^{-3})	Mass (g)	Magnetic field (G)
05:05 UT	0.22N, 1.06W	282°	3.34×10^5	1.24×10^8	2.68×10^{16}	2.40
05:15 UT	1.09N, 1.41W	308°	2.86×10^5	4.46×10^7	3.82×10^{16}	1.33
05:25 UT	2.03N, 1.81W	318°	1.92×10^5	2.81×10^7	4.42×10^{16}	0.86

the references therein], since the observed $T_b \geqslant 10^7$ K, is usually higher than that due to the emission from the background 'quiet' Sun ($\sim 10^6$ K) which is purely thermal in nature. But in the present case, the T_b of the enhanced radio emission observed at the location of the whitelight CME is less than the electron temperature (T_e) of the solar corona ($\approx 1.4 \times 10^6$ K, Fludra *et al.* 1999). Also no type IV emission was reported during our observing period on November 24, 2000 (Solar Geophysical Data, January 2001). We would like to point here that optically thin synchrotron radiation from the non-thermal electrons entrained in the magnetic field of the CME could also give rise to low values of T_b ($\sim 10^4 - 10^5$ K) as shown recently by Bastian *et al.* (2001) for the event of April 20, 1998. But we could not verify the above (through spectral index estimation) in the present case, since the radio imaging data is available at only one frequency. However it is to be noted that the CME event described here was not accompanied by any non-thermal continuum emission in the metric range unlike the event reported by Bastian *et al.* (2001) [see Solar Geophysical Data, June 1998 for details]. Therefore it is possible that the CME associated enhanced radio emission observed by us in the present case is most likely thermal in nature, and the excess emission observed off the limb in the north-west quadrant of the GRH images is due to bremsstrahlung from the extra electrons associated with the 'halo CME'. Its mass (M) is given by (Sheridan *et al.* 1978, Gopalswamy & Kundu 1992),

$$M = 2 \times 10^{-24} \left[5f^2 T_e^{1/2} T_b L^{-1} \right]^{1/2} V \quad \text{g} \tag{2.1}$$

where f (MHz) is the observing frequency and L (R$_\odot$) is the depth of the radio enhancement along the line of sight. The latter is unknown and is taken to be the same as the observed radial width. The volume (V) of the region of enhanced emission was determined by multiplying its radial and lateral width with the depth along the line of sight. We assumed that the coronal plasma is a fully ionized gas of normal solar composition (90% hydrogen and 10% helium by number), and each electron is associated with approximately 2×10^{-24} g of material. In addition to the above, we also derived the magnetic field strength (B) associated with the density enhancement assuming that the plasma $\beta \approx 0.05$, as found by Vourlidas *et al.* (2000) for some of the LASCO CMEs at about the same height range as the 'radio CME' described here. Table 1 lists the values of the different parameters of the latter obtained using the difference images with respect to that at 04:55 UT. The distance (s) travelled by a CME in a given time interval (t) can be found if its initial speed (u) and acceleration (a) are known. For the values of u = 1087 km/s, a = 157 m/s^2 and t = 29 min [time difference between the last and first height measurement using GRH (05:25 UT) and LASCO data (05:54 UT), respectively], we found that the centroid of the 'radio CME' should be located at a height of 5.78 R$_\odot$ from the center of the Sun, at 05:54 UT. According to the LASCO measurements, the leading edge of the CME was located at a height of 5.71 R$_\odot$ at 05:54 UT.

2.2. *The event of June 2, 1998*

There is no direct measurement of the parameters/kinematics of a CME along the line of sight. For calculations like the volume of a CME, one generally assumes that the depth of the CME along the line of sight is the same as its radial width. We present here a method using which one can estimate the parameters of a CME separately along each direction in a three-dimensional space. According to the CME catalog for the year 1998, the LASCO C2 coronagraph observed a massive CME on June 2, 1998 around 08:08 UT (the time at which it was first noticed in its field of view) The above CME event was also accompanied by an exceptionally bright prominence eruption. The left hand side of Figure 4 shows a composite of the radioheliogram obtained with the GRH at 109 MHz around 07:30 UT and the LASCO C2 image of the prominence eruption/CME obtained at 10:29:34 UT, on June 2, 1998. The leading edge of the CME was observed at a projected height of 2 $R_\odot$ around 07:40 UT in the field of view of the LASCO C1 coronagraph (Plunkett *et al.* 2000). This suggests that the extended, faint radio emission at the corresponding location in the composite picture on the left hand side of Figure 4 might be the counterpart of the former. The other bright whitelight feature in the southeast quadrant is probably associated with the CME that took place around 03:00 UT at a position angle (PA, measured counter clockwise from solar north) of $\approx 118°$, on that day. One can clearly notice its radio counterpart also at the corresponding location. The peak brightness temperature of the above two radio counterpart of the whitelight CME were $\sim 10^5$ K. This indicates that the emission is most likely thermal in nature. Also no nonthermal emission was reported during our observing period (Solar Geophysical Data, August 1998).

2.2.1. *Ray tracing method*

The brightness distribution of the 'quiet' Sun was computed theoretically from centimeter to meter wavelengths for the first time by Smerd (1950). He assumed a spherically symmetric corona to derive the solution by numerical integration of the radiative transfer equation for an ionized medium. The existence of density enhancements make the corona asymmetric and one has to take recourse in a more involved ray tracing technique to derive the brightness distribution. Such calculations were carried out by Newkirk (1961) and he derived the brightness profiles at short wavelengths. Sastry, Shevgaonkar & Ramanuja (1983) used a ray tracing technique similar to that of the latter to explain the one-dimensional brightness distribution observed by them in the decameter wavelength range. We have now extended the above scheme to the analysis of an observed two-dimensional thermal radio brightness distribution. In order to simulate the observed radio brightness distribution superposed on the LASCO C2 image in the left hand side of Figure 4, we calculated the brightness distribution of the solar corona using various types of electron density and temperature distributions. Though in principle one could use any standard electron density distribution model for the solar corona in which the rays are to be traced, we describe our technique using a model similar to that of Newkirk (1961). The electron density at any point in the corona is given by,

$$N_e(\rho) = N_0 \left[1 + C_n \exp(-\beta^2)\right] \quad \text{cm}^{-3} \tag{2.2}$$

where $N_0 = 4.2 \times 10^{4.32/\rho}$ (the spherically symmetric component of the background 'quiet' Sun) and ρ is the radial distance from the center of the Sun. The constant C_n is the strength of the density enhancement/depletion, and β depends on its location and

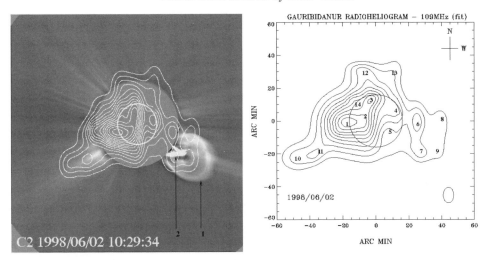

Figure 4. LEFT - A composite of the radioheliogram obtained with the GRH at 109 MHz around 07:30 UT and LASCO C2 image of the giant prominence eruption/CME at 10:29:34 UT on June 2, 1998. The peak brightness temperature in the radio map is $\sim 1.3 \times 10^6$ K. The contours are in interval of 8.6×10^4 K. The arrows labelled 1 & 2 indicate the CME and the prominence, respectively. **RIGHT -** Radio brightness distribution of the Sun at 109 MHz obtained through ray tracing calculations. The peak brightness temperature is $\sim 1.25 \times 10^6$ K. The contours are in interval of 10^5 K. The numbers 1-14 indicate the location of the centroid of the discrete sources used in the ray tracing calculations. The sources 6-9 correspond to the observed enhancement at the location of the CME in the south-west quadrant.

size. It is given by,

$$\beta^2 = \frac{(x - x_o)^2}{2\sigma_x{}^2} + \frac{(y - y_o)^2}{2\sigma_y{}^2} + \frac{(z - z_o)^2}{2\sigma_z{}^2} \tag{2.3}$$

where σ_x, σ_y, σ_z and x_o, y_o, z_o are the size (along the respective axes) and the location of the centroid of the density enhancement/depletion. Here x is towards the Earth (along the line of sight), and the xy-plane contains the axis of the localised region. y & z represent the longitudinal and lattitudinal directions on the Sun. All distances are in units of $R_\odot$. For the localised regions, we used a model in which the density falls off as a Gaussian function along the x, y & z directions, from their centroid. To determine the brightness temperature (T_b) at some point in the solar corona, rays initially directed towards that point are traced [using the technique described in Newkirk (1961)] from the Earth towards the Sun until the optical depth (τ) reaches a large value or the ray is moving away from the Sun and is atleast 5 $R_\odot$ from the Sun. The brightness temperature (T_b) for any ray path is evaluated using the following integral,

$$T_b = \int_0^\tau T_e \, e^{-\tau} d\tau \tag{2.4}$$

where T_e is the coronal electron temperature. The above procedure is repeated for different values of y & z, and the results are stored in a two-dimensional array. After trial and error, we were able to reproduce the observed brightness distribution by assuming a background corona of uniform brightness temperature $(T_e = 1.4 \times 10^6$ K) and a density profile equal to 0.35 times that given by equation (2.2). These values agree well with that reported recently by Fludra *et al.* (1999) for the 'quiet' Sun. The right hand side map in Figure 4 shows the brightness distribution obtained using the ray tracing technique described above. There is a good correspondence with the observed radio brightness

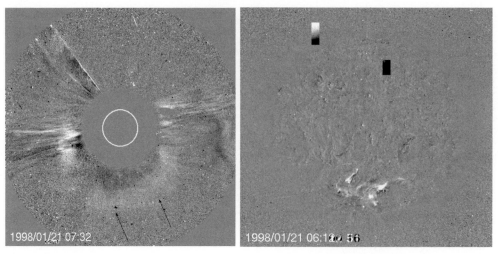

1998/01/21 07:32   1998/01/21 06:10 56

Figure 5. LEFT - Difference image (07:32-06:06 UT) of the 'halo' CME event observed with the LASCO C2 coronagraph on January 21, 1998. The CME can be noticed as a faint ring (indicated by the arrow marks) above the occulter of the coronagraph in the southern quadrant. **RIGHT** - A running difference image of the southern polar crown filament eruption of January 21, 1998 obtained at 06:10 UT with the Extreme ultra-violet Imaging Telescope (EIT) onboard SOHO at 195 Å.

distribution on the left hand side in Figure 4. The numbers 1-14 indicate the location of the centroid of the discrete sources used in our ray tracing calculations [see Kathiravan, Ramesh & Subramanian 2002 for details] It is to be noted here that we have not included scattering (by small scale density inhomogeneities in the solar corona) since their effects are more pronounced mainly at decameter wavelengths (Aubier, Leblanc & Boischot 1971). The sources 6-9 correspond to the observed enhancement at the location of the CME in the south-west quadrant in Figure 4 (left hand side) and their average electron density is $\approx 2.65 \times 10^7$ cm^{-3}. This is about 17 times greater than the ambient density at 2.8 R$_\odot$, the mean radial distance of the above sources. A knowledge of the width of the structure along the line of sight enabled us to calculate its volume and hence the mass in a straightforward manner. The estimated value of the mass is 2.02×10^{15} g.

2.3. *The event of January 21, 1998*

One of the results that came out from our ray tracing calculations described in section 2.2.1 was the possibility to obtain the location of the various discrete sources along the line of sight direction also, in a straight forward manner. This provides a technique to determine the three-dimensional position co-ordinates of the thermal radio counterpart of a density enhancement in the corona at a given epoch and estimate its space speed under situations where it shows noticeable displacement as a function of time. It is obvious that if the observed brightness distribution evolves with time, then the different parameters used in the ray tracing calculations will be unique to the radio map obtained at a particular epoch. According to the CME catalog for the year 1998, a full 'halo' event was observed on January 21, 1998. The event was first observed above the southern quadrant of the occulting disk of the LASCO C2 coronagraph around 06:37:25 UT. Its leading edge was located at a height of about 2.75 R$_\odot$ at that time. The left half of Figure 5 shows a difference image of the event obtained at 07:32 UT by subtracting a pre-event image taken at 06:06 UT. One can notice the CME as a faint ring like feature in the southern hemisphere. The estimated linear speed of the leading edge of the CME

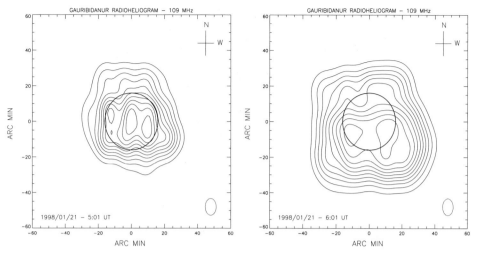

Figure 6. LEFT - Radioheliogram obtained with the GRH at 109 MHz on January 21, 1998 at 05:01 UT. The peak brightness temperature is $\sim 1.21 \times 10^6$ K and the contour interval is 1×10^5 K. **RIGHT -** Same as in the left hand side, but obtained at 06:01 UT. The peak brightness temperature is $\sim 1 \times 10^6$ K and the contour interval is 1×10^5 K.

in the plane of the sky was 361 km s^{-1}. The extrapolated onset time of the event from the center of the Sun was 05:09:09 UT. The event was associated with a H-alpha filament disappearance from location (S57 E19) on the solar disk between 04:00 - 06:03 UT on that day (right hand side of Figure 5; Solar Geophysical Data, July 1998). Figure 6 shows the radioheliogram obtained with the GRH on January 21, 1998 around 05:01:06 & 06:01:06 UT, respectively. The left hand side map in Figure 7 is the difference of the above two radioheliogram, i.e. 06:01:06 - 05:01:06 UT. One can notice enhanced radio emission at approximately the same location as the leading edge of the whitelight CME in the LASCO C2 difference image in Figure 5 (left hand side). The radio Sun was very 'quiet' and no non-thermal activity was reported, particularly during our observing period (Solar Geophysical Data, March 1998. This indicates that the emission seen in the left hand side map in Figure 7 is most likely due to thermal bremmstrahlung from the excess electrons (above the ambient) in the frontal loop of the whitelight CME, and is the radio counterpart of the latter.

2.3.1. *Analysis and Results*

The right hand side map in Figure 7 shows the brightness distribution obtained using the ray tracing technique described in section 2.2.1 for the observed radio map in the right hand side of Figure 6. There is a striking similarity between them. The numbers s1-s16 indicate the location of the centroid of the discrete sources used in our ray tracing calculations [see Kathiravan & Ramesh (2004) for further details]. A comparison of the two images in Figure 7 shows that there is a good correspondence between the location of the excess emission in the southern hemisphere of the latter and the sources s8, s9, . . ., s13 in the distribution obtained using ray tracing calculations. This suggests that the above set of discrete sources comprise the 'radio' CME. Its total mass (M_{radio}) is given by,

$$M_{radio} = m_8 + m_9 + m_{10} + m_{11} + m_{12} + m_{13} \qquad (2.5)$$

where m_8, m_9, . . ., m_{13} represent the mass of the structure s8, s9, . . ., s13 respectively. It was calculated by multiplying the volume of each one of them with the electron density at the location of their corresponding centroid. If $\vec{R}_{radio}$ indicates the position vector of

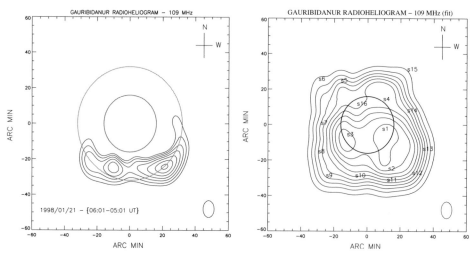

Figure 7. LEFT - Difference map obtained by subtracting the radioheliogram obtained at 05:01 UT from that at 06:01 UT in Figure 6. Contours with levels greater than 50% of the peak value are only shown here. The enhanced radio emission in the southern hemisphere correlates well with the leading edge of the whitelight CME in Figure 5 (left hand side). The outer bigger circle in the map corresponds approximately to the occulter size of the LASCO C2 coronagraph. **RIGHT** - Radio brightness distribution of the Sun obtained through ray tracing calculations of the GRH data obtained at 06:01 UT (Figure 6). There is a good correspondence between the two distributions. The numbers s1-s16 indicate the location of the centroid of the discrete sources used in the ray tracing calculations.

the 'radio' CME, then we have

$$\vec{R}_{radio} = \frac{1}{M_{radio}}[m_8\vec{r}_8 + m_9\vec{r}_9 + m_{10}\vec{r}_{10} + m_{11}\vec{r}_{11} + m_{12}\vec{r}_{12} + m_{13}\vec{r}_{13}] \qquad (2.6)$$

Here $\vec{r}_8$, $\vec{r}_9$, . . ., $\vec{r}_{13}$ represent the position vector of s8, s9, . . ., s13, respectively. Substituting for the different values in equations (2.5) & (2.6) we get $\vec{R}_{radio} = 1.16\hat{x} + 0.52\hat{y} - 1.54\hat{z}$. As pointed out earlier, the 'halo' CME of January 21, 1998 was closely associated with the disappearance of the Hα filament from the location (S57 E19) on the solar disk. Assuming the latter to be located on the solar surface, i.e. at a radial distance of 1 R$_\odot$ (in three-dimensional space) from the center of the Sun, we calculated its position vector as, $\vec{R}_{H\alpha} = 0.51\hat{x} - 0.18\hat{y} - 0.84\hat{z}$. According to the LASCO CME catalog, the lift-off time of the aforementioned 'halo' CME, projected to 1 R$_\odot$, is 05:41:17 UT. The above time was obtained by employing a linear fit to the LASCO C2/C3 height-time measurements of the leading edge of the whitelight CME in the plane of the sky, and is for the case where the CME is considered to travel at a constant speed. Note in the present case we have assumed that the source region of the CME on the solar disk is (S57 E19), i.e. the location of the Hα filament. This corresponds to a radial distance of $\approx$ 0.86 R$_\odot$ in the plane of the sky. We estimated the lift-off time of the CME from the above location by back-projecting its height-time curve, and it is 05:37:26 UT. This implies that the CME had travelled a distance of about 824,253 km [i.e. $\sqrt{(\vec{R}_{radio} - \vec{R}_{H\alpha})^2}$] in the three-dimensional space during the time interval 06:01:06 - 05:37:26 UT, i.e. at an average speed of $\approx$ 580 km s^{-1}. Note that this value should be treated with caution since we have assumed that the CME was propagating at a constant speed. We also separately calculated its speed along the line of sight and in the plane of the sky, and they are $\approx$ 318 & 485 km s^{-1}, respectively.

Acknowledgements

I thank Ch.V.Sastry and C.Kathiravan for their help in preparing this article. The staff of the Gauribidanur radio observatory are thanked for their help in data collection and maintenance of the antenna and receiver systems. The financial support provided by the International Astronomical Union and the Scientific Organizing Committee of the Symposium to attend the meeting and present this invited talk is acknowledged. The LASCO whitelight images presented are due to the kind courtesy of S.P.Plunkett, S.Yashiro and O.C.St.Cyr. The SOHO data are produced by a consortium of the Naval Research Laboratory (USA), Max-Planck-Institut fuer Aeronomie (Germany), Laboratoire d'Astronomie (France), and the University of Birmingham (UK). SOHO is a project of international cooperation between ESA and NASA. The CME catalog is generated and maintained by the Center for Solar Physics and Space Weather, the Catholic University of America in cooperation with the Naval Research Laboratory and NASA.

References

Aubier, M., Leblanc, Y. & Boischot, A. 1971, *Astron. Astrophys.* 12, 435.

Bastian, T. S. & Gary, D. E. 1997, *J. Geophys. Res.* 102(A7), 14031.

Bastian, T. S., Pick, M., Kerdraon, A., et al. 2001, *ApJ* 558, L65.

Brueckner, G. E., Howard, R. A., Koomen, M. J., et al. 1995, *Solar Phys.* 162, 357.

Dulk, G. A. 1980, in: M. R. Kundu & T. E. Gergely (eds.), *Radio Physics of the Sun*, (Reidel: Dordrecht), IAU Symposium 86, p. 419.

Fleck, B., Domingo, V. & Poland, V. (eds.), 1985, *The SOHO Mission* (Dordrect: Kluwer).

Fludra, A., Del Zenna, G., Alexander, D. & Bromage, B. J. I. 1999, *J. Geophys. Res.* 104(A5), 9709.

Gopalswamy, N. & Kundu, M. R. 1992, *ApJ* 390, L37.

Gopalswamy, N. 1999, in: T. Bastian, N. Gopalswamy & K. Shibasaki (eds.), *Solar Physics with radio observations*, Nobeyama radio observatory report No. 479, p. 141

Gosling, J. T., McComas, D. J., Phillips, J. L. & Bame, S. J. 1991, *J. Geophys. Res.* 96(A5), 7831.

Howard, R. A., Michels, D. J., Sheeley Jr. N. R. & Koomen, M. J. 1982, *ApJ* 263, L101.

Kathiravan, C., Ramesh, R. & Subramanian, K. R. 2002, *ApJ* 567, L93.

Kathiravan, C. & Ramesh, R. 2004, *ApJ* 610, 532.

Newkirk, G. 1961, *ApJ* 133, 983.

Plunkett, S. P., Vourlidas, A., Simberová, S., et al. 2000, *Solar Phys.* 194, 371.

Ramesh, R., Subramanian, K. R., SundaraRajan, M. S. & Sastry, Ch. V. 1998a, *Solar Phys.* 181, 439.

Ramesh, R. 1998b, *Ph.D. Thesis* Bangalore University.

Ramesh, R., Subramanian, K. R. and Sastry, Ch. V. 1999, *Astron. Astrophys. Suppl.* 139, 179.

Ramesh, R. 2000a, *Solar Phys.* 196, 213.

Ramesh, R. & Sastry, Ch. V. 2000b, *Astron. Astrophys.* 358, 749.

Ramesh, R., Kathiravan, C. & Sastry, Ch. V. 2001a, *ApJ* 548, L229.

Ramesh, R. & Ebenezer, E. 2001b, *ApJ* 558, L141.

Ramesh, R. & ShanmughaSundaram, G. A. 2001c, *Solar Phys.* 202, 355.

Ramesh, R., Kathiravan, C. and Sastry, Ch. V. 2003, *ApJ* 591, L163.

Sastry, Ch. V., Shevgaonkar, R. K. & Ramanuja, M. N. 1983, *Solar Phys.* 87, 391.

Sheridan, K. V., Jackson, B. V., McLean, D. J. & Dulk, G. A. 1978, *Proc. Astron. Soc. Australia* 3(4), 249.

Smerd, S. F. 1950, *Aust. J. Sci. Res.* A3, 34.

Solar Geophysical Data 643 (Part I), March 1998.

Solar Geophysical Data 646 (Part I), June 1998.

Solar Geophysical Data 647 (Part II), July 1998.

Solar Geophysical Data 648 (Part I), August 1998.

Solar Geophysical Data 677 (Part I), January 2001.
Vourlidas, A., Subramanian, P., Dere, K. P. & Howard, R. A.2000 *ApJ* 534, 456.

Discussion

BOTHMER : Does L~160.000 Mm for June 2, 98 event correspond roughly to length of
the arcade of loops ? L~160.000 Mm ≈ 20° in heliolog.

RAMESH: The June 2, 1998 radioheliogram presented was observed about 10 min prior
to the appearance of the associated CME in the LASCO field of view. So, it is possible
that the estimated line of sight depth of the discrete radio structure(s) would be the
location of the CME correspond to a pre-event coronal arcade system.

FILIPPOV : You showed the trajectory of a partial halo CME that was not the straight
line but the curved one. Is it real non-radial material motion or the result of brightness
redistribution within radially spreading CME body?

RAMESH: We followed the centroid of the discrete radio structure (whose position correlated well with the leading edge of the white light CME in the LASCO FoV) to obtain
the trajectory of the 'radio' CME. The estimated peak brightness temperature of the
structure did not change much during our observing interval. This indicates that the
trajectory is mostly real.

GRECHNEV : 1. Have you measured the temporal variation of the acceleration?
2. Can you comment on the brightness temperatures you've shown? They seem to be
rather low; is this due to calibration techniques?

RAMESH: 1. Yes, we have sufficient time resolution but I do not have a picture here.
2. The low brightness temperature reported are not due to any calibration errors. Note
that the radio brightness temperature due to a CME is always less than 10^6 K. This is
mostly because the CME is optically thin. Also it should be kept in mind that we have
refraction effects off the limb. This also could lead to a reduced brightness temperature.

JINGXIU WANG: What is the spatial resolution of your radioheliogram, if the resolution
is good enough to study the CME initiation process?

RAMESH: The spatial resolution of the Gauribidanur radioheliograph is approximately
5′. The instrument is better suited to study the global coronal changes associated with
the onset of a CME.

Coronal and Stellar Mass Ejections
Proceedings IAU Symposium No. 226, 2005
K. P. Dere, J. Wang & Y. Yan, eds.

© 2005 International Astronomical Union
doi:10.1017/S1743921305000207

A Special Flare-CME Event on April 21, 2002

Huang Guang-Li†

Purple Mountain Observatory
Chinese Academy of Science
Nanjing, 210008, P.R. China
email:glhuang@pmo.ac.cn

Abstract. The time and location of magnetic reconnection are indicated by radio (Nobeyama Radio Heliograph and Polarimeters, Hiraiso and Chinese radio spectrographs) and multi-wavelength (SOHO and TRACE satellites) data in a selected flare-CME event on April 21, 2002. Two hour radio burst started at high frequencies (maximum around 10 GHz). After that, a radio ejection at 17 GHz from one foot point was coincident with the expanded flare and post-flare loops. The reversal of polarization sense at the radio loop-top is associated with the strong coherent emissions around 2 GHz, which should be located above the loop-top at 17/34 GHz. The radio ejection and coherent emissions are also associated with a pair of moving type *IV* bursts at 0.2-2 GHz from high to low frequencies and 2.6-3.8 GHz from low to high frequencies, respectively. High time resolution (8 ms) data show three components of the frequency drifts at 2.6-3.8 GHz: very slow (5 MHz/s) of moving type *IV*, slow (50 MHz/s) of zebra strips, and fast (several GHz/s) of type *III*, which may represent respectively the speeds of flare loop or current sheet, outflows, and energetic electrons from the reconnection site.

Keywords. Sun: flares, CMEs, magnetic fields

---

1. Introduction

A special event on April 21, 2002 with X1.5 flare and very high-speed CME in AR9906 (S14W84) has attracted wide attention. The height, velocity and acceleration of CME are carefully fitted by Gallagher et al. (2003). Very rapid disruption of pre-CME streamer, very high Doppler shifts, and high temperature plasma in SOHO/UVCS are shown by Raymond et al. (2003). The spatial and temporal evolution of radio, X-ray and EUV data are carefully studied by Kundu et al. (2004).

One key problem to compare the observations with the models of flares-CMEs is how to get the location and time of magnetic reconnection (Forbs, 2004). The authors tried to give some evidences in this event as follow.

2. Radio ejection at 17/34 GHz with expanded flare and post-flare loops

Fig.1 shows the time evolution of TRACE images overlaid by 17 GHz contours of Nobeyama Radio Heliograph (NoRH). The main radio burst started at higher frequencies as shown in the first panel with a double source at two foot-points, which is confirmed by SOHO/MDI magnetograph in Fig.2a as well as 34 GHz contours of NoRH. After that, a radio ejection was detected from one foot point source (another one disappeared) at

---

† Present address: 2nd, West-Beijing Street, Nanjing, 210008, China

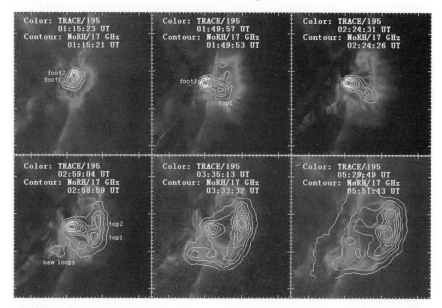

Figure 1. The TRACE images overlaid by 17 GHz contours of NoRH at different phase of the event

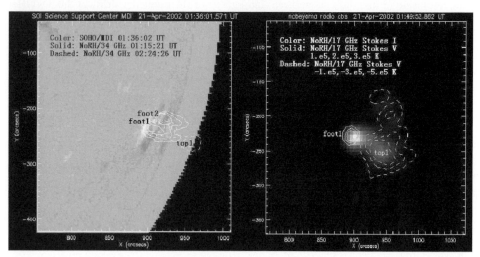

Figure 2. a. The SOHO/MDI magnetograph overlaid by 34 GHz contours at 01:15:21 (solid) and 02:24:26 (dashed) UT. b. The contours of Stokes V at 17 GHz are overlaid on Stokes I image at 17 GHz of NoRH (01:49:53 UT)

01:49 UT associated with the expanded flare and post-flare loops (Figs.1-2). The ejection was extended along the solar surface to form a mushroom shape well coincident with the EUV loops. The velocity of radio ejection and expanded loops keeps about 10 km/s until the post-flare phase. It is emphasized that the polarization sense of the radio ejection in loop-top source (LCP) is opposite to that at foot-point source (RCP) as in Fig.2b.

3. Associated with strong coherent emissions at 1-2 GHz

Fig.3 shows the Stokes I and V profiles at 1, 2, 3.75, 9.75, 17 and 35 GHz of Nobeyama Radio Polarimeters (NoRP) in this event. The maximum time is around 01:25 UT with

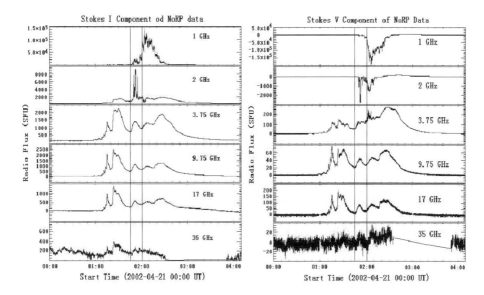

Figure 3. The Stokes I and V profiles at 1, 2, 3.75, 9.75, 17 and 35 GHz of NoRP in the event

turnover frequency of about 10 GHz corresponding to 17 GHz source at foot 2 in Fig.1. Almost during the same time as the radio ejection started (01:50 UT), a very strong emission at low frequencies (1-2 GHz) appeared with reversal polarization sense from RCP to LCP, which predicted as the coherent emissions by Kundu et al. (2004). Note that the polarization sense at 17 GHz of NoRP was not reversed as in NoRH image (Fig.2b), which may be explained by the high sensitivity and spatial resolution of NoRH.

4. A pair of moving type *IV* bursts at 0.2-2.0 and 2.6-3.8 GHz with fine structures

The coherent emissions are further analyzed with dynamic spectra of Hiraiso and Chinese Radio Spectrographs. There are a pair of moving type *IV* bursts respectively at 0.2-2 GHz from high to low frequencies and 2.6-3.8 GHz from low to high frequencies, respectively (Fig.4).

The fine structures with fast frequency drifts evidently exist in the type *IV* continuums in Fig.4. With high time resolution of 8 ms, the dynamic spectra at 2.6-3.8 GHz are selected in four time intervals as marked in Fig.4. The strong fluctuations at 2 GHz (Fig.3) mean that the type *IV* bursts are not real continuums. A group of quasi-periodic emission clusters are shown in the top panel of Fig.5. There are very slow frequency drift of moving type *IV* (5 MHz/s) and the fast drift of type *III* (several GHz/s) respectively in second and third panels. The 4th panel shows the multiple zebra strips with middle frequency drift (50 MHz/s). It is evident that the three components of frequency drifts commonly coexist in Fig.5.

5. Discussions

The fitting curve (Fig.6) of the velocity of the CME leading edge in this event (Gallagher et al., 2003) is well comparable with the typical falre-CME models, such as Lin & Forbes (2000). When the calculated velocity of the CME with average magnetic field

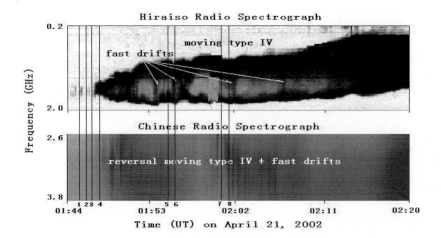

Figure 4. A pair of moving type *IV* bursts at 0.2-2.0 GHz and 2.6-3.8 GHz

of 100 Gauss in the active region is larger than the calculated local Alfvén velocity, the shock waves will be formed with radio type *II* bursts, which are reported in SGD by several spectrographs at meter bands around 01:20-01:30 UT, that consistent with the prediction in Fig.6.

On the other hand, the velocity of the radio ejection (top 1 in Figs.1-2) together with expanded flare and post-flare loops is comparable with the initial velocity of the CME. The reconnection site may be located in the 2 GHz source between two reversed type *IV* bursts (Fig.4), which should be above the loop-top of 17/34 GHz with the same velocity as the radio ejection and the velocity of moving type *IV* bursts. The different frequency drifts may correspond respectively to the speeds of current sheets, outflows, and accelerated electrons from the reconnection site.

Moreover, the coherent emissions as the radio signature of magnetic reconnection took place after the started time of the CME and the main flare phase, hence, it may refer to the second reconnection. The first reconnection may start in the pre flare- CME phase, such as the very rapid disruption of streamers, very high Doppler shifts, and high temperature plasma (Raymound, 2003), and the EIT ejection (Kundu et al., 2004). The fluctuations of coherent emissions with period of 10 seconds (typical Alfvén time) may be an intrinsic property of magnetic reconnection (Huang, 1990).

Acknowledgements

The author should thank Dr. Lin Jun for helpful discussions and the theoretical calculation in Fig.6. The author also thanks SOHO, TRACE, NoRH/NoRP, Hiraiso and Chinese radio spectrographs for the data usage. This study is supported by the NFSC projects with No.10273025 and 10333030, and "973" program with No. G2000078403.

References

T. G. Forbes, 2004, Proccedings of IAU Symposium 226 (invited)
P. T. Gallagher1, G. R. Lawrence, and B. R. Dennis, 2003, ApJ, 588, L53
G. L. Huang, 1990, Adv. Space Res., 10, 173
M. R. Kundu, V. I. Garaimov, S. M. White, and S. Krucker, 2004, ApJ, 600, 1052
J. Lin and T. G. Forbes, 2000, JGR, 105, 2375

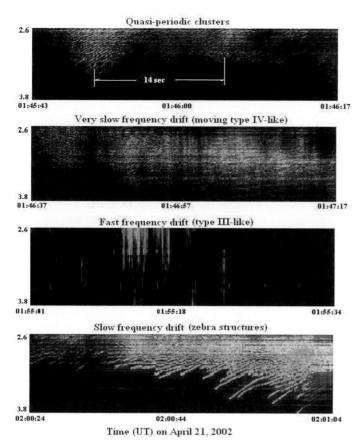

Figure 5. The quasi-periodic emission clusters and three components of frequency drifts from top to bottom in the dynamic spectra at 2.6-3.8 GHz

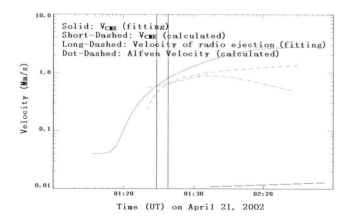

Figure 6. The fitting velocity of the CME and radio ejection in comparison with the calculated CME and Alfvén velocity

J. C. Raymond, A. Ciaravella, D. Dobrzycka, L. Strachan, Y.-K. Ko, and M. Uzzo, 2003, ApJ, 597, 1106

Discussion

SCHMIEDER: Some of the drifts that you show look like type III bursts. Are they associated with the jet features?

HUANG: The drifts of opposite directions are associated and mixed in these observations.

JUN LIN : The jet-like structure observed should correspond to the reconnection outflow along the current sheet.

HUANG: Yes, I agree with your comment.

Coronal and Stellar Mass Ejections
Proceedings IAU Symposium No. 226, 2005
K. P. Dere, J. Wang & Y. Yan, eds.

© 2005 International Astronomical Union
doi:10.1017/S1743921305000219

Occurrence of Solar Radio Burst Fine Structures in 1–7.6 GHz Range Associated with CME Events

Yihua Yan, Yuying Liu, Zhijun Chen, Qijun Fu, Chengming Tan, Shujuan Wang & Jian Zhang†

National Astronomical Obsevatories, Chinese Academy of Sciences, A20 Datun Road,
Chaoyang District, Beijing 100012, China
email: yyh@bao.ac.cn

Abstract. The solar radio bursts and accompanying fine structures recorded by spectrometers at Huairou, Beijing during 1999–2003 are presented. The spectrometers are with high temporal (5–10 ms) and spectral (4–20 MHz) resolutions. We found 91 radio burst events that occurred within half hour of the onset of the CME events which cause solar energetic particle events. The associations of radio fine structures with CME events are discussed.

Keywords. Sun: corona, Sun: coronal mass ejections (CMEs), Sun: Radio

1. Introduction

Radio observations at decimetric- and centimeter wavelengths provide important information for inferring fundamental processes of energy release, particle acceleration and particle transport in solar activities (Bastian *et al.* 1998). Temporal fine structures (FS) of solar radio emission in solar flares have been found in various wavebands for several decades (Allaart, *et al.* (1990), Bruggmann, *et al.* (1990), Benz, *et al.* (1992), Sawant, *et al.* (1994), Isliker & Benz (1994), Jiricka, *et al.* (2001), Fu, *et al.* (2004a)). They are considered to be related to primary energy release processes, etc. (Bastian *et al.* 1998). The Solar Radio Broadband Dynamic Spectrometer (SRBS) of China is the first instrument in microwave to acquire dynamic spectrums of solar bursts with the combination of wide frequency coverage (0.7–7.6 GHz), high temporal resolution, high spectral resolution, and high sensitivity (Fu *et al.* 2004b). We compare the occurrence of FSs in 1–7.6 GHz range with flare/CME events.

FSs in >1 GHz range are categoried by Isliker & Benz (1994), Jiricka, *et al.* (2001), Fu, *et al.* (2004a), etc., into different types. Kliem, *et al.* (2000) have found that the radio drifting pulsation structure are associated with CME initial process. Such structure are typical features for the active region NOAA 9077 during its passage over solar disk (Karlicky, *et al.* (2001), Wang, *et al.* (2001a)), and Wang, *et al.* (2001b) found that various radio FSs occurred during different phases for the Bastille Day event. The time sequence of the radio emission was analyzed by comparing with the hard X-rays (HXRs) and the soft X-rays (SXRs) in this flare. After the maxima of the X-rays, the radio emission in the range 1.0–7.6 GHz reached maxima first at the higher frequency, then drifted to the lower frequency. This comparison suggested that the flare included three successive processes: firstly the X-rays rose and reached maxima at 10:10–10:23 UT, accompanied by fine structures only in the lower altitude regime (range 2.6–7.6 GHz); secondly the

† Dept of Astronomy, Beijing University, Beijing, China

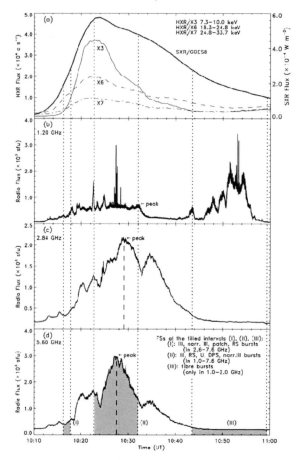

Figure 1. Time profiles of the HXRs and SXR and the radio emission at several typical frequencies during 10:00–11:00 UT on 14 July 2000. (a) HXRs by FY-2 satellite and GOES SXR; (b) 1.2 GHz; (c) 2.84 GHz; and (d) 5.60 GHz. The filled areas in (d) indicate three intervals only in which time interval many radio FSs occurred. During (I) 10:10–10:23 UT, accompanied by fine structures only in the range 2.6–7.6 GHz (II) many fine structures over the range 1.0–7.6 GHz at 10:23–10:34 UT (III) a decimetric type IV burst and its associated FSs (fibers) in the range 1.0–2.0 GHz appeared after 10:40 UT (Wang *et al.* 2001b).

microwave radio emission reached maxima accompanied by many fine structures over the whole altitude (range 1.0–7.6 GHz) at 10:23–10:34 UT; then a decimetric type IV burst and its associated FSs (fibers) in the high altitude regime (range 1.0–2.0 GHz) appeared after 10:40 UT. For the Oct–Nov 2003 flare/CME events, various radio FSs were also found appearing at different phases of the flare/CME process (Tan, *et al.* (2004)).

The temporal relationship between CMEs and associated flares is of great importance to understanding the origin of CMEs. Zhang, *et al.* (2001) have studied this issue using SOHO/LASCO and EIT observations. The association of radio type II bursts and CMEs have been found for decades (Cane (1984), Aurass, (1992)), and Chernov & Markeev (1997) discussed radio FSs in metric wave range associated with CMEs. Here the occurrence of FSs in 1–7.6 GHz with respect to the earth-effective CMEs, that cause solar energetic particle events, is statistically analyzed regardless the FS types. We first introduce the instruments and observations in §2. Then in §3 we present the statistical results and finally we draw our conclusions in §4.

Table 1. The Solar Radio Spectrometers at Huairou/NAOC

Frequency range:	1.0–2.0 GHz	2.6–3.8 GHz	5.2–7.6 GHz
Temporal resolution:	5ms (after June 2002) 20ms (before Dec. 2001)	8ms	5ms
Frequency resolution (MHz/chan.):	4/240 (after June 2002) 20/50 (before Dec. 2001)	10/120	20/120
Sensitivity:	3%,	2%,	2% $S_{quiet\ Sun}$
Dynamic range:	10 dB above 3 or 2% $S_{quiet\ Sun}$		
Polarizations:	LHCP, RHCP		
Observing time:	22-10UT (Summer), 0-8UT (Winter)		

Table 2. Observed solar radio bursts during 1999–2003 at Huairou/NAOC

Frequency range (GHz):	1.0–2.0	2.6–3.8	5.2–7.6
Number of Burst Events:	729	1616	1198
Number of Fine Structures:	110	131	48

2. Instruments and Observations

Since 1994, a broadband solar radio spectrometer had been developed in China, with a frequency coverage of 0.7–7.6 GHz, a frequency resolution of 1–10 MHz, and a temporal resolution of 1–10 ms (Fu, *et al.* (2004b)). This instrument is composed of multibands spectrometers and the three spectrometers at 1.0–2.0 GHz, 2.6–3.8 GHz, and 5.2–7.6 GHz are located at Huairou Solar Observing Station of National Astronomical Observatories, Chinese Academy of Sciences (NAOC). The radio environment has been measured and calibration techniques are developed to ensure reliable observations (Yan, *et al.* (2002), Sych & Yan (2002)).

The performance of the spectrometers is very powerful in detecting radio fine structures. Table 1 shows the description of the spectrometers at Huairou (Ji, *et al.* (2000), Ji, *et al.* (2003), Fu, *et al.* (2004b)).

Many events have been observed since the Chinese Solar Radio Broadband Spectrometers have been put into operation. The radio events observed at Huairou/NAOC during 1999 to 2003 is listed in Table 2.

The radio FSs have also been observed as shown in the above table. Figure 1 shows the occurrence of radio FSs associated with the flare/CME process for the Bastille Day event. For the Oct–Nov 2003 event we also observed FSs during flare/CME process (Tan, *et al.* (2004), Wang, *et al.* (2004)). Here we show the radio zebra pattern and spike fine structures during the rising phase of the 26 October 2003 event in Figure 2.

The CME list is obtained from SOHO/LASCO catalogue (http://cdaw.gsfc.nasa. gov/CME_list/) and the LASCO instrument is introduced in detail by Brueckner, *et al.* (1995).

3. Statistic Results

As mentioned above, we mainly analyze associations of the occurrence of FSs in 1–7.6 GHz with respect to the earth-effective CMEs, i.e., that cause solar energetic particle

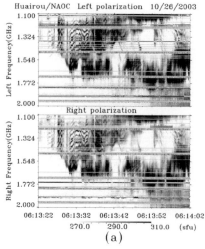

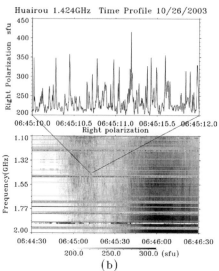

Figure 2. The radio (a) zebra pattern and (b) spike fine structures during the rising phase of the 26 October 2003 event.

events, regardless the FS types. Therefore we chose the radio burst events that occurred within half hour of the onset of the CME events.

During 1997–2003, there were 91 radio burst events in the 1–7.6 GHz range selected which are associated with the CME events, among which 86 were accompanying Type II/IV bursts as shown in Table 3. All these 91 events associated with GOES X-ray flares as listed in Table 3 as well. Among 91 events, 60 (66%) events have burst in all 3 bands employed by Huairou/NAOC spectrometers: 14/18 (78%) for X-class flares 39/57 (68%) for M-class flares and 7/16 (44%) for C-class flares. 78 out of 91 events (85.7%) were associated with Hα flares and the classification is shown in Table 4. If we look at the radio burst events and the GOES X-rays flares we can seen that most of them are radio complex events as shown in Table 5. The importance of the radio flux intensity at 2.84 GHz for these events are listed in Table 6. It can be seen that most are strong radio burst

Table 3. Radio bursts (in 1–7.6 GHz) associated with CMEs and Type II/IV bursts and SXR flares

CMEs	Type II	Type IV	X	M	C
91	49	37	18	57	16
(100%)	(54%)	(41%)	(20%)	(62%)	(18%)

Table 4. Radio bursts (in 1–7.6 GHz) associated with CMEs and Hα flares

Total event	3B	2B	1B	SF	1N	others	no flare
91	4	17	7	13	17	20	13

Table 5. Classification of soft X-ray flares

GOES flare	Event	47 GB	45 C	5 S	3 S	1 S	Other
X class	18	16	1	1	0	0	0
M class	57	15	27	4	9	1	1
C class	16	0	6	6	4	0	0
Total	91	31	34	11	13	1	1

events. This is agreeable to the selection that the CME events are all earth-directed and caused SEP events.

In all 91 events there are 38 (41.7%) containing FSs (or FS groups). The 26 FSs occurred during rising phase, 16 during maximum, and 8 in decaying phase (there are overlaps for FS occurrence). Note here a FS event may contain different types of FSs as we mentioned above. The detailed description of these events are listed in Table 7. We have also analyzed the time sequences of these flare/CME event and the results are listed in Table 8. It can be seen that most events are in radio burst-SXR flare maximum-CME time sequences. Please note that radio bursts and flares are measured at their peak time whereas the CME onsets are measured according to their C2/C3 onset time. Therefore it deserves careful analysis of these time sequence relationship.

4. Conclusions

In summary, during 1997–2003, there were 91 radio burst events in the 1−7.6 GHz range selected which are associated with the CME events that caused SEP events. The CME was simply chosen when they are half hour apart to the radio bursts and SXR flares.

(1) All the events are associated with soft X-ray flares (18 X class, 57 M class and 16 C class events), 78 are associated with Hα flares, 49 associated with Type II burst and 37 with Type IV burst.

(2) In the 91 CME-associated radio burst events, 60 (66%) have burst in all 3 frequency bands: 14/18 (78%) for X-class 39/57 (68%) for M-class and 7/16 (44%) for C-class. Most of them are of complex radio burst profiles (65 events) and with strong intensity (71 events with > 100 s.f.u. at 2.84 GHz). 50 events (65%) were in a time sequence of "radio burst" → "SXR flare" → "CME onset".

(3) There are 38 (41.7%) event containing FSs (or FS groups). Most radio FSs occurred during rising phase of flare/CMEs, and the occurrence of FSs decreases as frequency

Table 6. Solar Radio Flux at 2.84 GHz of the events

Flux range (s.f.u.)	< 100	100 − 500	500 − 1000	1000 − 5000	> 5000
Events	20	40	14	16	1

Table 7. Occurrence of radio FSs in different burst phases

Phases:	rising	peak	decay
1–2 GHz	19	15	4
2.6–3.8 GHz	14	3	3
5.2–7.6 GHz	5	2	1

Table 8. Time sequences of radio burst/CME/falres.

Radio burst maximum → SXR flare maximum → CME	59	(65%)
Radio burst maximum → CME → SXR flare maximum	8	(9%)
SXR flare maximum → Radio burst maximum → CME	12	(13%)
SXR flare maximum → CME → Radio burst maximum	2	(2%)
CME → Radio burst maximum → SXR flare maximum	7	(8%)
CME → SXR flare maximum → Radio burst maximum	3	(3%)

increases (19/14/5 FS events in rising, 15/3/2 at peak, and 4/3/1 when decay for 1–2 G/2.6–3.8 G/5.2–7.6 GHz regime).

(4) If we consider a standard flare/CME model the above statistic results are consistent with such a scenario and radio fine structures may manifest initial phase signature, which is important to diagnose coronal parameters.

Acknowledgements

The work is supported by CAS, NSFC grants 10225313, 10333030, and MOST grant G2000078403. We acknowledge SOHO/LASCO for providing CME catalog used in this study.

References

Allaart, M. A. F., van Nieuwkoop, J., Slottje, C., & Sondaar, L. H. 1990, *Sol. Phys.* 130, 183
Aurass, H. 1992, *Ann. Geophys.* 10, 359
Bastian, T. S., Benz, A. O., & Gary, D. E. 1998, *ARAA* 36, 131
Benz, A. O., Su, H., Magun, A., & Stehling, W. 1992, *A&A* 93, 539
Brueckner, G. E., Howard, R. A., Koomen, M. J., & et al. 1995, *Sol. Phys.* 162, 357
Bruggmann, G., Magun, A., Benz, A. O., & Stehling, W. 1990, *A&ASS* 240, 506
Cane, H. V. 1984, *A&A* 140, 205
Chernov, G. P., & Markeev, A. K. 1997, *IAU JD* 19, 16
Fu, Q., Yan, Y., Liu, Y., & et al. 2004a, *Chin. J. Astron. Astrophys.* 4, 176
Fu, Q., Ji, H., Qin, Z., & et al. 2004b, *Sol. Phys.* 222, 167
Isliker, II., & Benz, A. 1994, *A&A* 105, 205
Ji, II., Fu, Q., Liu, Y., & et al. 2000, *ACTA Astrophysica Sinica* 20, 209
Ji, H., Fu, Q., Liu, Y., & et al. 2003, *Sol. Phys.* 213, 359
Jiricka, T., & et al. 2001, *A&A* 375, 243
Karlicky, M., Yan, Y. H., Fu, Q. J., & et al. 2001, *A&A* 369, 1104
Kliem, B., Karlicky, M., & Benz, A. O. 2000, *A&A* 360, 715
Sawant, H. S., Fernandes, F. C. R., & Neri, J. A. C. F. 1994, *ApJS* 90, 689

Sych, R. A., & Yan, Y. H. 2002, Chin. J. Astron. Astrophys. 2, 183

Tan, C.-M., Fu, Q.-j., Yan, Y.-H., & Liu, Y-Y. 2004, *Chin. J. Astron. Astrophys.* 4, 205

Wang, S. J., Yan, Y. H., & Fu, Q. J. 2001a, *A&A* 370, L13

Wang, S. J., Yan, Y. II., Zhao, R. Z., & et al. 2001b, *Sol. Phys.* 204, 153

Wang, S. J., & et al. 2004, *IAU Symp* 226 (these proceedings)

Yan, Y., Tan, C., Xu, L., Ji, H., Fu, Q., & Song, G. 2002, *Since in China (Series A)* 45, 89

Zhang, J., Dere, K. P., Howard, R. A., Kundu, M. R., & White, S. M. 2001, *ApJ* 559, 452

Coronal and Stellar Mass Ejections
Proceedings IAU Symposium No. 226, 2005
K. P. Dere, J. Wang & Y. Yan, eds.

© 2005 International Astronomical Union
doi:10.1017/S1743921305000220

Observations of a Post-Eruptive Arcade on October 22, 2001 with CORONAS-F, other Spaceborne Telescopes, and in Microwaves

V.N. Borovik[1], V.V. Grechnev[2], O.I. Bugaenko[3], S.A. Bogachev[3], I.Y. Grigorieva[1], S.V. Kuzin[4], S.V. Lesovoi[2], M.A. Livshits[5], A.A. Pertsov[4], G.V. Rudenko[2], V.A. Slemzin[4], A.I. Stepanov[5], K. Shibasaki[6], A.M. Uralov[2], V.G. Zandanov[2] and I.A. Zhitnik[4]

[1]Main Astronomical Observatory, St. Petersburg, Russia, email: borovik@MK4099.spb.edu

[2]Institute of Solar-Terrestrial Physics, Irkutsk, Russia

[3]Sternberg Astronomical Institute, Moscow, Russia

[4]P.N.Lebedev Physical Institute, Moscow, Russia

[5]Institute of Terrestrial Magnetism, Aeronomy and Radiowave Propagation, Troitsk, Russia

[6]Nobeyama Radio Observatory, Japan

Abstract. Using multi-spectral data, we estimate plasma parameters in the post-eruptive arcade observed on October 22, 2001 at 100 Mm above the limb: the temperature is 6 MK and the plasma density is $(5-9) \cdot 10^9 \, \mathrm{cm}^{-3}$. We state a problem of the long-term equilibrium of the hot top of the arcade high in the corona: either the magnetic field surrounding the arcade well exceeds that one extrapolated in the potential approximation, or $\beta > 1$ both inside and outside the arcade. A downflow observed in soft X-rays can contribute to the equilibrium.

Keywords. Sun: activity, coronal mass ejections, UV radiation

1. Instrumentation, observations, and methods

The CORONAS-F space solar observatory was launched in August 2001. The SPIRIT complex aboard the CORONAS-F contains EUV and soft X-ray telescopes and spectro-heliographs providing full-disk solar images in 175 Å (FeIX–XI, $T_e \sim 1.5 \, \mathrm{MK}$), 284, 304, and 8.42 Å (MgXII, $T_e \sim 5$–15 MK, $T_{max} \approx 9 \, \mathrm{MK}$). A post-eruptive arcade on October 22, 2001 looks in MgXII channel like a 'ball' (Fig. 1) cospatial with the arcade observed with Yohkoh/SXT. The NoRH show the arcade at 17 GHz and the SSRT at 5.7 GHz. The RATAN-600 shows a polarized source cospatial with the arcade in one-dimensional scans at several frequencies from 2.8 to 15.6 GHz. Averaged NoRH images also show two oppositely polarized regions with the arcade in between. Yohkoh/SXT images shows dark features moving down onto the arcade from above.

2. Estimations and simulations

Temperature of the main part of the arcade including its brightest portion is 6 MK, and its upper part has a temperature of up to 8 MK, as obtained from Yohkoh/SXT data. The plasma density within the brightest part of the arcade is $5 \cdot 10^9 \, \mathrm{cm}^{-3}$. Optically thin free-free radio emission calculated from Yohkoh/SXT data fits microwave observations both visually and quantitatively to confirm the estimates of plasma parameters. The potential magnetic field extrapolation into the corona from the SOHO/MDI magnetogram corresponds to the magnetic configuration revealed from the microwave data. Calculated

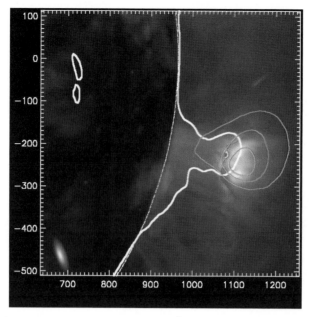

Figure 1. Arcade images. Background: EIT 195 Å (blue) and SXT (red-yellow) composite. Contours: white—NoRH (10^4 K), green—SPIRIT 8.42 Å. Green ellipse: NoRH beam.

magnetic field is ~ 7 G at ~ 100 Mm. Magnetic pressure at the arcade top turns out less than plasma pressure. The minimum required magnetic field strength of 20 G corresponds to the equality of the gas pressure inside the arcade to the outer magnetic pressure. As for the dark features flowing into the arcade top, we have found hints that the inflow regions are slightly cooler than their vicinity. This is possible if they are falling stuff, which can contribute to mass supply to the arcade and to its equilibrium.

3. Summary and conclusions

(*a*) Plasma parameters in the post-eruptive arcade: $N_e \sim 10^{10}$ cm^{-3}, $T_e \sim 6$ MK.

(*b*) The large-scale magnetic configuration surrounding the arcade high in the corona revealed from high-sensitivity NoRH and RATAN-600 data is in accord with the expected one from the potential magnetic fields extrapolation into the corona.

(*c*) Our results give rise to a problem of the long-term equilibrium of giant hot structures high in the corona: either the magnetic field surrounding the arcade well exceeds 20 G 100 Mm, or $\beta > 1$ both inside and outside the hot post-eruptive arcade.

(*d*) Dark features moving downward onto the arcade can have influence for the problem just mentioned. They likely show the downflow due to the falling down remnants of the filament pre-ejected. This downflow can contribute to the mass supply into the arcade and to its equilibrium.

Acknowledgements

We thank the instrumental teams of the NoRH, SSRT, RATAN-600; CORONAS-F & *Yohkoh* satellites for the open-data policies. This work was supported by the Russian Foundation of Basic Research under grants 02-02-16548, 03-02-16591, 03-02-17528, 03-02-17357, the Ministry of Industry and Science under grants 477.2003.2, 16 KI, OFN 18; and the Presidium of RAS under grant 7/2004.

Coronal and Stellar Mass Ejections
Proceedings IAU Symposium No. 226, 2005
K. P. Dere, J. Wang & Y. Yan, eds.

A Reconsideration of the Classification of Two Types of CMEs

A.Q. Chen, C.T. Yeh, J.X. Cheng and P.F. Chen

Department of Astronomy, Nanjing University, Nanjing 210093, China
email: jdye@nju.edu.cn

Abstract. Conventionally coronal mass ejections (CMEs) are categorized into flare-associated and filament-associated types. Since there are also many CMEs of the overlapping type, we classify CMEs into three types in order to compare their characteristics. It is found that the three types of CMEs have quite similar distributions of apparent speeds, with small difference in the average speeds.

Keywords. Sun: coronal mass ejections (CMEs), filament, flares

1. Introduction

Conventionally coronal mass ejections (CMEs) are categorized into two distinct types, i.e., slow CMEs which are associated with filament eruptions and fast CMEs which are associated with solar flares (Sheeley *et al.* 1999). However, it has been noted that some CMEs are associated with both filament eruptions and solar flares (Feynman & Ruzmaikin 2003). Therefore, it is worthwhile to reexamine the characteristics of CMEs of different types. In this paper, we divide CMEs into three types, i.e., (1) CMEs associated only with filament eruptions; (2) CMEs associated only with solar flares; (3) CMEs associated with both filament eruptions and solar flares, and investigate their characteristics in velocity distribution.

2. Observations

In this study, CME events observed by SOHO/LASCO (Brueckner *et al.* 1995) in 2002 are collected. The filament association is judged from the synthesis of the Hα images of Big Bear Solar Observatory, EUV images of SOHO/EIT (Delaboudiniere *et al.* 1995), and the core-like structures observed by LASCO. The judgment of the flare association is a little more complicated, which involves the backward extrapolation of the CME trajectory and some flare selection criteria. In this preliminary research, 257 out of the total 764 CMEs occurring in 2002 were selected into our sample, where 158 events belong to the first type, 77 events to the second type, and 22 events to the third type. The linearly-fitted apparent speeds of all the CMEs are presented in the CME catalog maintained by Gopalswamy et al. (http://lasco-www.nrl.navy.mil/cmelist.html).

3. Data Analysis and Results

The distributions of the three types of CMEs are shown in Figure 1. It is seen that the peaks of the three histograms do not differ significantly, though the tails at large speeds for the latter two types are enhanced compared to the filament-associated CMEs. The average speed of the three types is 476.6, 532.4 and 532.8 km s^{-1}, respectively.

The relative portions of each type of CMEs at every 100 km s^{-1} interval are depicted in Figure 2,which indicates that relative distributions of the three types are rather

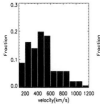 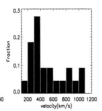 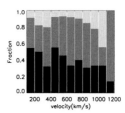

Figure 1. Histograms showing the relative distributions of velocity for the three types of CMEs. Left panel corresponds to events associated with filament eruptions only, middle panel to the events with solar flares only, and right panel to the events with both filament eruptions and solar flares.

Figure 2. Histogram showing the relative portions of the three types of CMEs at every 100 km s^{-1} velocity interval. Black columns correspond to the events associated with filament eruptions only, gray columns to the events with solar flares only, and white columns to the events with both filament eruptions and solar flares.

diffuse, though there is a slight trend that more events at high speeds belong to the flare-connected types.

4. Discussion

Phenomenologically it may not be appropriate to classify CMEs into two types, i.e., filament-associated slow CMEs and flare-associated fast CMEs. At least, there should be an overlapping type, which are associated with both filament eruptions and solar flares as in the classical CSHKP model. Therefore, we classify CMEs into three types, and investigate the distribution of apparent speeds for each type. It is found that the shapes of the distributions for the three types are quite similar, with somewhat stronger tails at small speeds and weaker tails at large speeds for the filament-associated type compared to the other two types. It is also noted that the average speeds for the three types of CMEs are not significantly different. Therefore, our results support the suggestion that there might not be intrinsic difference between various types of CMEs, and hence it is not physically significant to divide them into flare or filament associated. Their different velocity profiles, i.e., impulsive or acceleration types, are perhaps merely due to different time scale (Cliver 1999). In fact, during many CME events, which were previously thought not to be associated with flares, giant arcades or flares as weak as A-class are discerned (Wu *et al.* 2002; Zhang 2004).

Acknowledgements

The research is supported by FANEDD (200226), NSFC (10333040, 10221001 and 10403003) and NKBRSF (G20000784).

References

Brueckner, G.E., Howard, R.A., Koomen, M.J., *et al.* 1995, *Sol Phys.* 162, 357
Cliver, E.W. 1999, *J. Geophys. Res.* 104, 4743
Delaboudiniere, J.-P., Artzner, G.E., Brunaud, J., *et al.* 1995, *Sol Phys.* 162, 291
Feynman, J. & Ruzmaikin, A. 2003, *Sol Phys.* 219, 301
Sheeley, N.R., Walter, J.H., Wang, Y.-W., & Howard, R.A. 1999, *J. Geophys. Res.* 104, 24739
Wu, Y.Q., Tang, Y.H., Dai, Y., & Wu, G.P. 2002, *Sol Phys.* 207, 159
Zhang, J. 2004, *AAS.* 204, 7204

Coronal and Stellar Mass Ejections
Proceedings IAU Symposium No. 226, 2005
K. P. Dere, J. Wang & Y. Yan, eds.

A Possible Explanation for the Different Mean Speeds of Halo and Limb CMEs

J.X. Cheng, C.T. Yeh, D.M. Ding and P.F. Chen

Department of Astronomy, Nanjing University, Nanjing, 210093, China
email:chengjianxia@nju.org.cn

Abstract. In order to explain the different average speeds between halo and limb CMEs, we investigate the relationship between the brightness and speed for 17 halo CMEs. It is found that faster CMEs tend to be brighter, which implies that many halo CMEs with slow speeds are missed in observation owing to the limited sensitivity of LASCO detectors or identifications. As a result, the statistical average speed of halo CMEs turns to be much larger than that of limb CMEs.

Keywords. Sun: coronal mass ejections (CMEs)

1. Introduction

Coronal mass ejections are the most significant eruptive events in the solar atmosphere. In most cases, CMEs have an angular width of tens of degrees; however, CMEs appears to be fully or partially halo-like when their source regions close to the solar disk center. In this sense, CMEs can be categorized into limb and halo events. According to statistics, the average speed of halo CMEs is much higher than that of limb events (Webb *et al.* 1999; Michalek *et al.* 2003). It is generally assumed that halo CMEs are more energetic than limb events in order to explain the speed difference. However, such an explanation is not so justified since the occurrence rate should be the same for all longitudes. In this paper, we suppose that such a speed difference is possible due to an observation effect, as proposed by Andrews (2002).

2. Data analysis

We select 17 halo CMEs that are associated with flares for a statistical study. The brightness is derived by integrating over the whole area of each CME. For simplicity, the onset time of the CME is roughly considered as the peak time of the associated flare. The arrival time is determined by GOES observations as the time when the front loop propagates to the GOES satellite.

3. Our result

As shown in Fig. 1, the brightness and speed of CMEs are positively correlated with a correlation coefficient of 0.73. The faster the CME is, the larger the brightness is. This implies that faster CMEs are brighter that can be more easily observed while slower ones are less bright that are more difficult to observe. Since the emission of CMEs is from the Thomson scattering that peaks at a direction perpendicular to the ray path of photons from the photosphere, limb CMEs are easier to detect than halo CMEs, especially for those CMEs with a lower speed and brightness. Therefore, some slow halo CMEs are missed in observations for the limited sensitivity of the instrument while they

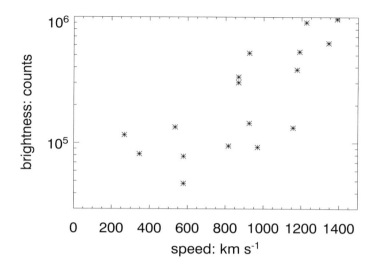

Figure 1. Brightness vs speed of CMEs showing a positive correlation. The faster the CME is, the larger the brightness is.

can possible be detected if appearing at the limb; as a result, the average speed of halo CMEs is larger than limb ones.

4. Conclusion

We investigate the relationship between the brightness and speed using a sample of 17 halo CMEs. It is found that faster CMEs are brighter. This can be understood as a pileup effect: CMEs with a larger speed can accumulate more coronal plasma during their propagation, and hence are brighter considering the effect of the Thomson scattering. Since the Thomson scattering is most significant when the CMEs propagate in the plane of the sky, halo CMEs are more difficult to detect than limb CMEs, especially for those CMEs with a slow speed. In other words, only fast halo CMEs can be easily detected while quite a number of slow halo CMEs are missed by the SOHO/LASCO coronagraph as their brightness is below the detection sensitivity. As a result, the average speed of halo CMEs is apparently larger than that of limb CMEs. Our result confirms the proposal by Andrews (2002).

Acknowledgements

We are grateful to Dr. K. Dere and A. Vourlidas for their useful comments. SOHO is a project of international cooperation between ESA and NASA. This work is supported by FANEDD (200226), NSFC (49990451, 10221001, 10333040,10403003) and NKBRSF (G20000784).

References

Andrews, M. D. 2002, *Solar Phys.* 208, 317
Brueckner, G.E., Howard, R.A., Koomen, M.J., *et al.* 1995, *Solar Phys.* 162, 357
Michalek, G., Gopalswamy, N., & Yashiro, S. 2003, *ApJ* 584, 472
Webb, D.F., St. Cyr, O.C., Plunkett, S.P., Howard, R.A., & Thompson, B.J. 1999, *BAAS* 31, 853

Coronal and Stellar Mass Ejections
Proceedings IAU Symposium No. 226, 2005
K. P. Dere, J. Wang & Y. Yan, eds.

© 2005 International Astronomical Union
doi:10.1017/S1743921305000256

The Three Dimensional Structure of CMEs from LASCO Polarization Measurements

Kenneth Dere[1], and Dennis Wang[2,3]

[1]George Mason University, USA email: kdere@gmu.edu
[2]Interferometrics, Inc., email: dennis.wang@nrl.navy.mil

Abstract. The degree of polarization of Compton-scattered photospheric light observed in a coronagraph is dependent on the distance of the scattering electrons from the plane of the sky. Measurements of the polarization of light scattered by CME structures have been observed by LASCO C2. We have reduced and analyzed a month long sequence of such measurements which were taken at a cadence of 1 hour. The CME brightness has been distributed throughout a 3 dimensional cube and visualized at a variety of angles. Several CMEs are found to have considerable fine-structure consistent with expanding loop arcades. The analysis is subject to a variety of assumptions such as a lack of knowledge of whether a source is before or behind the plane of the sky. Nevertheless, the results obtained to date are intriguing.

Keywords. polarization, Sun: coronal mass ejections (CMEs)

References

Dere, K. P., Wang, D., & Howard, R. A. 2005, ApJL, 620, 119

Coronal and Stellar Mass Ejections
Proceedings IAU Symposium No. 226, 2005
K. P. Dere, J. Wang & Y. Yan, eds.
© 2005 International Astronomical Union
doi:10.1017/S1743921305000268

Filament Eruption and Associated Partial Halo CME on 2001 September 17

Y. C. Jiang, L. P. Li, S. Q. Zhao, Q. Y. Li, H. D. Chen and S. L. Ma

National Astronomical Observatories/Yunnan Astronomical Observatory, CAS, Kunming
650011, China (email: jyc@ynao.ac.cn)

Abstract. We report the eruption of a small H_α filament and associated partial halo coronal mass ejection (CME) occurring in NOAA AR 9616. Accompanied by a M1.5 flare, the filament quickly erupted, a remote coronal dimming region far away from the eruption site was formed above quiet-sun area, and then a long H_α surge developed from the flare site. During the eruption, remote H_α and EUV brightenings appeared near the dimming, along the dimming boundary in EUV and in its interior in H_α, leaving behind EUV loops connecting the eruption source region and the remote EUV brightenings. Finnally, as a definite indication of the CME, a huge dark loop appeared to span the eruption region. These observations indicate that a much larger-scale rearrangement of the corona magnetic fields, eventually represented by the CME, was involved in the eruption of the small filament.

Keywords. Sun: activity, coronal mass ejections (CMEs), filaments, flares

---

On 2001 September 17, a partial Halo CME was observed by SOHO/LASCO. Using H_α data from Yunnan Astronomical Observatory (YNAO), EUV and white-light coronagraph data from SOHO/EIT and SOHO/LASCO, we will show that a filament eruption occurring in NOAA AR 9616 (S14E04) was closely associated with the CME.

In H_α observations (Figure 1(*a1-a4*)), a small filament around the preceding end of AR 9616, 'F', started to erupt violently at about 08:11 UT, followed by an X-ray class M1.5 flare with start, peak and end times at 08:18, 08:25 and 08:34 UT, respectively. Then a long surge appeared near the eruption F after the flare maximum. The flare consisted of three ribbons, 'R', 'Rb' and 'Rc'. The Rb showed spreading motion towards the south, and the Rc was a remote brightening containing a few less bright points. The Rb and Rc were located over quiet-sun regions with opposite magnetic polarities (Figure 1(*b*)). In EIT 195 Å observations (Figure 1(*c1-c4*)), a dimming , 'D', and a brightening, 'EB', appeared around the Rc site (see their 195Å light curves in Figure 2(*a*)). The Rc was inside the D, while the EB, along its northern boundary. Then EUV loops were formed to connect the EB and Rb. Similar to the situation in a halo event reported by Wang, *et al.* (2002), it seems that the R consisted of two unresolved flare ribbons resulted from the F eruption, while the Rb, EB and EUV loops were the product of the reconnection between the erupting F and outlying coronal loops. The dimming may be due to chromosphere evaporation involved near the footpoint of the remote flare ribbon.

A partial halo CME with a width of 166° was first seen in LASCO C2 images at 08:54 UT. The 195 Å difference images reveal that an expanding semicircular dark loop trailed the CME front. Figure 1(*d*) shows that its angular extent (see the white arrow) is approximately comparable with that of the CME front. Therefore, we believe that it is the definite indication of the CME start as a result of the F eruption. The CME front heights measured by Seiji Yashiro is shown in Figure 2(*b*), and the average velocity and acceleration from the first and second-order polynomial fits are 1010 km s^{-1} and -14.5 m s^{-2}. It is difficult to determine the true onset time of the flare-associated CME (Zhang

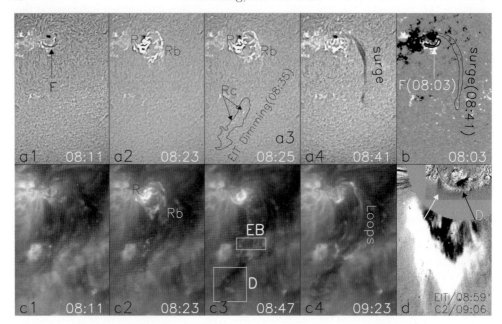

Figure 1. YNAO H_α images (a1-a4), MDI magnetogram (b), direct EIT 195 Å images (c1-c4) and a composite image of inner 195 Å with outer LASCO C2 difference images (d).

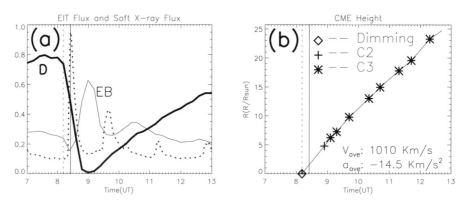

Figure 2. (a) GOES-8 soft X-ray flux (dashed line) and 195 Å light curves in the boxes (in panel c3 of Figure 1) centered on the EB and the D. (b) The CME front heights. The vertical dashed and solid bars indicate the dimming onset time and the flare maximum time.

et al., 2001), so we can only use the dimming process to infer the CME onset time since the coronal dimmings may be caused by the expansion of the coronal magnetic structure involved in the eruption. Figure 2(*a*) indicates that the dimming onset time was around 08:11 UT, so the CME was very likely to start to erupt at the same time.

Acknowledgements

The work is supported by the NSFC under grant 10173023.

References

Wang, H. M., Gallagher, P., Yurchyshyn, V., Yang, G. & Goode, P. R. 2002, *ApJ* 569, 1026

Zhang, J., Dere, K. P., Howard, R. A., Kundu, M. R. & White, S. M. 2001, *ApJ* 559, 452

Coronal and Stellar Mass Ejections
Proceedings IAU Symposium No. 226, 2005
K. P. Dere, J. Wang & Y. Yan, eds.

© 2005 International Astronomical Union
doi:10.1017/S174392130500027X

Hard X-ray Patterns of CME-Related Flares

Y. P. Li and W. Q. Gan

Purple Mountain Observatory, Chinese Academy of Sciences, Nanjing 210008, China
email: yplee@pmo.ac.cn

Abstract. The RHESSI hard X-ray spectra are investigated for a total of 23 CME-related flares. It is found that about 17%, 70%, and 13% of the samples can be attributed to type A, B, and C, respectively. These ratios are not significantly different from those obtained using the data of HXRBS/SMM, although the ratio of type C in our CME-related flares is a little higher. More samples are obviously necessary to study the difference of hard X-ray spectra between the CME-related flares and non-CME-related flares.

Keywords. Sun: CMEs, flares, X-rays

1. Introduction

The relationship between solar flares and coronal mass ejections (CMEs) is an important topic and has been extensively studied. At present it seems to be clear that part of flares has an associated CME. We call this kind of flares as CME-related flares. Harrison (1995) indicated that higher intensity events are more likely to be accompanied by a CME, but this is not always the case (Green et al. 2002). Recently, Andrews (2003) made a statistical study based on a total of 229 M-class and X-class flares occurred from 1996 to 1999, which had a corresponding LASCO observations. He found that all the 14 X-class flares in his samples have an associated CME, while there are 42% of M-class flares which have not any associated CME. The CME-related flares and non-CME-related flares seem to be independent of the positions where they happened on the solar disk.

The soft X-ray characteristics of CME-related flares and non-CME-related flares have been investigated by Kay et al. (2003). A comparison of flare parameters for a total of 69 flares indicates that there are systematic differences in the relationship between the peak intensity and duration, peak intensity and temperature for the two types of event. It was shown that higher intensity events are of longer duration in the case of CME-related flares, and that the CME-related flares tend to have lower peak temperatures than non-CME-related flares of the same intensity.

However, so far there does not seem to have any detailed work on the hard X-ray characteristics of CME-related flares. Different hard X-ray patterns would implicate different environment where the energetic electrons are accelerated. Do the CME-related flares tend to be in conjunction with the topology of the open magnetic field? Do the non-CME-related flares tend to be in conjunction with the topology of the closed magnetic field? Statistical study of hard X-ray spectra for the CME-related flares would be helpful to address this kind of problems.

The hard X-ray patterns of solar flares were divided into three types (e.g., Tanaka 1987): type A or hot thermal flares, type B or impulsive flares, and type C or gradual-hard flares. The detailed descriptions about this classification can also be found in Dennis (1988). Kosugi et al. (1988) made a statistical study for a total of 416 events observed with HXRBS/SMM. They found that among 416 events, 65 belonged to type A, 338 belonged to type B, and 13 belonged to type C. Cliver et al. (1986) noted that most type C flares are accompanied by CMEs.

Table 1. CME-related flares and their hard X-ray types

Types	Date of flares/time			
A	Feb-21-02/12:26	Feb-06-03/03:49	Apr-24-03/12:53	Apr-27-03/15:32
B	Feb-20-02/09:59	Apr-21-02/01:51	Aug-03-02/19:07	Aug-22-02/01:57
	Nov-09-02/13:23	Nov-10-02/03:21	Apr-23-03/01:06	Apr-26-03/08:07
	May-27-03/23:07	May-29-03/01:05	Jun-09-03/11:28	Jun-17-03/22:55
	Oct-19-03/16:50	Nov-01-03/22:38	Nov-02-03/17:25	Nov-03-03/09:55
C	Jul-23-02/00:35	Oct-24-03/02:54	Oct- 28-03/11:10	

The RHESSI (Lin *et al.* 2002) provides us an unprecedented chance to study the hard X-ray spectra with a high energy resolution. This paper is to study the hard X-ray spectra of CME-related flares using the RHESSI data.

2. Data analysis

The CME lists up to December 2003 were taken from the website of http://lasco-www.nrl.navy.mil/cmelist.html. Among the lists a total of 86 CMEs which have a corresponding flare were preliminarily selected. Then we checked the availability of RHESSI observations from these preliminary candidates. Excluding the events with incomplete time coverage and the relatively weak emissions up to 100 keV, at last we selected a total of 23 events with which we can properly analyze the hard X-ray spectra.

We used the standard code SPEX to fit the hard X-ray spectra. The fitted model constitutes a thermal component plus two power-laws. In order to avoid the influence of some factors like lower energy cutoff (e.g., Gan et al. 2001) on the spectrum, only the power-law index of the higher energy end is taken into account. Table 1 presents the selected samples of CME-related flares and the derived types of hard X-ray patterns. As a matter of fact, the so-called soft-hard-harder spectral variations of type C did not appear in our samples. The type C flare in Table 1 is characterized by its long duration (>10 min.) and harder spectral index (<3), the criterion as used by Kosugi *et al.* (1988).

From Table 1 we see that among a total of 23 CME-related flares, type A, B, and C are 4, 16, and 3, taking the ratio of about 17%, 70%, and 13%, respectively. In comparison with the results by Kosugi et al. (1988), the ratio of type C pattern in CME-related flares seems to be a little higher than that obtained in spite of CME-related or not, where the ratio is only 3%. However, both the small number of samples and the complex relationship between flares and CMEs prevent us to make a definite conclusion. Further studies should be based on more samples.

References

Andrews, M. D. 2003, *Solar Phys.* 218, 261.
Cliver, E. W. *et al.* 1986, *ApJ* 305, 920
Dennis, B. R. 1988, *Solar Phys.* 118, 49
Gan, W. Q., Li, Y. P., & Chang, J. 2002, *ApJ* 552, 858
Green, L. M. *et al.* 2002, *Solar Phys.* 205, 325
Harrison, R. A. 1995, *Astron. Astrophys.* 304, 583
Kay, H. R. M. *et al.* 2003, *Astron. Astrophys.* 400, 779
Kosugi, T. & Dennis, B. R. 1988, *ApJ* 324, 1118
Lin, R. P. *et al.* 2002, *Solar Phys.* 210, 3
Tanaka, K. 1987, *PASJ* 39, 1

Coronal and Stellar Mass Ejections
Proceedings IAU Symposium No. 226, 2005
K. P. Dere, J. Wang & Y. Yan, eds.

© 2005 International Astronomical Union
doi:10.1017/S1743921305000281

Statistics Analysis of Decimetric Type III Bursts, Coronal Mass Ejections and Hα Flares

Y. Ma[1,3]†, D. Y. Wang[2], M. Wang[1,3], Y.H. Yan[3]

[1]Yunnan Observatory, Chinese Academy of Sciences, P.O. Box 110, 650011 Kunming,
P.R. China
email: mayuanf@public.km.yn.cn

[2]Purple Mountain Observatories, Chinese Academy of Sciences, 210008 Nanjing, P.R. China

[3]National Astronomical Observatories, Chinese Academy of Sciences, 100012 Beijing China

Abstract. The Statistics analysis of decimetric type III bursts, coronal mass ejections (CMEs) and Hα flares are carried out. The relevant radio events observed from the 625-1500MHz spectrograph at the Yunnan Observatory during the 23rd solar cycle. It is found that the relation between the decimetric type III bursts and CMEs is not closer than that between the type II radio bursts and CMEs; All Hα flares generated decimetric type III bursts and correlated with CMEs are all gradual flares. The higher the energy of the flare correspond to the faster the initial velocity of the CMEs.

Keywords. Sun: activity, CMEs, radio radiation

---

1. Introduction

Is there any relation between the small-scale decimetric type III bursts and the large-scale CMEs? If any, how close is it? for this, the authors studied the relation between the type III bursts and CMEs and Hα flares, acquiring some physical information and approaching the cause of CMEs.

2. Data processing

264 decimetric type III bursts were recorded during the 23rd solar cycle from 625-1500MHz spectrograph at the Yunnan Observatory. If the comparison is made only from the angle of time, there are 138 type III bursts correlating to the CMEs, amounting to 52% of total bursts, and 90 type III bursts corresponding to the Hα flares. However, since the radio spectrograph has no spatial resolution one can not determine the active region. Therefore, if the relation between the type III bursts and CMEs is considered only from the angle of time, the statistical error would be very large because these type III bursts and CMEs do not come from the same source position though they occur simultaneously. For this, the authors only selected the type III bursts which occurred with the CMEs and Hα flares simultaneously for statistics. The identification and selection of the data are carried out according to the following steps: We choose the statistical data in the following procedure. An event of decimetric type III burst is taken firstly, and then to find CME data within the time interval (40 minutes) from LASCO. However, only the

† Present address:Yunnan Observatory, Chinese Academy of Sciences, P.O.Box 110, 650011 Kunming, P.R. China.

position of CME data is similar to the source of Hα flares , generated the type III burst, to be adopted.

3. Analysis Results and Discussions

1) The statistics of the events which occurred within 2 hours before the CMEs are collected, with 5 minutes being the time interval. The correlation between the decimetric type III bursts and the CMEs is 52% when the source position is not determined and is 28% after the source position is determined; most of the type III bursts occur in 45 minutes before the CMEs. 2) The Hα flares which generated decimetric type III and correlated to the CMEs are all the gradual flares. 82% Hα flares erupt before CMEs, with the occurrence being in first 33 minutes on average. Only 18% of Hα flares occur after CMEs, with the occurrence being in the last 29 minutes on average. 3) The initial frequencies of 64% of decimetric type III bursts occur at 700MHz; most of the frequency drift rates of the type III bursts corresponding to the CMEs occur at 100MHz/s and the next ones in number occur at 700 MHz/s. 4) The faster the initial velocity of the CMEs, corresponds to the higher the energy of the Hα, but the acceleration or de-acceleration of CME does not depend to the energy of Hα flare. 5) 25% of the type II bursts correspond to the type III bursts and 50% of the soft X-ray events correspond to the type III bursts.

It can be obtained from the above statistics analysis that the gradual flares correlating to the CMEs occur before and after the CMEs, however, most of CME occur later than Hα flares. As for the flares after the CMEs, the reason may be that the distance from the magnetic reconnection place to the photosphere, along which the electron beam propagated inwards is longer than that from the magnetic reconnection place to the corona, therefore the Hα flares occur later than the CMEs.

The authors hold that the high-energy electrons of the decimetric type III bursts should be accelerated in the process of the flares and they escape outwards along the open magnetic lines of force in the corona, thereby producing the type III bursts with extremely fast frequency drift while the type II radio bursts are accelerated by the shock waves produced when the CMEs propagate outwards. Thus, the type II bursts have the slow frequency drift of the propagating velocity of the shock waves (Nelson and Melrose, 1985). According to the above-mentioned points of view the correlation between the CMEs and the type II radio bursts should be far much greater than that between the CMEs and the type III radio bursts. The results obtained by the authors are that the correlation between the CMEs and the type III radio bursts is less than 40%, much less than the correlation, 60% or 80%. Between the CMEs and the type II radio bursts, which is in accordance with the above-mentioned model.

Acknowledgements

This research is supported by the Ministry of Science and Technology of China grant No. G2000078403, and also supported by the national NSF of China (grant Nos. 10473020 and 10333030).

References

Gopalswamy, Y. S. and Kaiser, M. L. 2001, *ApJ.* 548, 94

Jackson, B. V., Sheridan, K. V., Dulk, G. A. *et al.* 1978, *Solar Phys.* 3, 241

Michalek, G., Gopalswamy, N., Reiner, M. *et al.* 2001, *American Geophysical Union, Spring Meeting.*, 5

Nelson, G. J. and Melrose, D. B 1985, *Solar radiophysics.*, 333

Coronal and Stellar Mass Ejections
Proceedings IAU Symposium No. 226, 2005
K. P. Dere, J. Wang & Y. Yan, eds.

© 2005 International Astronomical Union
doi:10.1017/S1743921305000293

Eruption of a Large Quiescent Prominence and Associated CME on 2001 January 14

S. L. Ma, Y. C. Jiang, Q. Y. Li, S. Q. Zhao, L. P. Li, and H. D. Chen

National Astronomical Observatories/Yunnan Astronomical Observatory, CAS, Kunming
650011, China (email: jyc@ynao.ac.cn)

Abstract. We present observations of a spectacular eruption of a huge quiescent prominence, which was clearly associated with a coronal mass ejection (CME). The CME consisted of a typical three-part structure: a bright loop-like leading edge, a dark cavity and a bright core. The prominence exhibited a very symmetrical loop-like eruption in low corona and matched well with the bright CME core trailing the CME leading edge. By combining the H_α, 17GHz and EUV observations with white-light coronagraphs observations, the bright CME core was conclusively identified as the erupting cool, dense prominence material.

Keywords. Sun: activity, coronal mass ejections (CMEs), prominences

---

On 2001 January 14, a large quiescent prominence erupted in the north-western solar limb and an associated partial halo-type CME was well observed by the Large Angle and Spectrometric Coronagraphs (LASCO) on the *Solar and Heliospheric Observatory* (SOHO). In this paper, we investigate this prominence eruption and its relationship with the CME using H_α data from Yunnan Astronomical Observatory (YNAO), 17GHz microwave data from the Nobeyama Radioheliograph, EUV data from the Extreme-ultraviolet Images Telescope (EIT) on the SOHO, as well as the LASCO white light coronagraphs data.

The prominence activity started at approximately 03:00 UT, and began to ascend at about 04:30 UT. H_α, 17GHz and EIT 195 Å observations show that this was a very symmetrical loop-like eruption (see Figure 1(*a1-a3*), (*b1-b3*) and (*c1-c3*), respectively) . The whole prominence body lifted up while its two legs stayed anchored (see the inner EIT 304 Å image in Figure 1(*d*)). According to the EIT 195 Å observations, the height, speed and acceleration for the prominence are measured and plotted in Figure 2. It is noted that no GOES flare was recorded to associate with the eruption (see Figure 2(*b*)).

The CME with a central position angle of 356° and a width of 134° was first seen in LASCO C2 images at 06:54 UT as a bright loop. The CME consisted of a typical three-part structure: a bright loop-like leading front, a dark cavity and a bright core. There were two bright legs connecting the bright core back beneath the occulting disk. Figure 1(*d*) shows that the bright CME legs matched well with the legs of the erupting prominence (indicated by the arrows), definitely indicating that the CME core was the erupting cool, dense prominence material in the outer corona. This is consistent with the results of Dere, *et al.* (1997) and Plunkett, *et al.* (2000). The CME front heights measured by Seiji Yashiro are shown in Figure 2(*a*), and the average velocity and acceleration from the first and second-order polynomial fits are 945 km s^{-1} and 24.7 m s^{-2}, respectively. Therefore, this was a fast CME although it was correlated with the prominence eruption.

Acknowledgements

We thank the Nobeyama Radioheliograph group, the EIT and LASCO teams for data support. The work is supported by the NSFC under grant 10173023.

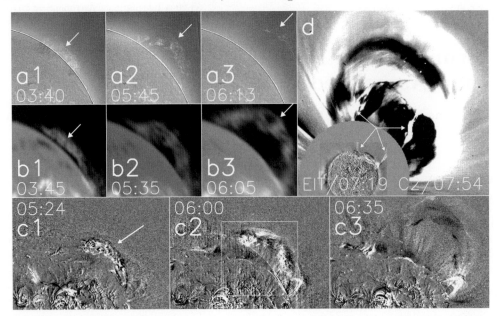

Figure 1. YNAO H_α (a1-a3), Nobeyama 17GHz microwave (b1-b3) and EIT 195 Å difference images (c1-c3) showing the prominence eruption (see the arrows); the white box in panel c2 indicates the field of view of H_α and 17GHz images. The panel d is a composite image of inner 304 Å with outer LASCO C2 difference images. The arrows indicate the legs of the erupting prominence and the CME core.

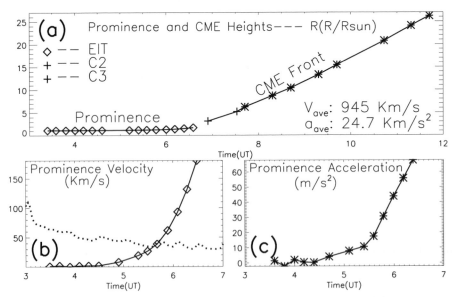

Figure 2. Plots showing heights for the prominence and CME (a), speed (b) and acceleration (c) for the prominence. The dashed line in panel b showing time profiles of GOES-8 soft X-ray.

References

Dere, K. P., *et al.* 1997, *Solar Phys.* 175, 601
Plunkett, S. P., *et al.* 2000, *Solar Phys.* 194, 371

Coronal and Stellar Mass Ejections
Proceedings IAU Symposium No. 226, 2005
K. P. Dere, J. Wang & Y. Yan, eds.

© 2005 International Astronomical Union
doi:10.1017/S174392130500030X

A CME and Related Phenomena on 2003 October 26

Zongjun Ning[1], C. Fang[1], M. D. Ding[1], C. -T. Yeh[1], H. Li[2], Y. N Xu[2], Y. Zhang[3] and C. M. Tan[3]

[1]Department of Astronomy, Nanjing University, Nanjing 210093, China
email: ningzongjun@hotmail.com
[2]Purple Mountain Observatory, Nanjing 210008, China
[3]Chinese National Observatory, Beijing 100021, China

Abstract.
We present the observational results of the solar bursts on the band of 1-80 GHz (NORH) associated with both a CME and a flare on Oct. 26 2003. This event shows two parts of radio bursts in the time profile. The first part is associated with an X1.2 flare. However, the following part seams related to both the flare and the CME, as the radio emission is enhanced while the Hα is decreasing. Thus, these two parts of radio bursts may originate from different physical processes, i.e., flare and CME shock. A primary study is performed on the difference between this two parts.

Keywords. Sun: flares, CMEs, radio radiation

1. Introduction

Solar radio microwave type IV bursts are generally thought to be produced by gyro-synchrotron radiation from energetic electrons trapped in the loops (e.g. Dulk 1970). It is widely accepted that magnetic reconnection is responsible for the acceleration of electrons during solar flares. The microwave type IV bursts are the radio bursts of solar flares above the frequency of 1 GHz. On the other hand, the CME shock is thought to be one of effective acceleration mechanism for the energetic electrons in the interplanetary space. It is an interesting question whether the shock can accelerate electrons in the corona in the initial phase? In this paper, we analyze a solar radio burst which takes place in association with a solar flare and a CME on 2003 Oct. 26. We try to seek the evidence of electrons acceleration from the CME shock in the corona.

2. Observations

The data include radio emission data observed by NORH, CME pictures by LASCO, Hα images by HUAIROU, and soft X-ray flux by GOES10. Figure 1 shows the time profiles of soft X-ray emission, solar radio emission at 19 GHz and Hα emission. The time profile of radio bursts can divided into two parts, the first from 06:10 to 06:35 UT, and the second from 06:40 to 7:35 UT (Tan *et al.* 2004). The start and end times of the radio bursts are defined respectively as the times when the intensity rises above and falls below 1.3 times the average intensity before the bursts. It is clear that the second part is much more intensive than the first one. Especially, the last peak in the second part is 3 times bigger than the maximum peak in the first part.

The soft X-ray is relatively gradual, which suggests that the coronal temperature (more exactly, the emission measure) is relatively stable when the radio emission changes

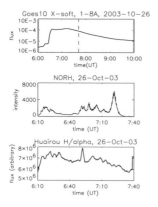

Figure 1. Top: soft X-Ray time profile from GOES10 on Oct. 26 2003. Middle: solar radio bursts from NORH on 34 GHz. Bottom: Hα emission from Huairou.

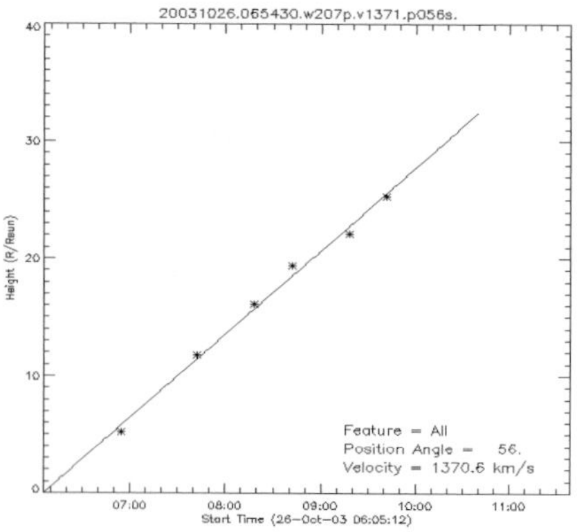

Figure 2. Linear extrapolation back to solar surface (photosphere) of CME event at 06:15 UT.

rapidly. On the other hand, this flare shows an Hα time profile with also two parts. The Hα intensity gradually decreases with time, especially during the second period, while the radio bursts shows a peak 3 times more intensive than the peak in the first part.

Figure 2 shows speed of the third CME event on 2003 Oct. 26. There are in total 6 CME events on that day (Bao *et al.* 2004). This CME is closely related to the flare, because EIT observations show a brightening in the corona during the flare. Later on, LASCO first detected this CME at 06:54 UT. Using linear extrapolation back to the solar surface, this CME event is estimated to start at 06:15 UT, which seams later than the flare.

3. Discussion

The radio bursts on 2003 Oct. 26 show two parts in the time profile. The first part is closely associated with the flare. In the second part, the radio emission increase significantly while the Hα emission decreases gradually. The increasing radio emission suggests that there are much more energetic electrons than before. However, the decreasing Hα intensity implies a gradual cooling of the chromospheric material. So the question is where the energetic electrons originate and contribute to radio emission. It is likely that the electrons are accelerated during the CME event. We propose that both the flare and the CME shock accelerate the electrons during the second part of radio bursts. In comparison, the CME shock may play the main role in the electron acceleration for this radio event (e.g. Caroubalos *et al.* 2004).

Acknowledgements

This work was supported by NKBRSF under grants G20000784, by NSFC under grant 10025315, 10221001 and 10333040, and a postdoctoral grant 0201003005.

References

Bao, X. M. Lin, J. 2004, submission
Caroubalos, C., *et al.* 2004, A&A, 413, 1125
Dulk, G. A. 1970, Proceedings of the Astronomical Society of Australia, 1, 372
Tan, C., Fu, Q., Yan, Y., & Liu, Y. 2004, Chinese Journal of Astronomy and Astrophysics, 4, 205

Coronal and Stellar Mass Ejections
Proceedings IAU Symposium No. 226, 2005
K. P. Dere, J. Wang & Y. Yan, eds.

© 2005 International Astronomical Union
doi:10.1017/S1743921305000311

Low Coronal Signatures of a Coronal Mass Ejection

K. P. Qiu, Y. Dai, and Y. H. Tang

Department of Astronomy, Nanjing University, Nanjing, 210093, P. R. China,
email: kpqiu@nju.edu.cn

Abstract. Using the observations of the EUV Imaging Telescope(EIT) and the Large Angle Spectrometric Coronagraph(LASCO) on the Solar and Heliospheric observatory(*SOHO*) and solar soft X-ray flux and radio bursts data, we study the low coronal signatures of a solar limb coronal mass ejection(CME) on November 4, 2003. The two prominent dimmings in EIT difference images were closely related to two large loops in this event. The onset time and height of the CME and the lower limit of the masses loss from dimming regions are estimated.

Keywords. Sun: corona, Sun: coronal mass ejections, Sun: flares, Sun: UV radiation

1. Introduction

Solar coronal dimmings observed in soft X-ray (See for example Sterling & Hundson 1997) and EUV (see for example Zarro *et al.* 1999; Delannée, Delaboudinière & Lamy 2000) are expected to be the very first signature of CMEs. Sometimes large coronal loops disappearing are associated to CMEs (see for example Delannée & Aulanier 1999). Here we study the EUV dimmings and loop structures of the 4 November 2003 event.

2. Observations

The great X28/3B flare occurred at 19:29 UT on 2003 November 4 in NOAA active region 10486 was associated to a fast halo CME. Observations of the EUV Imaging Telescope(EIT) and the Large Angle Spectrometric Coronagraph(LASCO) on the Solar and Heliospheric observatory(*SOHO*) and *GOES* soft X-ray flux and radio bursts data are used.

Figure 1 shows the large loops (denoted as 'L1' and 'L2') with one ends anchored in the flaring region (denoted as 'F') in EIT 195 Å and 284 Å, at 19:35 UT and 19:05 UT respectively. L1's height is about 0.35 $R_\odot$, or 2.5×10^5 km, from the solar surface. The loops disappeared in the next EIT 195 Å image at 19:47 UT and no longer appeared. In Figure 2 the EIT difference images by subtracting the preflare image at 19:25 UT show two dimmings (denoted as 'd1' and 'd2') at 19:47 UT, 19:59 UT and 20:11 UT. It is clear that d1 and d2 correspond to the footpoints of L1 and L2. A lower limit of the preflare masses from dimming regions is estimated to be about 6.6×10^{15} g and this is consistent with typical CME masses. The first CME observation of LASCO/C2 at 19:54 UT co-aligned with the EIT observation(the middle image in Figure 2) is shown in Figure 3. It can be seen that the double EUV dimmings approximately correspond to the CME footpoints. Using the CME velocity of ~ 2300 kms^{-1} inferred from the linear fit of LASCO observations we extrapolate that the CME time at the height of L1 was 19:44 UT. This is well in the loops disappearing time window of 19:35 UT – 19:47 UT.

From above observations we suggest that the CME initiated as the large loops expansion or opening, apparently disappearing, at about 19:44 UT. Figure 4 shows that the inferred CME onset time (denoted as the vertical line) is in the rising phase of the soft

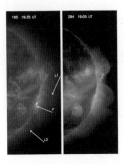

Figure 1. Large loops in EIT
195 Å and 284 Å channels

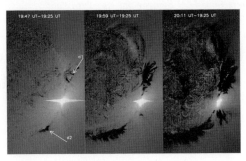

Figure 2. Double dimmings evolution in EIT
195 Å difference images

Figure 3. Co-alignment of LASCO/C2
and EIT 195 Å observations

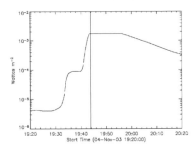

Figure 4. The CME onset time in *GOES* 1–8Å
soft X-ray flux

X-ray flare. The moving type IV radio burst (from Solar Geophysical Data) started at
19:47 UT and it is just after the CME onset time of ∼19:44 UT. According to plasma
emission mechanism, the upper frequency of the burst 180MHz gives a height of 0.38 $R_\odot$
under the corona density model of Sittler & Guhathakurta (1999) and this is consistent
with the loops height or the CME onset height of 0.35 $R_\odot$.

3. Conclusions

According to the observations and some estimations, we suggest that the CME initiated
as the large loops expansion or opening, leaving behind two prominent dimmings, in the
rising phase of the soft X-ray flare.

Acknowledgements

SOHO is a project of international cooperation between ESA and NASA. This work
is Supported by NKBRSF (G2000078404) and NSFC (No.10073005).

References

Delannée, C. & Aulanier, G. 1999 *Sol. Phys.* 190, 107
Delannée, C., Delaboudinière, J.-P. & Lamy, P. 2000 *A&A* 355, 725
Sittler, E.C. & Guhathakurta, M. 1999 *ApJ* 523, 812
Sterling, A.C. & Hudson, H.S. 1997 *ApJ* (letters) 491, L55
Zarro, D.M., Sterling, A.C., Thompson, B.J., Hudson, H.S., & Nitta, N. 1999 *ApJ* (letters) 520,
    L139

Coronal and Stellar Mass Ejections
Proceedings IAU Symposium No. 226, 2005
K. P. Dere, J. Wang & Y. Yan, eds.

© 2005 International Astronomical Union
doi:10.1017/S1743921305000323

Shock Wave Driven by an Expanding System of Loops

N.-E. Raouafi[1], S. Mancuso[2], S. K. Solanki[1], B. Inhester[1], M. Mierla[1] G. Stenborg[3], J. P. Delaboudinière[4], and C. Benna[2]

[1] Max-Planck-Institut für Sonnensystemforschung, 37191 Katlenburg-Lindau, Germany;
[2]OATo, Torino, Italy; [3]GSFC/CUA, Greenbelt, USA; [4]IAS, Orsay, France

Abstract. We report on a Coronal Mass Ejection (CME) observed on June 27 1999 by the UltraViolet Coronagraph Spectrometer (UVCS) telescope operating on board the SOHO spacecraft. The CME was also observed by LASCO (SOHO). Emission of hot material has been recorded by UVCS propagating in front of an opening system of loops generated by the CME. The evolution of the UVCS structure is highly correlated to the evolution of the opening loop. The data reveal excess broadening of the O VI doublet lines and an enhancement in the intensity of the Si XII $\lambda520.66$ and $\lambda499.37$ lines due to the motion of the expanding hot gas. The hot gas emission seems to be due to a shock wave propagating in front of a very fast gas bubble traveling along the opening loop system.

Keywords. plasmas, shock waves, Sun: coronal mass ejections (CMEs), UV radiation

1. Introduction

UVCS observations of CMEs usually show emission in low to moderate ionization stages. In a few cases UV spectra show emission from higher ionization states that can be interpreted in terms of coronal shock waves connected with the CME eruptions. The CME observed by UVCS on June 27 1999 produced a strong enhancement of the Si XII emission lines originating from hot plasma seen propagating along the slit.

2. Observations and data analysis

The results discussed in the present paper have been obtained from observations made on June 27 1999 by different instruments aboard the SOHO spacecraft. The analysis of this event also relies on radio spectra recorded by the decametric array of Nançay (France) and the Izmiran radio spectrograph (Russia). More details about the data analysis are given in Raouafi et al. (2004).

Three UVCS exposures taken at 2.55 $R_\odot$ reveal the propagation of hot plasma with highly enhanced emission in the Si XII lines, which is rare in CMEs (see Fig. 1). These lines trace the propagation of hot plasma along the UVCS slit. A co-spatial broadening of the O VI doublet lines is also remarkable (the same figure).

The observed broadening of the two O VI emission lines, together with the mentioned enhanced Si XII emission and the simultaneous detection of a metric type II radio burst by ground radiospectrographs, is suggestive of the passage of a coronal shock wave (e.g. Raymond et al. 2000; Mancuso et al. 2002).

The CME was observed initially by EIT to erupt as a system of loops. The white light images (Fig. 2) show a structure propagating toward the north pole that has the same general characteristics of propagation as the Si XII emission observed by the UVCS slit. At the same time, the curved shape of the white light brightening (see bottom frames of Fig. 2) is similar to that of the opening loop detected earlier by EIT.

127

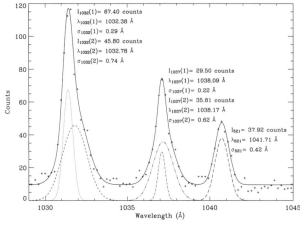

Figure 1. Spectra integrated over the spatial pixels where the hot gas emission (intensity enhancement of the Si XII lines) is detected in the UVCS data (+ signs). The multi-Gaussian fit of the data is also plotted (solid curve) together with the individual Gaussians. The O VI lines are fitted by two Gaussians each and the Si XII line by one. The fit parameters are also given (in counts for the amplitude and in wavelength unit for the center and width of each Gaussian).

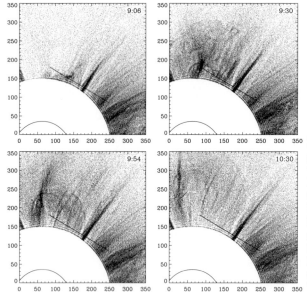

Figure 2. LASCO C2 images giving the evolution in space and time of the erupted system of loops as well as the CME material observed in white light. The black straight lines indicate the location of the UVCS slit. The opening system of loops observed in white light is highly correlated to the hot gas emission seen propagating along the slit of UVCS.

3. Discussion

The enhanced emission of the Si XII line and the line broadening observed for the O VI doublet indicates the presence of expanding hot material propagating along the UVCS slit. The EUV propagating structure is well correlated spatially and temporally to the evolution of the white-light LASCO observations of the expanding CME (see Fig. 2). The CME speed was high enough at low heights (considering an estimated projected speed of about 1200 km s^{-1} at the slit position) to be able to drive a coronal shock wave and produce local heating of the plasma that could be responsible for the observed enhanced emission and broadening of the EUV lines in higher ionization states. For more details, see Raouafi *et al.* (2004).

References

Mancuso, S., Raymond, J. C., Kohl, J., *et al.* 2002, A&A,383, 267

Raouafi, N.-E., Mancuso, S., Solanki, S. K., *et al.* 2004, A&A, 424, 1039

Raymond, J. C., Thompson, B. J., St. Cyr, O. C., *et al.* 2000, GRL, 27, 1439

Coronal and Stellar Mass Ejections
Proceedings IAU Symposium No. 226, 2005
K. P. Dere, J. Wang & Y. Yan, eds.

© 2005 International Astronomical Union
doi:10.1017/S1743921305000335

Characteristics of Solar Microwave Bursts Associated with CMEs: a Statistical Study

Chengwen Shao[1,2,4]† **,Min Wang**[1,2]**, Ruixiang Xie**[1,2]**, Qijun Fu**[2]**,**
Yu-ying Liu[2]**, Cheng-ming Tan**[2,4]**,Ying-na Su**[3,4] **and Yuan Ma**[1,2]

[1]National Astronomical Observatory/Yunnan Observatory,CAS,Kunming,China.650011
email: cwshaoanhui@etang.com,cwshaoanhui@163.com

[2] National Astronomical Observatory,Chinese Academy of
Sciences(CAS),Beijing,China.100012.

[3]Purple Mountain Observatory,Chinese Academy of Sciences(CAS),Nanjing,China.

[4] Graduate School of Chinese Academy of Sciences(CAS),Beijing,China.100012.

Abstract. We selected 133 solar microwave bursts (SMBs) recorded by SGD and 133 CMEs observed by SOHO/LASCO from November 1999 through September 2003. These SMBs are associated with CMEs and flares. We analyzed the characteristics of the SMBs, including duration, flux peak, burst type, and spectral index. Correlated events were distinguished by time and the location of flare associated with the SMB. We find that (1)The duration of SMBs associated with narrow($0° <$width$<20°$)/normal CMEs($20° <$width$<120°$) or slow CMEs is below 40 minutes. (2)The duration of SMBs associated with Halo-like CMEs($120° <$width$<360°$) or fast CMEs is from several minutes to 200 minutes. (3)The flux peak of SMBs associated with narrow CMEs/normal CMEs or slow CMEs is below 400sfu. (4)The flux peak of SMBs associated with Halo-like CMEs or fast CMEs is from several sfu to several thousands sfu. (5)the majority of SMBs,which are associated with Full Halo CMEs, are Complex/GB bursts. The majority of SMBs, which are associated with narrow CMEs/normal CMEs, are simple bursts. (6)U-shape spectra are observed. The spectra of SMB associated with CME is very flat when f>f_{max}. A statistical result suggest that CME/flares and SMBs is probably a different manifestation of the same physical process. CME/flare and SMB have intrinsically a physical relationship.

Keywords. Sun: coronal mass ejections (CMEs), radio radiation

1. Introduction

The aim of this paper is to study the observational characteristics of SMBs associated with CMEs, including duration, peak flux, burst type and the spectral index. Section II is data selection. Section III is statistical results. Section IV is conclusions and discussions.

2. Data Selection

The CMEs of this study are available on the world wide web*. SMB data are available in SGD. Spectral index is studied by using NoRP XDR data.

Firstly, temporal association: We recorded the time $T(R_s)$ when the CME initiated from $1R_s$. The CME is associated with a flare if the flare starts in the time window $T(R_s)\pm30$min (Zhou,G.P.*et al.* 2003). Secondly, spatial association: the CME is associated with a flare if the location of flare lies in the range of the CME span (M.Zhang 2002, Harrison *et al.* 1986).

† Present address: National Astronomical Observatory/Yunnan Observatory, Kunming, China.

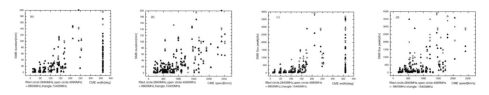

Figure 1. The relationship between SMB duration/flux peak and CME speed and width

The CME is associated with the flare if the CME and the flare satisfy the above two criteria simultaneously. The SMB, which is accompanied by a flare,is associated with the CME.

3. statistical results and discussions

According to the above two criteria, we selected 133 CMEs and their associated SMBs. We analyzed the relationship between the duration/flux peak of SMBs and CME width/speed. The burst type of SMBs and spectral index were analyzed.

(1)The relationship between duration/peak flux of SMBs and CME width/speed. As seen from the figure 1, both long-duration and short-duration SMBs are accompanied by CMEs. Both great/intense bursts and small bursts are accompanied by CMEs.

(2)The burst type of SMBs associated with CMEs. The simple burst associated with narrow/normal CME in 2840MHz, 4995MHz, 8800MHz, 15400MHz is 72%, 88%, 86%, 85% respectively. The complex burst associated with Full Halo CME in 2840MHz, 4995MHz, 8800MHz, 15400MHz is 68%, 72%, 65%, 69% respectively.

(3)Spectral index are studied. The spectral index is between 0 and 1.5. This is consistent with Fu's study (Fu2004).

This paper's conclusion is basically consistent with that of Dougherty(2002). The characteristics of SMBs also are studied in this paper. This paper's conclusions are as follows: Both long-duration and short-duration SMBs are accompanied by CMEs. Both great/intense bursts and small bursts are accompanied by CMEs. The majority of these SMBs accompanied by Full Halo CME are Complex bursts. The majority of SMBs accompanied by narrow/normal CME is simple,uncomplicated bursts. A statistical result suggests that CME/flares and SMBs are the different manifestations of the same magnetic process.

Acknowledgements

The work is supported by the National NSF of China grant(NO.19833050, 19973016 and 10333030), the 973 project and the foundation of the CAS.

References

Dougherty *et al.* 2002,ApJ,577,457
Fu,Qi-jun 2004,solar radio astronomy symposium,Chengdu,China/personal communication
Harrison *et al.* 1986,A&A,162,183
Zhang,M. *et al.* 2002,ApJ,574,L97-100
Zhou,G.P. *et al.* 2003,A&A,397,1057
*http://cdaw.gsfc.nasa.gov/cme_list

Coronal and Stellar Mass Ejections
Proceedings IAU Symposium No. 226, 2005
K. P. Dere, J. Wang & Y. Yan, eds.

© 2005 International Astronomical Union
doi:10.1017/S1743921305000347

Microwave Fine Structures in the Initial Phase of Solar Flares and CMEs

C.M Tan, Y.H. Yan, Q.J. Fu, Y.Y. Liu, H.R. Ji and Z.J. Chen

National Astronomical Observatories Chinese Academy of Sciences, Beijing 100012, China
email: Email: Tanchm@bao.ac.cn

Abstract. Solar radio fine structures (FSs) may be as an important diagnostics stool to draw the evolution map of the flare loop in the initial phase of solar flares. Also, it may be an important signature of the initial phase of CMEs. Here we analyzed a series of solar radio bursts with drift pulsation structures (DPS) and FSs during the former part of the 23rd solar activity cycle. Found they were associated with CMEs, and got some important statistic conclusions.

Keywords. Sun: coronal mass ejections (CMEs), flares, radio radiation

1. Introduction

Radio emission from solar flares offers a number of unique diagnostic tools to address a long-standing questions about energy release, plasma heating, particle acceleration, and particle transport in magnetized plasmas (Bastian *et al.* 1998). For example, drift pulsation structures (DPS) may be a signature of dynamic magnetic reconnection, the harmonic structure of zebra pattern structures (ZPS) indicate the localized region with the anisotropic distribution function of accelerated electrons, and the narrow-band dm-spikes with the MHD turbulence in the reconnection plasma out flows (Karlicky *et al.* 2000). Solar radio FSs was described in many papers (Benz 1986, 1992; Isliker *et al.* 1994; Karlicky 2000; Jiricka *et al.* 2001). DPS is a new phenomenon. It was firstly described in these papers (Karlicky 2000; Kliem *et al.* 2000; Jiricka *et al.* 2001).

2. observations and results

During the former part of 23rd solar activity cycle (1996-2003), 885 solar radio bursts were observed with Solar Radio Spectrometers of China in 1.0-2.0GHz waveband. 154 bursts have FSs. More than 40 events have DPSs. Here we select the 1.0-2.0GHz waveband to analyzed because the FSs and DPSs arose at low waveband more frequently than high waveband. Most of the DPSs are associated with CMEs, including the four significant LDEs (07/12/2000, 07/14/2000, 04/10/2001, 10/26/2003). The Oct.26, 2003 event was showed as Fig. 1

3. conclusion and discussion

Generally, DPS is a phenomenon that a groups of pulsations drifting negatively (from high frequency to low frequency) and slowly as a whole. They usually happened after or between the initial phase of flare, with a duration of ∼1min to ∼10min. The DPS might be consists of fast drifting burst (FDB), slowly drifting burst (SDB), type III burst, zebra pattern structures (ZPS), spikes or other types of FSs. The duration of

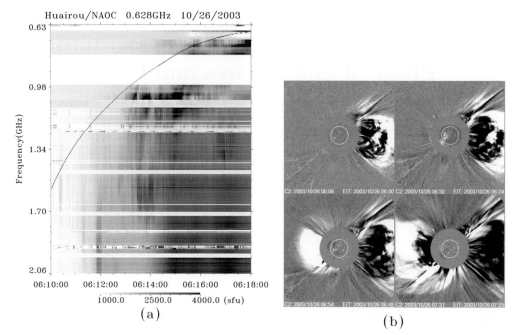

Figure 1. a is the DPS on Oct.26, 2003, the drift rate is -3~-7MHz/s, the black curve line describe the drift trend. b is the CME on the same day.

the individual pulsation is ~1sec to <1min. The global drift rate of the DPS is -2.0~-20.0MHz/s (corresponding velocity is ~100 to ~1000Km/s). In addition, the DPS might has harmonic structures (The Apr.10, 2001 event).

The four significant LDEs that have been mentioned above were associated with CME. Moreover, the corresponding four CMEs are fast and large angle width. In a same active region, the time sequence of the DPS, FSs, flare and CME might be as follows: Usually, DPS emerged after or between the pulse phase of flare, several minutes later was ZPS, spikes or other types of FSs, and 10~30 minutes later CME expanded. The time difference between DPS and the initial enhance of EIT is several minutes. The global drifting DPS, which happened during the initial phase of flare, is consistent with CME well.

Anyway, FSs and DPS are a series of important microwave spectrum burst phenomena. While DPS is a new phenomenon. It maybe caused by quasi-periodic particle acceleration episode that result from a dynamic phase of magnetic reconnection on a large-scale current sheet. And it plays a certain role in explaining the spatial structure and evolvement of flare loop, particle acceleration and shock wave acceleration.

Acknowledgements

This work was supported by CAS, NSFC, and MOST grants.

References

Benz A.O., Gudel M., Isliker H., Miszkowicz S., and Stehling W. 1991, Solar Phys, 133, 385–393.
Bastian T.S., Benz A.O., and Gary D.E. 1998, Ann. Rev. Astron. Astrophys. 36, 131–188.
Fu Q., Qin Z., Ji H., *et al.* 1995, Solar Physics, 160, 97–103.
Isliker H., and Benz A.O.. 1994. A&A Supplement series. 104, 145–160.
Jiricka K., Ksrlicky M., Meszarosova H., and Snizek V. 2001, A&A, 375, 243–250.
Ksrlicky M., Yan Y., Fu Q., Wang S., et al. 2001, A&A, 369, 1104–1111.

Coronal and Stellar Mass Ejections
Proceedings IAU Symposium No. 226, 2005
K. P. Dere, J. Wang & Y. Yan, eds.

© 2005 International Astronomical Union
doi:10.1017/S1743921305000359

SoHO/EIT Observation of a Coronal Inflow

D. Tripathi[1], V. Bothmer[1], S. K. Solanki[1], R. Schwenn[1], M. Mierla[1] and G. Stenborg[2]

[1]Max-Planck-Institut für Sonnensystemforschung, 37191 Katlenburg-Lindau, Germany
email: tripathi; bothmer; solanki; schwenn; mierla@mps.mpg.de

[2]NASA Goddard Space Flight Center, Washington DC, USA
email: stenborg@kreutz.nascom.nasa.gov

Abstract. A distinct coronal inflow has been discovered after ∼90 min of prominence eruption associated coronal mass ejection (CME) on 05-Mar-2000 by EIT (Extreme ultraviolet Imaging Telescope) aboard SoHO (Solar and Heliospheric Observatory). Evolution of the prominence seen by EIT was tracked into the LASCO/C2 and C3 field-of-view (FOV; 4-10$R_\odot$) where it developed as the core of a typical three-part CME. The speed of the inflow, which was only seen in EIT FOV, was 70-80 km/s at a height between 1.5-1.2 $R_\odot$ coinciding with the deceleration phase of the core of the CME in LASCO/C2. In contrast to dark inflow structures observed earlier and interpreted as plasma void moving down, the inflow reported here was bright. The inflow showed a constant deceleration and followed a curved path suggesting the apex of a contracting magnetic loop sliding down along other field lines

Keywords. Sun: corona, Sun: coronal mass ejections (CMEs)

1. Introduction

Detection of dark downflows after CME eruptions has been reported by several authors based on soft X-ray, LASCO and TRACE (Transition Region and Coronal Explorer) observations (McKenzie 2000; Wang *et al.* 1999; Innes *et al.* 2003). These downflows were interpreted as plasma voids. Based on the TRACE and RHESSI observations, non-thermal radiations were detected at the time of downflows which was interpreted as evidence for magnetic reconnection underneath the CME (Asai *et al.* 2004). In this paper we report an observation of a bright inflow identified by EIT at 195 Å in the course of a prominence associated coronal mass ejection.

2. Observations and Conclusions

(*a*) The prominence (P1, Fig. 1 left panel) was first accelerated with a constant acceleration of about 65-70 m/s^2 in EIT FOV (1.5 $R_\odot$), then decelerated with constant acceleration of 25 m/s^2 in the LASCO/C2 FOV (2-5 $R_\odot$) and finally achieved a constant speed of about 200 km/s higher up in the corona (Fig. 2, left panel).

(*b*) The inflow was first seen at 17:36 UT at a height of about 1.5 $R_\odot$, at the edge of the EIT FOV (Fig. 1, right panel). The vertical speed of inflow was initially about 70-80 km/s (Fig. 2, right panel). The speed of the inflow was comparable to the speed observed for downflows in white-light (Wang *et al.* 1999) and soft X-ray (McKenzie 2000) observations.

(*c*) The inflow started during the deceleration phase of the core of the CME (Fig. 2, left panel). The speed of the inflow decreased linearly with decreasing height, i.e., the downflow appeared to be decelerated (Fig. 2, right panel).

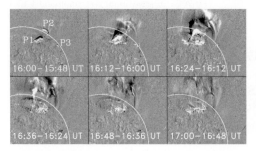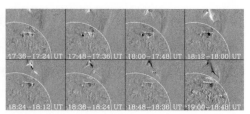

Figure 1. Left Panel: Running difference images taken by EIT at 195 Å showing simultaneous eruptions of the prominences marked by P1, P2 and P3. The eruption started on 05/03/00 at around 16:00 UT. In all images North is upward and West is to the right. Right Panel: Series of EIT 195 Å running difference images displaying the sunward moving plasma after the eruption. The inflow, visible as the bright feature in the NW part of the limb, started at around 17:36 UT in the EIT FOV.

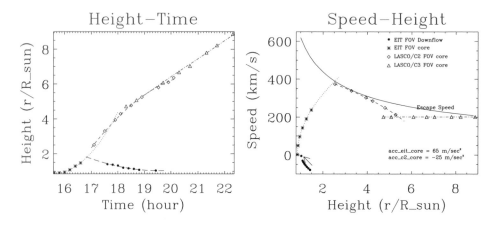

Figure 2. Left Panel: Height-time profile for the prominence (asterisks) in EIT, the core in LASCO/C2 (diamonds) & C3 (triangles) and the inflow (solid dots) in EIT. Right Panel: Speed height profile for all the features. Symbols have the same meaning as in the left panel. The solid line represents escape speed profile. Arrow marks the direction of motion.

(*d*) The inflow did not propagate vertically, but appeared to follow paths outlining the magnetic loops. Two branches emanating from the same apex were seen. The right branch showed a clear kink (Fig. 1, right panel).

Acknowledgements

This study is part of the scientific investigations of the project Stereo/Corona supported by the German "Bundesministerium für Bildung und Forschung" through the "Deutsche Zentrum für Luft- und Raumfahrt e.V." (DLR, German Space Agency) under project number 50 OC 0005. We also thank the SOHO/LASCO/EIT consortium for providing the data.

References

Asai, A., Yokoyama, T., Shimojo, M., & Shibata, K. 2004, *ApJ* (Letters) 605, L77
Innes, D. E., McKenzie, D. E., & Wang, T. 2003, *Sol. Phys.* 217, 247
McKenzie, D. E. 2000, *Sol. Phys.* 195, 381
Wang, Y.-M., Sheeley, N. R., Howard, R. A., St. Cyr, O. C., & Simnett, G. M. 1999, *Geophys. Res. Lett.* 26, 1203

Coronal and Stellar Mass Ejections
Proceedings IAU Symposium No. 226, 2005
K. P. Dere, J. Wang & Y. Yan, eds.

A Trans-equatorial Filament and the Bastille Day Flare/CME Event

Jingxiu Wang, Guiping Zhou, Yayuan Wen, Yuzong Zhang, Jun Zhang, Huaning Wang, and Yuanyong Deng

National Astronomical Observatories, Chinese Academy of Sciences, Beijing 100012, China
(email: wjx@ourstar.bao.ac.cn)

Abstract. The Bastille Day Event on July 14 2000, a major solar flare and a global coronal mass ejection (CME), is not a phenomenon of a single active region. Activation and eruption of a huge trans-equatorial filament is seen to precede the simultaneous filament eruption and flare in the source active region, AR9077, and the full halo-CME in the high corona. Evidence of reconfiguration of large-scale structures, manifested by SOHO EIT and Yohkoh SXT observations, is clearly seen. The large-scale magnetic composition related to the trans-equatorial filament and its sheared magnetic arcades appears to be the essential part of the CME parent magnetic structure. Estimations show that the filament-arcade system has enough magnetic helicity to account for the helicity carried by the related CMEs.

Keywords. Sun: coronal mass ejections (CMEs), Sun: filaments, Sun: activity

1. Activity of the Transequatorial Filament

To understand CME magnetism, the observations of Bastille Day event (Zhang *et al.* 2001; Deng *et al.* 2001) have been re-examined. It is found that one fact has been missing in the published literature: the pre-CME activation and eruption of a huge trans-equatorial filament. The active region filament in AR9077 could be considered as the extension, or rooting end of this huge filament.

A huge trans-equatorial filament was mapped by the Huairou full disk Hα filtergraph equipped with a 2k×2k CCD. Figure 1 shows the filament in comparison with its background magnetic field from MDI magnetogram and coronal structures from SOHO EIT and Yohkoh SXT images. The huge filament divided an extended bipolar region (EBR) which included the magnetic flux of AR9077, 9082 and several smaller ARs, as well as enhanced network.

The filament often twisted, knotted, kinked, or partial disappeared. Disturbances started from the filament in AR9082 on the southern hemisphere, was responded by the huge filament and propagated into AR9077. The global eruption of the trans-equatorial filament started approximately from 08:22 UT and earliest brightening in the huge filament was seen at 09:02 UT when the filament tied-up closely to its northern end. The filament in AR9077 began to break out and erupt since 10:02 UT, in coincidence with the beginning of Bastille Day flare.

2. Reconfiguration of Large-scale Magnetic Structure

The reconfiguration of large-scale magnetic structure appeared as followings. First, the arcades straddling over the trans-equatorial filament disappeared; Secondly, there appeared dimming at both EUV and soft X-ray wavelengths at 3 different locations

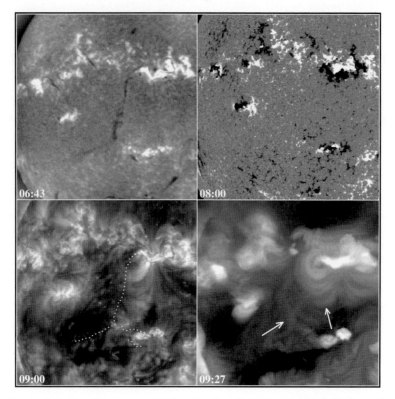

Figure 1. Upper-left: Hα filtergram showing the huge trans-equatorial filament; Upper-right: MDI magnetogram; Lower-left: EIT image showing the arcade system striding over the filament; Lower-right: SXT image showing the arcade system indicated by two arrows.

which were widely separated. It was not transient in nature, but manifested some long-term restructuring of the Sun's magnetic fields after the flare/CME event. The widely separated dimming seems to imply that not only the AR9077, but also the multiple flux systems covering a broad spatial range are involved in the physical process which led to the major activity.

Non-linear force-free extrapolations based on the MDI magnetograms show the magnetic arcades overlapping the trans-equatorial filament changed from reversed 'S'-shaped stressed structure before the flare/CME to more relaxed state. A gross estimation finds that the helicity of sheared arcades is approximate $3\times10^{43}Mx^2$. It seems to be enough to account for the helicity source of the CME.

Acknowledgements

The work is supported by the National Natural Science Foundation of China(10233050) and the National Key Basic Science Foundation(TG2000078404).

References

Deng, Y., Wang, J., Yan, Y., & Zhang, J. 2001, *Solar Phys* 204, 11–26.
Zhang, J., Wang, J., Deng, Y., & Wu, D. 2001, *ApJ* 548, L99–102.

Coronal and Stellar Mass Ejections
Proceedings IAU Symposium No. 226, 2005
K. P. Dere, J. Wang & Y. Yan, eds.

© 2005 International Astronomical Union
doi:10.1017/S1743921305000372

Diagnostics of Coronal Magnetic Field in Terms of Radio Burst and Fine Structures

M. Wang[1]†, C. Xu[1], R. X. Xie[1], and Y. H. Yan[2]

[1]National Astronomical Observatories/Yunnan Observatory, Chinese Academy of Sciences, Kunming 650011, China (email: wmynao@163.net)

[2]National Astronomical Observatories, Chinese Academy of Sciences, Beijing 100012, China

Abstract. A complex solar radio burst was observed on 19 October 2001 with the spectrometers of NAOC (National Astronomical Observatory of China) and Nobeyama Radioheliograph (NoRH). Basing on the analysis of brightness temperature spectra of radio sources and various fine spectral structures, we get a diagnosis of magnetic field of radio active region.

Keywords. Sun: coronal mass ejections (CMEs), magnetic fields, radio radiation

1. Introduction

Radio spectral observation is effective for studying the acceleration process, while the image observation is effective for studying the magnetic field configuration. Basing on a analysis of a solar radio burst observed by the spectrometer (0.65–7.60 GHz) of NAOC and NoRH, we give a diagnosis of magnetic fields in the radio active regions.

2. Observations and data analysis

2.1. NoRH observation and magnetic field diagnosis

In the radio image of NoRH, we find three sources, one is a pre-existing source named PS and the others are newly emerging sources named NES1 and NES2, respectively. From the polarization degrees of Sources NES1 and PS, we can calculate the range of θ (the angle between the line of sight and the magnetic fields), under the assumption that the emissions of these two sources are due to non-thermal gyro-synchrotron radiation. According to the Equation (38) of Dulk (1985), for the harmonic number $10 \leqslant s \leqslant 50$ and $\delta=3.5$ (for Source NES1) and 5.3 (for Source PS), θ should be between 83–158⁰ for Source NES1 and between 30–66⁰ for Source PS. We refer that the peak frequency is 9400 MHz for both Source NES1 and PS according to the data from SGD. According to the Equation (36), (37), and (39) of Dulk (1985), for the reasonable values of NL (about $10^{18-19}cm^{-2}$), we get that the values of magnetic field strength are 80-160 and 300-400 Gauss for Source NES1 and PS, respectively.

2.2. Spectral observation and magnetic field diagnosis

The highly polarized type III bursts in the frequency range of 2.6–3.8 GHz are due to fundamental plasma emissions. Using the method of Wang *et al.* (2003), we get the corresponding magnetic field should be about 120 Gauss.

Zebra patterns are due to upper hybrid waves or whistler waves. The frequency space between the Zebra patterns gives the diagnosis to the magnetic field of the source, because the frequency space is roughly equal to the local gyrofrequency. From the observed frequency space of about 30-40 MHz, we get the magnetic field is about 15 Gauss.

---

† Present address: PO Box 650011, Kunming, Yunnan, China.

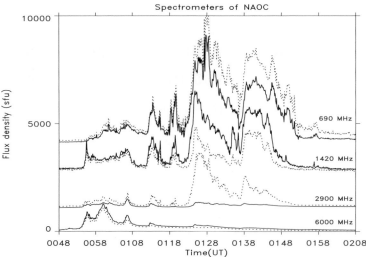

Figure 1. The spectrogram (time profiles) of radio bursts on 19 October, 2001 at selected frequencies, the dashed lines represent the right polarization, the solid lines left polarization.

We found some lower frequency fibers with normal frequency drifting rate of about 25 MHz/s and some higher frequency fibers with normal frequency drifting rate of about 90 MHz/s. The drift rates of fibers are relates to the local Alfvén speed (Treumann, Güdel, and Benz, 1990). For 0.65-0.8 GHz and 2.6-3.8 GHz, we can take density scale height λ_n be about 10^{10} and 0.5×10^{10} cm, respectively. We take frequency f_1=770 MHz and f_2=3000 MHz, electron density $n_1 = 1.8 \times 10^9\ cm^{-3}$, and $n_2 = 2.7 \times 10^{10}\ cm^{-3}$ (assuming harmonic emission), we get the Alfvén speeds $v_{A1} = 1400$km/s and $v_{A2} = 3500$ km/s, that means the local magnetic fields are 210 and 130 Gauss, respectively.

The observed 6-minute and 1-minute pulsations should be attributed to the standing Alfvén wave in the coronal loops driven by the photospheric velocity fields which modulate the radio emission and the kink fast magneto-acoustic mode, respectively. For 1-minute pulsations, according to Aschwanden *et al.* (1999), for the observed period about 60 s, assuming second harmonic emission, n=$2.8 \times 10^{10}\ cm^{-3}$ and L=0.94×10^{10} cm, we get B=170 G. For 6-minute pulsations, L= 5.0×10^{10} cm, according to Strauss, Kaufmann, and Opher (1980), we get the Alfvén speed V$_A$=2550km/s. At the height of 3 GHz emission, the electron density n=$1.3 \times 10^9\ cm^{-3}$, so the magnetic field B= 50 gauss; for the height of 17 GHz emission, n = $1.0 \times 10^{11}\ cm^{-3}$, we get B= 380 Gauss.

Acknowledgements

This work is supported by the Ministry of Science and Technology of China (No. G2000078403) and the national NSF of China (Nos. 19833050,10333030, and 10473020).

References

Aschwanden, M. J., Fletcher, L., Schrijver, C. J., & Alexander, D. 1999, *ApJ* 520, 880
Dulk, G. A. 1985, *ARAA* 23, 169
Strauss, F. M., Kaufmann, P., & Opher, R. 1980, *Solar Phys.* 67, 83
Treumann, R. A., Güdel, M., & Benz, A. O. 1990, *AA* 236, 242
Wang M., Duan, C. C., Xie R. X., & Yan, Y. H. 2003, *Solar Phys.* 212, 401

Coronal and Stellar Mass Ejections
Proceedings IAU Symposium No. 226, 2005
K. P. Dere, J. Wang & Y. Yan, eds.

© 2005 International Astronomical Union
doi:10.1017/S1743921305000384

Multi-Wavelength Radio Features Associated with Large CMEs on Oct. 26–28, 2003

S. J. Wang†, Y. Yan, Q. Fu, Y. Liu and Z. Chen

National Astronomical Observatories of China, Beijing 100012, China
email: wsj@bao.ac.cn

Abstract. We study in focus to the multi-wavelength radio spectra associated with quite strong CME events observed in the time interval of October 26–28, 2003. Using multi-wavelength observations recorded by WIND/WAVES experiment, Learmouth Spectrograph (Australian), PHOENIX-2 (Switzerland), Hiraiso (Japan), DAM/Nançay (France), Izmiran (Russia) and Huairou/NAOC (China), an analyzing of radio bursts was performed for the those events over a large coronal region across the microwave, the decimetric wavelength, the metric wavelength and even to much longer wavelength. The composite spectra indicate there were many complicated structures of radio bursts, including type II bursts, type IV bursts, type III bursts, drifting pulsation structures (DPS) and many radio fine structures (FS).

Keywords. Sun: coronal mass ejections (CMEs), radio radiation

1. Introduction

There is evidence that many CMEs launch from the low corona and accompanying with rich radio bursts (St. Cyr *et al.*, 1999, Delannèe *et al.*, 2000, Wang *et al.*, 2001), but relatively few imaging data are available to investigate the spatial relationships between bursts and ejections. Therefore a global radio spectra from microwave to metric wavelength across over a large coronal region would be a basic study, and they are essentially needed and useful for the associated faster and/or halo-like CMEs to provide base features and to identify the cause of this association.

2. Results and Discussions

In the late 2003, many violent solar activities were observed. We focus to the large CMEs in Oct. 26–28, 2003 which involved large spanning region over more than 140 degree, even up to 360 degree. Also very strong flare activities were recorded in the NOAA 0486 and 0484 during this period with importance of M2.7/2F to X17.2/4B.

An overview of solar radio bursts on 26 and 27 over 1 MHz-7.6 GHz is given in Fig. 1. The composite radio spectrum on 28 Oct. resembles the modality in Fig. 1, but cannot be put here because of the page limit. The radio emission on 26 started at about 6:13 UT with a group of fast type III bursts which extended over all of the radio frequency range at 6:15 UT. Meanwhile some DPS were found in the range of 25 MHz-1.5 GHz. After 6:49 UT several interplanetary type II bursts were found which continued to about 9:12 UT. There were type IV bursts abroad in the global radio range lasting more than 2.5 hours. In summary, this event included globally complicated structures such as type IV bursts, type III bursts, type II bursts and DPS. For the event on 27 Oct., some fast type III bursts were found after 7:58 UT over almost the whole radio range. At 8:07 UT several DPS appeared first near 1.0 GHz, then they drifted to low-frequency nearly 20

† This work supported by NSFC grants 10225313 & 10333030 and MOST grant G2000078403.

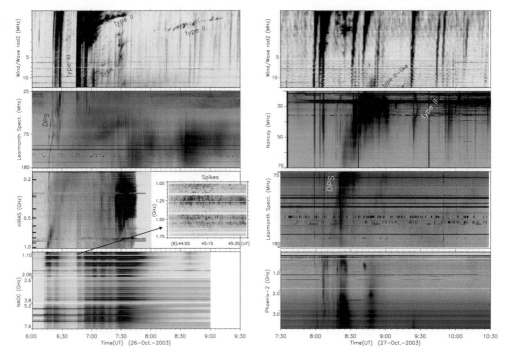

Figure 1. Composite radio spectra on 26 (a group of spikes in the inset) and 27 Oct., 2003

MHz until to 9:03 UT. Afterwards a type II burst was recorded by DAM/Nançay after 9:20 UT. Also in the range 10–14 MHz there seemed to be two slower drifting structures at about 8:45 UT and 9:15 UT (type II-like bursts). Then for the event on 28 Oct., the global radio spectrum was also very complicated including type IV bursts, type III bursts, type II-like bursts and DPS. Meanwhile using our millisecond data, many fine structures were also detected including narrow band drifting lines, zebra pattern, fiber bursts, patch and many spikes (an example showed in the inset of the left of Fig 1.).

Recently, a kind of slowly negatively drifting pulsation structures (DPS) was presented and interpreted as a signature of dynamic magnetic reconnection by Kliem *et al.*, (2000) and Karlicky *et al.*, (2001). The slow negative drift could caused by the upward motion of the plasmoid of the whole reconnection region to one of lower plasma density. DPS are closely associated with CME. It is possible that DPS in decimetric frequency band manifests the initial phase of the CME. In our events, DPS were found to be the typical signatures of those radio bursts associated with CMEs. Also the composite spectra indicate there were many complicated structures of radio bursts accompanying with the large CMEs including type II burst, type IV burst, type III burst and DPS, and many kinds of radio fine structures. Therefore it could be indicated that the strong solar activity associated with complicated magnetic structures and magnetic reconnection processes.

References

Delannèe, C., Delaboudiniére, J. -P & Lamy, P. 2000, *Astron. Astrophys.* 355, 725
Karlický, M., Yan, Y., Fu, Q. *et al.* 2001, *Astron. Astrophys.* 369, 1104
Kliem, B., Karlický, M. & Benz, A. O. 2000, *Astron. Astrophys.* 360, 715
St. Cyr, O. C., Burkepile, J. T., Hundhausen, A. J. *et al.* 1999, *J. Geophys. Res.* 104, 12493
Wang, S. J., Yan, Y., Zhao, R. *et al.* 2001, *Solar Physics* 204(1), 153

Coronal and Stellar Mass Ejections
Proceedings IAU Symposium No. 226, 2005
K. P. Dere, J. Wang & Y. Yan, eds.

© 2005 International Astronomical Union
doi:10.1017/S1743921305000396

A Type I Noise Storm and the Bastille Day CME

Yayuan Wen and Jingxiu Wang

National Astronomical Observatories, Chinese Academy of Sciences Beijing 100012, China
email: yayuanwen@ourstar.bao.ac.cn

Abstract. Based on Nançay Radioheliograph (NRH) observations, we have identified 3 Type I noise storm continua sources associated with the Bastille Day flare/CME event. Two of them were stable and closed to active regions. Their outskirts covered AR9077 and 9082, respectively. One source was over the south-west limb and in the middle corona, it was stable for hours. All the Type I storm sources weren't observed simultaneously before 10:20 UT at the onset of the global CME, which indicated the intrinsic association of Type I noise storm and CME initiation. The wide span of the Type I storm sources and burst sources clearly implied that the Bastille Day flare/CME involves large or even global magnetic interaction.

Keywords. Sun: coronal mass ejections (CMEs), radio radiation

1. Introduction and observation

In this paper, we analyze the relationship between the sources of the Type I noise storm and the Bastille Day CME. The type III noise storm can't be observed by NRH. The radio burst sources after the X5.7 flare are also discussed.

In the Bastille Day event of 2000, besides NOAA AR9077, there are several smaller active regions relating to the event, such as AR9073, AR9081, AR9082 and so on (see Solar-Geophysical Data). NOAA AR9077 rotated to the east limb of the Sun on about 7 July 2000. Noise storm activity was detected in the Nançay data, well before the eruption. Most of the active regions that can be identified on the sun on 14 July 2000, have weak noise storms associated with them (Maia *et al.*, 2001). Figure 1 shows the development of the sources of the Type I noise storms and radio bursts. The sources marked by numbers $(1, 2, 3)$ are storm sources, while marked by letters $(A, B, \cdots, F)$ are burst sources. 1a and 1b mean that two positions are seen close to the position 1, so we call them 1a and 1b. At 10:18:17 UT, storm source and radio burst source 'A' are observed on the disk. It suggests that the flux intensity starts to rise and the structure of the radio emission regions become increasing complex. After 10:18:17 UT, we can see that the storm source 1, 2, and 3 were not observed, they may in fact disappear but more likely are simply masked by the much brighter outburst. While the burst sources were moving. At about 10:27:27 UT, four sources appear on the solar disk, they cover nearly all the visible range. Their moving features show that a large-scale halo CME is ejecting to all directions. At about 10:40 UT, the source region 1a appears the new burst source. After 11:00 UT, Sources 1 and 2 do in fact disappear. A new source appears at an intermediate position. Noise storm 3 seems to survive although slightly shifted. The features of storm sources and burst sources are indicated in table 1.

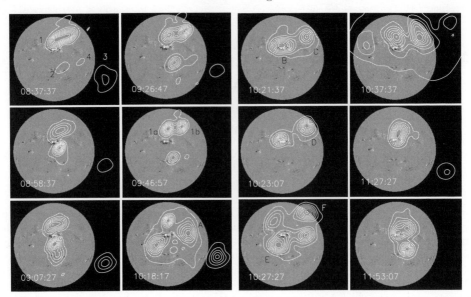

Figure 1. The intensity contours of the storm sources(left) and the radio burst sources(right) at 164MHz of the NRH superposed on the MDI magnetograms at the closest time.

Table 1. The features of the storm sources and burst sources

Source	Position	Interval	Remark
1	N0.53E0.06	08:27-09:25	stable, offset to AR9077
1a	N0.47E-0.13	08:27-10:20	stable, offset to AR9077
1b	N0.53E0.27	08:37-09:46	instable
2	N-0.33E-0.02	08:26-10:19	stable, offset to AR9082
3	N-0.5E1.25	08:26-10:21	stable, in middle corona
A	N0.17E0.38	10:16-10:19	moving toward north west
B	N0.28E0.25	10:21-10:23	moving slowly
C	N0.31E0.31	10:19-10:22	moving toward north west
E	N-0.14E-0.34	10:27-10:28	lasting for a very short time

2. Discussion and conclusion

In this study, we have showed that the development of the Type I noise storm sources before the Bastille Day flare/CME. Some sources are relative to the active regions, but only their outskirts cover the active regions. All the Type I noise storm sources were not observed before the event at the onset of the global CME, it indicated the intrinsic association of Type I noise storm and CME initiation. At the same time, it clearly implied that the Bastille Day flare/CME involves large or even global magnetic interaction.

Acknowledgements

The work is supported by the National Natural Science Foundation of China(10233050) and the National Key Basic Science Foundation(TG2000078404). The authors are grateful to Dr Maia for guiding the NRH procedure.

References

Maia, D., Pick M & Hawkins III, S. E. 2001, *Solar Phys.* 204, 199

Coronal and Stellar Mass Ejections
Proceedings IAU Symposium No. 226, 2005
K. P. Dere, J. Wang & Y. Yan, eds.

© 2005 International Astronomical Union
doi:10.1017/S1743921305000402

Correlation of the Observational Characteristics of Microwave Type III Bursts with CMEs

Y. Ma[1]†, R.X. Xie[1], X.M. Zheng[1], Y.Y. Liu[2], Y.H. Yan[2], and Q. J. Fu[2]

[1]National Astronomical Observatories/Yunnan Observatory, Chinese Academy of Sciences,
P.O.Box 110, 650011 Kunming, P.R.China
email: mayuanf@public.km.yn.cn

[2]National Astronomical Observatories, Chinese Academy of Sciences 100012, Beijing,
P.R.China.

Abstract. A total of 266 type III bursts observed with the 2.6 - 3.8 GHz high temporal resolution dynamic spectrometer of NAOC during the 23rd solar cycle (from April in 1998 to January in 2003) are statistically analyzed in this present paper. The frequency drift rates (normal and reverse slop), durations, polarizations, bandwidth, starting and ending frequencies are analyzed in detail. From the statistical results of starting and ending frequencies we show that the regions of starting frequencies are very large, which are from less than 2.6 GHz to greater than 3.8 GHz; but the ending frequencies regions are relative concentration, which are from 2.82 GHz to 3.76 GHz. These phenomena mean that the sites of electrons acceleration are quite scatter, while the cutoff regions of the radio type III bursts are in the limiting domain. The bursts number with positive and negative drift rates are nearly equal. This correlation may interpret the suggest that a proportional number of electron beams in the directions of upward and downward are accelerated in the range of 2.6 - 3.8 GHz. The other statistical results are similar to those of decimetric type III bursts as statistics in previous literature. The emission mechanisms of microwave type III bursts are mainly caused by the plasma radiation and electron gyro-maser radiation.

From the statistics of microwave type III bursts and associated coronal mass ejections (CMEs), it is found that the 36% of type III bursts (97) are corresponding to the CMEs for occurring time and site. The correlation between the type III bursts and CMEs is not close, and most type III bursts are occurred in the time regions of 26 – 30 minutes before CMEs. This means that the partial microwave type III bursts may be a precursor of the CMEs.

Keywords. Sun: coronal mass ejections (CMEs), radio radiation

---

1. Introduction

Since type III bursts play a very important role for electron acceleration in solar flares, it is of great importance to observe and research type III bursts at different frequencies. The researches on type III bursts are mostly those on metric and decimetric type III bursts and only no more authors have reported on the statistical studies of microwave type III bursts (e.g., Aschwanden *et al.*, 1995; Benz *et al.*, 1983). A coronal mass ejection (CME) is one of the highest energy activities on the solar surface, and the connection between the CMEs on the solar disk and other active phenomena is one of the open question. Previous studies show that CMEs have close relation to Hα flares, X-ray, type

† Present address:Yunnan Observatory, Chinese Academy of Sciences, P.O.Box 110, 650011 Kunming, P.R.China.

II and type IV bursts, there would be about 60% or 80% of the events accompanied by CMEs (Chertok, 1997; Chernov, 1998 Xiaoma, *et al.*, 2003; Kejun, *et al.*). However, up to now it has not been clear whether CMEs are related to type III bursts in radio bursts or not.

2. Results

(1). We have found 189 type III bursts appeared in the range of 2.6 - 3.8 GHz, in which 86 are negative drift rates, 103 are positive drift rates. In addition, 77 bursts are starting beyond the frequency band of instrument, but ending in this region of 2.80 - 3.76 GHz. (2). The most (185) of type III bursts are occurred in the impulsive phase of the flares. Only 53 bursts are in the decade phases. In 266 bursts contain 9 narrow-band ($\Delta f \leqslant$ 100 MHz) and 114 broad-band ($\Delta f \geqslant$ 500 MHz) bursts. (3). The half-power duration lies between 100 ms and 1620 ms, with the average of 613 ms. The type III bursts with long durations are much less than those with short durations. (4). 61% of bursts have polarization degree of 6% – 90% with the average of 35%. (5). The correlation between the type III bursts and the CMEs is 36%. Most type III bursts occurred 26 – 30 minutes before the CMEs.

3. Discussions

(1). 189 type III bursts are started in the region of 2.62 - 3.76 GHz, and 41 in the regions less than 2.6 GHz and 36 in the regions greater than 3.8 GHz. It has known that the starting frequencies of type III bursts are co-spatial with the acceleration regions (Aschwanden, 2002), so the electrons acceleration sites are very scatter. The acceleration regions are up to the centimeter wavelength, which is greatly beyond the region of 400 - 1000 MHz (Bastian, Benz and Gary, 1998). (2). From the statistics of type III bursts and CMEs, we may known that some acceleration electrons have escaped before the CMEs. This suggests that some type III bursts may be the precursor of the CMEs.

Acknowledgements

This research is supported by the Ministry of Science and Technology China under grant No. G20000784, and also supported by the national NST of China (grant Nos. 10333030 and 10473020).

References

Aschwanden, M.J. 2002, *Space Science Rev.*. 101, 1
Aschwanden, M.J. & Benz, A.O. 1995, *ApJ.* 455, 347
Bastian, T.S., Benz, A.O. & Gary, D.E. 1998, *ARAA* 36, 131
Benz, A.O., Bernold, t.e.x. & Dennis, B.R. 1983, *ApJ* 271, 351
Chertok, J. 1997, *Moscow Phys.* 7, 31
Chernov, G.P. 1998, *Astrophysics Reports* (Special Issue) 3, 9
Isliker, H. & Benz, A.O. 1994, *A&A Suppl.* 104, 145
Li, K.J., Schmieder, B., Malherbe, J.M., *et al.* 1998, *Solar Phys* 90, 37
Swant, H.S., Lattari, C.J.B. & Benz, A.O. 1990, *Solar Phys.* 130, 57
Gu, X.M. & Ding, M.D. 2002, *ChJAA* 2, 92

Coronal and Stellar Mass Ejections
Proceedings IAU Symposium No. 226, 2005
K. P. Dere, J. Wang & Y. Yan, eds.

© 2005 International Astronomical Union
doi:10.1017/S1743921305000414

The Successive Ejection of Several Halo CMEs from NOAA AR. 652 July 2004, a Physical Study

Shahinaz Yousef[1], Mostafa M. El-Nazer[1] and Aisha Bebars[2]

[1] Astronomy& Meteorology Dept, Faculty of Science, Cairo University, email:
shahinazyousef@yahoo.com

[2] National Research Institute of Astronomy and Geophysics, Cairo, Egypt

Abstract. Although we are on the descending branch of solar cycle 23, a very strong active region, NOAA 652, crossed the solar visible hemisphere during the second half of July 2004. Its very large sunspot was of beta- gamma- delta type. This active region was a source of numerous X-ray flares and coronal mass ejections. Being in a favorable position not far from the equator, it represented a threat to planet Earth particularly when near its central meridian passage.

Using EIT images, it was possible to locate the position of the 2004 July 25 CME ejection as dimming of several loops to the SW of AR 652. As usual, the CME lift up was accompanied by type IV burst and followed by type II burst and a proton flare.

Keywords. Sun: coronal mass ejections (CMEs), flares, particle emission

---

1. The Ejection Conditions of The 2004 July 25 CME

A list of coronal mass ejections (CMEs) produced by active region, NOAA 652 is given in ftp://lasco6.nascom.nasa.gov/pub/lasco/status/LASCO_CME_List_2004.

Since space is limited, only the July 25 CME will be studied here. It was preceded by two brief X-ray flares. The sequence of events that lead and accompanied the CME are; a type IV burst, a M1.1 X-ray flare,a 1F chromospheric flare and a Type II radio burst respectively as seen below.

Type	start	max	end	location
C2.1	13:18	13:25	13:32	N04W29
M2.2	13:37	13:49	13:55	N04W30
M1.1	14:19	15:14UT	16:43	N08W33
1F Flare	1433	1448	1643	N08W33
Type IV/2	1415	////	1731	
Type II	1521	////	1526	

Shocks generated during the CMEs are detected as type II radio bursts while ejected plasmoids are observed as type IV bursts(Gopalswamy *et al.* 1999).The plasmoid was perhaps lifted over around the time of type IV start at 1415.

A front side halo CME is usually apparent in EIT images as dimming (Zhang *et al.* 2003). This dimming is actually apparent in fig. 1 images at 1348 and 1500. In the first image a faint bright loop seen first at 1326 was dimmed. The second important dimming of several coronal loops seen at 1500 is the actual site of CME ejection. However the CME was actually ejected earlier. It was first seen in C2 around 1430 and in C3 in SW limb at 1518 UT. Plane-of-sky speed at PA 145 was 1280 kms^{-1} (ftp://lasco6.nascom.nasa.gov/pub/lasco/status/LASCO_CME_List_004)

145

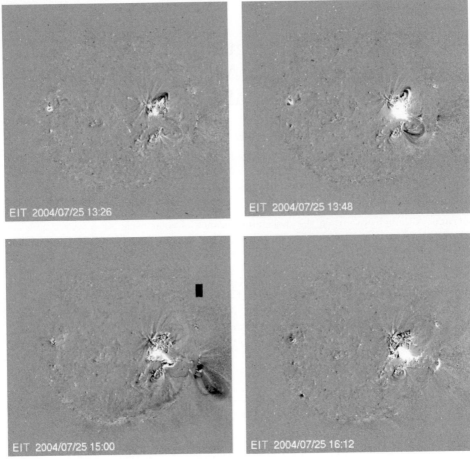

Figure 1. A sequence of EIT images showing the site of coronal mass ejection of 2004/07/25. Note the faint bright loop at 1326 that was dimmed on 1348. Several dimmed loops to the SW of AR NOAA 652 seen on 1500 image and disappeared later, are the most probable site of initiation of coronal mass ejection. The wave front to the SE of those dark loops is disturbed on 1500 due to type IV radio burst. Two C2.1 and M2.2 flares were near their maximums on 13.26 and 13.48 images respectively. Another M1.1 flare was going on during the images at 1500 and 1612 between the two ARs NOAA 652 and 653 and magnetic loops were seen later to connect them. Images are reproduced from ftp://ares.nrl.navy.mil/pub/lasco/halo/20040725/

2. Conclusion

The 25th of July 2004 CME was ejected from a site to the SW of ARs NOAA 652 and 653. Apparent dimming of several coronal loops in this location was evident in EIT images. The ejection was preceded by two brief X-ray flares. A type IV radio burst started four minutes ahead of a third long duration M1.1 X-ray flare accompanied the lift off of the CME. A type II radio wave and a proton flare followed.

References

Gopalswamy, N., Lazio, T. J. W., Kassim, N. E., and Erickson, W. C. 1999, BAAS, 194, 848
Zhang, J., Dere K. P., Howard, R. H., and Bothmer, V. 2003, ApJ 582, 520

Session 3

Coronal mass ejections source regions

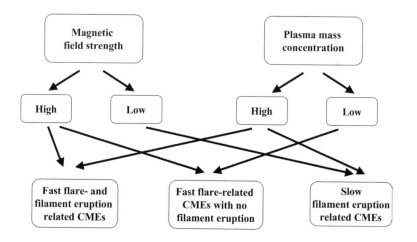

Coronal and Stellar Mass Ejections
Proceedings IAU Symposium No. 226, 2005
K. P. Dere, J. Wang & Y. Yan, eds.

© 2005 International Astronomical Union
doi:10.1017/S1743921305000426

Source Regions of Coronal Mass Ejections

Brigitte Schmieder[1,2] and L. van Driel-Gesztelyi [1,3,4]

[1] Observatoire de Paris-Meudon, LESIA, France email: brigitte.schmieder@obspm.fr
[2] ITA, P.O.Box 1029, Blindern, N-0315, Oslo, Norway
[3] Mullard Space Science Laboratory, University College London, Holmbury St. Mary, Dorking, Surrey, RH5 6NT, UK
[4] Konkoly Observatory, P.O. Box 67, H-1525, Budapest, Hungary

Abstract. The majority of flare activity arises in active regions which contain sunspots, while CME activity can also originate from decaying active regions and even so-called quiet solar regions which contain a filament. Two classes of CME, namely flare-related CME events and CMEs associated with filament eruption are well reflected in the evolution of active regions, flare related CMEs mainly occur in young active regions containing sunspots and as the magnetic flux of active region is getting dispersed, the filament-eruption related CMEs will become dominant. This is confirmed by statistical analyses.

All the CMEs are, nevertheless, caused by loss of equilibrium of the magnetic structure. With observational examples we show that the association of CME, flare and filament eruption depends on the characteristics of the source regions: (i) the strength of the magnetic field, the amount of possible free energy storage, (ii) the small- and large-scale magnetic topology of the source region as well as its evolution (new flux emergence, photospheric motions, canceling flux), and (iii) the mass loading of the configuration (effect of gravity). These examples are discussed in the framework of theoretical models.

Keywords. Sun: coronal mass ejections (CMEs), flares, magnetic fields, sunspots

1. Introduction

Coronal mass ejections are episodic expulsions of mass and magnetic field from the corona into the interplanetary medium. They reflect a high level of activity in the solar atmosphere. It would be very important to understand where they come from and what the relationship is between CMEs and other, related, sources of activity (flare activity, filament eruption). In the old days Dodson and Hedeman (1970) postulated that 93% of the flare activity arises in active regions which contain sunspots. However, the span of CME activity is much longer and extends well into the decay phase of active region evolution when the magnetic field is dispersed and the region is classified as quiet solar region containing filament (van Driel-Gesztelyi *et al.* 1999). Subramanian and Dere (2001), based on a sample of 32 CMEs, found that 85% of them were associated with active regions and 15% with so-called quiet regions. They found that 44% of these CMEs were associated with eruption of an active region filament and the 15% coming from "quiet" regions were all associated with filament eruption. These results are in good agreement with other statistical analyses of more numerous events having a narrower multi wavelength coverage (Saint Cyr and Webb, 1991; Webb, 1998; Zhou *et al.* 2003).

The main characteristics of the source regions have been intensively described in the review of Wang (2002). He also emphasized that CMEs are large scale events. An average-size CME is as large as five active regions. Therefore CMEs are frequently related to large scale magnetic structures, such as giant magnetic loops or transequatorial loops (Delannée and Aulanier, 1999), giant filaments or filament channels (Wang *et al.* 2002), or super magnetic configuration consisting of alignments of sunspots (Avignon, Martres and Pick, 1964). Falconer and co-workers found increasing correlation between CME productivity and global non-potentiality measures in the source region (Falconer, Moore and Gary, 2002). CMEs tend to appear in a small-scale core field in a complex magnetic region, where persistent flux cancellation (Deng *et al.* 2001; Zhang *et al.* 2001), new emerging flux or moving magnetic features and consequent activation of filaments are present.

In this paper we review the principal conditions for CME and then discuss what the different conditions are for flare- and/or filament eruption related to CMEs.

2. Conditions for CMEs

2.1. *General conditions*

There is a consensus that the presence of significant magnetic stresses in the source region is a necessary condition for CME. The magnetic helicity quantifies how the magnetic field is sheared or twisted compared to its lowest energy state (potential field). Observations provide plenty of evidence for the existence of such stresses in the solar magnetic field and their association with flares and CME activity. The first quantitative estimate of the helicity injection into an observed solar active region was made by Wang (1996), who deduced a 10^{43} Mx^2 of helicity increase in an emerging flux region (AR 6233) over a period of just a few hours. He computed change in magnetic helicity density using vector magnetograms and tracing the change of α, the force-free parameter. In young active regions magnetic stresses are increased by (i) twisted magnetic flux emergence and the resulting magnetic footpoint motions (e.g. Chae 2001; Chae *et al.* 2002; Kusano *et al.* 2002; Moon *et al.* 2002; Nindos and Zhang, 2002; Nindos *et al.* 2003; Démoulin and Berger, 2003) as well as (ii) torsional Alfvén waves which bring up helicity from the sub-photospheric part of the flux tube and replenish coronal helicity after CME events (Longcope and Welsch, 2000; Démoulin *et al.* 2002a; Green *et al.* 2002). A possible manifestation of such torsional Alfvén wave is sunspot rotation, which has indeed been observed (e.g. Brown *et al.* 2003). In old, decayed active regions twist can be redistributed through cancellation events transferring helicity from small- towards the large scales, it can be increased by large scale photospheric motions (differential rotation) (DeVore, 2000; Démoulin *et al.* 2002b; Berger and Ruzmaikin, 2000) or brought up by torsional Alfvén waves (Démoulin *et al.* 2002a; Green *et al.* 2002).

Whether the eruption is possible or not, the strength of the magnetic field overlying a sheared arcade or a twisted flux tube may play an important role. Since such overlying field has a stabilizing effect, a strong field can actually prevent the eruption (Török *et al.* 2004; Roussev *et al.* 2003). However, in a magnetically complex configuration such overlying field can be (at least partially) removed by reconnection. The so-called breakout scenario can lead to partial opening of the overlying field which stabilizes the increasingly sheared core field, which can burst open leading to a CME (e.g. Antiochos, 1998; Antiochos, DeVore and Klimchuk, 2000; Aulanier *et al.* 2000; Wang, 2002). Therefore magnetic complexity of the magnetic field in the source region and in its environment is another important factor leading to CME.

2.2. *Conditions for flare-associated CME*

The correlation between a CME and a solar flare depends on the energy that is stored in the relevant magnetic structure, which is available to drive the eruption: the more energy that is stored, the better the correlation is; otherwise, the correlation is poor (Svetska, 1986; Lin, 2004). The correlation between solar flares and CMEs depends on the strength of the magnetic field in the source region - strong fields obviously can store more free energy. It is well-known that large, complex magnetic regions with strong and increasing non-potentiality are highly CME productive. Therefore it is important to evaluate the amount of the free energy storage versus time.

2.3. *Conditions for filament eruption associated CME*

The correlation between a CME and eruptive prominence, on the other hand, depends on the plasma mass concentration in the configuration prior to the eruption. If the mass concentration in the source region is significant, CME will be associated with prominence/filament eruption, otherwise a CME develops without an apparent associated eruptive prominence (Lin, 2004).

These results confirm that solar flares, eruptive prominences and CMEs are different signatures of a single physical process that is related to the energy release in a disrupted coronal magnetic field. The impact of gravity on CME propagation is shown to be important in a low background magnetic field (around 20 G). The gravity of a filament or prominence may play an important role in the process of slow CMEs by prohibiting the catastrophe to occur at the very beginning (Isenberg, Forbes and Démoulin, 1993). There are threshold values (mass and magnetic field strength) defining whether CME is possible or not.

2.4. *Summary*

The field strength, the presence of stress in the magnetic field, the mass of prominence are important ingredients to get CME or not. We have also to consider the pre-eruption magnetic environment of the CMEs. The small as well as the large scale magnetic topology of the source region is an important factors. Furthermore, the strength of the magnetic field overlying a sheared arcade, appears to be very important. The overlying field has a stabilizing effect, and may prevent eruption. The highly stressed (sigmoid) regions appear in large and also in small scale configurations, which poses the question whether the size of the source region influences or not its eruptive nature. We shall review these different aspects of the CME process through different examples of CME observations and modeling studies.

3. Review of examples

3.1. *Large strength of magnetic field and high stress*

Commonly, active regions with strong magnetic field and high stresses are sites of large energetic flares producing fast CMEs (e.g. 15 June 2000, the Bastille day flare and the twelve X-ray class flares during the period of October–November 2003 in three complex active regions NOAA 10484, 10486 and 10488). All three source regions are delta-spot regions formed through numerous episodes of flux emergence. As an example, the active region NOAA 10486 had an overall quadrupolar magnetic configuration on October 28 2003. The magnetic field strength was still high although the active region had entered into its decaying phase, when an X17 GOES class flare and associated CME occurred (Figure 1). The magnetograms of Huairou indicate strong shear along the two parallel inversion lines which squeezed a bridge of negative polarities between two positive polarities (Zhang *et al.* 2003).

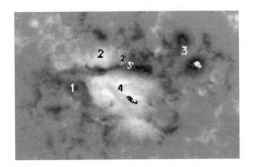
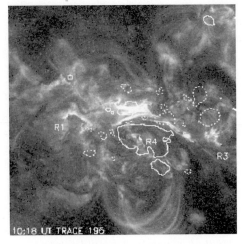

Figure 1. X-ray 17 Flare of October 28, 2003, right panel: TRACE observations of the four ribbons signature of a quadrupolar reconnection before the flare , left panel: Complexity of the magnetic topology of the active region 10486, where the main X17 flare occurs (Schmieder et al. 2005)

3.2. Evolution of magnetic flux density, stresses and activity through the life of two active regions

Comprehensive analyses of the long term evolution of two active regions confirm that evolving magnetic flux density plays an important role in flare and CME activity and its evolution must influence all the activity signatures (van Driel-Gesztelyi et al. 1999, 2003; Green et al. 2002). The complex analysis concerns an isolated active region (NOAA 7978 July-December 1996). The magnetic field of the AR was clearly distinguishable for at least seven solar rotations. It was found that flares mainly occur when the magnetic field of the AR has the highest complexity and magnetic flux density during the two main flux emergence phases (1st two rotations in July and August 1996) and while the number of CME is sustained and may be more closely related to the magnetic stresses in the region, since the value of the linear force-free α parameter was found to be roughly proportional with the CME rate (Table 1). During the first observed flux emergence phase 10^{22} Mx flux surfaced. During the decay phase magnetic flux gradually spreads over an ever increasing area, flare activity shows a sharp decrease and practically ceases with the disappearance of sunspots, while CME activity remains on a relatively high level (3-4 CMEs per rotation). During the decaying phase of the active region cancelling flux along the major inversion line were identified and could participate to the redistribution of the twist. The filament which was lying along the inversion line erupted after a sustained period of cancellations. Similar evolutionary pattern was found in NOAA AR 8100 by Green et al. (2002).

3.3. Sigmoids of small and large scale

Sigmoids, in many cases, indicate the presence of high magnetic stresses and have been linked to CMEs (Gibson et al. 2002; Manoharan et al. 1996). Though the sigmoid-CME connection is still statistically ambiguous (Canfield et al., 1999; Glower et al. 2000) it has been suggested that the magnetic helicity content in S-shaped magnetic configuration may reach a threshold leading to instability and eruption (van Driel-Gesztelyi et al. 2000; Török and Kliem, 2003). In the same way as CMEs can range from large-scale

Table 1. Evolution of flare- and CME activity and of the magnetic stresses in AR 7978 Note, that above the horizontal line sunspots are present in the active region, while under it it becomes a spotless 'plage' region.

No. of rotation	Date of CMP	Flares (GOES class)				CMEs	α $10^{-2}Mm^{-1}$
		X	M	C	B		
1st	07/07/96	1	2	14	11	11	1
2nd	02/08/96	-	-	-	16	5	0.3-0.75
3rd	30/08/96	-	-	1	8	3	0.9-1
4th	25/09/96	-	-	-	-	5	1-1.4
5th	23/10/96	-	-	-	-	4	0.9-1.4
6th	18/11/96	-	-	-	-	3	0.9
Total		1	2	15	35	31	

to very narrow events, the scale of erupting structures can range from interconnecting transequatorial loop size to the size well below of a typical AR.

The well known geoeffective CME event of January 6, 1997, produced a large magnetic cloud that damaged a satellite. For two-three days prior to the CME MDI observations of the active region show persistent cancellation of magnetic flux along the magnetic inversion line in the center of the AR (van Driel-Gesztelyi *et al.* 2002), that, through magnetic reconnection, is thought to lead to a reorganization of the magnetic configuration, increasing the twist in the large-scale magnetic structure of the overlying filament. In soft X-rays the region had a sigmoidal shape. Van Driel-Gesztelyi *et al.* (2000) proposed a model involving reconnection in a sheared arcade, which lead to the formation of short, highly sheared loops in the center and long sigmoidal loops connecting the outer edges of the bipole. The long sigmoidal loops may become unstable leading to a CME. As the sigmoid expands, a current sheet is formed under it and a cusp structure appears.

A similar event was observed in the center of the disk but related to a very small dipole, which had a two orders of magnitude less magnetic flux than a usual active region. Mandrini *et al.* (2004), carrying out a multi-wavelength analysis of a sigmoidal coronal bright point, found a high level of non-potentiality in the magnetic field and provided several independent evidence for its eruptive behavior (flaring followed by dimming and appearance of cusped loops, change of shape of the sigmoid from elongated prior to and compact after the eruption) (Figure 5 and Figure 8 in Mandrini *et al.* 2004). Using a linear force-free model of the pre- and post-eruption coronal loops they computed the change in coronal helicity due to the eruption (CME). Analyzing in-situ data obtained by the WIND spacecraft, they found a small magnetic cloud, which could be linked to the small solar eruption by timing, spatial magnetic orientation and field direction. Modeling the magnetic cloud and having constraints on the length of its flux tube from the short lifetime of the solar source region, they calculated its helicity. The helicity change in the corona and the helicity content of the MC were very similar, -2.7±0.4 x 10^{39} Mx^2 in the corona and 2.3 ±0.8 x 10^{39} Mx^2 in the cloud, which, given the unusually small size of the source region, can be regarded as a lower bound for the helicity loss due to CME.

4. Role of magnetic topology

Do CME source regions have bipolar or quadrupolar magnetic structure?

Forbes and Isenberg (1991) and Isenberg, Forbes and Démoulin (1993) developed a model where the eruption is preceded by converging motions towards the magnetic inversion line of a bipole. A loss of equilibrium leads to the rise of a flux rope situated along the inversion line. Reconnection occurring in the current sheet formed under the flux rope allows the flux rope to escape. Converging motion of opposite polarities and consequent cancellation along the magnetic inversion line before CMEs have indeed been observed (e.g. van Driel-Gesztelyi et al. 2002, Schmieder et al. 2005).

However, the highly CME-active source regions are magnetically complex! What is the main role of a multipolar magnetic structure in the CME process?

The analysis of the magnetic field evolution, topology and multi-wavelength data before and during the X 17 flare on October 28, 2003, shows how a large twisted flux tube supporting the long filament lying along the main inversion line was first formed then erupted. The slow build-up of magnetic stress through converging motions along the magnetic inversion line begins well before the eruption. If we try to classify the X 17 flare within the scenario of "storage and release" models (Klimchuk, 2001) i.e. tether cutting or tether straining, it appears that two mechanisms could be present: firstly, twist in the flux rope is built up as small-scale cancellation events (reconnection) occurs in a sheared arcade aligned along the main neutral line. The flux tube starts to rise slowly, as more and more tethers are being cut. During that time the twist in the tube increases. Before the X 17 flare, two episodes of large-scale quadrupolar reconnection occur in the active region. The second episode, being more intense, implies more important field-rearrangements. These quadrupolar reconnection events remove stabilizing field lines from above the flux rope (filament) which succeeds to break out. Reconnection under the erupting flux rope starts after the lift-off, forming a post-flare loop arcade. This evolution is similar to the breakout model in quadrupolar configurations proposed by Antiochos, DeVore and Klimchuk (1999). These models require reconnection above the erupting arcade prior to the CME eruption. Multiwavelength observations combined with modeling provided one of the first convincing observational evidence that breakout model is indeed a viable model for CME (Aulanier et al. 2000).

In the breakout models the quadrupolar magnetic configuration is a necessary condition. However, it is still a question whether topological complexity is indispensable for CMEs or not. Is the loss of equilibrium of a flux rope, in which twist exceeded threshold level (see Török, Kliem, and Titov (2004) and references within), provides sufficient condition for CME? The latter seem to work even in a simple bipolar magnetic configuration. Since the Sun is complex, we must leave open the possibility that more than one CME model might be correct!

It has been a long-standing question which comes first: flare or CME? Both observations and models suggest that an inflation of the magnetic structure due to increasing magnetic stresses is the first step - this makes observers say that CME starts before flare. The main flare energy release occurs during the reconnection of field lines extended by the eruption, which, again, points towards that CME should come first. However, both in the tether cutting (flux rope) and the tether straining (break-out) models pre-eruption reconnection is required under or above, respectively, of the erupting twisted/sheared magnetic structure (filament). In the break-out model this pre-eruption reconnection is expected to occur high in the corona in a region of weak field, releasing too little energy to be observed. However, if the reconnection occurs in a strong field region, like in a complex active region, where tethers are not external weak fields overlying the sheared core field but they are part of the active region having a quadrupolar configuration, the released magnetic energy can be high enough to qualify as flare, therefore we may find that (an impulsive, quadrupolar) flare precedes the CME, while the latter includes a filament eruption and a related two-ribbon flare representing the post-eruption arcade. Similar scenario was proposed for the 15 July 2002 CME by Gary and Moore (2004), and for the 28 October 2003 CME by Schmieder et al. (2005).

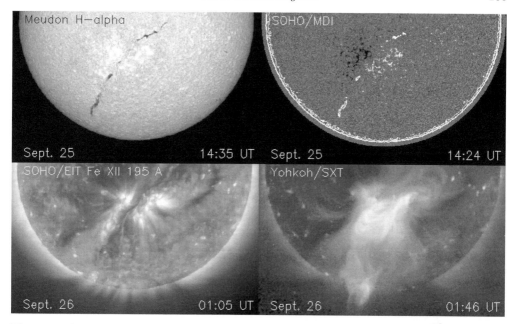

Figure 2. CME of September 26, 1996 due to the eruption of a filament in a non active region

In any case, in the models most relevant to observations (storage and release models) the build-up of magnetic stresses, i.e. strong shear and twist are necessary conditions for CME to occur. In observations twisted CME structures are frequently seen (Fig. 4) and there is amounting evidence that considerable amount of twist is being carried away from the Sun by CMEs. These emphasize the importance of magnetic helicity in the CME process.

5. Mass loading in prominences

The free energy stored in a stressed magnetic structure prior to the eruption depends on the strength of the background field. The stronger the background field is, the more free energy can be stored, and thus the more energetic the eruptive process is (Lin 2004). This eruptive process refers to any disruption of the coronal magnetic field that causes either a flare, or a filament eruption or CME or all of them. In the case of CMEs related to quiet sun region, there is a critical strength of the background magnetic field (<27 G) where the effect of gravity becomes significant enough to prevent the CME from progressing (Isenberg, Forbes and Démoulin 1993). If the field is larger than this value, the system evolves smoothly in response to the slow change in the boundary conditions and can end up with a slow CME (Schmieder *et al.* 2000). The gravity of the filament or prominence may play an important role in the process of slow CMEs by prohibiting the catastrophe to occur at the very beginning. A prominence rises slowly before erupting and can reach altitude as high as 200 Mm (Figures 2 and 3). The mass of CMEs inferred from SOLWIND observations from 1979 to 1981 (Howard *et al.* 1985) ranged from 2×10^{14} to 4×10^{16} g. The amount of mass and the mass concentration should be important for the CME-eruptive prominence association.

Two aspects derived from the observations have to be taken into consideration after this statement. On one hand a filament appears darker in chromospheric images before eruption. This indicates a higher density of plasma or new condensation of material.

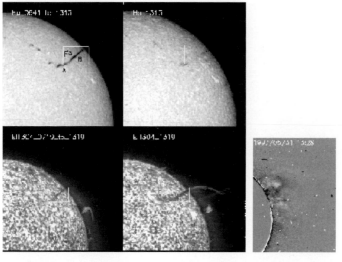
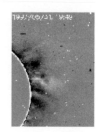

Figure 3. Quiet region source of a CME due to a filament eruption on May 31, 1997
(Spectroheliograms of Meudon, EIT images, LASCO C1 images (Schmieder *et al.* 2000)

A strength of the stresses in the magnetic field supporting the prominence can result of this increasing mass. The models of magnetic support of prominences show a higher complexity of the field lines as the stress increases (Aulanier, DeVore, and Antiochos, 2002). Another possibility to explain this increasing density is the following. Before the eruption the plasma in prominences is commonly heated, the optical thickness of Hα increases and it is in favor to see darker filaments before the eruption (Schmieder, Tziotziou, and Heinzel, 2003). This latter mechanism does not imply increasing magnetic stresses due to mass loading.

On the other hand it has been recently found that solar filaments observed in EUV lines are much more extended than their Hα counterparts (Heinzel *et al.* 2001, Schmieder, Tziotziou, and Heinzel, 2003a). This was explained by a large difference between the hydrogen Lyman-continuum and Hα opacities. Two different MHD models were suggested to explain the EUV filament extensions: the model based on parasitic polarities (Aulanier and Schmieder, 2002) and the model with twisted flux tubes (Anzer and Heinzel, 2003). The latter model tentatively describes the possibility of the EUV extension to be located relatively high in the atmosphere (Schmieder *et al.* 2004). These heights can be computed using a new spectroscopic model of EUV filaments (Heinzel *et al.* 2003a). The mass which is loaded into the EUV filament extensions is then estimated on the basis of non-LTE transfer calculations. The total filament mass is larger than that derived for the Hα filament itself by a factor 1.5 or 2 and this may have consequences for the structure and the mass loading of CMEs (Heinzel *et al.* 2003b).

6. Conclusion

The driver behind solar flares, prominence eruptions and CMEs is the instability of stressed coronal magnetic fields which have been driven towards a highly stressed state by photospheric motions and twisted flux emergence. From a theoretical point of view (Lin, 2004) and observations of CME source regions it appears that some parameters are important to determine which type of CMEs the region is able to expel. The strength of magnetic field and the complexity of the magnetic configuration prior to the eruption determines the correlation between flares and CMEs, and the impact of the gravity on

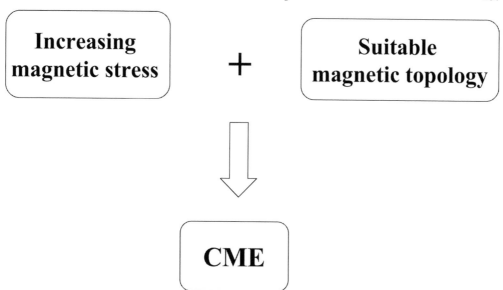

Figure 4. What are the magnetic conditions for getting CMEs?

the above correlation can be important (mass and concentration of mass in prominence) if the background magnetic field is weak. The size of the region is not an important parameter, even very small magnetic regions are able to produce CME. An increase of stress in the source region is a crucial factor, and the strength of the field determines how much free energy can be stored in it. However, the strength of the overlying field, which has an important stabilizing effect, appears to play an important role in whether eruption occurs or not. The correlation between CME and eruptive prominence depends on the amount and the concentration of the plasma mass in the related magnetic configuration. The CME may commence with an apparent prominence eruption if the mass inside the structure is larger than 4.5×10^{15} g. Our conclusions are summarized in two charts (Figs. 4, 5).

Acknowledgments

LvDG is supported by the Hungarian Government grant OTKA T-038013. This research was supported by the European Commission through the RTN programme ESMN (contract HPRN-CT-2002-00313).

References

Antiochos, S. K. 1998, Astrophys. J., 502, L181
Antiochos, S. K., DeVore, C. R., and Klimchuk, J. A. 1999, Astrophys. J., 510, 485
Anzer, U. and Heinzel, P. 2003, Astron. Astrophys., 404, 1139
Aulanier, G. and Schmieder, B. 2002, Astron. Astrophys., 386, 1106
Aulanier, G., DeLuca, E. E., Antiochos, S. K., McMullen, R. A., and Golub, L. 2000, ApJ, 540, 1126
Aulanier, G., DeVore, C. R., and Antiochos, S. K. 2002, Astrophys. J., 567,. L97-L101
Avignon, Y., Martres, M. J. and Pick, M., 1964, An. Astro., 27, 23.
Avignon, Y., Martres, M. J., and Pick, M. 1964, An.Astro., 27, 23
Berger, M. A. and Ruzmaikin, A. 2000, J. Geophys. Res., 105, A5, 10481.
Brown, D. S., Nightingale, R. W., Alexander, D., *et al.* 2003, Solar Phys. 216, 79.
Canfield, R. C., Hudson, H. S., and McKenzie, D. E. 1999, Geophys. Res. Lett. 26, 627.

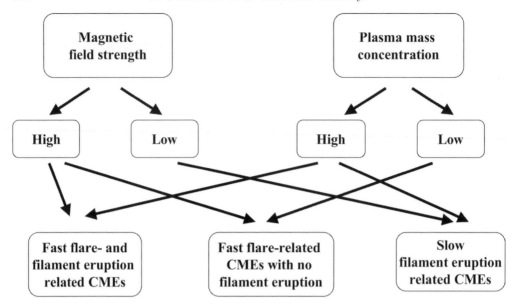

Figure 5. Summary of the different characteristics of the source regions producing flares, filament eruptions, CMEs

Chae, J. 2001, Astrophys. J., 560, L95.

Chae, Jongchul, Moon, Yong-Jae, Wang, Haimin, and Yun, H. S. 2002, Solar Phys., 207, 73

Chae, J., Wang, H., Qiu, J., Goode, P. R., Strous, L., and Yun, H. S. 2001, Astrophys. J., 560, 476.

Delannée, C. and Aulanier, G. 1999, Solar Phys., 190, 107

Deng, Y., Wang, J. X., Yan, Y., and Zhang, J. 2001, Solar Phys., 204, 13

Démoulin, P., Mandrini, C. H., van Driel-Gesztelyi, L., et al. 2002a, Astron. Astrophys., 382, 650.

Démoulin, P., Mandrini, C. H., van Driel-Gesztelyi, L., et al. 2002b, Sol. Phys., 207, 87.

Démoulin, P. and Berger, M. A. 2003 Solar Phys., 215, 203.

DeVore, C. R. 2000, Astrophys. J., 539, 944.

Dodson, H. W. and Hedeman, E. R. 1970, Solar Phys. 13, 401

Falconer, D. A., Moore, R. L., and Gary, G. A. 2002, Astrophys. J., 569, 1016

Forbes, T. G. and Isenberg, P. A. 1991, Astrophys. J., 373, 294

Gary, G. A. and Moore, R. L. 2004, Astrophys. J., 611, 545

Gibson, S., Fletcher L., DelZanna G., et al. 2002, Astrophys. J., 574, 1021

Glover, A., Ranns, N. D. R., Harra, L. K., and Culhane, J. L. 2000, Geophys. Res. L., 27, 2161

Green, L. M., Lopez-Fuentes, M. C., Mandrini, C. H., Démoulin P., van Driel-Gesztelyi, L., and Culhane, J. L. 2002, Solar Phys. 208, 43

Heinzel, P., Schmieder, B., and Tziotziou, K. 2001, Astrophys. J., 561, L223

Heinzel, P., Anzer, U., and Schmieder, B. 2003a, Solar Phys., 216, 159

Heinzel, P., Anzer U., Schmieder, B., and Schwartz, P. 2003b, ESA SP 535, 447-457

Howard, R. A., Sheeley, J. N. R., Koomen, M. J., and Michels, D. J. 1985, J. Geophys. Res. 90, 1356

Isenberg, P. A., Forbes, T. G., and Démoulin, P. 1993, Astrophys. J., 417, 368

Klimchuk, J. A. 2001, in Space Weather (Geophysical Monograph 125), ed. P. Song, H. Singer, & G. Siscoe (Washington: Am. Geophys. Un.), 143-157

Kusano, K., Maeshiro, T., Yokoyama, T., and Sakurai, T. 2002, Astrophys. J., 577, 501.

Lin, J. 2004, Solar Phys. 219, 169

Longcope, D. W. and Welsch, B. 2000, Astrophys. J., 545, 1089

Mandrini, C. H., Pohjolainen, S., Dasso S., Green, L. M., Démoulin, P., van Driel-Gesztelyi, L., Copperwheat, C., and Foley, C. 2004, Astron. Astrophys., in press

Manoharan, P. K., van Driel-Gesztelyi, L., Pick, M., and Démoulin, P. 1996, Astrophys. J, 468, 73

Moon, Y.-J., Chae, J., Choe, G. S., Wang, H., Park, Y. D., Yun, H. S., Yurchyshyn, V., and Goode, P. R. 2002, Astrophys. J., 574, 1066.

Nindos, A. and Zhang, H. 2002, Astrophys. J., 573, L133.

Nindos, A., Zhang, J., and Zhang, H. 2003, Astrophys. J., 594, 1033.

Roussev, I. I., Forbes, T. G., Gombosi, T. I., Sokolov, I. V., DeZeeuw, D. L., and Birn, J. 2003, ApJ., 588, 45.

Schmieder, B., Delannée, C., Yong, D. Y. Vial, J. C., and Madjarska, M. 2000, Astron. Astrophys., 358, 728

Schmieder, B., Tziotziou, K., and Heinzel, P. 2003, Astron. Astrophys. 401, 361.

Schmieder, B., Yong Lin, Schwartz, P., and Heinzel, P. 2004, Solar Phys. 221, 297

Schmieder, B., Mandrini, C. H., Démoulin, P., Pariat, E., Berlicki, A., and DeLuca, E. 2005, Adv. Space Res., in press

Subramanian, P. and Dere, K. D. 2001, Astrophys. J., 561, 372

St. Cyr, O. C. and Webb, D. F. 1991, Solar Phys. 136, 379

Svetska, Z. 1986, in D.F. Neidig (ed) The lower Atmosphere of Solar Flares, NSO/Sac Peak Publ., 332

Török, T. and Kliem, B. 2003, Astron. Astrophys., 406, 1043

Török, T., Kliem, B., and Titov, V. S. 2004, Astron. Astrophys., 413, L27

van Driel-Gesztelyi, L., Mandrini, C. H., Thompson, B., Plunkett, S., Aulanier, G., Démoulin, P., Schmieder, B. and de Forest, C. 1999, Third Advances in Solar Physics Euroconference: Magnetic Fields and Oscillations, ASP Conference Series . Eds. B. Schmieder, A. Hofmann, J. Staude,vol.184, p. 302

van Driel-Gesztelyi, L., Manoharan, P. K., Démoulin, P., *et al.* 2000, J. Atm. Solar-Terr. Phys., 62/16, 1437

van Driel-Gesztelyi, L., Schmieder B., and Poedts S. 2002, Proc. SOLSPA-2001 Euroconference 'Solar Cycle and Space Weather'; ESA SP Series (SP-477), ISBN 92-9092-749-6, 47.

van Driel-Gesztelyi, L., Démoulin, P., and Mandrini, C. H. 2003a, Second Franco-Chinese Meeting on Solar Physics, 'Understanding Active Phenomena, Progress and Perspectives', eds. J.-C. Hénoux, C. Fang and N. Vilmer, World Publishing Corporation, 37.

van Driel-Gesztelyi, L., Démoulin, P., and Mandrini, C. H. 2003b, Adv. Space Res., 32, No. 10, 1855.

Wang, J. X. 1996, Solar Phys., 63, 319.

Wang, J. X. 2002, proceedings of the second French-Chinese meeting eds. J.C. Hénoux, C. Fang and N. Vilmer, World Publishing Corporation, 145.

Wang, Tongjiang, Yan, Yihua, Wang, Jialong, Kurokawa, H., and Shibata, K. 2002, Astrophys. J., 572, 580

Webb, D. F. 1998, in IAU Colloquium 167, Webb D., Rust D. and Schmieder B. (eds), APS Conference Series, Vol 150, 463

Zhang, H. Q., Bao, X. M., Zhang, Y., Liu, J. H., Bao, S. D., Deng, Y. Y., *et al.* 2003, Chinese Journal of Astronomy and Astrophys., Vol 3, N6, 491

Zhang, J., Dere, K. P., Howard, R. A., Kundu, M. R., and White, M. 2001, Astrophys. J., 559, 452

Zhou, Guiping, Wang, J. X., and Cao, Z. L. 2003, Astron. Astrophys., 397, 1057

Discussion

STERLING: Do you think the apparent darkening of filaments before eruption could be due to a slow rise of the filament rather than due to an increase of filament mass?

SCHMIEDER: Yes, it could be motion combined with increased turbulence. I think that turbulence is more likely than mass increase.

SCHWENN: Assume you had been able to follow the evolution on Oct. 28 in real-time and analyze it fast enough: at which time would you have predicted the big flare to occur? In other words: will your analysis lead to better flare/CME predictions?

SCHMIEDER: I was the PI of the campaign (Oct 20-30, 2003) and I chose this active region on Oct 26 because of its complex topology due to continuous emerging flux. This is a necessary condition to get flares and CMEs. Of course it is not sufficient to know at what time these events will occur.

KOUTCHMY: From white light observations we know that an important component of CMEs is the cavity part, including the magnetic field associated with the cavity. Considering quiescent filaments, you described observations bringing even more material around filaments, so what about the cavities?

SCHMIEDER: The cavities observed around filaments in coronal lines ($\lambda < 912$ Å) could contain cool material not observed in Hα but optically thick enough to get absorption of the coronal emission by Lyman continuum.

Coronal and Stellar Mass Ejections
Proceedings IAU Symposium No. 226, 2005
K. P. Dere, J. Wang & Y. Yan, eds.

Formation of Non-Potential Magnetic Field and Flare-CMEs Eruption

Hongqi Zhang

National Astronomical Observatories, Chinese Academy of Sciences, Beijing 100012, China
email:hzhang@bao.ac.cn

Abstract. We analyze the formation process of delta configuration in some well-known super active regions based on the photospheric vector magnetogram observations. It is found that the magnetic field in the initial developing stage of some delta active regions shows the potential-like configuration in the solar atmosphere, the magnetic shear develops mainly near the magnetic neutral line with the magnetic islands of opposite polarities, and the large-scale photospheric twisted field forms late gradually. Some results are obtained: (1) The analysis of magnetic writhe of whole active regions cannot be limited in the strong field of sunspots, because the contribution of the fraction of decayed magnetic field is non-negligible. (2) The magnetic model of kink magnetic ropes, proposed to be generated in the sub-atmosphere, is not consistent with the evolution of large-scale twisted photospheric transverse magnetic field and the relationship with magnetic shear in some delta active regions completely.

The photospheric current helicity density is a quantity reflected the local twisted magnetic field and relates to the remain of transfered magnetic helicity in the photosphere, even if the mean current helicity density brings the general chiral property in a layer of solar active regions. As the emergence of new magnetic flux in active regions, the changes of photospheric current helicity density with the injection of magnetic helicity into the corona from the sub-atmosphere can be detected. Because the injective rate of magnetic helicity and photospheric current helicity density contain the different means in the solar atmosphere, the injected magnetic helicity probably is not proportional to its remain (current helicity density) in the photosphere. A evidence is that the rotation of sunspots does not synchronize with the twist of photospheric transverse magnetic field in some active regions (such as, delta active regions) completely, as one believes that the rotation of sunspots reflects the magnetic one and connects with the injection of magnetic helicity. They represent different aspects of magnetic chirality. The synthetical analysis of the observational magnetic helicity parameters actually provides a relative complete picture of magnetic helicity and its transfer in the solar atmosphere.

Keywords. Sun: coronal mass ejections (CMEs), magnetic fields

---

The vector magnetograms in active region NOAA 5354- 5395-5441-5470 in 1989 are shown in Fig. 1, which were observed at the Huairou Solar Observing Station. It is found that a large dipole active region formed in February 1989 and the tilt angle of active region was about $11°$ on Feb. 10 obtained from Solar Geophysical Data. It was a regular magnetic configuration in the early stage of active region. The proceeding magnetic pole of negative polarity broken gradually, as this region appeared again from the eastern limb of the Sun in March. In March it became a typical delta region (Wang et al. 1991; Zhang 1995; Ishii et al. 1998). The main polarity of this region was positive and the magnetic shear of transverse field formed in the vicinity of magnetic neutral line in the eastern side of the magnetic main pole of positive polarity. It is found that the delta configurations only occurred in some evolution stages of the active region and the large-scale twisted magnetic field above the photosphere related to the highly sheared emerging magnetic flux in the active region.

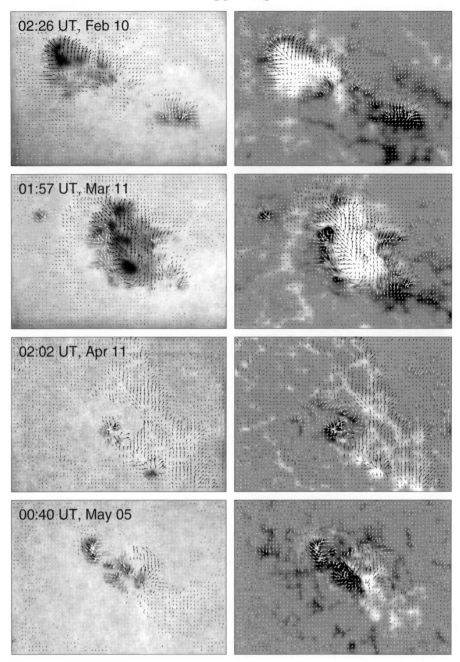

Figure 1. Photospheric images (left) and corresponding vector magnetograms (right) in active region NOAA 5354-5395-5441-5470 in 1989. The white (black) indicates positive (negative) polarity in magnetograms. The arrows indicate the transverse components of field. North is at the top, west is at the right. The size of images is $5.23' \times 3.63'$.

The electric current increased with the formation of magnetic shear in the active region. By investigating the evolution of magnetic field in the active region, it is found that the formation of local strong current in the eastern side of active region on March 11 was caused by the emergence of magnetic flux with opposite arrangement of magnetic

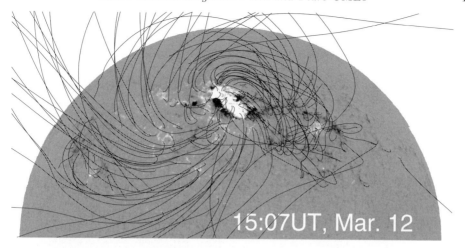

15:07UT, Mar. 12

Figure 2. Spatial configuration of magnetic field extrapolated from linear force free field in the north hemisphere based on the MDI full disk magnetogram on 1989 March 12.

polarities. In this active region flares tended to occur near the highly sheared magnetic neutral line, where the transverse field is almost parallel to it, and small scale magnetic features of opposite polarities moved out on either side of the magnetic main pole of positive polarity and curled around it in curved trajectories in the active region (Wang et al. 1991). This means that the magnetic shear and electric current in the solar atmosphere relate to the emergence of magnetic flux of opposite polarity. The flares near the highly sheared magnetic neutral line provide evidence on the reconnection between the emerging magnetic flux and overlying field. In April and May the magnetic main pole of positive polarity broken and the total flux of the active region decreased gradually, while there was also the left handedness of helical magnetic configuration in the active region dominantly. The α_{best} of force free field is $-1.6 \times 10^{-8} m^{-1}$ on Feb. 10, $-4.2 \times 10^{-8} m^{-1}$ on Mar. 11, $-3.9 \times 10^{-10} m^{-1}$ on Apr. 11 and $-3.0 \times 10^{-8} m^{-1}$ on May 5, respectively. It is consistent with that the mean sign of current helicity density h_{cz} in the active is negative. Figure 2 shows the magnetic lines of force in the solar atmosphere extrapolated by the linear force free field with $\alpha = -3.8 \times 10^{-8} m^{-1}$. It is of a similar order of α_{best} inferred from photospheric vector magnetogram on Mar. 11. We can find that the most lines of force connect the magnetic main pole of positive polarity and the enhanced network of negative one in the active region. This means that, as the main pole of active region becomes the enhanced network field, the field of active region also keeps the basic bipolar topology obviously in the higher solar atmosphere, even if the field probably is force free, i.e. the current exists in the atmosphere.

The formation of highly sheared magnetic field can be inferred by a series of photospheric vector magnetograms in delta active region NOAA 9077 (Liu and Zhang, 2001; Zhang, 2002). For comparison of the evolution of the highly sheared vector magnetic field, the soft X-ray configuration is shown in Figure 3, the possible formation process of the highly sheared magnetic field above the photosphere in the active region can be inferred. As one believes that the soft X-ray loops at 01:53:04UT on July 11 showed the potential-like field and connected the photospheric footpoints of opposite polarities (even if the prospective effects is significant), while the post soft X-ray loop of "Bastille Day" powerful flare at 17:09:21UT on July 14 provides the relaxed process of magnetic lines of force above the photosphere. The inverse "sigmoid" configuration formed with the decrease of the magnetic shear of photospheric transverse field near the magnetic neutral

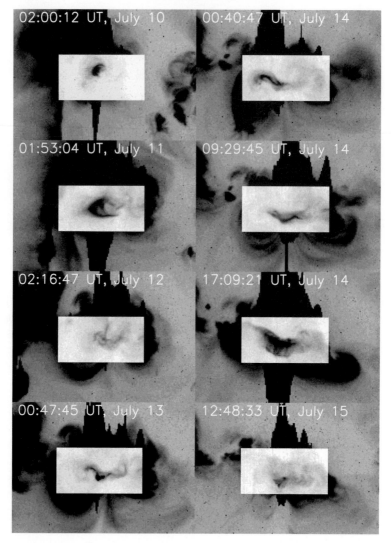

Figure 3. A series of soft X-ray images of active regions NOAA 9077 in 2000 July.

line in the active region on 12 - 14 gradually before the flare. The significant developed inverse "sigmoid" configuration can be found at 00:40:47UT on July 14. It provides an evidence of intense non-potentiality and helicity of magnetic field above the photosphere (Rust and Kumar, 1996; Pevtsov et al., 1997). The mean value of force free field parameter α is about $2.0 \times 10^{-8} m^{-1}$ in the corona, as the estimation of Pevtsov et al. (1997) on the soft X-ray loops has been used. We can conclude that the evolution of photospheric vector magnetic field and corresponding relationship with TRACE 171Å and soft X-ray features in active region NOAA 9077 provide the formation of that the highly sheared magnetic field transfers from the lower solar atmosphere into the corona in the active region before the "Bastille Day" powerful flare-CME (Yuchyshyn et al., 2001).

The proper motion of sunspot features provides some information on the evolution of magnetic field as the magnetic frozen-in state in the photosphere. One normally can inferred the evolution of magnetic field by the sunspot features. The comparison between the accumulation of magnetic helicity above the photosphere inferred from the

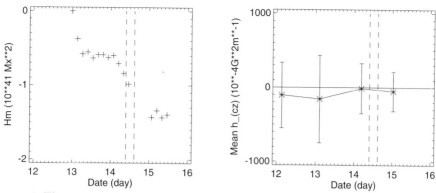

Figure 4. The time profile of the accumulated change of magnetic helicity from the estimated dH/dt (left) and the mean current helicity density h_{cz} (right) in active region NOAA 9077 . The vertical dashed lines mark the begin and end times of "Bastille Day" flare on 2000 July 14.

MDI longitudinal magnetograms by the local correlation tracking techniques and mean photospheric current helicity density h_{cz} in the whole of active region NOAA 9077 is shown in Figure 4. It is found the significant injection of magnetic helicity and the decrease of mean photospheric current helicity density before the "Bastille Day" flare in the active region. It was demonstrated by Zhang (2002) that the high intense current helicity density in the photosphere formed in the vicinity of magnetic neutral line in the active region, on July 12 and 13, and the "Bastille Day" flare-CME on 2000 July 14 occurred after the decay of current helicity density in the photosphere. It is consistent with the injection process of magnetic helicity discussed above, as a whole of the active region. It reflects that the decrease of positive photospheric current helicity density relates the injection of the magnetic helicity from the sub-atmosphere, as comparing the distribution of current helicity density on July 13 and 14 in active region NOAA 9077.

We briefly summarize the main results on the magnetic helicity in the active regions in the following: (1) The magnetic and current helicity commonly provide some similar information on the handedness of twisted magnetic field. It is found that the current helicity evolves with the magnetic helicity in the emerging flux process in active regions. (2) The injection rate of magnetic helicity from the sub-atmosphere and current helicity density in the photosphere contain some different information on the helical properties of magnetic field. The former reflects the transfer process of magnetic helicity, while later does the message of the helical degree of magnetic field and relates to the remain of injected magnetic helicity in the photosphere only. The non-synchronism between them can be found in their evolution process in active regions.

References

Ishii T., Kurokawa H., Takeuchi T., 1998, *ApJ*, 499, 989
Liu Y., Zhang H., 2001, *A&A*, 372, 1019
Pevtsov A., Canfield C., Zirin H., 1997, *ApJ*, 481, 973
Rust D. and Kumar A., 1996, *ApJ*, 464, L199
Wang H., Tang F., Zirin H., Ai G., 1991, *ApJ*, 343, 489
Yurchyshyn V. B., Wang H., Goode P.R. and Deng Y., 2001, *ApJ*, 563, 381
Zhang H., 1995, *A&AS*, 111, 27
Zhang H., 2002, *MNRAS*, 332, 500

Discussion

SCHMIEDER: Does the flux tube emerged twisted or does it become twisted in the upper atmosphere?

ZHANG: Yes, the magnetic flux probably twists in the upper atmosphere, but it is also hard to get obvious evidence how the twist of the magnetic field in the sub-atmosphere from observations of vector magnetic fields.

KUSANO: Is the measured magnetic helicity sufficient to activate the kink instability?

ZHANG: The measured magnetic helicity above the solar photosphere in some active regions probably can be used to demonstrate the model of magnetic kink instability. While there is still the problem of finding the evidence that the magnetic ropes formed kinked in the sub-atmosphere before they emerge.

Coronal and Stellar Mass Ejections
Proceedings IAU Symposium No. 226, 2005
K. P. Dere, J. Wang & Y. Yan, eds.

Large-Scale Activity Initiated BY Halo CMEs

I. Chertok[1] and V. Grechnev[2]

[1]IZMIRAN, Troitsk, Russia, email: ichertok@izmiran.troitsk.ru

[2]Institute of Solar-Terrestrial Physics, Irkutsk, Russia, email: grechnev@iszf.irk.ru

Abstract. We summarize results of our recent studies of CME-associated EUV dimmings and coronal waves by 'derotated' fixed-difference SOHO/EIT heliograms at 195 Å with 12-min intervals and at 171, 195, 284, 304 Å with 6-h intervals. Correctness of the derotated fixed-difference technique is confirmed by the consideration of the Bastille Day 2000 event. We also demonstrate that long narrow channeled dimmings and anisotropic coronal waves are typical of the complex global solar magnetosphere near the solar cycle maximum. Homology of large-scale dimmings and coronal waves takes place in a series of recurrent eruptive events. Along with dimmings coinciding entirely or partially in all four EIT bands, there exist dimmings that appear different, mainly in the transition-region line of 304 Å and high-temperature coronal line of 284 Å.

Keywords. Sun: coronal mass ejections (CMEs), corona, UV radiation

1. Introduction

Strong reconfiguration of the magnetic fields during coronal mass ejections (CMEs) are known to be accompanied, in particular, by such large-scale phenomena as dimmings and coronal waves (e.g., Thompson *et al.*, 1998; Zarro *et al.*, 1999; Gopalswamy & Thompson, 2000; Webb, 2000; Hudson & Cliver, 2001). Dimmings, or transient coronal holes, are regions of temporary depressions of soft X-ray and EUV emissions formed after a CME near the eruptive center, for example, at the periphery of pre-eruptive sigmoid structures. Analysis of heliograms recorded with the EUV Imaging Telescope (EIT; Delaboudinière *et al.*, 1995) aboard SOHO and with the Soft X-ray Telescope (SXT; Tsuneta *et al.*, 1991) aboard *Yohkoh* showed that dimmings can exist during many hours and cover a large portion of the solar disk. Coronal waves are emitting fronts that are observed fairly often ahead the developing dimmings and propagate from the eruptive center with a speed of several hundreds km/s.

Usually, both the dimmings and coronal waves are studied with so-called running-difference images obtained by consecutive subtracting of each current heliogram from the following one. Such running-difference images emphasize changes of the brightness, location, and configuration of observed structures between two subsequent heliograms, but inevitably result in some artifacts. In this paper, we present some results of our recent studies of dimmings and, to a lesser degree, coronal waves, obtained with SOHO/EIT data by means of the method of 'derotated' fixed-difference images (Chertok & Grechnev, 2003a,b, 2005a,b). Those images are formed in two stages: first, the solar rotation is compensated using three-dimensional rotation of all images ('derotation') to the time of a pre-event base heliogram, and then this heliogram is subtracted from all following ones.

In the next section, the method of data processing is outlined. Then we demonstrate reasonableness of the derotated fixed-difference technique with an example of one of the famous eruptive events. Section 4 deals with anisotropic channeled dimmings characteristic of a complex topology of the global solar magnetosphere near the maximum of the solar activity. Homology of large-scale dimmings and anisotropy of coronal waves are described in Section 5. Section 6 is devoted to dimming manifestations in four SOHO/EIT different-temperature bands of 171, 195, 284, and 304 Å. Some concluding remarks are given in Section 6.

2. Data processing

First, we process all SOHO/EIT images using the IDL† routine EIT_PREP.PRO. This routine performs background subtraction, degridding images, and the flat-field correction. Next, the images are properly oriented, centered, and variations of the exposure time are corrected. We also resize all the images to 512×512 pixels, which is sufficient to study such large-scale features as dimmings, but reduces the time consumption in data processing.

In the 512×512 pixel EIT images with a pixel size of $5''.24$, the solar rotation effect appears to be almost inessential when difference images (as running ones) are obtained by subtracting heliograms taken 12 min apart. However, the solar rotation becomes important when the interval between frames exceeds tens of minutes, as in the case of the fixed-difference images. In particular, the displacement of any feature due to the solar rotation produces in the difference image a false eastern edge of the opposite contrast.

To avoid the appearance of false darkenings and brightenings, before the formation of the difference images, we compensate the rotation of the Sun by means of the 'derotation' of all heliograms to the same selected time, usually to the time of the reference frame before the event. In the software used, the solid rotation is applied to the whole visible solar hemisphere at the photospheric radius. Such a rotation is not perfect, because in the reality the solar rotation is not solid, and 3D structures observed in the EUV range comprise some altitude interval at larger heliocentric distances. However, possible errors due to the solid rotation of the sphere at $1R_\odot$ are small and cannot change noticeably the observed picture of dimmings and other features, especially for events occurring in the central part of the solar disk.

To 'rotate' a solar image around the solar polar axis, we first convert the rectangular sky-plane coordinates of pixels on the visible solar surface into the longitude and latitude. Then we transform the longitudes according to the solar rotation during the required time interval, and convert the longitudes and latitudes to the sky-plane coordinates again. Finally, we transform the solar image to the last coordinates. The coordinate-conversion IDL routines were developed by M. J. Aschwanden.

When displaying the difference images, we limit their brightness with symmetric bipolar thresholds. In this way, show up negative dimmings, which are faint (typically 100–200 counts) with respect to very bright flare regions (some 1000 counts) observed simultaneously in the same frames.

To study dimmings quantitatively, we compute time profiles in a few small regions on the solar disk in the same way as Zarro et al. (1999) did. These light curves allow estimating the depth of the dimmings and show the evolution of their various parts as well as their temporal relation.

3. Reasonableness of derotated fixed-difference images

Correctness of the derotated fixed-difference images in comparison with the running-difference images, especially for dimmings, can be illustrated by the famous Bastille Day 2000 event referring to the paper by Andrews (2001). The author of this paper considers,

† IDL is the trademark of the *Research Systems, Inc.*

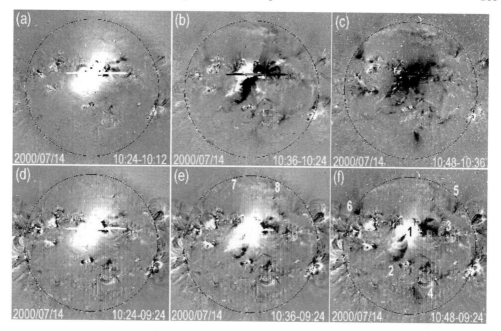

Figure 1. SOHO/EIT 195 Å running-difference (a–c) and derotated fixed-difference (d–f) full-disk images of the Bastille Day 2000 event. The real picture of the dimmings is displayed by the fixed-different images, but not running-difference ones.

in fact, running-difference EIT images at 195 Å similar to those presented in the upper row of Figure 1, but interprets them as fixed-difference images. This brings him to wrong conclusions. In particular, he correctly notes that a large dark region in frame (c) roughly corresponds to the area that was bright at 10:24 UT (frame (a)). However, his conclusion that this dark region is a real dimming, is erroneous. As a matter of fact, the greater central part of this darkening that appears in the running-difference image (c) is due to the intensity decrease of the flare brightening happened after 10:24 UT, and therefore represents a false dimming.

The real picture of dimmings displayed by the fixed-difference images shown in Figure 1 (d–f) is quite different (Chertok & Grechnev, 2005a). Two main pronounced dimmings extend southward and westward of the bright structure above the eruptive center. The south transequatorial J-like dimming 1–2 is observed in conjunction with several emitting pre-event loops which increased their brightness as a result of the flare. The western composite dimming 1–3 covers an area between the eruptive center and remote active region. Besides, there are several additional dimming elements extending particularly to the southwestern region 4 as well as to the northwestern (5) and northeastern (6) limbs. The north emitting structure 7–8 in frame (e) can be identified as a coronal wave front.

4. Channeled dimmings

There were halo CME events in which both coronal waves and dimmings developed more or less symmetrically in all directions from the eruption center as quasi-isotropic disturbances. It has been noted that such events occurred mainly in 1997–1998, i.e., at the ascending phase of the 23rd cycle when the global solar magnetosphere was comparatively simple. For example, one of well-known events of that kind, May 12, 1997, was observed when a single active region was present at the disk (Thompson *et al.*, 1998). On the

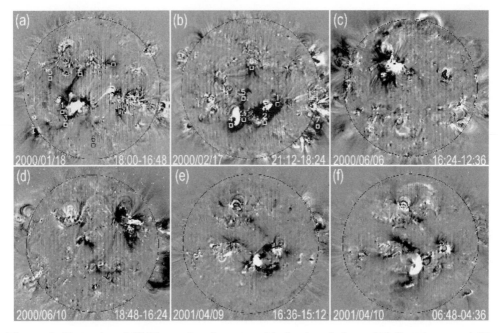

Figure 2. Examples of CME-associated events with developed channeled dimmings as visible at 195 Å SOHO/EIT derotated fixed-difference images.

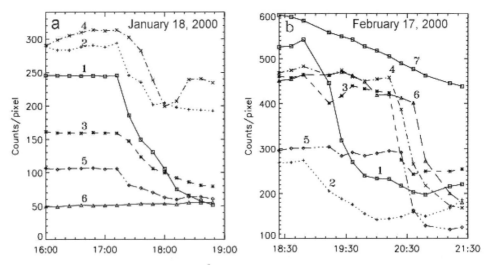

Figure 3. Brightness variations at 195 Å during events of January 18, 2000 and February 17, 2000 for several $53'' \times 53''$ areas labeled correspondingly in Figure 2 (a,b).

contrary, our analysis of halo CME events for 2000–2002 (Chertok & Grechnev, 2003a) revealed that strongly anisotropic dimmings stretching along narrow extended features (channels) are typical of the complex solar magnetosphere at the cycle maximum epoch (cf. Delannée, 2000).

Several examples of such channeled dimmings are given in Figure 2 where derotated fixed-difference images with multi-hour intervals between the reference and current heliograms are shown. One can see that in each event, the channeled dimmings have a global character, with a length exceeding the solar radius; they are transequatorial and extend

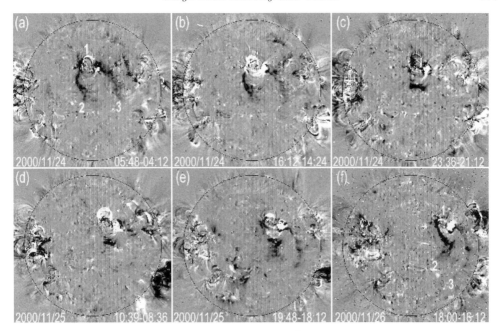

Figure 4. SOHO/EIT fixed-difference images at 195 Å illustrating homology of large-scale dimmings in events of the November 2000 series.

between remote active regions. By their characteristics, such as the intensity depletion and lifetime, the channeled dimmings do not differ significantly from compact dimmings adjoining to eruptive centers. This is illustrated at Figure 3 where the depth and temporal changes in some selected $53'' \times 53''$ areas are presented for two events shown in Figure 1 (a,b). The corresponding areas and time profiles are labeled by the same digits. It follows from these data that the depth of the channeled dimmings reaches several tens per cent, they develop for several tens of minutes, and their lifetimes are, at least, several hours. In channeled-dimming events, coronal waves are also anisotropic, and they propagate within a restricted sector of the disk (see next Section).

5. Homology of dimmings and coronal waves

Two events shown in Figure 2 (e,f) occurred from the same eruptive center with a time interval of about 14 h. They show rather obvious similarity in the location and topology of the large-scale channeled dimmings. Such homology of CME-associated disturbances appears to be a general property of seriate eruptive events.

Chertok *et al.* (2004a) studied large-scale activity on the solar disk associated with a November 24–26, 2000 series of five X-class and one M3.5-class recurrent flares and halo CMEs. The analysis showed that those eruptive events were homologous not only by their flare and CME characteristics, as Nitta and Hudson (2001) described, but also in terms of their large-scale CME-associated manifestations. SOHO/EIT 195 Å derotated fixed-difference images (Figure 4) show that two channeled transequatorial dimmings 1–2 and 1–3, extending southward of the eruptive center, predominated in practically all events of this series. These two dimmings appear to outline footpoints of a coronal arcade observed above an Hα filament, with the eastern dimming 1–2 stretching along the western boundary of a narrow preexisting coronal hole (CH).

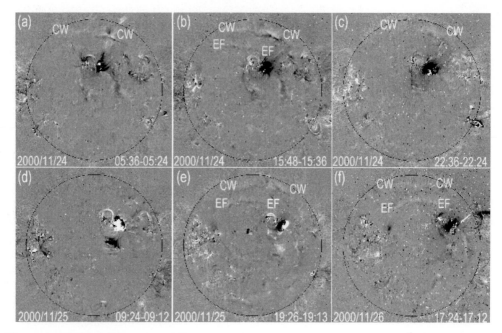

Figure 5. SOHO/EIT running-difference images with 12-min intervals at 195 Å showing homologous coronal waves (CW) and quasi-stationary emitting fronts (EF) in events of the November 2000 series.

Large-scale homology of the events of this series was manifest in accompanying coronal waves. Because they are relatively weak, we use for the illustration running-difference images keeping in mind the above remarks. As one can see from Figure 5 (and especially with corresponding movies), all of the events (except perhaps event d) were accompanied by similar coronal waves (CW). In these running-difference images, coronal waves are visible as bright fronts, and dark fronts behind them mark their locations in preceding images. The main property of these disturbances is that in the different events they propagated anisotropically inside the same restricted angular sector. It is essential that, whereas the main dimmings described above developed southward of the eruptive center 1, all observed coronal waves propagated mainly in the northern direction. Moreover, the second and also similar emitting fronts (EF) were observed, at least, in events b, e, d, and f. A peculiarity of these emitting fronts also seen with movies is that they almost do not propagate and, instead, occupy more or less stationary positions on the disk (cf. Delannée, 2000).

Note that remarkable large-scale homology, especially for dimming patterns, also took place in a series of extremely powerful flares and large CMEs associated with a complex of three active regions in October–November 2003 (Chertok & Grechnev, 2005b).

6. Dimmings in four different-temperature bands

In the standard mode of SOHO/EIT observations, full-disk heliograms are produced almost every 12 min at 195 Å, but only 4 times per day near 01, 07, 13, and 19 UT at 171, 284, and 304 Å. Let us remind that three coronal channels of 171, 195, and 284 Å with predominating lines FeIX/X, FeXII, and FeXV are sensitive to plasma temperatures of 1.2, 1.5, and 2.0 MK, respectively, while the 304 Å filter passes the transition-region line HeII (0.02–0.08 MK) and much weaker coronal component SiXI (1.6 MK) (Delaboudinière

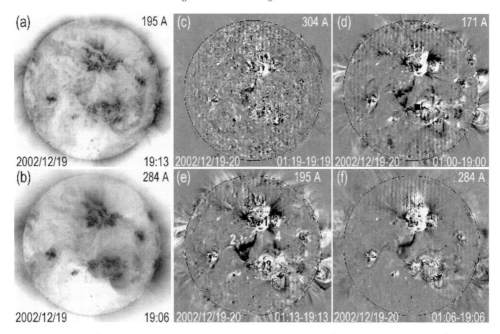

Figure 6. SOHO/EIT images related to the eruptive event of December 19, 2002: (a,b) Negative pre-event heliograms at 195 and 284 Å; (c–f) Derotated difference images with a 6-hour interval illustrating dimmings in four EIT bands.

et al., 1995). Due to the derotation procedure, we are able to form difference images with 6-h or 12-h intervals in all four bands simultaneously, and therefore to study dimming manifestations in different-temperature lines (Chertok & Grechnev, 2003b).

The eruptive event of December 19, 2002 (Figure 6) represents an example of CME-initiated global dimmings, some of which are almost identical in all three coronal lines, with a pronounced dimming manifestation also in the transition-region line. The eastern transequatorial dimming 1–2–3 extending from the northern eruptive center to the remote southern active region coincides at 171, 195, 284 Å and is clearly visible at 304 Å. At the same time, the western fragmentary transequatorial dimming is very similar in two moderate-temperature coronal lines 171 and 195 Å, but is practically absent in high-temperature coronal line 284 Å as well as in the transition-region line 304 Å.

In June 10, 2000 CME event (Figure 7), again, very similar large-scale channeled dimmings extending through the disk are observed in both the moderate-temperature coronal lines 171 and 195 Å. One transequatorial dimming structure 1–2 stretches from northeastern outskirts of the eruptive center in the southwestern direction. The second similar structure is formed by dimmings located along the transequatorial line 3–4–5 interconnecting remote active regions. The most remarkable feature of this event is that both the transequatorial dimming structures visible at 171 and 195 Å practically are not manifest in the high-temperature coronal line 284 Å. The dimming manifestations are comparatively poor also in the transition-region line 304 Å. Only some small-area analogs of the coronal dimmings, located near the eruptive center 1 and partly near other involved active regions, are distinguished on the disk in this low-temperature line. A sole exception is the northern high-latitude dimming ribbon 6–7 also visible at two moderate-temperature coronal lines of 171 and 195 Å.

During the April 29, 1998 event (Figure 8), the eruptive center 1 was, in fact, the only active region on the disk, but it was adjacent to a large CH that occupied nearly the

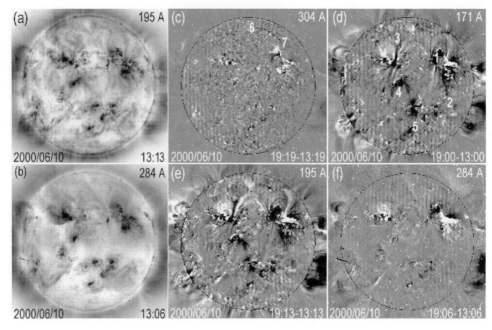

Figure 7. Same as in Figure 6 for the CME event of June 10, 2000.

whole central zone of the disk. At 195 Å, the main dimmings developed in the direction of the northeastern limb: the deepest transequatorial dimming 1–2 formed in the immediate vicinity of CH extends toward northern region and joins with the eastern meridional dimming 3–4. As usually, similar overall dimming structure is also observed at 171 Å. One of the minor differences from the 195 Å pattern is that dimming 1–2 is somewhat broadened in the longitudinal direction in the immediate vicinity of the eruption center 1, and that only individual fragments are visible at the place of the eastern meridional dimming 3–4. The main transequatorial dimming 1–2 is also present in the high-temperature 284 Å coronal line. A remarkable peculiarity of this event is that a large dark area was observed in 304 Å band within the pre-event CH without any counterparts in coronal lines. This pronounced dimming 1–5–6 starts from the eastern side of the eruptive center, covers an extensive portion of CH stretching westward to point 5, then bends and extends southeastward to point 6 located near the southern boundary of CH. It is reasonable to suppose that the two-component structure of the CME observed in this case was connected with the described dimming pattern: perhaps the bright northeastern CME component corresponds to the coronal dimming, while the dimming-like feature visible in 304 Å band corresponds to the southwestern CME component.

7. Concluding remarks

Thus, due to the derotation technique, we are able to form fixed-difference EUV images with diverse time intervals between the base and current heliograms and in different-temperature lines. This allows us obtaining real pictures of long-living dimmings and detecting some interesting features of CME-associated large-scale disturbances.

The analysis of derotated fixed-difference images has revealed, in particular, that when the global solar magnetosphere is complex, which is typical of the solar cycle maximum, dimmings are often anisotropic and spread along narrow extended channel-like structures. In halo CME events, these channeled dimmings often have a global character, and

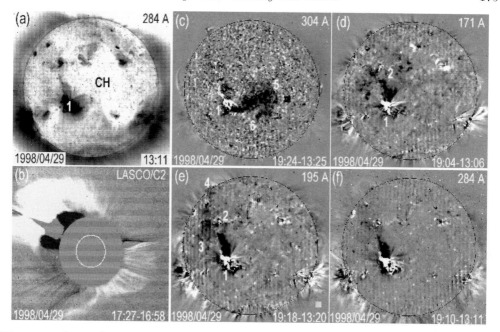

Figure 8. SOHO/EIT & LASCO images related to the eruptive event of April 29, 1998: (a) Negative pre-event heliogram at 284 Å showing a large central CH; (b) Two-component CME; (c–f) Derotated difference images with a 6-hour interval illustrating dimmings in four EIT bands.

can embrace almost entire visible solar disk extending between remote active regions located on different sides from the helioequator. The EUV intensity depletions in the channeled dimmings are comparable to or somewhat less than those in the isotropic and patchy dimmings. Herewith, coronal waves are also observed as anisotropic disturbances propagating within a restricted angular sector.

Large-scale homology of dimmings and coronal waves in repetitive events originating recurrently from the same eruptive center has been found. Dimming homology means that the same large-scale structures, exposed to a strong CME-caused restructuring, can restore their plasma characteristics and emissivity at a time scale of 10–15 h. As for observed coronal wave homology and anisotropy, they seem be determined by the structure of the global solar magnetosphere and to confirm that the corresponding disturbances tend to avoid strong magnetic fields, particularly active regions (e.g., Ofman and Thompson, 2002).

The analysis of dimmings in four SOHO/EIT channels based on derotated fixed-difference images taken at intervals of 6 h and sometimes of 12 h indicates that dimmings are normally strongly pronounced and have similar large-scale structures in the moderate-temperature 171 and 195 Å coronal lines. The dimmings can be also visible in the higher-temperature 284 Å line, although there are cases when some coronal dimmings, particularly similar to transequatorial interconnecting loop structures (see Khan and Hudson, 2000), are not pronounced in this line. In addition, some coronal dimmings are visible in the transition-region line 304 Å, especially those adjacent to the eruptive center. Sometimes 304 Å dimming-like features are observed without coronal counterparts.

The coinciding dimmings in three coronal lines or sometimes in all four spectral bands support their interpretation as a result of full or partial opening of magnetic field lines in some areas and structures in the course of CME processes accompanied by evacuation of plasmas from those structures (e.g., Thompson *et al.*, 1998; Zarro *et al.*, 1999; Harrison *et al.*, 2003; Harra & Sterling, 2001). At the same time, different appearance of dimmings, mainly, in the transition-region line 304 Å and in the highest-temperature coronal line 284 Å, suggests that temperature variations can also play an important role in the formation of some dimming structures.

The fact that dimmings, both compact and extended, are also observed sometimes at 304 Å shows that the transition region plasma is also involved in CME processes. Transition-region dimmings result from the opening of the magnetic fields and plasma outflow from originally closed, low-lying structures, i.e. a transient 'hole' is formed down to the transition region. One cannot also rule out occultation effects in absorbing material of erupting filaments for short-term and moving dimmings (Slemzin *et al.*, 2005). Such an effect is likely responsible for the 304 Å dimming observed without coronal counterparts.

On the other hand, long evolution times of many dimmings along with gradual fade of emitting structures sometimes observed suggest the possibility of other mechanisms, e.g., CME-caused suppression of either heating of coronal structures, or mass supply into them (Uralov & Grechnev, 2004).

The outlined technique of derotated fixed-difference images has also been applied to the analysis of EUV data from the SPIRIT telescopes aboard the CORONAS-F spacecraft (Chertok *et al.*, 2004b; Slemzin *et al.*, 2004).

Various difference images and corresponding movies for many CME events, including ones described in this paper, are presented at the Web site

http://helios.izmiran.troitsk.ru/lars/Chertok/.

Acknowledgements

We are grateful to SOHO/EIT team members for data used in this research. SOHO is a project of international cooperation between ESA and NASA. This work is supported by the Russian Foundation of Basic Research (grants 03-02-16049, 03-02-16591) and the Ministry of Education and Science (grants 447.2003.2, 1445.2003.2).

References

Andrews, M. D. 2001, *Solar Phys.*, 204, 181
Chertok, I. M. & Grechnev, V. V. 2003a, *Astron. Reports* 47, 139
Chertok, I. M. & Grechnev, V. V. 2003b, *Astron. Reports* 47, 934
Chertok, I. M. & Grechnev, V. V. 2005a, *Solar Phys.*, accepted
Chertok, I. M. & Grechnev, V. V. 2005b, *Astron. Reports* 49, 155
Chertok, I. M., Grechnev, V. V., Hudson, H. S., & Nitta, N. V. 2004a, *J. Geophys. Res.* 109, doi:10.1029/2003JA010182
Chertok, I. M., Slemzin, V. A., Kuzin, S. V. et al. 2004b, *Astron. Reports* 48, 407
Delaboudinière, J.-P., Artzner, G. E., Brunaud, J. et al. 1995, *Solar Phys.* 162, 291
Delannée, C. 2000, *Astrophys. J.* 545, 512
Gopalswamy, N. & Thompson, B. J. 2000, *J. Atmos. Sol-Terr. Phys.* 62, 1457
Harrison, R. A., Bryans, P., Simnett, G. M. et al. 2003 *Astron & Astrophys.* 400, 1071
Harra, L. K., Sterling, A. C. et al. 2001 *Astrophys. J. Lett.* 561, 216
Hudson, H. S. & Cliver, E. W. 2001 *J. Geophys. Res.* 106, 25199
Khan, J. I. & Hudson, H.S. 2000, *Geophys. Res. Lett.* 27, 1083
Nitta, N. V. & Hudson, H. S. 2001, *Geophys. Res. Lett.* 28, 3801
Ofman, L. & Thompson, B. J. 2002, *Astrophys. J.* 574, 440
Slemzin, V., Zhitnik, I. A., Ignat'ev, A.P. et al. 2005, this volume, 119
Thompson, B. J., Plunkett, S. P., Gurman, J. B. et al. 1998 *Geophys. Res. Lett.* **25**, 2465–2468.

Tsuneta, S., Acton, L., Bruner, M. 1991, *Solar Phys.* 136, 37
Uralov, A. M. & Grechnev, V. V., Hudson, H. S. 2005, *J. Geophys. Res.*, accepted
Webb, D. F. 2000, *J. Atmos. Sol-Terr. Phys.* 62, 1415
Zarro, D. M., Sterling, A. C., Thompson, B.J. et al. 1999 *Astrophys. J.* 520, L139

Discussion

GOPALSWAMY: You mentioned several possible mechanisms of dimming. Especially, there were two possibilities (wave process and plasma out flow) now debated. In my opinion, a CME process contains both. The plasma outflow is the CME and the wave part surrounds the CME.

GRECHNEV: I listed various possible mechanisms to explain different darkening phenomena. It is possible that the magnetic configuration opens, and then some part is taken away by CME, and some by waves. However, 1) long-term deepening of dimmings with impulsive launch of CMEs and 2) observations of fading closed structures suggest also suppression of either heating, or mass supply into closed coronal structures or both. So different mechanisms can contribute to dimming phenomena.

NINDOS: How did you correlate the EIT channeled dimmings with the large-scale patchy channeled features sometimes observed at 17 GHz with Nobeyama Radioheliograph?

GRECHNEV: The bright chains in the microwave emission were really found by the first author of this talk, I. Chertok, before the finding channeled dimmings. However, we have not yet studied associations of channeled dimmings with other extended structure, but we want to do it. The only thing I can state now is that dimmings appear along features between active regions, sometimes along transequatorial loops, and often embrace almost all the visible solar disk.

SCHMIEDER: 1. Comments: The anisotropy of the EIT waves during the maximum solar cycle could be explained by the presence of active regions and inversion magnetic field lines.
2. The dimming observed in 304 Å can be due to the absorption of the EUV emissions. At this wavelength the optical thickness of cool material is larger than the optical thickness in Hα. This implies that cool material is not visible in Hα but could be efficient for the absorption mechanism by He Lyman continuum.

GRECHNEV: 1. Yes, it is definitely connected with the structure of the global solar magnetosphere, as a said.
2. I just put it No.1 in my Concluding Remarks: occultation by absorbing material. This is definitely caused by some absorption, but it is a question what kind of absorption works: non-resonant or resonant. Of course it is an interesting question to study.

ZHUKOV: 1. The first comment is about your statement that dimmings could be due to Doppler effect. EIT is not a spectrometer, its bandpasses are wide and contain a lot of spectral lines, so I don't think this explanation is plausible.
2. I think that the difference of dimming patterns in coronal and transition region bandpasses is caused by the fact that you use low-cadence (6 hours) observations in 304 Å bandpass.

GRECHNEV: 1. I only mentioned the possible interpretation that was proposed.

2. No, there is another reason, I think. The dimming in the 304 Å band is likely due to absorption in remnants of an ejected filament, but this absorption is not so efficient in coronal emission lines (171 and 195 Å).

KOUTCHMY: When de-rotating your images from a time sequence, did you assume a rigid rotation and what about using some "differential" rotation?

GRECHNEV: Yes, we apply solid rotation. The differential rotation is hardly important for intervals of several hours. Another related question is that we rotate the sphere of a fixed radius, but the coronal structures have some extent in altitude. However, according to our estimation, both these circumstances do not significantly affect such relatively long-scale features, as dimmings.

DELABOUDINIERE: Dimmings in 304 Å are due to occultation by filament ejected material and are totally different from coronal dimmings (not at the same location!!).

GRECHNEV: Yes, this is just the first point from my slide "Nature of dimmings": Occultation by absorbing material. I list various possible reasons for various darkening effects.

JINGXIU WANG: After the careful data deduction, your results seem to not support the wave interpretation for dimming? Is this true?

GRECHNEV: No, waves definitely are present in several eruptive events. We were against mixing methodical artifacts with waves and dimmings which really exist, and their real structure with that one which appears in running-difference images, which are actually the temporal derivative of the structure observed.

Coronal and Stellar Mass Ejections
Proceedings IAU Symposium No. 226, 2005
K. P. Dere, J. Wang & Y. Yan, eds.

© 2005 International Astronomical Union
doi:10.1017/S1743921305000451

The Evolution of Photospheric Source Regions of CMEs

K. Muglach[1] and K. Dere[2]

[1]Artep, Inc., Naval Research Laboratory, Washington, DC, USA
email: muglach@nrl.navy.mil

[2]George Mason University, Fairfax, VA, USA
email:kdere@gmu.edu

Abstract. In this presentation we determine the source regions of CMEs that were observed with SoHO/LASCO during times of solar activity maximum (Feb./Mar. 2000) and during the declining phase of the solar cycle (Nov./Dec. 2002). The CMEs were traced back onto the disk and EIT EUV images were used for identifying the sources. With the help of MDI synoptic magnetograms we follow the evolution of the photospheric magnetic flux about 24h before and 12h after the event. We find that about 87% of the identified CME source regions show small–scale flux changes before the event, usually flux emergence and/or flux disappearance. In 13% of the cases we find no signature of photospheric flux changes.

Keywords. Sun: magnetic fields, Sun: photosphere, Sun: coronal mass ejections (CMEs)

1. Introduction

The solar magnetic field is thought to play a fundamental role in the process that leads to coronal mass ejections (CMEs). Ideally one would wish to know the complete large–scale 3d structure of the magnetic field involved. Coronal magnetic fields can not be directly measured at the present time although a first successful attempt has recently been carried out by Lin et al. 2004. Coronal fields are usually indirectly inferred e.g. from an extrapolation of the field at the solar surface. In this contribution we study the evolution of the photospheric magnetic field in active regions (ARs) and around filaments that produce CMEs. These photospheric fields represent the lower boundary of the 3d structure. Field changes like flux emergence, submergence, reconnection or shearing have been found associated with various solar eruptive phenomena like flares or disappearing filaments. The current work is the continuation of an earlier analysis by Subramanian & Dere (2001).

2. Observations

We examined data of two months of CME events in 2000 (Feb./Mar.) and 2002 (Nov./Dec.) during the peak and the declining phase of the solar cycle. We make use of several SoHO instruments: the LASCO/C2 coronograph, EIT (195 Å coronal images) and MDI (longitudinal magnetograms). Event and source detection followed the steps outlined below. We selected events where we could unambiguously determine the source region and study the short–term photospheric magnetic activity related to the CME. E.g. multiple sources were rejected, when several ARs showed coronal signatures at the projected time and location. Sources very near the limb were also rejected as the line–of–sight component of the magnetic field is reduced near the limb and weak field changes

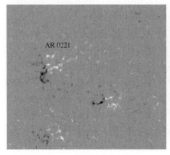 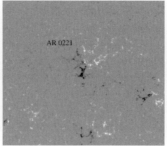 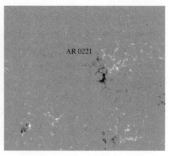

Figure 1. event A, magnetogram taken on 12th Dec. 2002, 16:00UT.

Figure 2. event A, magnetogram taken at 17:41UT, event onset in EIT at 17:36UT, magnetic flux of both polarities disappears in the northern part of AR 0221.

Figure 3. event A, magnetogram taken on 13th Dec. 2002, 04:53UT.

can not be detected (e.g. in filaments channels). In addition flux changes due to solar rotation and due to the evolution of the AR itself can not be disentangled near the limb.

In the first step we used the CME lists and catalog which are available on the LASCO public web-site (http://lasco-www.nrl.navy.mil/cmelist.html). All events in the catalog were checked regardless of their size. Among others the catalog gives position angle and size of the CME and the time of its first signature in C2. We also compared the list with the daily on-line C2 movies to identify coronal outflows as CME signatures. These movies have a temporal resolution of about 20 min.

Then we used coronal images in EIT 195 Å to search for possible source regions. These synoptic full-disk images are taken about every 12 min. Various signatures of activity were used like brightenings, dimmings and EIT waves. We allowed for a time difference of up to 2 h before the first C2 signature and assumed an approximately radial outflow of the plasma. Dark filament channels were also taken into account, usually in addition to checking Hα images from various public web–sites.

Finally MDI line–of–sight magnetograms were used to investigate photospheric flux changes. We used 96 min cadence full–disk synoptic data and studied a period of about 24h before and 12h after the event. Some events had to be rejected as there were not sufficient MDI data available.

3. Results

Out of several hundred CMEs listed in the catalog we ended up with 32 events with an identified source region. 28 (87%) of them showed changes in the magnetic flux. For four events (13%) no flux changes could be detected. Figures 1–9 give three examples to demonstrate the variety of flux changes that are present in the data. The magnetogram in Fig. 1 of event A is taken on 12.Dec. 2002 at 16:00UT. The event onset in EIT was on 13.Dec. 2002 at 17:36UT, the magnetogram in Fig. 2 was taken at 17:41UT. In the northern part of AR 0221 patches of opposite polarity flux move closer and disappear as can be seen in Fig. 3 (flux cancellation). Event B in Figs. 4–6 (AR 8904) is a case of flux emergence: in the southern part of the AR white polarity flux appears at the edge of the black polarity flux of the bipolar region. This flux emergence continues after the CME for many more hours (although it is eventually masked by solar rotation effects) and the AR has two more CMEs in the following 20h. Finally, the flux changes in event

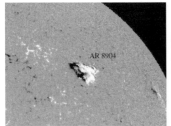 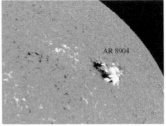 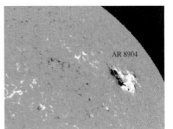

Figure 4. event B, magnetogram taken on 10th Mar. 2000, 23:59UT.

Figure 5. event B, magnetogram taken on 11th Mar. 2000, 20:47UT, event onset in EIT at 21:23UT, white polarity flux emerges in the southern part of AR 9804, flux emergence continues over many hours after the event and the AR shows two more CMEs.

Figure 6. event B, magnetogram taken on 12th Mar. 2000, 09:41UT, flux emergence seems to continue after the first event but the AR gets too near to the limb and solar rotation masks actual flux changes.

C in Figs. 7–9 are rather complex: in the northern part we can see flux emerging (white, Fig. 8) and at the same time the main white polarity structure gets weaker and decays. The decay continues after the CME (Fig. 9).

From the 28 events that showed flux changes we can summarize the following points:

• Flux changes are all small-scale. They can be the size of a small dipole (a few pixels in MDI) up to the changes seen in event B (Figs. 4–6).

• Flux changes happen several hours before the CME onset, some continue after the event as can be seen in the examples given here.

• Flux emergence always seems to happen in an already existing AR. There are several cases of flux emergence in the MDI data where a new AR is formed away from any pre-existing AR. This leads to enhanced coronal activity (as seen in EIT) but not to a CME.

• Flux changes happen right at or very near the onset location as determined by EIT.

4. Discussion

Subramanian & Dere (2001) surveyed 32 CMEs during the rising phase of the solar cycle. They also found that flux emergence and cancellation are associated with the CMEs. The flux changes happen in time–scales of several hours. The initiation of the events happens much faster, probably faster than the cadence of EIT (12 min) which we can also confirm from our current data set. Thus, we consider the flux changes as a trigger of the CME.

Flux emergence in connection with CMEs has been noted before by several authors, like Feynman & Martin (1995), Wang & Sheeley (1999) and Chen & Shibata (2000) for CMEs due to filament eruptions, and Lara et al. (2000) for flaring ARs.

Note that with the line–of–sight magnetograms we can only describe the flux changes that can happen during CME events (e.g. Wang, 2001). To provide a physical interpretation we would need to know the full magnetic field vector, preferably derived from full Stokes spectro–polarimetry. E.g. if flux cancellation is associated with the submergence of a flux tube which connects the two polarities, then this would have clear signatures in the magnetic field vector (e.g. in an increase of the horizontal component of the field). Recent spectro–polarimetric observations of flux cancellation indicate such a scenario (Chae et al. 2004). Also for a proper correction of solar rotation the magnetic field

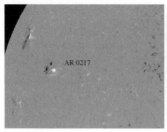

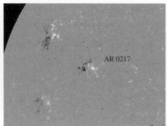

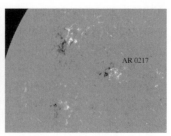

Figure 7. event C, magnetogram taken on 9th Dec. 2002, 12:48UT.

Figure 8. event C, magnetogram taken on 10th Dec. 2002, 16:05UT, event onset in EIT at 15:48UT, flux changes are complex, white polarity flux emerges in the northern part of AR 0217, while the central flux region decays (seen also Fig.9).

Figure 9. event C, magnetogram taken on 11th Dec. 2002, 04:53UT.

vector is essential, as the assumption of a radial magnetic field is often not valid, e.g. in a sunspot penumbra and in case of flux emergence or retraction.

This study would also benefit a lot from better resolution, spatial resolution and in particular temporal resolution. Although the flux changes happen over several hours the 96 min synoptic data cover that period only roughly. In some cases 1 min cadence MDI magnetograms were available and five minute averages produced the best magnetograms but usually they did not cover a period of 24h before the event. The up–coming Solar–B mission will have the ability to provide this kind of data.

Acknowledgements

K.M. acknowledges financial support from the NRL LASCO project The SoHO data used in this work are courtesy of SoHO/LASCO, SoHO/EIT and SoHO/MDI consortium. SoHO is a project of international cooperation between ESA and NASA.

References

Chae, J., Moon, Y–J., & Pevtsov, A.,A., 2004 *ApJ* 602, L65
Chen, P.F., Shibata, K. 2000, *ApJ* 545, 524
Feynman, J., & Martin, S.F. 1995, *JGR* 100, 3355
Lara, A., Gopalswamy, N., & DeForest, C. 2000, *GRL* 27, 1435
Lin, H., Kuhn, J.R., & Coulter, R. 2004 *ApJ* 613, L177
Subramanian, P., & Dere, K.P. 2001, *ApJ* 561, 372
Wang, J. 2001, *SSR* 95, 55
Wang, Y.–M., & Sheeley, N.R. 1999, *ApJ* 510, L157

Discussion

GOPALSWAMY: You have classified AR8904 as a case of flux emergence. However, at the same location where the white polarity emerges, there is black polarity disappearing. So, the classification of this must be "complex" rather than emergence or cancellation.

MUGLACH: You are correct, some of the black polarity flux disappears. But the increase of white polarity flux clearly dominates, so we have classified this event as flux emergence. In event C (complex) some of the white polarity flux seems to emerge next to the black one and the main white polarity flux decays. But it might also be that the flux that seems to dispersed flux from the decaying region. The current 96 min MDI data does not allow one to differentiate between these two possibilities. I also do not want to overemphasize (sub) categories here. We just want to show the variety of flux changes in our sample of source regions.

JUN LIN: Corresponding to these features observed in EIT & MDI, what does the evolution in those observed in Hα look like?

MUGLACH: We have checked Hα data for some events where a filament eruption seemed to be involved. This was primarily done to verify that an Hα filament was involved. A few co-temporal Hα movies confirmed the filament eruption, in some other cases the pre-event filament had disappeared (in images taken several hours before and after the event).

KOUTCHMY: You are talking about changes of the "flux" of the longitudinal component of the magnetic field, not the changes of the magnetic fluxes. Did you look at the possibility to explain your observations by assuming only change of the direction (topology) of the field?

MUGLACH: MDI measures magnetic flux, a combination of $|\vec{B}|$, $\cos\phi$ and α (ϕ = angle between the line of sight and $\vec{B}$, α magnetic filling factor). Flux changes can be due to any one of these three or a combination of them. We avoided events too close to the limb as they are dominated by line-of-sight effects. Part of the observed flux changes are nevertheless probably due to changes in $\cos\phi$: e.g., in event A flux disappears by flux cancellation. If we assume a small loop connecting the two polarities submerges, then this would be an example of flux changes due to topological changes. To determine the physical process involved in the flux changes one needs the vector field $\vec{B}$, ideally derived from spectro-polarimetry which can determine $\vec{B}$ and α.

VOURLIDAS: Flux emergence occurs always on the solar photosphere. How can we be sure that your statistics are not just a coincidence?

MUGLACH: The fact that these flux changes are almost always at the onset location makes me think that they are not random. In the case of flux emergence, a pre-existing large-scale magnetic structure seems also to be necessary. I have seen several cases of flux emergence in the quiet sun (away from active regions) that did not result in a CME (within 24 hours). Thus, flux emergence seems to be a necessary but not sufficient condition for a CME.

NINDOS: Comment: Sometimes flux emergence occurs not only within an old AR but also in quiet sun regions resulting in CME-productive ARs.

MUGLACH: I can only comment on the sample I have currently available, and I did not see this.

SCHWENN: Did you measure the emerging flux quantitatively? MDI allows that?

MUGLACH: No, I did not do any quantitative calculations. To do this, I have to correct for projection effects and solar rotation. Accurate correction can only be carried out when having the vector field available. E.g., the assumption of a radial field that is often used is not valid in a sunspot penumbra and many of the active regions in the sample have a complex flux distribution.

Coronal and Stellar Mass Ejections
Proceedings IAU Symposium No. 226, 2005
K. P. Dere, J. Wang & Y. Yan, eds.

© 2005 International Astronomical Union
doi:10.1017/S1743921305000463

The Evolution of Vector Magnetic Fields and the Origin of Coronal Mass Ejections (CMEs)

Jun Zhang, Guiping Zhou and Jingxiu Wang

[1]National Astronomical Observatories Chinese Academy of Sciences, Beijing 100012, China
email:wjx@ourstar.bao.ac.cn; zhougp@ourstar.bao.ac.cn; zjun@ourstar.bao.ac.cn

Abstract. Coronal mass ejections (CMEs) are intrinsically associated with magnetic structures and evolutions in the solar photosphere. Based on the analysis of vector magnetic field data, we found that: 1, magnetic flux cancellation is the most universal magnetic change in the course of CME onset; 2, new flux emergence also plays an important role in CME origination; 3, interaction and reconnection of flux systems with opposite sign helicity is another key element in the magnetism of CME initiation.

Keywords. Sun: magnetic fields, Sun: coronal mass ejections (CMEs)

1. Introduction

Coronal mass ejections (CMEs) are known as the most spectacular form of solar activity. An equivalent phenomenon has just begun to be identified in the other astrophysical objects. Despite of the significant improvement of coronagraphic and non-coronagraphic (see Hudson & Cliver 2001) observations of CMEs in recent years, the physics of their initiations remain unsettled. The low coronal observations (in X-rays and EUV) are very easily associated with a limb CME, but missing good measurements of surface magnetic field. Magnetic flux erupting in CMEs are believed to connect (at least initially) to the photosphere without question. Therefore CMEs' initiations can not be understood properly without reference to the surface magnetism (Feynman & Martin 1995).

CMEs always are associated with filaments' eruptions. A filament will erupt when new magnetic flux emerges within or adjacent to the unipolar magnetic fields astride the filament in an orientation favorable for reconnection. Flux cancellation in the vicinity of a neutral line is suggested to be a necessary condition for a filament's formation and eventually for its eruption. The association of flares and canceling magnetic fields was first noted by Martin, Livi , & Wang (1985) and was discussed later by Livi et al. (1989) and(Wang, & Shi 1993).

Many CMEs contain helical magnetic structures (Dere et al. 1999), which has been proved by *in situ* observations near 1 AU in magnetic clouds (Burlaga & Behannon 1982). CMEs are thought to originate from an over-accumulation of magnetic helicity and to take off the magnetic helicity from the Sun into the interplanetary space (Rust & Kumar 1994; Low 1996; van Driel-Gesztelyi et al. 1999). Thus, to understand the buildup processes of magnetic helicity in CMEs becomes a central issue in CMEs' studies.

Magnetic helicity quantifies the topology complexity which constrains the minimum energy status of a given flux system during its evolution. Thus, it has been taken as one of the necessary ingredients in flare/CME models. Many other elements, such as the magnetic connectivity, magnetic non-potentiality (vertical currents, magnetic shear), magnetic flux evolution (flux emergence and cancellation), and dynamics (coherent pore and sunspot motion, sunspot rotation), describe other, more or less, independent ingredients of the flare/CME magnetism (Wang, 2002).

2. Magnetic flux cancellation and CME.

Magnetic flux cancellation is described as the mutual flux disappearance in closely-spaced magnetic fields of opposite polarities. The opposite polarity flux involved in the flux cancellation is referred to as a Canceling Magnetic Feature. Reconnection in the lower atmosphere was implied.

On July 14, 2000, a great solar flare with X-ray importance of X5.7 launched near the disk center in active region, NOAA 9077. The flare was accompanied by a giant filament eruption and an extended Earth-directed CME. The arrival at Earth of the massive electrified gas cloud from this CME caused vivid aurora on July 16. Figure 1 shows the general appearance of the active region, NOAA 9077, and the various manifestations of the event. The arrow at Panel 'A' indicates a filament. Seen from the TRACE 1600 Å movie, we found that the brightening first appeared from the region shown by a window at Panel 'B', then the bright material (the arrow at Panel 'B') moved along the channel (shown by a dot curve) of the filament to the right. Several hours before the filament eruption, some bright points (or patches) had already appeared on both sides of the filament. The arrow at Panel 'C' indicates one main bright patch. It enlarged along the filament to form a bright ribbon. This bright patch was also identified in TRACE 195 Å and 1600 Å images (see Panels 'A' and 'B'). Comparing the magnetogram with the Hα filtergram in the figure, we noticed that the bright patch was located at the region (indicated by the arrow at Panel 'D') where a pair of opposite polarity fields was closely contacting and canceling. The eruption of the filament and the onset of the flare were accompanied by an extended halo CME, see the running difference image of LASCO C2 in Panel 'F'.

To illustrate the process of the filament eruption and the flare onset, we present in Figure 2 the time sequence of TRACE 195 Å images. The filament (indicated by the arrow at 09:30:15 UT) was apparently consisted of two twisting threads (indicated by the upper two arrows at 09:46:06 UT), one was thicker and diffusive, and the other, thinner and compact. At an inflection point of the filament, the filament appeared to be bifurcated (see the lower arrow at 09:46:06 UT). At first, the thinner one of the two threads was cut off at the point, indicated by an arrow in the frame of 09:48:17 UT, then the two threads broke off. At the broken point, a bright flare patch (shown by the arrow at 10:09:19 UT) appeared while the filament began to erupt. The filament was torn into two pieces from the broken point. The upper piece seemed to be stable, while the lower one rose from one end closed to the broken point with another end fixed in place. Several minutes later, another flare patch (see the arrow at 10:11:39 UT) appeared near the fixed end of the filament. This flare patch became larger and larger, meanwhile the whole body of the lower piece of filament started to rise and erupt. The erupted piece appeared rotating, and two threads (see arrows at 10:15:03 UT) were clearly seen to untwist during the eruption. Near the maximal phase of the flare, only a small segment of this piece of the filament was remained.

In Figure 3, the straight line at 00:08 UT indicates a piece of magnetic neutral line. The transverse field alignments show that the positive magnetic field (above the line) and the negative field were a single couple of magnetic features. It is identified from the history of flux evolution from MDI and HSOS magnetograms that they represented an emerging flux region (EFR) in this superactive region. The negative magnetic field squeezed upward to the left of the positive magnetic flux. During the 10 hour observations

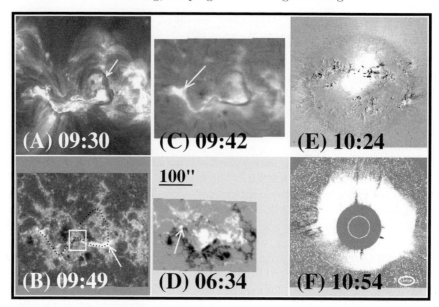

Figure 1. The appearance of the active phenomena in the region NOAA 9077 on July 14, 2000. A: a TRACE 195 Å image; B: a TRACE 1600 Å image; C: an Hα filtergram; D: the corresponding line-of-sight magnetogram at 06:34 UT; E: a running difference of SOHO/EIT image. There is a bright patch of the X5.7 class flare; F: a running difference of LASCO C2 image shown a halo coronal mass ejection (Zhang et al. 2001).

from 23:32 UT July 13, the negative polarity field moved about 7.1×10^3 km (related to the negative polarity field shown by arrow '1' in Figure 3). The mean speed was 0.2 km s^{-1}. The positive polarity field moved downward and canceled with a nearby negative field (indicated by a open square bracket at 01:01 UT) until the disappearance of negative flux. We have also noticed the flux cancellation when some positive magnetic patches (the two arrows at 01:01 and 04:14 UT) slid and intruded into the negative magnetic fields to its south. As a result of the shearing motion of opposite polarities in the EFR and related flux cancellation, the orientation of the magnetic neutral line altered obviously. During the period from 00:08 UT to 08:12 UT, its alignment changed 70° (see the solid and dot lines at 08:12 UT, they represent the neutral lines at 08:12 and 00:08 UT, respectively).

Flux cancellation seen in vector magnetograms hints the magnetic reconnection in the photospheric level. Opposite polarity fields come from independent flux systems Transverse fields change alignment. Zhang et al.(2001) presented the first evidence that the slow magnetic reconnection in the lower atmosphere (in AR9077), which is manifested as observed flux cancellation, is of overwhelming importance in leading to the global instability responsible for the major magnetic activity.

3. New flux emergence and CME origination

Flux cancellation is always accompanied by flux emergence. Emerging flux regions (EFRs) are elementary building bricks of magnetic fields of ARs, and play a central role in explosive activity. Wang (2002) study the magnetic field evolution of AR 8100, the active region associated with four flare-CME events by vector magnetograms taken at Huairou. Some representative magnetograms are shown in Figure 4. At least five EFRs appeared since November 1, and they added large amount of the magnetic flux to the main bipole. EFRs 1 and 2 emerged in the northern periphery of the region. EFR 2

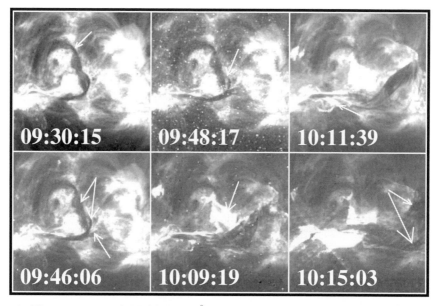

Figure 2. Time sequence of TRACE 195 Å images shown the evolution of the filament. The field of view is about 290 by 290 square arcsec. The arrows in this figure are described in the text. (Zhang et al. 2001)

showed an exceptionally long duration and large separation. Its continuous growth and cancellation with the pre-existing negative flux in the north periphery (indicated by two arrows) were a central evolutionary feature of AR 8100. The flux cancellation was signified by an obvious decrease of negative flux on the eastern side of the magnetic neutral line, and an increase of EFR's positive flux on the western side. Note that the interface of EFR 2 and its impacted negative flux was characterized by the discontinuity in transverse field alignment. This is often seen when two topologically unconnected flux systems interact with each other. Flare-CME associated active regions always show obvious new flux emergence. In the example active region, AR 8100, all the CME-associated flares occurred close to EFR 2 that had exceptionally long duration and large separation.

4. Helicity and CME initiation

Emerging flux regions usually carry magnetic helicity. CMEs are thought to originate from the over accumulation of magnetic helicity (Low, 1996). Recent studies (Chae et al. 2001; Nindos & Zhang, 2002; Demoulin et al. 2002, Green et al. 2002; Moon et al. 2002) revealed the incompetence of AR fields in creating enough helicity for CMEs. CME is a large-scale, or global scale activity. Its magnetic helicity must have been maintained in large or global scale fields.

Wang, Zhou, & Zhang (2004)have tried to examine if particular helicity patterns are retained by CME-associated active regions (ARs). 9 ARs are selected and present complicated helicity patterns. Both helicity signs are seen in each AR as previously reported by Pevtsov, Canfield & Metcalf (1994). For the 5 ARs with definitively identified key EFRs, it is found that the EFRs were born with the helicity of the sign opposite to the dominant helicity sign of the ARs. For the 2 ARs with somewhat uncertain EFRs, the results are the same.

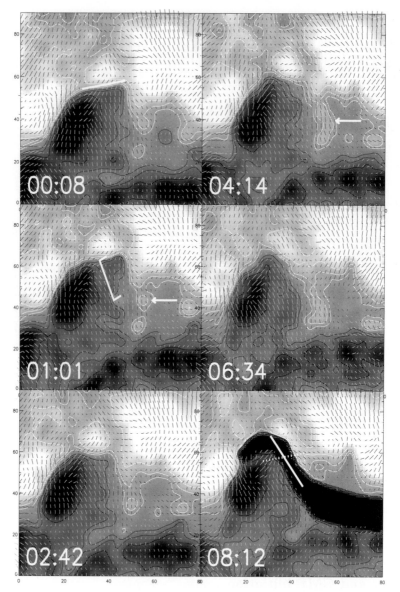

Figure 3. Time sequence of vector magnetograms showing the magnetic field evolution. On x axis, 1 unit = 0.613 arc sec, on y axis, 1 unit = 0.425 arc sec (Zhang et al. 2001).

A famous CME-prolific AR in this solar cycle, AR8100, exhibited exactly particular helicity pattern. Liu et al.(2002) presented a helicity charging picture for this AR. They demonstrated that the magnetic reconnection between EFRs and pre-existing flux played a role in the formation of the sigmoidal structure of the AR. This appears to be contradictory to the findings in this work. Fortunately, very good time sequence of vector magnetograms were available to resolve this apparent contradiction. The EFR that was exemplified by Liu et al. (2002) was a smaller EFR emerged within the main positive sunspot and lasted only one day (see the lower-left panel of Fig.1 and the upper-right panels of Fig.2 of Liu et al. 2002). Wang, Zhou, & Zhang (2004) show this EFR in the upper panel of our Figure 5 by a long arrow. It exhibited marginally the same sign of

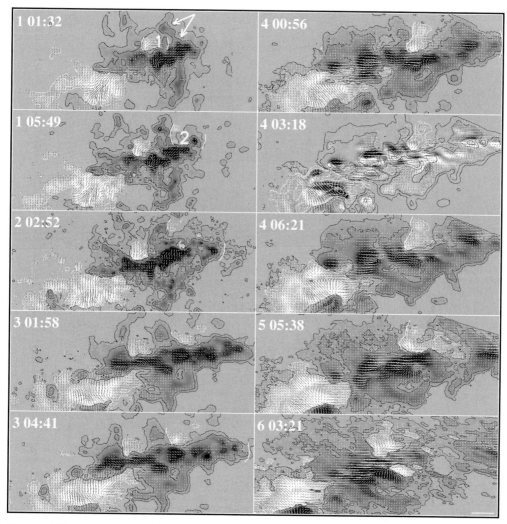

Figure 4. Vector magnetic field evolution of AR8100. Huairou vector magnetograms, which have been de-projected into the heliographic plane, are shown in time sequence. The longitudinal fields are given in gray-scale, i.e., lighter (darker) colors for positive (negative) polarity, and also in contours, whose levels are $\pm 50, 250, 1000, 1500$G. The transverse fields are presented by short line segments with length proportional to field strength and alignment parallel to the field direction. Two EFRs are numbered as '1' and '2' in brackets which embrace opposite polarities of each EFR, respectively. For clear identification, EFR 2 was marked by brackets in several magnetograms. An image of current helicity $B_z J_z$ (at November 4, 03:18 UT) is also shown. The white bar at the right bottom corner indicates the scale of 30 arcsec (Wang 2002).

helicity of the AR, and triggered a sizable, but confined flare. Moreover, it is also recognized that part of the magnetic flux in the main sunspot might come from EFRs which showed dominant negative helicity and played some roles in the formation of the complex topology of AR 8100. However, the key EFR (marked by the bracket in the magnetograms of Figure 5) which triggered explosive flares and homologous CMEs was an extraordinarily large and long-duration EFR. It grew for continuous 6 days. There were 6 homologous flare-CME events which appeared in the immediate interface between the positive flux of this EFR and pre-existing negative flux in the AR (Delannée & Aulanier 1999). The

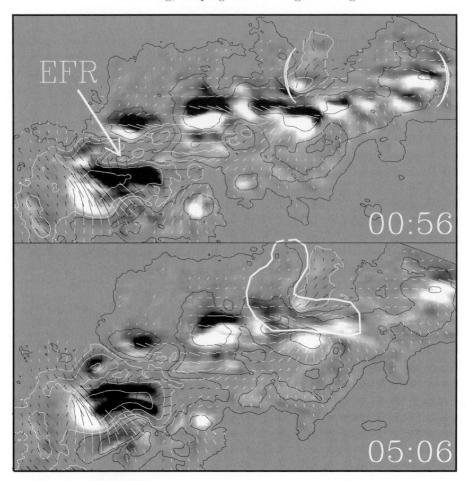

Figure 5. Helicity images of AR8100 scaled in the range of $\pm 20\ A^2 m^{-3}$, superposed by vector magnetograms with contours representing the flux density and short bars indicating the field azimuth. Light(dark) colors refer to positive(negative) helicity. The contour levels are $\pm 100, 500, 1500$ G. A long arrow indicates an EFR that was described by Liu et al.(2002). A bracket marks the key EFR; heavy white contours represents the flare ribbons at 05:52 UT of November 4 (Wang, Zhou, & Zhang 2004).

flare ribbons of the major flare-CMEs that initiated from 05:52 UT is contoured in the lower panel of the figure. This key EFR exhibited predominant positive helicity during the whole flux emergence of 6 days. Thus, although it can not be excluded that some other EFRs may load the same sign helicity to the AR and caused confined flares, for the onset of flare-CME events the key EFR with opposite sign helicity seems to be necessary.

For 9 sampling ARs, contrary to the helicity charging picture, Wang, Zhou, & Zhang (2004) find evidence that the new emerging flux often brings up the helicity with sign opposite to the dominant helicity of the ARs. This support the paradigm that interaction of topology-independent flux systems is a key ingredient in flare/CME magnetisms. Counter-helicity interaction causes the largest amount of magnetic free energy release, while co-helicity interaction results in the highest final energy state of the flux system. This idea is consistent with the 3D MHD simulation by Linton et al. (2001) and the helicity annihilation model by Kusano et al. (2003)

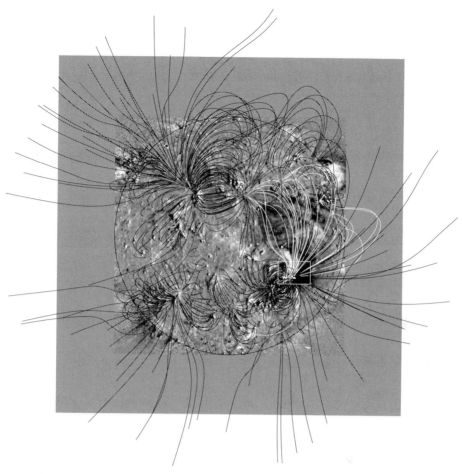

Figure 6. Parent magnetic structure of CMEs. Extrapolated 3-D magnetic lines of force superposed on an EIT difference image (06:15-05:41 UT) that shows EUV dimming of CME 3. This potential field extrapolation uses as the boundary condition an MDI magnetogram in the longitudinal range of -60 to 70 degrees and latitudinal range of -60 to 60 degrees. The vertical component is assumed to be equal to the longitudinal component in the data domain, outside which the flux density is set to zero (Wang 2002).

5. Discussion

In both the flare-associated and the flare/CME-associated active regions, evolving magnetic features are always the same. Why some flare-associated active regions have no CMEs? The magnetic field of AR 8100 as a whole showed a marked dominance of negative polarity. The total flux of the region in the observed field-of-view (see Figure 4) increased from 3×10^{22} to 6×10^{22} during November 1–6, while the net flux varied in the range of $-(0.5 - 1.5) \times 10^{22}$ Mx. There must have been some large-scale magnetic loops connecting this region to other regions dominated by positive flux. This expectation was substantiated by potential field lines as extrapolated (using a boundary element method) on a MDI magnetogram of November 4 in Figure 6. They are plotted on an EIT difference image showing global dimming. The magnetic lines of force with height lower than 0.5 $R_\odot$ and footpoint separation larger than 0.4 $R_\odot$ are highlighted in white. Spreading out from the vicinity of the flare site(see the box in the figure), they almost covered the whole dimming area. Wang (2002) identify these field lines with giant magnetic loops

(GMLs). Similar large-scale interconnecting loops are reproduced by extrapolations on MDI data taken at other times during November 3–6. This leads us to suggest that the pre-CME magnetic skeleton be characterized by the GMLs that were vigorously impacted by smaller-scale magnetic loops (especially EFR 2) in AR 8100, in which the magnetic field was much stronger than that at the other end of the GMLs (see the volume density of magnetic lines of force). The global connectivity manifested by GMLs is also supported by the helicity distribution of AR 8100. The negative flux in the north periphery, canceling with EFR 2, had negative helicity, $B_z J_z$. Moreover, on average, AR 8100 had a negative current helicity (see the bottom part of Figure 5), which indicated a connection with the northern hemisphere (Canfield, Pevtsov, & McClymont 1996). The structure of coronal magnetic fields over the active regions may be predominant! CME-associated coronal magnetic fields exist: 1. Large-scale magnetic connectivity; 2. Global magnetic non-potentiality and complexity, e.g., the total magnetic (current) helicity.

Acknowledgements

The work is supported by the National Natural Science Foundation of China (10233050) and National Key Basic Research Science Foundation (TG2000078404).

References

Burlaga, L. F., & Behannon, K. W. 1982, *Sol. Phys.* 81, 181

Canfield, R.C., Pevtsov, A.A. & McClymont, A.N. 1996, in Bentley, R.D. and Mariska, J.T. (eds.) Magnetic Reconnection in the Solar Atmosphere, *PAS Conf. Series* 111, p.341.

Chae, J., Wang, H., Qiu, J., Goode, P. R., Strous, L., & Yun, H. S. 2001, *ApJ* 560, 476

Delannée, C., & Aulanier, G. 1999, *Sol. Phys.* 190, 107

Démoulin, P., Mandrini, C. H., van Driel-Gesztelyi, L., Thompson, B. J., Plunkett, S., Kövari, Zs., Aulanier, G., & Young, A. 2002b, *A&A* 382, 650

Dere, K. P., Brueckner, G. E., Howard, R. A., Michels, D. J., & Delaboudiniere, J. P. 1999, *ApJ* 516, 465

Feynman, J., & Martin, S. F. 1995, *J. Geo. Res.* 100, 3355

Green, L. M., Lopez Fuentes, M. C., Mandrini, C. H., Démoulin, P., van Driel-Gesztelyi, L., & Culhane, J. L. 2002, *Sol. Phys.* 208, 43

Hudson, H. S., & Cliver, E. W. 2001, *J. Geo. Res.* 106, 25199.

Kusano, K., Yokoyama, T., Maeshiro, T., & Sakurai, T. 2003, *Adv. Space Res.* 32, 1931

Linton, M.G., Dahlburg, R. B., & Antiochos, S. K. 2001, *ApJ* 553, 905

Livi S. H. B., Martin S. F., Wang H, & Ai G. 1989, *Sol. Phys.* 121, 197

Liu Y., Zhao, X. P., Hoeksema, J. T., Scherrer, P. H., Wang, J., & Yan, Y. 2002, *Sol. Phys.* 206, 333

Low, B. C. 1996, *Sol. Phys.* 167, 217

Martin S. F., Livi S. H. B., & Wang J. 1985, *Australian J. Phys.* 38, 929

Moon, Y.-J., Chae, J., Wang, H., Choe, G. S., & Park, Y. D. 2002, *ApJ*, 580, 528

Nindos, A., & Zhang, H. 2002, *ApJ* 573, L133

Rust, D.W., & Kumar, A. 1994, *Sol. Phys.* 155, 69

van Driel-Gesztelyi, L., Mandrini, C. H., Thompson, B., Plunkett, S., Aulanier, G., Démoulin, P., Schmieder, B., & DeForest, C. 1999, *ASP Conf. Ser.* 184, 302

Wang, J., & Shi, Z. 1993, *Sol. Phys.* 143, 119.

Wang, J. 2002, In Understanding Solar Active Phenomena (eds. J.-C. Henoux, C. Fang, & N. Vilmer), *Int. Sci. Publ.* & World Publ. Corp., Beijing, China, p.143

Wang, J., Zhou, G., & Zhang, J. 2004, *ApJ* 615, 1021.

Zhang, J., Wang, J., Deng Y., & Wu, D. 2001, *ApJ* 548, L99

Discussion

JUN LIN: How much does the magnetic flux need to change in order to trigger a typical eruption?

ZHANG: About 5~10% of the total flux of the active region.

SCHMIEDER: How do you relate the low atmosphere reconnection and the flare?

ZHANG: As at the site where flux cancellation takes place, brightening always appears, we consider the flux cancellation as low atmosphere reconnection. Now the direct connection between low atmosphere reconnection and a flare is still one of the unresolved problems.

GOPALSWAMY: You remarked that pre-eruption reconnection takes place at the photospheric level. What is the spatial and temporal relationship of pre-eruption reconnection and the reconnection in the corona associated with actual eruption?

ZHANG: The pre-eruption reconnection and the reconnection in the corona is always co-spatial, but the reconnection in the corona is several hours later than the pre-eruption reconnection.

YOUSEF: When you have two magnetic loops of opposite field direction come in contact that will lead to magnetic annihilation and release of energy and would lead to disappearance of the photospheric magnetic field. So the magnetic annihilation may start high up in corona leading to field disappearance in photosphere. Can you comment on that?

ZHANG: Yes, sometime reconnection in corona leads to field disappearance in photosphere. However, flare onset and other solar activities always appear several hours later than flux cancellation observed in photosphere. This implies that flux cancellation may trigger coronal activities.

KOUTCHMY: Regarding the topology of the reconnection site, especially the case of the Bastille day 2000 flare, can you say something about the behavior of the horizontal component of the magnetic field near the site of apparent reconnection?

ZHANG: The horizontal component of this magnetic field is high shear, meaning almost parallel to the neutral line of the two reconnecting magnetic elements.

SHIBATA: What is the time scale of low atmospheric reconnection leading to global eruption? Have you observed velocity field associated with low atmospheric reconnection? If so, how large is the velocity?

ZHANG: Several tens of hours of low atmospheric reconnection lead to global eruption! We have not observed velocity field associated with low atmospheric reconnection!

Coronal and Stellar Mass Ejections
Proceedings IAU Symposium No. 226, 2005
K. P. Dere, J. Wang & Y. Yan, eds.

© 2005 International Astronomical Union
doi:10.1017/S1743921305000475

The Association of Big Flares and Coronal Mass Ejections: What is the Role of Magnetic Helicity?

A. Nindos[1] and M.D. Andrews[2]

[1]Section of Astrogeophysics, Physics Department, University of Ioannina, Ioannina GR–45110, Greece
email: anindos@cc.uoi.gr

[2] Computational Physics Inc, NRL, Code 7660, Washington DC 20375, USA
email:michael.andrews@nrl.navy.mil

Abstract. In a recent study Andrews found that approximately 40% of M-class flares between 1996 and 1999, classified according to GOES X-ray flux, are not associated with Coronal Mass Ejections (CMEs). Using 133 events from his dataset for which suitable photospheric magnetograms and coronal images were available, we studied the pre-flare coronal helicity of the active regions that produced big flares. The coronal magnetic field of 78 active regions was modeled under the "constant α" linear force-free field assumption. We find that in a statistical sense the pre-flare value of α and coronal helicity of the active regions producing big flares that do not have associated CMEs is smaller than the coronal helicity of those producing CME-associated big flares. A further argument supporting this conclusion is that for the active regions whose coronal magnetic field deviates from the force-free model, the change of the coronal sign of α within an active region is twice more likely to occur when the active region is about to produce a confined flare than a CME-associated flare. Our study indicates that the amount of the stored pre-flare coronal helicity may determine whether a big flare will be eruptive or confined.

Keywords. solar-terrestrial relations – Sun: activity –Sun: flares – Sun: coronal mass ejections –Sun: magnetic fields –Sun: corona

1. Introduction

Recently, Andrews (2003) considered the complete list of the X- and M-class GOES soft X-ray flares observed during the years 1996-1999. He identified possible CME candidates for the 229 flares of his list with good LASCO coverage and concluded that 40% of the M-class flares do not have associated CMEs. The probability of finding a CME candidate did not depend on the solar location of the flare which supports the conclusion that the lack of observed CMEs was not an observational selection effect. In this paper we shall try to understand why some M-class flares do have associated CMEs while other M-class flares do not. Our data set consists of the events studied by Andrews (2003). For such task, one needs to study in detail the properties of the active regions (ARs) which produce the big flares. Here we shall compute the coronal magnetic helicity of the corresponding active regions prior to the flare onset. Our study will demonstrate that the coronal magnetic helicity of the ARs plays an important role concerning the association (or the absence thereof) of big flares with CMEs.

2. Coronal Magnetic Helicity

The magnetic helicity of a field $\mathbf{B}$ within a volume V is defined as $H_m = \int_V \mathbf{A} \cdot \mathbf{B} dV$ where $\mathbf{A}$ is the magnetic vector potential. The magnetic helicity is physically meaningful only when $\mathbf{B}$ is fully contained inside V. When this condition is not satisfied (for example in the solar atmosphere), we define a gauge-independent relative magnetic helicity (hereafter referred to as

Table 1. Active Regions and LFFF Extrapolations

Extrapolation	Active Regions (Flare-CME)	Active Regions (Flare, no CME)	Total Number
Acceptable	47	31	78
Both signs of α	15	25	40
Uniform α sign, large deviation	10	5	15

helicity) of $\mathbf{B}$ with respect to the helicity of a reference field $\mathbf{B_p}$ having the same distribution of vertical magnetic flux on the surface S surrounding V: $H = \int_V \mathbf{A} \cdot \mathbf{B} dV - \int_V \mathbf{A_p} \cdot \mathbf{B_p} dV$. Being a potential field it is a convenient choice for $\mathbf{B_p}$. The quantity $\mathbf{A_p}$ is the corresponding vector potential satisfying $\nabla \cdot \mathbf{A_p} = 0$ and being horizontal on S. Then the term $\int_V \mathbf{A_p} \cdot \mathbf{B_p} dV$ vanishes (Berger 1988).

In an open volume like the solar atmosphere, helicity can change either because of the emergence of new twisted field lines that cross the photospheric surface or/and by shearing motions on the photospheric surface. Such motions include differential rotation and/or transient flows. On the other hand when a CME is launched, it carries away part of the helicity of its source magnetic field. Demoulin et al. (2002) and Green et al. (2002) developed a method to compute the coronal helicity H_c of ARs. A photospheric magnetogram is used as boundary condition for linear force-free field (lfff) magnetic extrapolations ($\nabla \times \mathbf{B} = \alpha \mathbf{B}$ with α being constant over the AR). The extrapolated field lines are fitted with the AR's coronal loops. The value of α giving the best overall fit between the models and observations is adopted for the computation of the coronal helicity. Then one follows Berger (1985) and after linearizing the derived expression in order to avoid helicity enhancements close to the resonance values, the resulting coronal helicity is

$$H_c = 2\alpha \sum_{n_x=1}^{N_x} \sum_{n_y=1}^{N_y} \frac{|\tilde{B}_{n_x,n_y}^2|}{(k_x^2 + k_y^2)^{3/2}} \tag{2.1}$$

where $\tilde{B}_{n_x,n_y}$ is the magnetic field's Fourier amplitude of the (n_x, n_y) harmonic, $k_x = 2\pi n_x/L$, $k_y = 2\pi n_y/L$ with L being the horizontal extension of the computation box used for the force-free field extrapolations.

3. Results

From the 229 flares studied by Andrews (2003), we select those which originate from ARs located within $\pm 50°$ from the central meridian at the time of the flare. For the time interval which starts 1.5 hours prior to the flare start time (as defined in the GOES catalogs) we require the availability of at least one MDI magnetogram and EIT 195 Å images obtained with cadence higher than 25 min. The above selection criteria yield 133 events for further analysis. For each case we use the MDI magnetogram taken 25 min prior to the flare start time as boundary condition for linear force-free field extrapolations. This is possible when 1-min-cadence MDI magnetograms are available. When they are not available, we create a magnetogram for the time we need, taking into account the solar rotation and interpolating the two magnetograms obtained closest to the desired time.

For each event, the extrapolated field lines are fitted with the corresponding AR's EIT coronal loops. We determine the best value of α iteratively following basically the procedure developed by Green et al. (2002). The interested reader is referred to their §2.6 for details. Here, we summarize the technique briefly: (1) We calculate the mean distance d_{mean} between a given EIT loop and the computed field lines resulted from a given α. (2) Through successive steps we select the value of α which gives the lowest d_{mean} for the loop. (3) The same procedure is repeated for all loops appearing in the EIT image. The value of α giving the best overall

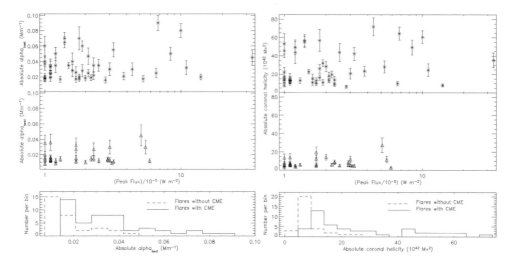

Figure 1. *Left column, top*: Scatter plot of the pre-flare absolute values of α_{best} as a function of the flare's peak X-ray flux for the ARs producing CME-associated flares. *Left column, middle*: Same as top panel, but for the ARs producing flares that do not have associated CMEs. See text for details concerning the error bars. *Left column, bottom*: Histograms of the values of α_{best} appearing in the top and middle panels. The solid line represents the histogram of α_{best} of the ARs which give CME-associated flares while the dashed line is the histogram of α_{best} of the ARs which produce flares that do not have CMEs. *Right column*: The absolute coronal helicity of the 78 ARs appearing in the left panel. The format is identical to the format of the left column.

fit between the models and observations (α_{best}) is the one which minimizes $< d_{mean} >$. The derived α_{best} is considered satisfactory and used in the subsequent analysis if two conditions are met: (1) the derived values of α for individual loops should all have the same sign and (2) $< d_{mean} > \leqslant 1.9$ Mm which is close to the pixel size of the high-resolution EIT images. The above criteria have been implemented because the constant value of α above an active region is a simplification. The values of α_{best} which survive the two criteria are associated with mean deviations that never exceed 25%-30% of the corresponding α_{best}. In Table 1 we give the number of ARs which satisfy both conditions, the number of ARs that do not pass the first condition, and the number of ARs that pass the first condition but do not satisfy the second. In Table 1 we give separately the numbers of ARs producing CME-associated flares and the numbers of ARs producing flares that do not have CMEs.

In fig. 1 (left column) we show the absolute values of α_{best} of the 78 ARs which passed our two conditions as a function of flare's peak flux. Each error bar denotes the mean deviation to the value of α_{best} over the AR. Also in the left column of fig. 1 we give the histograms of α_{best}. A visual inspection of the scatter plots indicates that there is a weak correlation between the values of α_{best} and the corresponding X-ray peak flux; the correlation coefficient is 0.21. Similar small correlation coefficients have been derived between the values of α_{best} and the total X-ray flux and duration of the flares (the corresponding scatter plots are not given for the sake of brevity). The average of all values of α_{best} of fig. 1 is 0.028 ± 0.017 Mm^{-1}. This is about a factor of 4 larger than the average photospheric α_{best} derived by Pevtsov, Canfield & Metcalf (1995) who studied 69 diverse ARs with varying level of flare activity. The large difference between the two studies is due to selection effects: our sample consists of ARs observed a few minutes before powerful flares. In the right column of fig. 1 we give the scatter plots and histograms of the absolute coronal helicity, H_c, using the values of α_{best} and eq. (2.1). The average of all values of H_c is $(19.5 \pm 17.0) \times 10^{42}$ Mx^2.

Fig. 1 shows that several ARs which give big flares without CMEs have smaller values of α_{best} and H_c than those producing CME-associated flares. This result shows better in the histograms and it is statistically significant. We have computed the average α_{best} and H_c separately for the ARs which give flares that do not have CMEs and for the ARs which give CME-associated flares. We find: $< \alpha_{nocme} >= 0.018 \pm 0.010$ Mm^{-1}, $< \alpha_{cme} >= 0.035 \pm 0.018$ Mm^{-1} and $< H_{nocme} >= (8.3 \pm 5.2) \times 10^{42}$ Mx^2, $< H_{cme} >= (26.8 \pm 18.1) \times 10^{42}$ Mx^2. From the scatter plots, and taking into account the error bars, we find that 45% of the events without CME come from ARs with smaller values of α_{best} than the values of α_{best} of each and every AR which gave CME-associated flare. A similar percentage (52%) has been found for the coronal helicities of the ARs without CMEs with respect to the coronal helicities of all ARs producing CME-associated events. The analysis of our results for the $n = 78$ ARs appearing in fig. 1 shows that the ARs with $\alpha_{best} > 0.02$ Mm^{-1} are a factor of 2.25 more likely to produce flare with CME than the ARs with $\alpha_{best} \leqslant 0.02$ Mm^{-1}. We have used the ϕ coefficient for evaluation of statistical significance of the above result. This coefficient is related to chi-square values through $X^2 = n\phi^2$, which can be compared to tabulated chi-square values with one degree of freedom. For our dataset we find $X^2 = 16.4$, which means that the null hypothesis (i.e. that there is no association between the initiation of CME-associated flare and whether the AR's α_{best} is bigger or smaller than 0.02 Mm^{-1}) can be rejected at better than the 99.5% confidence level. By the same measures, the ARs with $H_c > 15 \times 10^{42}$ Mx^2 are a factor of 2.4 more likely to produce flare with CME than the ARs with $H_c \leqslant 15 \times 10^{42}$ Mx^2. Here we obtain $X^2 = 23.2$ and again the null hypothesis can be rejected at better than the 99.5% confidence level.

The EIT images show low-lying, relatively cool loops. In order to prove that our results are accurate, our best-fit magnetic extrapolations should be checked against Yohkoh SXT and TRACE images. From the 78 ARs appearing in fig. 1, 52 of them have been observed simultaneously by EIT and SXT and 7 of them by EIT and TRACE. For these ARs the extrapolated field lines are fitted with the corresponding AR's SXT and TRACE coronal loops. In 80% of the cases, the difference between the derived value of α_{best} and the value of α_{best} derived from EIT is less than $\pm25\%$-30% of EIT's α_{best}. More importantly, the statistical results presented earlier do not change. Furthermore in one event, vector magnetograms from Huairou Solar Observatory are available and the α_{best} derived from the vector magnetogram data is also consistent with the corresponding EIT's α_{best}.

Another aspect of our study is that about 40% of our ARs (see Table 1) show coronal structures that cannot be fitted with a uniform value of α over the AR, indicating that the linear force-free approximation cannot represent their coronal magnetic field satisfactorily. Burnette, Canfield & Pevtsov (2004) have argued in favor of the uniformity of the coronal value of α of the ARs they studied. The difference between the two studies may be due to two reasons: (1) part of our coronal dataset consists of images with better spatial resolution than the full-frame SXT images they used and (2) their dataset was dominated by mature ARs with relatively simple bipolar topologies and areas being either constant or decreasing. It is also interesting that most ARs whose coronal field deviates from the linear force-free approximation show both signs of α within them (see Table 1). Several such cases become obvious simply by visual inspection of the EIT images: for example in some images both S-shaped and reversed S-shaped structures appear. Such structures may correspond to positive and negative sign of α, respectively (e.g. Rust & Kumar). Furthermore, the change of the coronal sign of α is more frequent in ARs producing flares without CMEs than in those ARs producing CME-associated flares: it happens in 41% of the ARs giving flares without CMEs and only in 21% of the ARs giving CME-associated flares. Recently, Kusano et al. (2004) proposed that magnetic reconnection between oppositely sheared loops works as a trigger mechanism of solar flares. In their calculations, however, it is not clear whether the ejected flux escapes into infinity accounting for CME. Their model predicts that the position of flare brightenings should coincide with the magnetic field's shear reversals. The fact that we have used only pre-flare images, makes a direct comparison of our results with their model somewhat difficult.

4. Conclusions and Summary

While there is no doubt that CMEs eject helicity from the Sun, its role in the initiation of transient activity is a subject of hot debate. Some argue (e.g. Antiochos & DeVore 1999) that the global helicity by itself yields little information on coronal evolution while others (e.g. Low 1996) argue that the accumulation of helicity into the corona is at the origin of CMEs. Recently, the theoretical work by Amari et al. (2003) supports that a large enough helicity seems to be a necessary condition for an ejection to occur but not a sufficient one. In this paper we investigated whether the coronal helicity has anything to do with the fact that some big flares are associated with CMEs while other big flares do not have associated CMEs. Our starting point was the dataset of big flares studied by Andrews (2003). From his dataset we selected 133 events for which suitable pre-flare photospheric magnetograms and coronal images were available. Our dataset was in a statistical sense similar to the complete Andrews's (2003) dataset because 46% of the events we analyzed did not have associated CMEs. Our computations yielded 78 ARs whose coronal magnetic field could be approximated satisfactorily under the force-free assumption and subsequently their coronal helicity was computed. From the 78 ARs, 40% produced flares without CMEs.

A key conclusion of our study is that the pre-flare coronal helicity of the ARs producing big flares that do not have CMEs is smaller, in a statistical sense, than the coronal helicity of the ARs producing CME-associated big flares. Overall, our study indicates that the amount of the stored pre-flare coronal helicity may determine whether a big flare will be a confined event (i.e. flare without CME) or an eruptive event (i.e. CME-associated flare). The findings supporting this conclusion are:

- The average values of α_{best} and coronal helicity are 0.035 ± 0.018 Mm^{-1} and $(26.8 \pm 18.1) \times 10^{42}$ Mx^2 for the ARs producing eruptive events but only 0.018 ± 0.010 Mm^{-1} and $(8.3 \pm 5.2) \times 10^{42}$ Mx^2 for the ARs producing confined events.

- About 45%-52% of the ARs producing confined events are associated with values of α_{best} and coronal helicities H_c that are smaller than the values of α_{best} and H_c of all ARs producing eruptive flares.

- ARs with $\alpha_{best} > 0.02$ Mm^{-1} and $H_c > 15 \times 10^{42}$ Mx^2 are likely to produce confined flare with probabilities of only 29% and 16%, respectively.

- In the ARs where the linear force-free model is not acceptable, the change of the coronal sign of α within an AR occurs more often in those ARs producing confined flares (in 41% of them) than in the ARs producing eruptive flares (only in 21% of them). This finding may indicate that the distribution of coronal helicity in CME-productive ARs is more coherent than in ARs giving events that do not have associated CMEs.

Finally, a word of caution is needed. Our study does not necessarily imply that the amount of coronal helicity stored in a pre-flare configuration is the *only* factor which determines whether the flare will be confined or eruptive. A detailed study of the pre-flare magnetic topology is also required in order to settle this issue. However, such analysis was beyond the scope of this paper and it will be carried out in the future.

References

Amari, T., Luciani, J.F., Aly, J.J., Mikic, Z., & Linker, J. 2003, *ApJ*, 595, 1231
Andrews, M.D. 2003, *Sol. Phys.*, 218, 261
Antiochos, S.K. & DeVore, C.R. 1999, *in Magnetic Helicity in Space and Laboratory Plasmas, ed. M.R. Brown, R.C. Canfield, & A.A. Pevtsov, AGU: Geophysical Monograph 111*, 187
Berger, M. A. 1985, *ApJS*, 59, 433
Berger, M. A. 1988, *A&A*, 201, 355
Burnette, A.B., Canfield, R.C., & Pevtsov, A.A. 2004, *ApJ*, 606, 565
Demoulin, P., et al. 2002, *A&A*, 382, 650
Green, L., M., Lopez Fuentes, M. C., Mandrini, C.H., Démoulin, P., van Driel-Gesztelyi, L., & Culhane, J.L. 2002, *Sol. Phys.*, 208, 43
Kusano, K., Maeshiro, T., Yokoyama, T., & Sakurai, T. 2004, *ApJ*, 610, 537
Low, B.C. 1996, *Sol. Phys.*, 167, 217.
Pevtsov, A.A., Canfield, R.C., & Metcalf, T.R. 1995, *ApJ*, 440, L109
Rust, D.M., & Kumar, A. 1996, *ApJ*, 464, L199

Discussion

HAISHENG JI: For a specific active region, did you check the α values and helicity values before and after flare/CME event?

NINDOS: I didn't consider the values of α and helicity after the flare.

ZHUKOV: The appearance (or not) of a flare and/or a CME depends on the availability in a given place of free magnetic energy, which is converted into the radiated energy (flare) and/or kinetic energy (CME). As several previous speakers mentioned, the emergence of new magnetic flux is an important property of flares/CMEs, which may provide this additional free energy. My question is: what is the additional information provided by the helicity balance in comparison to the energy balance consideration?

NINDOS: 1. Helicity, according to the Taylor hypothesis, dissipates much slower than magnetic energy. It is the only quantity which is conserved under reconnection.
2. Sometimes flux emergence contributes to the increase of the absolute value of the AR's helicity. But this is not always the case, as previous speakers demonstrated.
3. I don't imply that helicity is the only parameter controlling whether a flare will be eruptive or confined.
Definitely we need to study helicity and topology together.

KOUTCHMY: You were considering cases of flares which did not produce CMEs. We know that there are also filament eruptive case which do not produce CMEs. The mass loading process and filament formation/eruption is an important aspect of this CME physics. Do you see any relationship between the helicity behavior with respect to the production of CMEs and the filament eruption phenomenon?

NINDOS: I didn't consider the topology of the eruption. I only computed the global pre-flare coronal helicity.

Coronal and Stellar Mass Ejections
Proceedings IAU Symposium No. 226, 2005
K. P. Dere, J. Wang & Y. Yan, eds.

© 2005 International Astronomical Union
doi:10.1017/S1743921305000487

The Large-Scale Source Regions of Coronal Mass Ejections

Guiping Zhou, Jingxiu Wang, Jun Zhang and Chijie Xiao

National Astronomical Observatories, Chinese Academy of Sciences, Beijing 100012, China
email: zhougp@ourstar.bao.ac.cn

Abstract. Using the observations of LASCO aboard SOHO in the interval from Mar. 1997 to Dec. 2003, 301 earth-directed halo CMEs are selected and the source regions are located in MDI synoptic charts. A statistical analysis has been made with the emphasis on the CMEs' large-scale source regions as well as the correlation between CMEs and solar surface activity. The statistics show that CMEs are intrinsically related to surface activity. Four groups of CMEs' large-scale source structures are identified on the photosphere. They are: I, Extended bipole regions (EBRs) with long magnetic neutral line; II, Closely packed active regions (ARs); III, Large-scale magnetic flux of the same polarity runs through the opposite hemisphere, along the boundary there appears transequatorial filaments; IV, Between two EBRs with long filament. The result shows that CME-associated source activity is closely related to the types of large-scale magnetic structures.

Keywords. Sun: coronal mass ejections (CMEs), Sun: activity, Sun: magnetic fields

1. Introduction

CMEs arise in large-scale closed coronal structures. Their average masses and kinetic energies are a few times 10^{15} g and 10^{31} erg. It is driven by the magnetic fields(Webb, 2000). To understand CME initiations and onset mechanisms, one should know the magnetic environment and the characteristics of magnetic evolution, or else meaningful results can never be gotten in the studies of CME mechanisms (Wang et al. 2002).

The studies of CMEs' source regions have progressed very much recently, but the problem is still far from being solved. By far, there still have no clear identification and classification about CME's source regions. Many studies about CMEs' source regions are focused on smaller scale, such as flares or filament eruptions (e.g. Sheeley et al. 1983, Hudson et al. 1995, Subramanian & Dere 2001, J.Lin 2004). However, CMEs, as a large-scale solar activity, their sources may be disproportional to such small-scale activity.

Some case studies about CMEs' sources are also attempted around the destabilization of large-scale, but these are limited in coronal structures at higher altitudes, such as coronal streamer(e.g. Hundhausen, 1993), X-ray loop(Nitta & Akiyama 1999), loop arcades (Chertok, 2001), transequatorial filaments (Wang et al, 2004, submitted) and so on. Up to now, there are no systematic studies established to investigate CMEs' source regions in large-scale magnetic structures at lower latitude. Large-scale magnetic structures are intrinsic components of solar magnetism. Their destabilization, expansion, and eruption into the interplanetary space are the basic processes which lead to the CMEs (Wang et al. 2002). To understand CME magnetism, it seems necessary to investigate the CMEs' source regions from large-scale magnetic structures at lower altitudes.

After the successful mission of SOHO and Yohkoh, the disk observations of CME initiation become possible, particularly for the earth-directed CMEs. To avoid the ambiguity in locating CME source regions, as argued by Feynman & Martin (1995), only

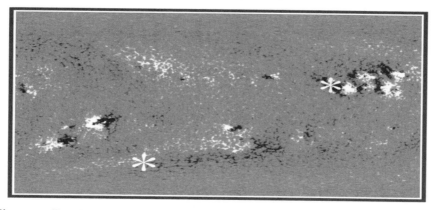

Figure 1. Two halo CMEs' source regions (indicated by asterisks) are located in the corresponding synoptic chart.

earth-directed halo CMEs are selected in this statistics. A CME with span angle greater than 130^0 is referred to as a halo CME in our approach.

In this work, the CMEs' sources are considered as the large-scale magnetic features on the photosphere. The statistics are emphasized on the CMEs' large-scale source regions, as well as the correlation between CMEs and surface activities in terms of flares and filament eruptions as Zhou et al, (2003, paper1 hereafter). The database are enlarged to 301 earth-directed events from 1997 to 2003.

The database are presented in section 2. In section 3, the correlation between CMEs and surface activity are shown. The categories of the large-scale CMEs' source structures are described in section 4. In the last section, the conclusion and discussion are given.

2. Database and source region locations

The primary database for this study is SOHO LASCO and EIT time-lapse observations. All the CMEs are from CME Category (see http://cdaw.gsfc.nasa.gov/CME list/). The GOES X-ray counts and Yohkoh SXT images are used to identify flares. $H\alpha$ filaments are identified from the observations from Big Bear $H\alpha$ filtergrams, Huairou Solar Observing Station (HSOS), Hiraiso Solar Terrestrial Research Center (HSTRC) and Holloman Air Force Base (HAFB).

To identify the earth-directed halo CMEs, two criteria are applied. One is the associated surface activity happen in the time interval (CME's initial time) ± 30 min. The second criterion is that the surface activity's position (PSSA) identified in the EIT images is under the span of the associated CME or just near the span's edge. If both criteria are satisfied, the CME is thought as earth-directed. 302 earth-directed halo CMEs from Mar. 1997 to Dec. 2003 are well identified. The associated surface activity are considered as the CMEs' source regions, which are located in the corresponding MDI synoptic charts as shown in Fig. 1.

In addition, according to the MDI daily magnetograms, the relationship between CMEs and ARs were also checked. If the difference between SRPA and the central position angle or the position angle (for 360^0 halo CMEs) of a CME is not greater than 20^0, we define it as symmetric, or else it is asymmetric.

Table 1. the correlation between front-side halo CMEs (1997.3-2003.12) and the associated surface activity

Assoc. Surface activity	CME Num.	Percent
Flares	270	89%
Filaments	280	93%
ARs	268	86%
Asymmetry	153	51%
Total CMEs	302	

3. Correlation between CMEs and surface activity

Studying correlation of CMEs with other solar activity will help us to understand the physical links between the very large scale activity (e.g. CME) and the rather small scale phenomena (e.g. flares and filament eruptions).

In this study, only two primary forms of associated surface activity, flares and filament eruptions, are taken into account. Some other active manifestations, e.g. EIT wave and/or dimming, are considered as secondary active phenomena. The correlations are shown in table 1. It is found that 89% halo CMEs are associated with flares, while more than 93% are related to filament eruptions. The CMEs are intimately related to the other surface activity, as either flares, or filament eruptions, or both.

In previous work, CMEs were considered to be associated with ARs. Based on SOHO's successful observations, the relationship is checked again. The statistic shows that as many as 86% earth-directed halo CMEs are related to ARs.

In addition, we also check the asymmetry or symmetry between CMEs and associated surface activity. It is found that about 51% of the CME source regions are asymmetric with corresponding CMEs, which are contradictory to present CME models. All the results listed in table 1.

4. Categorize the CMEs' large-scale source regions

After identifying the earth-directed halo CMEs' source regions, it is found that each CME has a large-scale structure counterpart on the photosphere. Such large-scale structures are referred to as the CMEs' sources. They are distinguished by magnetic structures with distinct characteristics. The total 301 earth-directed CMEs are associated with 204 large-scale source regions, categorized as four groups. Usually, there is more than one CME corresponding to one source structure. The magnetic structures of each category are shown in the Fig. 2, in which all the plus signs denote CMEs' source locations. The statistic results are listed in Table 2.

Category I: the CMEs' source locates in one extended bipole region (EBR). EBRs are brought forward as Evolving Magnetic Structures (EMS) by Feynman (1997). It can be identified as a large-scale bipole structure in synoptic charts, whose polarity distributions follow the Hale law. 25% CMEs are associated with this kind of structures. As shown in Fig. 2I, the source of a CME on August 30 1997 located in an EBR in the northern hemisphere.

61% CMEs are associated with Category II that their source regions lie in closely packed ARs, which include more than two ARs. This category is differentiated by that the threshold of the edge magnetic intensity of each case is greater than 100 G, in the range of which the average magnetic flux is greater than 40 Maxwell/cm^2. Fig. 2II shows

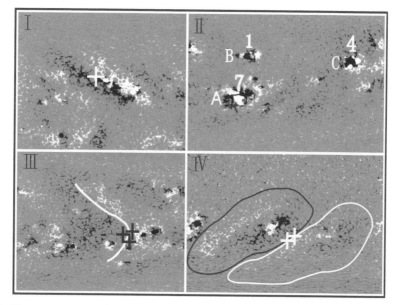

Figure 2. Four categories of CMEs' large-scale magnetic source structures: I, one EBR; II, closely packed ARs; III, single polarity that runs through two hemisphere accompanying by transequatorial filaments; IV, between two EBRs with long filament.

Table 2. statistics of large-scale magnetic structures of CMEs' source regions

Assoc. sour. struc.	Sour. num.	Sour. per.	CME num.	CME per.
I:	59	29%	73	25%
II:	107	52%	183	61%
III:	11	5%	17	6%
IV:	27	13%	29	9%
Total sour.	204			
Total CMEs	302			

three cases (shown as A, B & C) in this category, and 7,1 & 4 are the numbers of associated CMEs.

In Category III, all the CMEs' source regions lie along the single polarity that runs through two hemispheres. Such magnetic structures are always accompanied by transequatorial filaments, whose eruptions are related to CMEs. As indicated in Fig. 2III, along the boundary of the negative polarity straddling northern and southern hemisphere, there is a transequatorial filament denoted by the white curve. 6% CMEs are related to this category.

As indicated by Fig. 2IV, the CMEs' sources located between two EBRs with long filaments. Such magnetic structures are considered as Category IV, whose associated CMEs that occupy 9% are always related to the long filaments' eruptions.

From Table 2, it can be found that the predominant two kinds of magnetic structures are closely packed ARs and EBRs, with which most of CMEs are associated. It doesn't means that the other two kinds of categories are unimportant or illogical, but indicates that the physics behind the fore two categories are more easy to cause CMEs or they appear more frequently in the sun. The reason is that the surface activity related to CMEs can be distinguished from their corresponding magnetic structures on the photosphere.

For Category I, flare and filament eruption often happen together. For Category II, flares are the main activity. Filaments' eruptions with long neutral lines are often related to Category III & IV.

5. conclusion and discussion

Using the observations from LASCO, EIT, GOES X-ray, Hα and synoptic & daily MDI magnetograms, following the former work in paper 1, we examined the relationship between halo CMEs and the surface activity by enlarged sample from Mar. 1997 to Dec. 2003. All the large-scale magnetic structures on the photosphere, as the CMEs' sources, are studied statistically.

The large-scale magnetic structures are categorized into four groups: (I), one EBR; (II), closely packed ARs; (III), the single polarity that runs through two hemispheres, which are always accompanied by transequatorial filaments; (IV), long neutral lines between two EBRs with long filaments. They four present the large-scale characteristics of CMEs' source on the photosphere.

The main results about the correlations of CMEs with surface activity are presented in Table 1. Table 2 lists the details about the categories of the CMEs' large-scale source structures. 301 earth-directed halo CMEs are identified. Their sources locate in 204 large-scale source structures on the photosphere.

The results show that CMEs are intimately linked to surface activity, which behaved as flares or filament eruptions or both. Half of the CMEs are offset to their source that is not consistent with the present CME models. More than 80% of the earth-directed CMEs have the sources inside ARs. However, all of these CMEs can be found large-scale source counterparts on the photosphere. It implies that CMEs' sources are correlated with large-scale structures more than small-scale ones, e.g. AR scale.

In addition, since both of the structures in different scales are closely related to CMEs simultaneously. It suggests there be physical interactions between them. Recently, Zhang et al. (2001) have identified that the interaction between the large-scale structures and active region scale (or small scale) magnetic field in the form of flux cancellation may transport the magnetic energy and complexity into large-scale magnetic loops. This may cause the large-scale structure to be destabilized, partially opened and erupted into interplanetary space, which often appear as CMEs.

The further work of us is to search large-scale phenomena at higher altitude, such as propagating EUV waves, large-scale magnetic loop and so on. The aim is to well understand the CME initiations and onset mechanisms, and finally establish the physics-based prediction models for CMEs.

Acknowledgements

The work is supported by the National Natural Science Foundation of China(10233050) and the National Key Basic Science Foundation(TG2000078404). We are grateful to all members of the SOHO EIT, LASCO and MDI teams, as well as BBSO team, Yohkoh team, HSTRC team, HAFB team and for HSOS station providing the wonderful data. SOHO is a project of international cooperation between the ESA and NASA.

References

I.M. Chertok 2000, *Sol. Phys.* 198, 367
Feynman, J., & Martin, S. F. 1995, *J. Geophys. Res.* 100, 3355

Joan Feynman 1997, in: Nancy Crooker, J. A. Joselyn, J. Feynman (eds.), *Coronal mass ejections*, Evolving Magnetic Structures and Their Relation to Coronal Mass Ejections (Washington, DC : American Geophysical Union), vol. 99, p. 299

Hundhausen, A.J. 1993, *J. Geophys. Res., 98* 13, 177

Hudson, H., Haisch, B. & Strong, K.T. 1995, *J. Geophys. Res.,* 100, 3473

J. Lin 2004, *Sol. Phys.* 219, 169

Liu, Y., Zhao, X.P. & Hoeksema, J.T. 2004, *Sol. Phys.* 219, 39

Nitta, N. & Akiyama, S. 1999, *ApJ* 525, 57

Ograpishvili, N.B. 1998, *Sol. Phys.* 115, 33

Sheeley Jr., N.R., Howard, R.A. *et al.* 1983, *ApJ* 272, 349

Subramanian, P. & Dere, K. P. 2001, *ApJ* 561, 372

Wang, J.X., Zhang J., Deng, Y.Y. *et al.* 2002, *Science in China (Series A)* 45, L57

J.X. Wang, G.P. Zhou, Y.Y. Wen *et al.* 2004, *Ch.J.A.A.* submitted

Webb D. 2002, *J. of Atmospheric and solar-terrestrial phys.,* 62, 1415

Zhou, G.P., Wang, J. & Cao, Z.L. 2003, *A&A* 397, 1057

Zhang J., Wang, J., Deng, Y.Y. *et al.* 2001, *ApJ* 548, L99

Discussion

PENGFEI CHEN: You got a correlation between halo CME speeds and flare flux, while halo CMEs suffer seriously from projection effects. So, we might be cautious about the measurement of halo CME speed.

ZHOU: Yes, you are right, but the results would be meaningful if projection effects were not predominant.

NINDOS: What do you mean by the term "closely packed ARs?"

ZHOU: Some people once referred to such magnetic structures as active region (AR) nests. In our work, the "closely packed ARs" is similar to the "AR nests".

Coronal and Stellar Mass Ejections
Proceedings IAU Symposium No. 226, 2005
K. P. Dere, J. Wang & Y. Yan, eds.
© 2005 International Astronomical Union
doi:10.1017/S1743921305000499

On the Radio Signatures Associated with the Development of Coronal Mass Ejections

Dalmiro Jorge Filipe Maia

Observatorio Astronomico Prof. Manuel de Barros da Faculdade de Ciencias da Faculdade do
Porto Alameda do Monte da Virgem, 4430-148 Vila Nova de Gaia Portugal
email: dmaia@fc.up.pt

Abstract. The methods of radioastronomy are in important observational tool to explore magnetic energy releases in the solar corona. When combined with the useful diagnostics provided by observations in other wavelengths, namely with data from space missions such as Yohkoh, SOHO, and more recently RHESSI, these datasets allow us to track the progression of solar eruptive events from the low corona into the interplanetary medium. One of the most dramatic forms of solar activity, coronal mass ejections (CMEs) encompass a large range of spatial scales in a question of a few minutes. These go from the very small like current sheets, to small like active regions, to the very big like trans-equatorial loops and the transient seen in white light images (with angular extents in excess of 100 degrees for some events). Hence, in order to understand the CME phenomenon, its origin, and early development, we need a set of observations able to image the whole Sun with time cadences of the order of the second. Radio observations can do that presently. Multifrequency radio observations of the solar corona in the metric domain provide diagnostics of a wide variety of phenomena that occur in association CMEs. Radio imaging instruments can follow the processes leading to CME initiation, follow the expansion of the CME in the low corona, both on disk, and above the solar limb, and as such make the link with coronagraphic data. The characteristic signatures of the many CME related phenomena go from thermal emission of the eruptive cavity in the low corona, to direct imaging of the CME loops from synchrotron emission, to radio continua and shock associated emissions, recent progress on the understanding of the early development of CMEs, and on the coronal restructuring in the aftermath of the mass ejection, based on solar radio imaging from the Nancay Radioheliograph, is reviewed here.

Keywords. Sun: coronal mass ejections (CMEs)

Discussion

JIE ZHANG: Given the high cadence of radio observation in the inner corona, can you see the acceleration or velocity change of the CME in the inner corona?

MAIA: We are limited by resolution of the instrument. Above ~ 1.5 $R_\odot$ the velocity seems to be the same as in LASCO C2. Below that there could be some acceleration but I can not estimate it reliably.

YOUSEF: If you have acceleration of electrons at the top of the loop coming down you get X-ray emission at the 2 footpoints of the loop. Is this the case in radio event you have shown?

MAIA: I don't have X-ray emission available for these events. But the bursts are due to the plasma emission mechanism, that is, we are seeing only a slice in density.

GOPALSWAMY: 1. What is the nature of the intense radio emission at the place where the synchrotron loop connects to? May I suggest that it is the reconnection region from which non thermal electrons are injected?

2. In that case, the loop should be the same as the traditional moving type IV burst.

MAIA: 1. Maybe. The origin of the electrons is not clear.

2. Yes, the CME radio loops are one of the categories of the traditional moving type IV burst.

SCHWENN: Comment to the question about backward extrapolation. It works only for limb CMEs. If it comes from disk center, you will be off by several minutes. Careful! Also, the apparent lateral loop expansion could be due to a projection effect of a non-limb CME.

MAIA: Yes, that is right. Extrapolating CMEs to the limb only makes sense for very fast CMEs (2000Km/s means $1R_\odot$ in about 6min); or for the ones you are sure that are very close to the limb. I agree also that "lateral" expansions must take care about projection effects. In the event presented (Nov 6, 1997), it is a "true" expansion because we see that the small loop systems which disappear are affected at the time the CME flanks reach them - For this event it's not simply a large structure emerging from behind the limb. In general Dr. Schwenn is right, it is more likely that we would be seeing projection effects.

RUFFOLO: I'd like to ask more about the dangers of extrapolation. When we study solar energetic particles, from very fast CMEs, we want to compare with a CME lift-off time, which we extrapolate from LASCO C2 and C3. How bad is the timing error, and can radio people give us a more accurate time?

MAIA: On average it's probable not very important but for a particular event you can be in error for something like 30min for these fast events ($\sim$1000Km/s). It depends greatly on the height of the first image. I found events whose velocity higher than $5R_\odot$ is about 70% less than in the low corona. And there is also the problem of knowing the longitude of the CME source.

Coronal and Stellar Mass Ejections
Proceedings IAU Symposium No. 226, 2005
K. P. Dere, J. Wang & Y. Yan, eds.

© 2005 International Astronomical Union
doi:10.1017/S1743921305000505

Solar Cycle Variation of the Internal Magnetic Field Structure of CMEs

Volker Bothmer

[1]Max-Planck-Institut für Sonnensystemforschung, Katlenburg-Lindau, D 37191, Germany

Abstract. The internal magnetic field structure of CMEs and the field structure of the solar source regions were systematically investigated during different phases of the solar cycle in cycles 19-23 based on plasma and magnetic field measurements sampled by various satellites and through multi-wavelength remote sensing observations. It is found that: 1. To first order, the internal magnetic structure of CMEs varies systematically from one solar cycle to the next with respect to the prevailing hemispheric magnetic patterns of bipolar regions following the law of hemispheric helicity dependence. 2. To second order, the field structure in CMEs varies with respect to the complex spatial evolution of the magnetic flux in the photosphere in both hemispheres over the course of the cycle itself. The two effects can naturally explain the cyclic behavior of the SN, NS variations of the internal magnetic fields in CMEs in the solar wind as well as intermittent periods of mixed distributions.

Keywords. Sun: coronal mass ejections (CMEs)

Discussion

KOUTCHMY: To talk about a magnetic cloud I suspect that you need a lot of measurement made in 3-D to get the topology of the cloud and say if the magnetic cloud is really disconnected from the Sun?

BOTHMER: Yes, I fully agree. It is very dangerous to make conclusions on overall topology. That's also why we can't get proper values for helicity.

NINDOS: Comment on Dr. Koutchmy's comment: when we observe bi-directional flows in a MC, it is probable that the MC is rooted on the solar surface.

BOTHMER: Yes, but there is no unique interpretation. Flare particles (after impulsive onsets) suggest that some parts are still connected to the Sun.

YOUSEF: In a Beijing meeting 8 years ago I published a paper predicting that cycle 23 and the following 3 cycles would be weak cycles like those occurring around 1800 cycles 5, 6, 7 and 1900 cycles 12, 13. As a matter of fact, cycle 23 became weak and solar induced climate change occurred in 1997. Can you find evidence from your work on cycle 23 that it is different from the previous cycles 22 and 21?

BOTHMER: I have not studied this issue so far. So I can't give you a good answer right now.

Coronal and Stellar Mass Ejections
Proceedings IAU Symposium No. 226, 2005
K. P. Dere, J. Wang & Y. Yan, eds.

Study of Filament Eruption and its Relationship to a CME on 2003 August 25

H. D. Chen, Y. C. Jiang, Q. Y. Li and S. Q. Zhao

National Astronomical Observatories/Yunnan Observatory, CAS, Kunming 650011, China
email: hdchen@ynao.ac.cn

Abstract. By means of multiwavelength data, we study the eruptions of two filaments and the relationship between the first filament eruption and a subsequent CME. The main results are: (1) The disturbances of the two filaments showed different features, indicating that their eruptive mechanisms were possibly different. (2) A subsequent CME was well correlated with the first eruption in both time and space.

Keywords. Sun: activity, coronal mass ejections (CMEs), filaments, flares

It is well known that CMEs are closely associated with filament eruptions and flares (Hudson & Cliver (2001)). On 2003 August 25, two filaments, 'F1' and 'F2' successively erupted at the boundary of NOAA AR 100442 (S09E40). The F1 eruption was accompanied by an X-ray class C3.6 two-ribbon flare, and then a CME was observed to span the eruption region. Using H_α data from Yunnan Astronomical Observatory (YNAO), EUV and white-light coronagraph data from TRACE and SOHO/LASCO, we find that the eruptive processes of F1 and F2 were different. It seemed that the F1 eruption, the flare and the CME could be regarded in a same eruptive process framework.

The F1 and F2 were located on the neutral polarity inversion zone at the boundary of AR 100442. Their general eruptive processes are showed by the H_α and TRACE 171 Å images in Figure 1. The F1 started to erupt at about 02:10UT. By 02:32UT, it completely disappeared in H_α (Figure 1(a2)) but can be visible in TRACE 171 Å image (Figure 1(c2)). The two-ribbon flare with start, peak and end times at about 02:30, 02:59 and 03:35 UT (see Figure 2((a))) occurred in the course of the F1 eruption. Its two ribbons, 'R1' and 'R2', clearly showed spreading motion. When the R1 came near the F2, the F2 became instable and then erupted. The detailed H_α movie shows that the activation state and eruption processes of F1 and F2 were obviously different. The F1 first became darker and thicker, then bifurcation appeared in its body, and finally it showed a whiplike eruption. However, one part of the F2 first disappeared, then the rest part underwent horizontal axial motion, and eventually it erupted wholly. The different eruption processes probably indicated the different eruption mechanisms. We also note that the F1 eruption was possibly associated with new emerging magnetic flux in the photosphere as shown by MDI magnetograms, whereas the F2 eruption was obviously correlated with the separate motion of the flare ribbons as a result of the F1 eruption and the possible interaction between F2 and the R1.

After the F1 eruption and the flare, the CME was first seen in the field of view of the LASCO C2 at 03:26 UT. Figure 1(d) shows the CME situation at 04:26 UT. Using TRACE/171 Å images and the measures of Seiji Yashiro, we plot the projection heights of F1 and CME front in Figure 2((a)). The filament started to rise about one hour before the occurrence of the flare ribbons, and reached a velocity of several tens of km s^{-1}. After the flare onset (around 02:30 UT), however, it was rapidly accelerated to over 200 km s^{-1} and reached at least 500 km s^{-1} when it moved beyond the TRACE field of view around

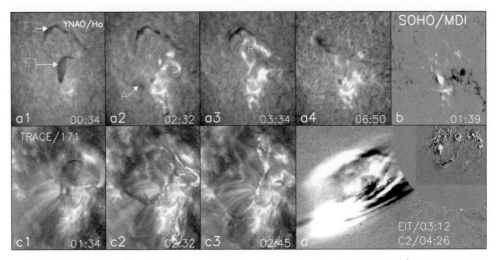

Figure 1. YNAO H_α images (a1-a4), MDI magnetogram (b), TRACE 171 Å images (c1-c3) and a composite image of inner 195 Å with outer LASCO C2 difference images (d).

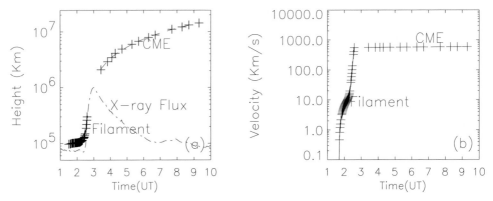

Figure 2. (a) GOES-8 soft X-ray flux (dash dot line), F1 and CME height (b) The velocity of F1 and CME front.

02:35 UT. The velocities of F1 and CME are shown in Figure 2((b)). Obviously, the CME front velocity was basically a constant, and the average acceleration from a second-order fit was only about 1.5 m s^{-2}. However, the F1 height and velocity profiles show a fast acceleration during the impulsive phase of the flare. Thus, the major acceleration of the CME should take place at its early stage (Zhang, et al. (2001)). It seems that the F1 eruption, the flare and the CME could be regarded in a same eruptive process framework (Wang et al. 2003).

Acknowledgements

We thank the TRACE team, the LASCO and MDI teams for data support. The work is supported by the NSFC under grant 10173023.

References

Hudson, H. S. & Cliver, E. W. 2001, *J. Geophys. Res.* 106, 25199
Wang, H. M., Qiu, J., Jing, J. & Zhang, H. Q. 2003, *ApJ* 593, 564
Zhang, J., Dere, K. P., Howard, R. A., Kundu, M. R., & White, S. M. 2001, *ApJ* 559, 452

Coronal and Stellar Mass Ejections
Proceedings IAU Symposium No. 226, 2005
K. P. Dere, J. Wang & Y. Yan, eds.

Variability of Coronal Mass Ejections

Probhas Raychaudhuri

Department of Applied Mathematics, Calcutta University, Calcutta-700009, INDIA,
E-Mail: *probhasprc@rediffmail.com*

Coronal mass ejections (CME) from the solar corona are the most spectacular phenomena of solar activity. Solar physicists are tried to relate CME with other forms of solar activities. CMEs are the result of a large scale rearrangement of solar magnetic field and they are often observed as an eruption of twisted magnetic fields from the solar atmosphere. SOHO/LASCO detected (*http://cdaw.gsfc.nasa.gov/CME_list*) more than 7500 CMEs during 1996-2003 June. The catalog contains all the CMEs with primary characteristics e.g. linear speed, central position angle, and the angular width. We will use these characteristics to study the variations of CME within these periods. The period starts from the sunspot minimum to entire sunspot maximum range where the solar activity is high. Solar proton events ($E > 10MeV$) were collected from NOAA website (*http: /www.lep.gsfc.nasa.gov/waves*) of the associated CMEs with halo CMEs. We find from CMEs data that the occurrence of average CME rate is 121.51 per month during June 1999 to June 2003 (sunspot maximum range) whereas the occurrence of average CME rate is 41.24 per month during January 1996 to May 1999 (sunspot minimum range), although during the year 1996 (when the average sunspot number is 8.6 per month) occurrence of average CME rate is 18.16 per month. The CME occurrence rate is also correlated with the sunspot numbers with high statistically significant level. The CME number is highest in 2002 but CME is higher in 2000 than in 2001. There is an overall similarity between sunspot number and CME rates but there are differences particularly from June 1999 which is the beginning of the sunspot maximum range. The CME rate peaks in September 2001 to October 2002, which is about 1.25 year after the sunspot maximum. Similarly the average speed of CME at the time of sunspot maximum range and sunspot minimum range are 575 km/sec. and 266 km/sec. respectively. This means that the average speed of CME increases from 1996 to June 2003. The CME speed is also correlated with the sunspot numbers with less significant level than the average rate of CME occurrence. The maximum monthly average speed is about 677.3 km/sec. at the time of April 2001, which is about 5 months earlier than the second sunspot maximum. From the preliminary list of halo CME events from SOHO/LASCO during January 1996 to June 2003 we find that the occurrence rate of average halo CME events during January 1996 to May 1999 is about 1.10 per month whereas during June 1999 to June 2003 is about 4.00 per month, during the year 1996 only two halo CMEs is occurred. We also find that the average speed of halo CME events during sunspot minimum range is 838 km/sec, whereas average speed of the halo CME events during sunspot maximum range is 1000 km/sec. Although during the year 1996 the average speed of halo CME events is 451 km/sec. From the characteristics of halo CMEs in years we find that the number of halo CME increases from 1996 to 2001 and the number of halo CME is maximum in the year 2001, after that number of halo CME decreases. In the 23rd solar cycle maximum solar activity occurred during June to September 2001 we call the time as 2nd sunspot maximum time. We also find that number of high speed ($> 1000\,km/sec$.) halo CME is highest during 2nd sunspot maximum range (i.e., during 2001-2002). We find from the halo CME data that average halo CME speed increases from 1996 to 1998 and then decreases from 1998 to 2000 and again increases from 2000 to 2003 and we expect that the average speed of halo CME will decrease after 2003. We find 78 solar proton events ($E > 10MeV$) from CME and about 43 of them are from halo CME during 1996 to 2003. We noticed that the maximum solar proton events occurred at the second sunspot maximum, which is occurred after $1\frac{1}{2}$ sunspot maximum in the 23rd solar cycle. We find there exist 5 phases of solar proton events ($E > 10MeV$) data in the 23rd solar cycle. The first phase is at the sunspot minimum, 2nd phase is after two years from the sunspot minimum, 3rd phase is at the time of sunspot maximum and 4th phase occurs just one and half year (usually it is about 2/3 years) after the sunspot maximum and 5th phase occurs 2/3 years before the sunspot minimum. We find six solar proton events ($E > 10MeV$) data within 1999 to 2003 with 12900 to 31700 pfu which produced strong geomagnetic storms and all of them are very high-speed halo CME. It is known that very fast CMEs ($V_p > 1000km/sec.$) are capable of causing extremely intensive geomagnetic storm when D_{st} index $< -300nT$. We find that there is a significant correlation between the speed of the CME and solar proton events ($E > 10MeV$) data. Solar radius measurement at Rio de Janeiro from 1997-2000 shows that the solar radius varies in phase with the solar cycle. Astrolabes of Antalya, Rio de Janeiro and Santiago suggest that the solar radius varies in phase with the solar cycle. From the detection of solar radius variations with MDI on board SOHO it is found that the solar radius increases with the number of sunspots[l]. It appears that solar radius variation and solar neutrino flux variation with the solar cycle is due to the variation of solar core pulsations and is mainly responsible for the variation of CME and its speed that is in phase with the solar cycle. We suggest that the above-mentioned characteristics are interrelated and that a pulsating solar core may be their common origin [2].

Keywords. Sun: activity, coronal mass ejections (CMEs), magnetic fields, particle emission

---

References

Noel, F. 2002, *Astronomy and Astrophysics* 396, 667

Raychaudhuri, P. 1994, *Solar Phys.* 153, 445.

Coronal and Stellar Mass Ejections
Proceedings IAU Symposium No. 226, 2005
K. P. Dere, J. Wang & Y. Yan, eds.

© 2005 International Astronomical Union
doi:10.1017/S1743921305000530

Cross-correlations between CMEs and other Solar Activity Indices

W.B. Song† and J.X. Wang

National Astronomy Observatories, Chinese Academy of Sciences, Beijing 100012, China
email: wenbin@ourstar.bao.ac.cn

Abstract. Using the list of CMEs observed by SOHO/LASCO, we compile a daily CME counts from January 1996 to December 2003. Cross-correlations between the CME counts and other three solar activity indices, i.e., flare index, sunspot number, and photospheric magnetic flux, are examined in both real and Fourier spaces. We find that correlations are all significant in real space, but only photospheric magnetic flux has good correlation with CME counts in Fourier space. Typical periods of CME occurrence are presented and discussed.

Keywords. Sun: activity, coronal mass ejections (CMEs), flares, magnetic fields

1. Data Analysis

A study on CMEs is an important topic that relates directly to space environments. Here we take our emphasis on correlations between CME occurrence and other solar activity indices.

Figure 1*a* has shown four daily sequences: CME counts (CC), sunspot number (SN), flare index (FI) and photospheric magnetic flux (PMF). The former three ones come from Websites, The last one is constructed by connecting NSO/Kitt Peak synoptic charts (Stenflo & Güdel, 1988) and calculating magnetic flux like Ballester & Oliver (2002). About five months long gaps of CC are filled with a method introduced by Fahlman & Ulrych (1982).

To assess the degree of correlation we compute cross-correlation coefficient ω at zero lag defined as $\omega = \frac{\sum_{j=1}^{M}(x_{j-1}-\bar{x})(y_{j-1}-\bar{y})}{\sqrt{[\sum_{j=1}^{M}(x_{j-1}-\bar{x})^2][\sum_{j=1}^{M}(y_{j-1}-\bar{y})^2]}}$. To test for significance level we use F-test. Its corresponding function is $R_c = \sqrt{\frac{F_c(1,M-2)}{(M-2)+F_c(1,M-2)}}$, where $F_c(1, M-2)$ can be looked up in F-list at one definite level ($\alpha = 0.99$). If $\omega > R_c$, we can say that this correlation is significant, otherwise it can be considered due to chance.

2. Results

We examine the correlations both in real and Fourier spaces. In real space, from Table 1 we can find that three correlation coefficients are all positive and very significant. Besides, the coefficients between CC and SN, PMF are both rather high. In Fourier space, Figure 1*b* has shown the normalized Fourier power spectra of these four sequences after zero-meaned, we compute the correlation coefficients in $0.005 - 0.05/day$ frequency domain and get that $\omega(CC-FI) = 0.0015$, $\omega(CC-SN) = -0.011$, $\omega(CC-PMF) = 0.35$. Their corresponding R_c are all about 0.23. Therefore, only PMF has good correlation with CC in Fourier space. This is very different from what happened in real space. Moreover, from

† Present address: A20 Datun Rd., Chaoyang Dist., Beijing 100012, China

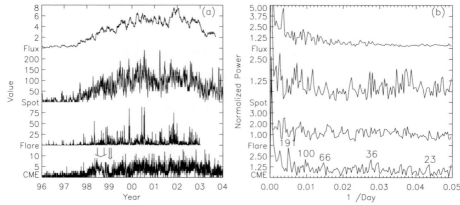

Figure 1. (*a*) Daily variations of CC, FI, SN and PMF from 1996 to 2003. Arrows point to the gaps of CMEs list. Here the PMF unit is 10^{23} Mx; (*b*) Fourier power spectra of the four sequences shown in Figure 1*a*.

Figure 1*b* we can find some typical periods in CC, including 191-day period, 100-day period, 66-day period, and so on. Many of them are also mentioned by Lou *et al.* (2003).

Table 1. Cross-correlations between CC and FI, SN, PMF

Solar Activity Indices	$M(days)$	ω	R_c
Flare Index	2557	0.28	0.067
Sunspot Number	2922	0.49	0.062
Photospheric Magnetic Flux	2812	0.55	0.063

3. Discussion

We find that the correlations between CC and FI, SN, PMF are highly significant in real space. While in Fourier space we find that only the flux sequence has such good correlation. Therefore, we consider that pulses in PMF can place a premium on CME occurrence. This may due to the fact that they both belong to global (or large scale at least) solar indices. It is really elusive that there is almost no correlation between CC and FI in Fourier space. First, the CME and flare may have very different physical mechanisms, for example, the flare basically an active region phenomenon. Second, different definitions and measurements of such two indices cannot be ignored. That the fewer correlation in real space and non-correlation in Fourier space between CC and SN would be expected because their difference is obvious except solar cyclic changes.

Acknowledgements

This work is supported by National Natural Science Foundation of China (10233050) and National Key Basic Research Foundation of China (G2000078404).

References

Ballester, J. L., Oliver, R. & Carbonell, M. 2002, *ApJ* 566, 505
Fahlman, G. G. & Ulrych, T. J. 1982, *MNRAS* 199, 53
Lou, Y. Q., Wang, Y. M., Fan, Z. H., Wang, S. & Wang, J. X. 2003, *MNRAS* 345, 809
Stenflo, J. O. & Güdel, M. 1988, *A&A* 191, 137

Coronal and Stellar Mass Ejections
Proceedings IAU Symposium No. 226, 2005
K. P. Dere, J. Wang & Y. Yan, eds.

© 2005 International Astronomical Union
doi:10.1017/S1743921305000542

CMEs and Flux Appearance in the Periphery of Two Unipolar Sunspots

X.L. Yang† W.B. Song, G.P. Zhou, J. Zhang, and J.X. Wang

National Astronomy Observatories, Chinese Academy of Sciences, Beijing 100012, China
email: yxl@ourstar.bao.ac.cn

Abstract. A class of large-scale magnetic compositions have been identified to be CME-prolific, which is characterized by a huge unipolar sunspot appearing in a large-scale extended bipolar region in synoptic magnetic charts. To understand the CMEs' origin and the nature of flux appearance, we scrutinize the long time-sequence of MDI magnetograms of high-resolution mode for super active region AR9236. Two types of magnetic features are clearly identified. They are moving magnetic features (MMFs) emanated radially from the penumbral boundary and emerging flux regions (EFRs) whose growing opposite polarities rotate out from the inner boundary of sunspot moat along helical paths in opposite directions. The interaction between the MMFs and EFRs often creates multi-fold magnetic neutral lines where the flare/CMEs initiated.

Keywords. CMEs,moving magnetic feature,emerging flux region

1. Introduction

A huge monopolar sunspot, which located in a favorite large-scale magnetic configuration and presented persistent flux appearance, is often flare/CMEs creative. For example, in NOAA AR9236 there initiated 5-6 homologous flare/CMEs from the sunspot periphery. However, there are controversies in interpreting the nature of flux appearance in the sunspot periphery. Nitta & Hudson (2001) considered the fact of this region undergoing a dynamic restructuring due to flux emergence in the form of EFRs. Zhang & Wang (2002) identified that the main magnetic changes are flux emergence in the form of MMFs. A same type but less active sunspot was seen in NOAA AR8375, for which the magnetic evolution and activity were described by Yurchyshyn & Wang (2001), Zhang, Solanki & Wang (2003), and Yang *et al.* (2004). In this contribution we focus on the flux appearance in AR 9236, and to address: 1) what types of flux appearance can be seen in the sunspot periphery, 2) are they distinct from flux appearance in other environment, 3) how do they associate with flare/CMEs initiated from in the AR?

2. Two types of flux appearance and CME initiation

We identified two typical forms of flux appearance in AR9236. They are MMFs and EFRs. Figure 1 shows MDI magnetograms in time sequence. Unlike in AR8375, the MMFs in the sunspot moat did not show much in pairs of opposite polarities. The average magnetic flux of EFRs and MMFs is 10^{19-20} and 10^{18} Mx respectively. Their horizontal velocity is about 0.4 km/s although the MMFs were always showing a bit faster. EFRs in the sunspot periphery are peculiar in the sense that grow and separate along the circle encircling the sunspot, their magnetic orientation is always the same. For AR9236, the negative flux rotated clockwise, while the positive flux, anti-clockwise. They emerged successively and interact one another, making the magnetic configuration

† Present address: A20 Datun Rd., Chaoyang Dist., Beijing 100012, China

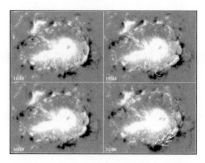

Figure 1. Time sequence of MDI magnetograms of AR9236 on Nov.24 2000. Two pairs of MMFs and one peculiar EFR are marked by polygon and brackets, respectively. Arrows indicate many folds of the magnetic neutral lines which is created by interacting EFRs.

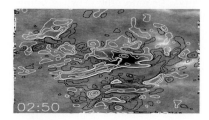

Figure 2. H_β filtergram of AR9236 taken in the earliest phase of flare/CME at 02:50/03:30 UT on November 26, 2000. Superposed are contours of current helicity. Darker (light) contours indicate the negative (positive helicity).

very complicated. Often there appear multi-fold magnetic neutral lines (see arrows in Figure 1). To understand the CME initiation in such magnetic environments, we select the flare/CME at 02:50/03:30 UT of Nov.26 shown in Figure 2 for this study. The earliest flare patches were closely correlated with the interaction of EFRs, which created multi-fold magnetic neutral lines and strong magnetic shear. Such structures are much easier to place a premium on the flare/CMEs. For this flare/CME event, EFRs are, indeed, more responsible. It can not be excluded that MMFs also play some role in the magnetic evolution leading to the flare/CME. As revealed by Wang *et al.* (2004), the flare/CME initiation site is characterized by the close contacting and canceling of magnetic flux of opposite helicities EFRs identified in AR8375 and AR9236 have the same helicity sign of the sunspot, predominately; while the MMFs have opposite helicity sign to the sunspot usually. MMFs brought up the opposite helicity to the AR, which may help with triggering the opening of the overall fields. Other flare/CMEs in AR9236 seem to share the common magnetic configuration and evolution.

Acknowledgements

This work is supported by National Natural Science Foundation of China (10233050) and National Key Basic Research Foundation of China (G2000078404).

References

Nitta, N. V. & Hudson, H. S. 2001, *GRL* 28, 3801
Yang, X. L., Song, W. B., Zhou, G. P., Zhang, J. & Wang, J. X. 2004, *ASR* submitted
Yurchyshyn, V. B. & Wang, H. M. 2001, *Solar Phys.* 202, 309
Zhang, J., Solanki, S. K. & Wang, J. X. 2003, *A&A* 399, 755
Zhang, J. & Wang, J. X. 2002, *ApJ* 566, L117
Wang, J., Zhou, G. P. & Zhang, J. 2004, *ApJ* 615, 1021

Coronal and Stellar Mass Ejections
Proceedings IAU Symposium No. 226, 2005
K. P. Dere, J. Wang & Y. Yan, eds.

Parametric Survey of Emerging Flux for Triggering CMEs

X. Y. Xu, P. F. Chen, C. Fang and M. D. Ding

Department of Astronomy, Nanjing University, Nanjing, 210093, China
email: xyxu@nju.edu.cn

Abstract. Observations suggest that solar coronal mass ejections (CMEs) are closely associated with reconnection-favored flux emergence, which was explained as the emerging flux trigger mechanism for CMEs by Chen and Shibata (2000) based on numerical simulations. This paper presents a parametric survey of the CME-triggering environment. Our numerical results show that whether the CMEs can be triggered depends on both the amount and the location of the emerging flux. The results are useful for space weather forecast.

Keywords. Magnetohydrodynamics (MHD), Sun: coronal mass ejections

1. Introduction

Recent observations suggest that coronal mass ejections (CMEs) are strongly correlated with reconnection-favored flux emergence. As suggested by observations (Feynman & Martin, 1995), two types of emerging flux with reconnection-favored direction can trigger filament eruptions (and then CMEs): one is within the filament channel, and the other is on the outer edge of the channel. This was regarded as the emerging flux trigger mechanism for CMEs and further verified by numerical simulations (Chen & Shibata, 2000). Later on, using a simple analytic model, Lin, Forbes, and Isenberg (2001) investigated the circumstances under which CMEs may be triggered, and found that there is no universal relation. Extending the work by Chen, Shibata, and Yokoyama (2001), we present in this paper a parametric survey of the emerging flux in relation to the CME triggering using the model by Chen and Shibata (2000). As the first attempt, we assume that the emerging flux has the same polarity orientation as the pre-existing magnetic field overlying the filament.

2. Numerical method

With the gravity and heat conduction being omitted, two-dimensional time dependent compressible resistive MHD equations are numerically solved with a multi-step implicit scheme (Hu, 1989). The initial conditions are the same as that in Chen and Shibata (2000), while the amount and the location of the emerging flux are free parameters. The length scale L_0 (2×10^4 km) for the horizontal and vertical coordinates is equal to the half width of the filament channel.

3. Numerical results

The evolutions of temperature (*gray*, in unit of 10^6 K), velocity (*arrows*), and magnetic field (*lines*) in three scenarios are shown in Figures 1 to 3, with the former two cases corresponding to non-eruption type, and the third one to eruption type. The results are

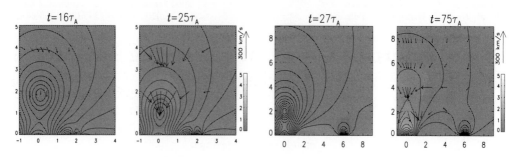

Figure 1. A non-eruption example where the emerging flux is very close to the magnetic neutral line.

Figure 2. A non-eruption example where the emerging flux is very far from the magnetic neutral line.

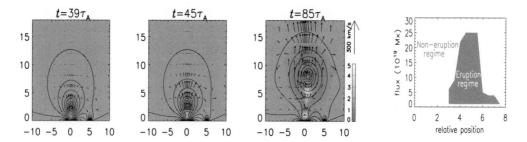

Figure 3. An eruption example where the flux rope loses its equilibrium after the reconnection between the emerging flux and the pre-existing field, and then is ejected as reconnection proceeds in the current sheet below.

Figure 4. The parameter space of emerging flux for triggering CMEs.

depicted in Figure 4, which indicates that the emerging flux with the parameters in the shaded area can trigger the onset of CMEs.

4. Conclusions

We numerically investigated the response of a flux rope system to flux emergence. With the same polarity direction as the background field, the results indicate that whether a CME can be triggered depends on both the amount and the location of the emerging flux, which may provide useful information for space weather forecast. Similar study with opposite polarity orientation of the emerging flux will be conducted soon, and the detailed comparison with observations is devoted to another paper.

Acknowledgements

This work was funded by NSFC under grants 10221001, 10333040 and 10403003, by NKBRSF under grant G20000784 and FANEDD under grant 200226.

References

Chen, P.F. & Shibata, K. 2000, *ApJ* 545, 524
Chen, P.F., Shibata, K. & Yokoyama, T. 2001, *Earth, Plants and Space* 53, 611
Feynman, J. & Martin, S.F. 1995, *JGR* 100, 3355
Hu, Y.Q. & Low, B.C. 2000, *ApJ* 342, 1049
Lin, J., Forbes, T.G. & Isenberg, P.A. 2001, *JGR* 106, 25053

Coronal and Stellar Mass Ejections
Proceedings IAU Symposium No. 226, 2005
K. P. Dere, J. Wang & Y. Yan, eds.

© 2005 International Astronomical Union
doi:10.1017/S1743921305000566

The Evolving Features of the Source Region of a Fast Halo CME with Strong Geo-effects

Zhang Guiqing†

National Astronomical Observatories, Chinese Academy of Sciences, Beijing 100012
email: zgq@bao.ac.cn

Abstract. A fast halo CME and correspondingly strong solar burst that occurred in NOAA9684 led to intense geo-effectiveness. The analysis and study in this paper were focused on the evolution of this active region.

Keywords. Sun: coronal mass ejections (CMEs), magnetic fields, sunspots

1. Introduction

The study of Coronal mass ejections (CMEs) have been one of the most significant topics since they were first observed in the 1970s. The studies of CMEs covered were very extensive. This work tries to analyze and study the evolution of NOAA9684 where a fast halo CME and a strong flare occurred. NOAA9684 was not a high-productive region of remarkable CME and flare. It only produced one remarkable fast Halo CME and an X/3b flare during its all-life. The CME that accompanied the X1.0/3B flare erupted on 4 November 2001. This strong eruption led to Sudden Ionospheric Disturbance, proton event with the peak flux of 31700 PFU (E$\geqslant$ 10 Mev) and a geomagnetic disturbance with a significant Bz deflection of -292 nT.

2. Evolving Features of Source Region

AR9684 appeared on the east limb on 27 October, 2001. It was located at N 06° L 136°. The leading spot of N polarity in the active region dominated and its following spot was weak and scattered when it went onto the visible disk (Figure 1(a)).

The Tilt Angle of the Magnetic Axis (TAMA) in a sunspot group has been defined in the studies of Tian (2002). Furthermore, the sign of TAMA and the relation between TAMA and the writhe of a flux tube that formed an active region has been deduced by Fan (1999) and Tian (2002). That is to say, in the northern/southern hemisphere, an Hale active region should rotate clockwise/counter-clockwise and the writhe of magnetic flux tube should be right/left-handed if the rotation appeared in it.

The TAMA of NOAA 9684 was roughly $+2°$ on 30 October (Figure 1(a)). The TAMA was roughly $-23°$ on 3 November. This case implied that the magnetic flux tube forming the active region rotated 25° counter-clockwise (left-handedness, (Figure 1(a) and (b)).

We know that the positive/negative values of α_{best} correspond to the direction of the twist of magnetic lines in right/left-handedness (Tian 2002). We calculated the force-free parameter α_{best} of NOAA9684 when it was on the heliocentric longitude. It was $-0.027 \pm 0.002 Mm^{-1}$. This result indicated that the twist of field lines was left-handed in this region.

† The work is supported by NSFC through grants 10073013, 10233050 and 4999-0451.

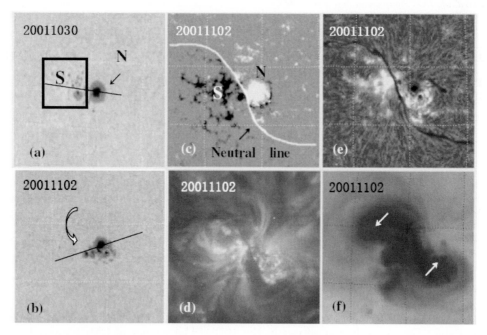

Figure 1. The rotation of sunspot group and "Counter-sigmoid" structure on various layer in NOAA9684. The white and bent arrows indicate the rotating direction of sunspot group and writhing direction of the filament and the loops. The black line indicates the magnetic axis

Figure 1(c) displayed that the magnetic neutral line of NOAA9684 superposed on the neutral line of background magnetic field. Figure 1(d),(e) and (f) are the images of 195$\mathring{A}$, H_α and soft X-ray. These images showed that the torsion of "Counter-sigmoid" structure of the filament and the magnetic loops are consistent to Counter-clockwise rotation of the active region and the trend of the neutral line of background field.

3.   Discussion and Conclusion

NOAA9684 was a non-Hale region that disobeyed *Joy*'s law and its maximum area was only 550 μh. Only one fast Halo CME that accompanied a big flare occurred in it and led to violent geo-effectiveness.

One of the features of the active region was the rotation. The writhe of the magnetic flux tube forming the active region and the twist of the magnetic line were all left-handed, which revealed that the kink instability existed in the active region.

The second feature of NOAA9684 was that its magnetic neutral line superposed on the neutral line of background field, which should indicate that the CME and large flare wonldlead to geo-effectiveness because this location was advantageous to energetic particle flux propagate toward the earth.

References

Fan, Y., Zweibel, E. G., Linton, M. G. & Fisher, G. H. 1999, *ApJ* 521, 460
Tian, L., Liu, Y. & Wang, J-X. 2002, *Solar Phys.* 209, 361

Coronal and Stellar Mass Ejections
Proceedings IAU Symposium No. 226, 2005
K. P. Dere, J. Wang & Y. Yan, eds.

© 2005 International Astronomical Union
doi:10.1017/S1743921305000578

Correction of Large-Spread-Angle Stray Light for Measurements of Longitudinal Magnetic Signals

Jiangtao Su[1]† and Hongqi Zhang[1]

[1]National Astronomical Observatories, Chinese Academy of Sciences, Beijing 100080, China
email: sjt@sun10.bao.ac.cn

Abstract. To examine the stray light in magnetograph observations, we determined the point spread function of the Video Vector Magnetograph mounted on the Solar Magnetic Field Telescope (SMFT) installed at the Huairou Solar Observing Station (HSOS). Then we obtained the curve on large-spread-angle (LSA) stray light intensity as a function of distance from disk center. A new way to correct LSA stray light is proposed.

Keywords. Instrumentation, magnetic fields

1. Determination of LSA Stray Light intensity

The observed Stokes I is

$$I^{obs} = a_{non}I + \sum_j c_j \Phi(r, b_j) + c_l \Phi_l(r, b), \qquad (1.1)$$

where the contribution functions $\Phi(r, b_j)$ and $\Phi_l(r, b)$ are numerically calculated by using the formula given by Martinez Pillet (1992). All the coefficients of this linear combination are to be obtained. To the end, we make a linear-square fit to the aureole data. The data we use in the present study were obtained with the HOSO Video Vector Magnetograph on 31 May 2004. We adopted the research method of Chae, *et al.* (1998) how to select the data to be fitted. Figure 1 shows the intensity profile across the limb and the result of model fitting. In the process of fitting, two Gaussians were used and the corresponding width parameters are $b_1 = 2.5''$ and $b_2 = 12.5''$. The width parameter of Lorentzian is $b = 100''$. The LSA stray light intensity integral is presented as a dot-dashed curve. Since it slowly increases with distance from the disk center, the same integral might be used in observations from different days. In the next section, we will use the curve (for convenience, named correcting curve) of it vs. the distance to correct LSA stray light.

2. Correction of LSG Stray Light

In routine observations, what we should do first is to make telescope point to the Sun limb to measure background noise signal (BNS) coming from stray light and instruments, then subtract them from polarized light intensity. These BNS are called as Black Level (BL). Since we subtract off a constant background, this method should be suitable only to correct LSA stray light. However, it is clear that the stray light intensity near an active region (AR) is not equal to that of Sun limb. Now, based on the results of section 1, we propose a new method to correct a constant BNS, which is a LSA stray light in

---

† Present address: 20A Datun Road, Chaoyang District, Beijing, China.

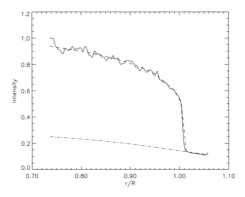

Figure 1. The observed intensity profile across the Sun limb (solid line). The model fit is drawn as dashed line, and the LSA stray light integral is presented as a dot-dashed curve.

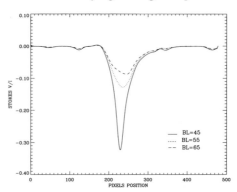

Figure 2. Stokes V/I observed with BLs of 45, 55, 65. vs. pixel location.

the instrument. If BL of Sum limb symbol as $L_{r=1}$, we can get a formula for BL near an AR on the solar disk by the correcting curve in Figure 1, as

$$L_r = L_{r=1}(I_r/I_{r=1}) - (I_r - I_{r=1})L_{zero}/I_{r=1}, \tag{2.1}$$

where $L_{zero} = 110$. We apply it to the AR 10218 to obtain the 'real' BNS (its distance from disk center being $r \sim 0.5$), which was observed with different BLs from 35 to 80, in step of 5. The BL of Sun limb is 65 known from the real time monitor. The rough ratio of LSA stray light intensity of $r = 1$ to that of $r = 1$ is 1.93 according to the correcting curve. So BL of $r = 0.5$ is ~ 43 by equation (2). Figure 2 shows the distributions of Stokes V/I (observed with BL of 45, 55, 65) VS. pixel location. We think Stokes V/I with BL of 45 is closer to the 'real' V/I than the others. However, some artificial signal may be added into it. This mainly ascribes to the correcting data for AR 10218 observations obtained in different days.

Acknowledgements

The authors gratefully thank Yuanyong Deng and Shudong Bao for constructive discussion. We also thank Guoping Wang and Hongwei Qi for their laborious observations.

References

Chae, J., Yun, H. S., Sakurai, T., and Ichimoto, K. 1998, *Solar Phys.* 183, 229
Martinez Pillet, V. 1992, *Solar Phys.* 140, 207.

Coronal and Stellar Mass Ejections
Proceedings IAU Symposium No. 226, 2005
K. P. Dere, J. Wang & Y. Yan, eds.

© 2005 International Astronomical Union
doi:10.1017/S174392130500058X

Four Corona Mass Ejections and their Associated Surface Activity Observed on 26 October 2003

Xingming Bao[1], Hongqi Zhang[1], and Jun Lin[2]

[1]National Astronomical Observatories of China, Chinese Academy of Sciences, Beijing 100012, China
email: baoxm@sun10.bao.ac.cn

[2]Harvard-Smithsonian Center for Astrophysics, 60 Garden Street, Cambridge, MA 02138, USA

Abstract. Four coronal mass ejections (CMEs) occurs successively from the solar disk near the west limb on October 26, 2003. They, together with the associated activities of the solar surface, were observed by various instruments both in space and on ground, such as the Large Angle and Spectrometric Coronagraph Experiment (LASCO), the Extreme Ultraviolet Imaging Telescope, and the Michelson Doppler Imager on board the *Solar and Heliospheric Observatory,* as well as the Huairou Solar Observing Station and the Big Bear Solar Observatory. These four events start with a filament eruption that manifests a two-ribbon flare in a spotless region, destroyed a helmet streamer, and give rise to an X1.2 flare in the active region AR0484. We notice that these eruptions occur either in active region, or in quiescent region, or in the region without any precursors. The time profiles of the CME (filament) heights show that the main acceleration takes place within one solar radius ($R_\odot$) from the solar surface, and that all the CMEs almost propagate at constant speeds as they appear in the field of view of LASCO C2. We conclude that the most dynamical process of each of these CMEs happens at the altitudes lower than one $R_\odot$ from the surface. Among the four activities, the fourth one comes from AR10484 and shows the largest speed projected on the sky plane, which is about 1500 km s^{-1}; and the first filament shows the largest acceleration, ~ 50 m s^{-2}.

Keywords. Sun: activity, coronal mass ejections, filaments, flares

1. Introduction

Both observations and theories suggest that CMEs be involved in the reorganizations of large scale magnetic fields in the corona, be closely related to the eruptive activities (e.g., solar flares and eruptive prominences) on the solar surface (see Forbes 2000; Priest & Forbes 2000; Lin et al. 2003 for reviews), and often be related to disruptions of helmet streamers over the solar limb (e.g. Raymond 2003 and the references therein).

A series of eruptions occurring on October 26, 2003 provides us a set of nice samples that allows us to investigate various CMEs discussed above: two CMEs starting with the eruptive prominences in the spotless region, a CME from an active region, and a CME that develops from a helmet streamer and has the magnetic structure of the helmet streamer as one of its legs. We give the results deduced from these data in next section, discuss our results and finally summarize this work in Section 3.

2. Results

The active region AR0484 is located on the west solar disk (N10, W50) on 2003 October 26 . A long filament and a short filament, which are denoted as F1 and F2 respectively, stayed at southwest to AR0484. Filament F1 erupts at 00:12 UT and gives rise to a CME, of which the leading edge enters the field of view (FOV) of LASCO C2 at 01:30 UT, and a two-ribbon flare starts to develop from 01:48 UT at the location where F1 used to sit. As both CME1 and the associated two-ribbon flare are in progress, filament F2 begins to take off at around 01:25 UT. The eruption of F2 yields a CME with a complex double-loop structure that enters FOV of LASCO C2 from 5:00 UT and also leads to a two-ribbon flare very close to the previous one.

The third CME undergoes with the destruction of a helmet streamer located at position angle $\sim 45°$ and is north to the previous two CMEs. In the processes of CME1 and CME2, the helmet streamer is slightly deflected. Unlike the other CMEs that grow from the helmet streamer (Raymond 2003), CME3 of the present case spans around 90° over the north pole and the helmet streamer consists of the legs of the corresponding expanding arcade. Due to the projection effect, the real span of CME3 must be wider than 90°. Because it is a slow CME, no apparent surface activity can be correlated to CME3.

The fourth CME is obviously related to AR10484. EIT 195 movie shows a series of successive precursors starting from 11:00 UT until 16:48 UT when an eruption-like activity occurs in the region south to AR0484, then a significant brightening in EIT 195 commences at 17:24 UT when an X1.2 flare is observed. Correspondingly, a CME front begins to appear in FOV of C2 at 17:54 UT.

The height-time profiles of the four CMEs and the eruption of the long filament have been plotted. The four CMEs show a nearly constant velocity after appeared in LASCO C2 and the first filament shows the largest acceleration, ~ 50 m s^{-2} before appeared in LASCO C2.

3. Discussions and Conclusions

We have selected four CMEs in this study with various origin and propagation which appeared above the west limb on October 26, 2003. Though the four CMEs are not enough to account for characters of all CMEs, we believed that the four CMEs represent various CMEs, either with different origins or without, in active region or in quiescent region, speed ranging from 200km/s to nearly 2000km/s. We try to link the eruptive phenomenon occurred in difference height to a single physical process by using the data from chromosphere to out corona. The height-time profile show that most CMEs propagate in a nearly constant speed since it appeared in C2. The height-time curve of filament and its derived velocity-time and acceleration-time curve show a significant acceleration $(50\,\mathrm{m/s^2})$ within 1R$_\odot$. This reveal that the dynamical process of eruptive events mainly happen within 1 $R_\odot$ (from solar disk).

Our study suggests that the CME related to the disruption of streamer maybe caused by impact of CMEs leading edge originated beside streamers and may pull out large mount of plasma which trapped within their closed loop. Our study suggests that such a CME belongs to a different class of CMEs to those associated with flare and filament eruption, both in triggering mechanism and propagation.

References

Forbes, T. G. 2000, *J. Geophys. Res.* 105, 23153
Lin, J., Soon, W., & Baliunas, S. L. 2003, NewA Rev., 47, 53
Priest, E. R. & Forbes, T. G. 2002, A & A Rev., 10, 313
Raymond, J. C., Ciaravella, A., Dobrzycka, D., Strachan, L., Ko, Y.-K., Uzzo, M., & Raouafl, Nour-Eddie 2003, ApJ, 597, 1106

Coronal and Stellar Mass Ejections
Proceedings IAU Symposium No. 226, 2005
K. P. Dere, J. Wang & Y. Yan, eds.

© 2005 International Astronomical Union
doi:10.1017/S1743921305000591

A Test of the Tanaka Model with NOAA 10488

Hui Zhao[1], Yuzong Zhang[1,2], Jie Jiang[1] and Jingxiu Wang[1]

[1]National Astronomical Observatories, Chinese Academy of Science, Beijing 100012, China
email: zhaohui@ourstar.bao.ac.cn

[2]Department of Astronomy, Beijing Normal University, Beijing 100875, China

Abstract. We investigated the complex subsurface magnetic rope structure of a super-active region NOAA 10488. With the set of twisted magnetic loop, knot and bifurcate configuration ,we could explain the complicated flux emerging, developing and disappearing by following Tanaka model (Tanaka, 1991). Based on Huairou photospheric vector magnetograms, we calculated the current helicity and found the dominant helicity sign is positive. We deduced that the whole active region might be one twisted magnetic rope.

Keywords. Sun: magnetic fields, sunspots

1. Introduction

The details and origin of magnetic rope structure under the photosphere are still not known clearly. In 1991 Tanaka firstly sought complex magnetic rope structures under the surface as inferred from detailed evolution data and analyzed the August 1972 region successfully. Kurokawa, Wang, & Ishii (2002) constructed a model of an emerging twisted flux rope to explain the outstanding features of the sunspots evolution in NOAA AR9026 (Figure 2(a)). Fan *et al.* (1999) performed three-dimensional MHD simulations of the rise of twisted magnetic flux tubes in an adiabatically stratified model solar convection zone. In this paper we try to use a set of the complex subsurface magnetic rope structure to explain NOAA AR10488's emerging, evolving and disappearing and test the model of Tanaka.

2. Data Analysis

During two weeks in late October and early November 2003, there is a series of large solar activities. The active region NOAA 10488 emerged on October 26 and developed into a large AR rapidly on October 27, then stepped into its phase of decay when approaching the west edge of the sun on November 3. With the MDI longitudinal magnetograms we traced its continuous evolution. Hence we can pick up pairs of bipolar appearance in the same time and identify the topological connectivity of magnetic field lines. And the sub-photospheric structures can also be identified through the movement, rotation and decay of longitudinal magnetic field.

By photospheric vector magnetograms and images of Hα we obtain more details of transverse field alignments. The 180 degree ambiguity in the observed field azimuth was carefully corrected by both theoretical and empirical methods (Wang, 1996). We calculated current helicity based on the vector field and found the dominant sign of helicity is positive (Figure 1(b)). So we could consider the whole active region may be one twisted magnetic rope like the model of Tanaka. The Figure 2(b) is one of possible structures which satisfy the evolvement of NOAA AR 10488.

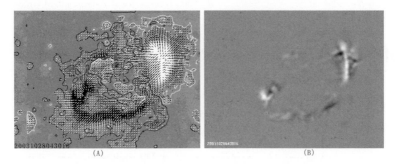

(A)   (B)

Figure 1. (a) One of the photospheric vector magnetograms from Huairou (which corresponding to the second longitudinal magnetogram employed in Fig. 3.). (b) The current helicity for (a).

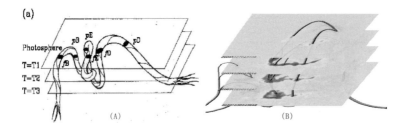

(A)   (B)

Figure 2. (a) Schematic drawing of an emerging twisted flux rope. The photospheric surfaces are drawn at three successive times(Kurokawa, Wang, & Ishii (2002)). (b) The schematic drawing of an emerging twisted flux rope. The background is the Huairou's longitudinal magnetograms at the four close time with (a). The sections of the flux rope cut by each longitudinal magnetogram are the locations of sunspots of that time.

3. Conclusion and Discussion

Tanaka model can reasonably explain the emerging and disappearing of sunspots and corresponding phenomena. However, to a given active region, the structure of flux rope drawn only by this model is not unique. The calculation of the current helicity distribution of NOAA AR 9026 basically accords with Kurokawa's flux rope structure of the same region. In our work the loop and hook twist structure, the bifurcate configuration and their evolution with time could explain the increasing, decreasing and disappearing of the sunspots, with the twist structure of knot, it is easy to understand the magnetic shearing. In addition, we try to model the complex magnetic configuration of the AR 10488 following the minimum energy principle.

Acknowledgements

The work is supported by the National Natural Science Foundation of China(10233050) and the National Key Basic Science Foundation(TG2000078404).

References

Hiroki Kurokawa, Tongjiang Wang, & Takako T. Ishii 2002, *ApJ* 572, 598–608
Katsuo Tanaka 1991, *Solar Physics* 136, 133–149
Y.Fan, E. G., Zweibel, M. G. Linton, & Fisher, G. H. 1999, *ApJ* 521, 460–477

Coronal and Stellar Mass Ejections
Proceedings IAU Symposium No. 226, 2005
K. P. Dere, J. Wang & Y. Yan, eds.

© 2005 International Astronomical Union
doi:10.1017/S1743921305000608

The Relationship Between Magnetic Helicity and Current Helicity

J.H. Liu, Y. Zhang and H.Q. Zhang

[1]National Astronomical Observatories, Chinese Academy of Sciences, Beijing 100012, China
email: ljh@sun10.bao.ac.cn

Abstract. We have studied the magnetic helicity transport rate and the current helicity for solar active region (AR) NOAA 10488 and find a complex relationship between them. We further extend this study to a statistical one, and find that 33 among the selected 57 ARs show same sign for the two parameter.

Keywords. Sun: activity, magnetic fields

1. Observation and method

With the vector magnetograms from Huairou Solar Observing Station (HSOS) of the National Astronomical Observatories of China, longitudinal current helicity $B_\| \cdot (\triangledown \times B_\|)$ had been calculated. And using the 96 minute cadence longitudinal magnetograms taken by the Michelson Doppler Imager (MDI) on board the Solar and Heliospheric Observatory (SOHO), we calculated the magnetic helicity transporting rate $dH_{corana}/dt = -2 \oint_{photohpere} (A_p \cdot V_\perp) B_n d^2 x$ (Démoulin and Berger 2002) with the local Correlation Tracking method (LCT, Chae *et al.* 2001). We have gathered 5 days of observing data for each AR. Before applying the LCT method to the MDI magnetograms, we do the nonlinear mapping, flux density interpolating and the geometrical foreshorten correcting works according to Chae *et al.* (2001).

2. Result

We calculate the magnetic helicity transport rate and the current helicity of AR NOAA 10488 who has passed across the solar disk from Oct 26 to Nov 3 (Zhang, *et al.* 2003). Figure 1 shows the comparison between the spatial distribution of $-2(v \cdot A_p)B_z$ and the current helicity density. The former is a measure of the local contribution of the foot point motion to the rate of the magnetic helicity transport, and the latter is a measure of the photospheric helicity state. So the figure implies that the current helicity at the photosphere changes with the transfer of magnetic helicity from the sub-atmosphere in the emerging process of magnetic flux. Figure 2 shows the temporal variations of dH/dt and average current helicity in AR 10488. The data is gathered from Oct 27 to 30 when the AR was not far from the disk center. We can see that the magnetic helicity transport rate was negative, and the average current helicity decreased weakly with time.

Figure 3 shows the statistical relationship between magnetic helicity transport rate and current helicity imbalance $\rho_h = \sum h_c / \sum |h_c|$ (Bao and Zhang 1998) of 57 ARs. We see a weak same-sign trend. It should be pointed that the ARs we selected were bi-pole or multi-pole ARs and showed not obviously hemispheric preference for both helicity.

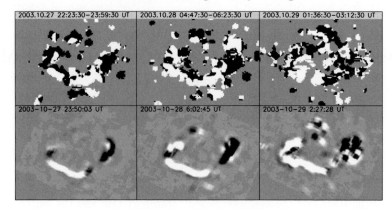

Figure 1. Gray-scale map of $-2(v \cdot A_p)B_z$ (upper) and $B_{||} \cdot (\nabla \times B_{||})$ (lower) of AR10488.

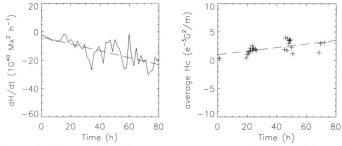

Figure 2. Measured helicity transport rate as a function of time (left) and time variation of average current helicity (right) of AR 10488

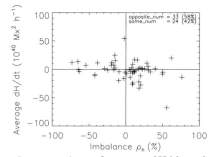

Figure 3. comparison of average dH/dt and ρ_h for 57 ARs

That the two parameters of 10488 and other ARs have same-sign means that the rotation of sunspots synchronize with the twist of vector magnetic field. It reflects the complexity of the helicity injection and the interaction of magnetic configuration.

Acknowledgements

We thank the staff in Huairou group for their discussions. We are grateful to all members of the SOHO teams for providing the wonderful data.

References

Bao, S.D. and Zhang, H.Q. 1998, *ApJ* 496, L43

Chae, J. 2001, *ApJ* 560, L95

Démoulin, P. and Berger, A. 2002, *SoPh* 215, 203

Zhang, H.Q., Bao, X.M., Zhang, Y., Liu, J.H., Bao, S.D., Deng, Y.Y., *et al.* 2003, *ChJAA* 3, 491

Coronal and Stellar Mass Ejections
Proceedings IAU Symposium No. 226, 2005
K. P. Dere, J. Wang & Y. Yan, eds.

The Relationship Between Photospheric Magnetic Field Evolution and Major Flares

Y. Zhang, J.H. Liu
and H.Q. Zhang

[1]National Astronomical Observatories, Chinese Academy of Sciences, Beijing 100012, China
email: zhy@sun10.bao.ac.cn

Abstract. On 28 Oct 2003, one of the biggest flares (4B/X17.2) seen in recent years occurred in Active Region (AR) NOAA 10486 associated with a violent halo coronal mass ejection. It was a complex $\beta\gamma\delta$ region. After studying the evolution of the AR and the phenomena of this powerful flare, we obtained the following result. (1) Highly sheared transverse field was formed gradually on both sides of the neutral line by squeeze during the AR development; (2) Rotations of penumbra of main polarities were discerned, and the average horizontal velocities was as large as 0.55 km/s; (3) The spiral transverse field of main positive polarity was diffused after the large flare; (4) Some magnetic features submerged or emerged in the vicinity of the flare onset point. The emergence of this rotational and complex magnetic topology implies a transport of magnetic energy and complexity from the low atmosphere to the corona. Moreover, the rapidly submergence (emergence) and movements of the small magnetic features which represent the enhancement (cancellation) and squeeze of the magnetic field play a key role in the onset of the flare.

Keywords. sun: flare —sun: magnetic field—sun: activity

1. Observation

For our current study, the data are from the following source: 1. complex sets of vector magnetograms from Huairou Solar Observing Station (HSOS) of the National Astronomical Observatories of China; 2. white light and EUV images from Transition Region and Coronal Explorer (TRACE) satellite.

2. Result

2.1. Long Term Evolution

The AR NOAA 10486 was a complex $\beta\gamma\delta$ configuration groups during its passage across the solar disk from Oct 22 to Nov 2 (Zhang, et al. 2003). The most important evolutionary mechanisms in this region are: 1. gradually formed highly sheared transverse field and longitudinal grad on both sides of the neutral line by squeeze during the AR development (Fig. 1); 2. different rotational pattern of sunspots (Fig. 2). The spiral magnetic structure which covered the whole P2 on Oct 28 diffused after the flare. The penumbra shared by P1 and P2 had a counterclockwise rotation pattern, while N4 and P3 had clockwise rotation pattern. N3 was squeezed by the different rotational polarities on both sides and developed as slender negative region.

2.2. Onset of the Flare

We present the time sequence of TRACE 171 Å images in Fig. 3. Seen from 171 Å movie, the AR showed various manifestations of activation before the powerful flare. It slowly

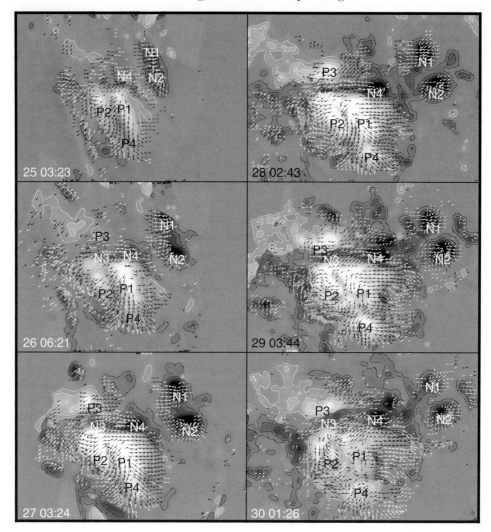

Figure 1. HSOS vector magnetograms in AR 10486. The longitudinal component of the magnetic field is presented by gray-scale patches and isogauss contours with levels of ±200, 500, 900 and 1400 G. White patches represent positive polarity fields and black patches represent negative fields. The transverse components of field is shown with short lines, with lengths proportional to the relative field strength. The size of images is $4.75' \times 3.34'$.

rose, meanwhile brightening appeared at part of the filament. Then it split into threads and each thread rose and fluctuated. Some filament materials ejected from 09:50 UT. At 10:17 UT part of the filament became brighten and rose rapidly, finial it erupted partially. At 11:01 UT, the filament erupted and a powerful two-ribbon flare developed rapidly.

2.3. Rapid Change Association with Flare

Disappearing sunspots which were located in the vicinity of highly sheared neutral line are outlined by windows 1 in Fig. 4. These spots were merged and canceled with nearby opposite polarity P3 and totally disappeared around 08:00 UT. After the flare, shear angle here was declined from 60° to 23°. Arrows 1 and 2 mark the emerging sunspots. Arrow 1 represents a penumbral feature which continuously emerged and developed as a little spot around 06:00 UT. The spot indicated by arrow 2 was already developed as

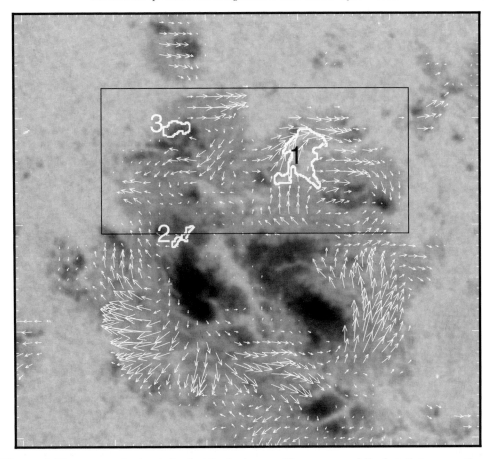

Figure 2. The average horizontal velocities is shown with arrows, and the length is proportional to the relative value of the velocity. The thick contours are the site of the flare onset. See details in the text.

a little spot on Oct 28 01:44 UT as shown in Fig. 4. It moved with spot 1 in the same direction rapidly. And before the powerful flare, they were already on the north of N4.

Acknowledgements

The authors would like to thank the staff in Huairou group for their comments and discuss. We are grateful to all members of TRACE teams for providing the wonderful data.

References

Zhang, H.Q., Bao, X.M., Zhang, Y., Liu, J.H., Bao, S.D., Deng, Y.Y., *et al.* 2003, *ChJAA*, 3, 491.

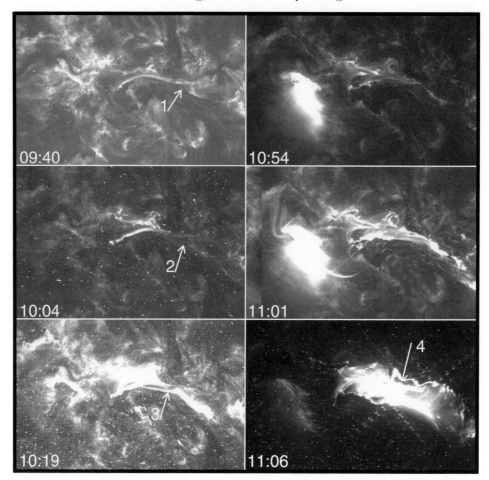

Figure 3. A series of 171 Å images obtained by TRACE. Arrows indicate: (1) the filament; (2) the ascend of filament; (3) brightening of the filament; (4) the post flare loops;

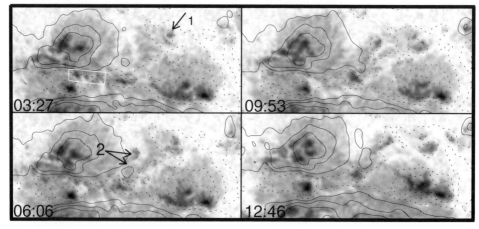

Figure 4. The rapid evolution of the sunspots. The window and arrows are described in the text.

Coronal and Stellar Mass Ejections
Proceedings IAU Symposium No. 226, 2005
K. P. Dere, J. Wang & Y. Yan, eds.

Spectral Features in Solar Microwave Emission Preceeding CME Onset

Olga A. Sheiner[1]and Vladimir M. Fridman[2]

[1,2]Radiophysical Research Institute (NIRFI),
25 Bol'shaya Pecherskaya Street, Nizhny Novgorod, 603950 Russia,
[1]email: rfj@nirfi.sci-nnov.ru
[2]email: fridman@nirfi.sci-nnov.ru

Abstract. The sporadic solar radio emission of patrol solar radio observations within the periods 1980, 1984–1989 and of observations with high temporal and spectral resolution in 1989 are used to find the manifestation of pre CMEs activity.

Keywords. Sun: radio radiation, coronal mass ejections (CMEs), evolution

1. Introduction

Studies of the phenomena in the microwave radio emission, which precede CMEs registration, are based on the concept of their formation and initial propagation in the lower layers of solar atmosphere, inaccessible to observations on coronagraphs. Such phenomena cover the significant high-altitude three-dimensional scales of solar atmosphere. In this case for determining the mechanisms of emission in the radio-frequency band it is important the spectral approach to a study of CMEs phenomena.

2. Observational Data

For this purpose we used as the observational data of sporadic radio emission obtained using regular observations as special observations with high spatial resolution. The first one was carried out over a wide range of wavelengths (3cm–3m) at the Radio Astronomical Observatory NIRFI "Zimenki" within the periods 1980, 1984, and 1989.

Special observations of microwave emission were made by RT–22 array of the Crimea Astrophysical Observatory (angular resolution about 3–4′) in August 1989. Sweeping spectrograph in 14–17 GHz was used to obtain the spectrum each second with 100 MHz spectral resolution and the average one during 1 min. The results of high spatial resolution presented are concerned to observations of solar radio emission from Active Region (AR) NOAA/USAF 5638 (S18W90) on the 12th of August. The observational time (5:00–14:00 UT) is coincide with the time of CMEs formation and propagation (SMM gives 12:59–14:32 UT as the time of CMEs observations).

3. Results

Spectra of radio emission have stable smooth behavior with flux increasing in frequency range. Approximately 5 hours before CMEs detection one can see very small spectral feature on the right side of spectra in 14–17 GHz range, which is less than 10% of AR flux and about 100–150 MHz in frequency range. The closer the time of CMEs detection the larger amplitude of feature is. The value of its flux is about 1 sfu and spectral interval varies in 200–400 MHz frequency interval just 1.5 hours before CMEs registration.

Spectra become flatter in the average: mean spectral indexes over the frequency range observed are 0.65–0.7 for a long time increasing to 0.75–0.8 in 2–3 h before CMEs registration.

Flattening spectrum can be connected with several reasons, one of which is the shift of the maximum of emission to short waves. The fact of the retention of emission value in the long-wave part of the range during the observations testifies in favor of this assumption. The latter can be caused by both the warming up of low layers of chromosphere and corona and by increase in the magnetic field because of the floating up new magnetic fluxes.

One more special feature of the dynamics of radiation spectrum into this period, is observed: the presence of the narrow-band component of emission with the frequency width 1–1.5 GHz. It appears in the band of the analysis of spectrograph for 1.5–3 seconds and moves from the low frequencies to the high, which composes the speed of the motion of about 1–2 GHz/s. The presence of such narrow-band component have been discovered by us with the earlier observations in the range 8–12 GHz and found satisfactory explanation within the framework of the transformation of plasma emission into the electromagnetic during the motion of thermal fronts in the flare loops.

Analyzing monitoring solar radio emission coincided CMEs onset we looked at so called isolated CMEs events. It is the situation when during the 8-hour time interval before and 6-hour time interval after these CMEs there were no other recorded CMEs events. It corresponds to well-known fact that it takes many hours for the atmosphere of active region to recover after the CMEs pass.

We find the presence of definite class of non stationary radio events before CMEs appearance during 2-hour intervals. During this interval there are observed sharp decay or variations of noise storm in meter emission, simultaneous appearance of microwave bursts of C- or S-types in cm–dm region. In most cases CMEs formation is accompanied by non stationary events in radio emission (about 80% of all events registered on SMM coronagraph). Non stationary radioemission corresponding to CME formation is observed in a greater frequency range than the radioemission without CME. 50% of nonstationary events preceding the CMEs onset are broad-band, they are observed at least in 3 parts of cm–dm frequency region.

CMEs that are not accompanied by non stationary events in solar radio emission mostly have high speed (mean value is about 700 km/s) and narrow width (mean width is less than 40 ang.deg). There are no CMEs Loop type among them.

4. Conclusions

The studies performed allow us to obtain results on the spectral-time dynamics of the preflare development of high-power solar events and demonstrate the promising nature of the spectral studies of microwave emission before CMEs onset. The high sensitivity and stability of the data on the fluxes of radio emission of the Sun at "Zimenki" facility show their effectiveness for statistical studies of weak events of solar activity.

Acknowledgements

The work is carried out with the support of Russian Fund for Basic Research (grant 03-02-16691), Ministry of Education (grant E02-11.0- 27) and FPSTP (Astronomy).

O. A. Sheiner would like to thank IAU for the supporting participation in IAU Symposium 226.

Coronal and Stellar Mass Ejections
Proceedings IAU Symposium No. 226, 2005
K. P. Dere, J. Wang & Y. Yan, eds.

Quasi-Periodic Components of Solar Microwave Emission Preceeding The CME Onset on 19 October, 2001

Olga A. Sheiner[1] and Vladimir M. Fridman[2]

[1,2]Radiophysical Research Institute (NIRFI),
25 Bol'shaya Pecherskaya Street, Nizhny Novgorod, 603950 Russia,
[1]email: rfj@nirfi.sci-nnov.ru
[2]email: fridman@nirfi.sci-nnov.ru

Abstract. The results of solar microwave observations in the Radio Astronomical Observatory NIRFI "Zimenki" are examined. Data analysis shows the presence of periodic component, that arose prior to burst connected to CMEs onset, and its absence after burst. Obtained data are compared with the dynamics of the development of activity on the solar disk. Results can be considered as the illustration of the dynamics of wave motions in the periods of flare activity.

Keywords. Sun: radio radiation, oscillations, coronal mass ejections (CMEs), evolution

1. Introduction

The laws governing the wave and fluctuating motions in the structures of solar atmosphere in the periods of flare activity and shaping of the coronal mass ejections are being established. We investigated the rapidly changing quasi-periodic components of solar radio emission with the use of data of the patrol observations of solar flux.

We would like to emphasize the effort in obtaining the data. The physical processes that can describe such a phenomena are not analyzed consciously. It is well known that there is a lot of research on the diagnostic possibilities of flare loops parameters, using parameters of the periodic oscillations of radio emission in the bursts.

Obtained by us results can be considered as a step in direction in development of diagnostic possibilities under the conditions for the monitoring observations of solar radio flux with the increased time resolution.

2. Observation Data

The situation in the sun during the observation day 19.10.01. was that there were 2 active regions on the Sun. It is evident that the analyzed dynamics of radio emission can be attributed to the changes in most developed active region AR 9658 (S14 W47) and CMEs event, which relates to this period. Peculiar 1B/M5.7 flare of 09:40 UT and radio emission and CMEs are discussed.

Data of radio emission was obtained in the period of patrol observations of the general flux of solar emission carried out in the period of the high solar activity during October 2001. Observations were conducted on 4 waves of cm and dm ranges in the Radio Astronomical Observatory "Zimenki" of Radiophysical Research Institute (NIRFI).

In this project we examine observation results at wavelengths 3 and 10 cm (time resolution on the radiometers of the corresponding wavelengths was 50 and 200 milliseconds, respectively). There was relatively high discretion of the registration of signals (0.5 s).

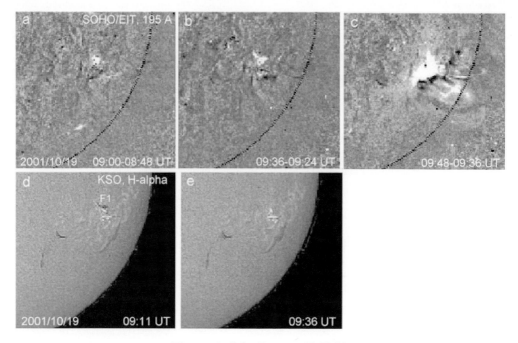

Figure 1. Solar Images. 19.10.01.

For the analysis we used parts of the record both directly preceded and after the burst, when the signal level did not change substantially within the limits of record.

3. Results

We used for this analysis the realizations of the records, that made it possible to analyze periods in the interval of 2–30 seconds. For the spectral treatment the program of Fast Fourier Transform was used which made it possible to estimate the presence of periodic components and their relative value.

Data analysis in 3rd cm emission shows the presence of periodic component (~ 14 s), that arose approximately 15 minutes prior to burst connected to CMEs onset, and its absence after burst.

In 10 cm emission we look at 2 components: about 6 seconds and 14–17 seconds. The first one exists prior to the burst during a whole 25-minute interval, while the component of ~ 14–17 seconds becomes steady 15 minutes prior to burst. The latter result coincides with ones obtained for 3rd cm emission. 6th seconds component remains also at the stage after burst.

The event in question preceded appearance a compact (size $\sim$ several angular minutes) UV brightening in the line 195Å in the northern environment of AR 9658 between 08:48 and 09:00 UT (Fig. 1a). As show higher sensitive difference images, at this time in the south there was observed the less intensive UV brightening, which was connected with the first weakly luminous loop, elongated westwards. The manifestation of this activity was observed as steel weak pulse dm-bursts at 900 and 600 MHz frequencies, and also the group of type III bursts (see Gnezdilov, Gorgutsa, Sobolev, *et al.* (2002)), which covered entire range of spectrograph from 270 to 25 MHz, and it was most intensive in interval of 08:55–08:58 UT. This followed the formation (or ejection) of the small transient H_α fiber F1 (Fig. 1d), which coincided in the localization with the basic UV brightening and was observed during 09:02–09:20 UT. To the time interval 09:36–09:38 UT we can see that UV brightening was renewed and increased in the size (Fig. 1b), and H$_\alpha$ fiber F1 ceased to be visible (Fig. 1e). From the LASCO data (http://cdaw.gsfc.nasa.gov/CME_list/) it is evident that CMEs took the form of the bright, clearly outlined, compact (angular dimensions of 15–20 deg.) ejection, which is extended in the corona with an uncommonly low speed of V$\sim$ 240 km/s. It should be noted that localization on the position angle, form and angular dimensions of this CMEs correspond well to the loop structure, which is visible into 09:48 UT on the UV difference images in the line 195 Å above the bright region of eruption (Fig. 1c).

The comparison of all data shows, that the appearance of fluctuations coincides with the appearance of dark fiber and the brightening. Fluctuations with periods (about tens of sec) disappear after the destruction of fiber, damage of the structure of flare loop.

4. Conclusions

Thus, the obtained results can be considered as the illustration of the dynamics of wave motions in the periods of pre flare and pre CMEs activity. It is an important step in increasing the possibilities of the stationary monitoring observations of the general flux of solar radio emission. The increased time resolution for the analysis of the dynamics of wave and fluctuating motions and development of diagnostic ideas is also vital.

Acknowledgements

The work is carried out with the support of Russian Fund of Basic Research (grant 03–02–16691), Ministry of Education (grant E02–11.0–27) and FPSTP (Astronomy).

O. A. Sheiner would like to thank IAU for the supporting the participation of IAU Symposium 226.

References

Gnezdilov, A.A., Gorgutsa, R.V., Sobolev, D.E., Fridman, V.M., Chertok, I.M., Sheiner, O.A., and Podstrigach, T.S. 2002, *Active Sun, Proc. Intern. Conference*, 66 (in Russian)

Session 4

Theoretical models of coronal mass ejections

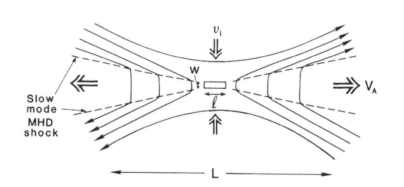

Coronal and Stellar Mass Ejections
Proceedings IAU Symposium No. 226, 2005
K. P. Dere, J. Wang & Y. Yan, eds.

Theories of Eruptive Flares

K. Shibata

Kwasan and Hida Observatories, Kyoto University, Kyoto Japan

Abstract. Recent progress of theories of eruptive flares (and CMEs) is reviewed within a framework of reconnection model with emphasis on development of basic idea and concept.

Keywords. Sun: coronal mass ejections (CMEs), magnetohydrodynamics: MHD

1. Introduction

Recent development of space solar observations have revealed various type of evidence of magnetic reconnection, not only for large scale flares and CMEs but also for small scale flares. Observations have also revealed that the association of mass ejections (plasmoids or flux rope) with these flares is much more common than previously thought, and have led to develop theories of flares and CMEs in a unified way (e.g., Shibata 1999). On the other hand, rapid development of supercomputers has developed MHD modeling of flares and CMEs greatly: we can now calculate realistic model of eruptive flares including various physical processes, such as reconnection, heat conduction, radiative cooling, and evaporation (see Shibata 2003 for a short review on this subject). Keeping these development in mind, we review recent progress of theories of eruptive flares and CMEs with emphasis on the development of basic concept and idea.

It should be noted that this paper is not a comprehensive review, and there are many important papers which are not cited in this paper. From that point of view, the reader should be referred to review papers by Priest and Forbes (2002) and Aschwanden (2002) for more complete citation and discussion of related papers.

2. Present Status of Reconnection Theory

At first, we should note that basic physics of magnetic reconnection has not yet been established (e.g., Hoshino *et al.* 2001). Fundamental puzzles of magnetic reconnection are summarized as follows:

1) What determines the reconnection rate? Or, what is the condition for fast reconnection?

2) What is the structure of reconnection region? Sweet-Parker type or Petschek type or others?

3) How much fraction of energy goes to nonthermal particles?

Hence, now is the stage that laboratory, space, and solar plasma physicists are collaborating to solve this basic physics. It should be emphasized that solar physicists have a lot of chances to contribute to solving this basic physics of reconnection, using excellent imaging data of solar flares and flare-like phenomena.

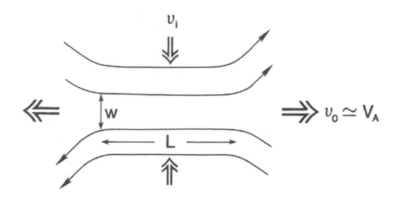

(a) Sweet-Parker reconnection

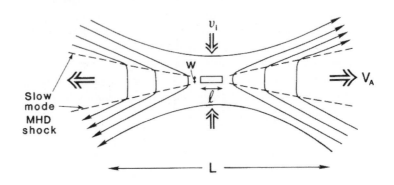

(b) Petschek reconnection

Figure 1. Reconnection models

Furthermore, we have following key questions on the origin of flares and CMEs.

4) How is energy stored? Is it the shearing motion or emergence of twisted flux tube?

5) What is the triggering mechanism for flares/CMEs?

6) What is the role of magnetic helicity in flares/CMEs?

As for the role of magnetic helicity, the reader should refer to Kusano (2005) and Hu (2005).

3. The Standard Model

The standard model of eruptive flares has been developed by the following pioneering researchers : Carmichael (1964), Sturrock (1966), Hirayama (1974), and Kopp-Pneuman

(1976). Hence the standard model is often called CSHKP model. Here we note the brief history how the term 'CSHKP' model appeared. At first, US people called the standard model 'Kopp-Pneuman' model in 80's. Since this was not fair, Shibata (1991) proposed to change it to 'SHKP' model, respecting pioneering work by Sturrock (1966) and Hirayama (1974). At that time, Sturrock himself was fair, and added 'C' just in front of 'SHKP' in 1992, noting the real pioneering work by Carmichael. Svestka and Cliver (1992) also used the term 'CSHKP' model in the same proceedings book.

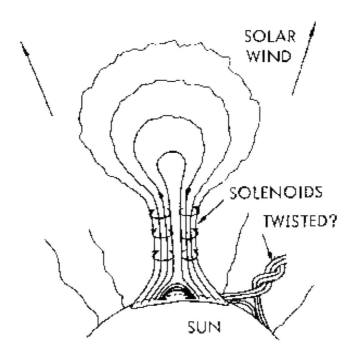

Figure 2. Carmichael (1964)

The standard model of eruptive flares has evolved significantly after 1976, especially because of development of understanding basic physics of reconnection region, as summarized in a nice review paper by Priest and Forbes (2002). Kopp and Pneuman (1976) considered that after reconnection of open field line, the solar wind along open field line collide to form shock inside the reconnected closed field, which heat the coronal plasma to flare temperature. However, Cargill and Priest (1982) correctly pointed out that we should consider the role of slow mode shock associated with Petschek type reconnection. Forbes and Priest (1984) noted the formation of fast shock (termination shock) due to reconnection jet above the reconnected loop, and Forbes and Malherbe (1986) pointed out that the slow shock is dissociated to isothermal slow shock and conduction front in solar flare condition.

Yokoyama and Shibata (1997) carried out for the first time the self-consistent MHD simulation of reconnection including heat conduction, and confirmed that the adiabatic slow shock is dissociated to isothermal slow shock and conduction front as predicted by Forbes and Malherbe (1986). This is very important to understand the structure of cusp-shaped flares observed by Yohkoh (Tsuneta *et al.* 1992). Yokoyama and Shibata (1998, 2001) succeeded to perform 2D MHD simulation of reconnection with heat conduction

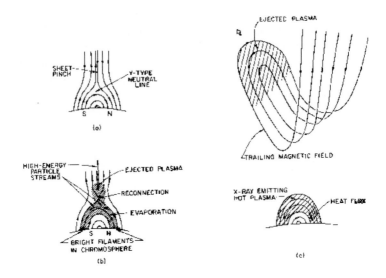

Figure 3. Sturrock (1966)

and chromospheric evaporation, and this is the most advanced model of eruptive flares. Yokoyama and Shibata (1998, 2001) further discovered the following scaling law for flare temperature:

$$T \sim 10^7 K (B/50G)^{6/7} (L/10^9 cm)^{2/7} (n_0/10^9 cm^{-3})^{-1/7} \qquad (3.1)$$

where B is the magnetic field strength, L is the reconnected loop length, and n_0 is the preflare electron density. This enabled the further unification of solar and stellar flares (Shibata and Yokoyama 1999, 2002).

4. Aly-Sturrock Theorem

It has often been discussed that the Aly-Sturrock conjecture (Aly 1984, 1991, Sturrock 1991) is a difficulty for modeling of eruptive flares and CMEs. Is this a real difficulty? We briefly outline a history of this conjecture.

Barnes and Sturrock (1972) calculated nonlinear evolution of force free field, and obtained the result that the energy stored in the nonlinear force free field is greater than that of the open field. At first, this result was thought to be consistent with observations, because the closed force-free field with energy higher than the open field can easily evolve to open field, and then magnetic reconnection occur in such open field current sheet to produce flares. More than 10 years later, however, Aly (1984) presented a conjecture 'the energy of any smooth force free field occupying a 'half coronal space' should be smaller than the energy of the so-called open field having the same flux distribution on the plane photospheric boundary'. Yang, Sturrock & Antiochos (1986) recalculated the Barnes-Sturrock problem, and reached the conclusion that 'Our new results differ from the earlier results of Barnes and Sturrock and we conclude that the earlier article was in error.' Furthermore, using the analytical approach, Sturrock (1991) showed 'the Aly conjecture is valid'. This is why the conjecture was called the Aly-Sturrock conjecture, and people began to think that the conjecture is not consistent with standard model (CSHKP model) of eruptive flares, since people thought the "open" vertical current sheet is necessary for the standard model whereas it is not easy to have such "open" vertical current sheet on the basis of the Aly-Sturrock conjecture.

However, it should be noted that the Aly-Sturrock conjecture is based on very simplified assumptions, such as force free (gas pressure and gravity can be neglected), Cartesian geometry, and that all magnetic field lines are connected to the boundary. So there are many ways out of this dilemma as follows.

- True opening of field line is not necessary for reconnection and mass ejections (Aly 1991).
- Non-force free (e.g., gas pressure, gravitational) (Sturrock 1991).
- Initially partly open, partly closed field (Sturrock 1991).
- Cylindrical axisymmetric geometry with spherical boundary (Lynden-Bell and Boily 1994).
- Resistive process (Mikic and Linker 1994).
- Quadrupole magnetic field (e.g., Biskamp and Welter 1989, Antiochos *et al.* 1999, and many).
- Two bipole sources (Choe and Cheng 2002).

Break out model by Antiochos *et al.* (1999) has been thought to be the promising way out of the dilemma. However, we should remember that there are many flare models with quadrupolar or multipolar magnetic field (e.g., Biskamp and Welter 1989, Uchida *et al.* 1999, Chen and Shibata 2000; also classical emerging flux model by Heyvaerts *et al.* 1977, Sweet (1958)'s model all belong to this category), and it is not fair that only break out model is discussed.

Why multipolar flux system is favorable? The reason is simple. If the magnetic field is bipolar, large energy is necessary for plasma to escape from closed bipolar flux system since plasma has to stretch many field lines. If the flux system is multipolar, small energy is enough for plasma to escape from the closed field region, since the number of field lines that plasma has to stretch is much smaller than that for bipolar flux system.

5. Current Sheet Formation (Energy Storage) Model

Traditionally, the following models have been considered to be applied to small scale flares or non-eruptive flares: converging flux model (Sweet 1958, Uchida *et al.* 1999, Priest *et al.* 1994), emerging flux model (Heyvaerts *et al.* 1977, Forbes-Priest 1984, Shibata *et al.* 1992, Yokoyama-Shibata 1995), sheared or converging arcade model (Mikic *et al.* 1988, Biskamp and Welter 1989, Forbes 1990, Kusano *et al.* 1995, Choe and Lee 1996, Magara *et al.* 1997, Hu 2000, Choe-Cheng 2001). However, there is no essential difference between small and large scale flares. It is also theoretically possible to unify these models and the standard model for eruptive flares, since the plasmoid (flux rope in 3D) is easily created in the current sheet (Shibata 1997).

Hence Shibata (1998, 1999) proposed the unified model, which he call plasmoid-induced reconnection model, in which plasmoids play following two roles 1) to store energy by inhibiting reconnection, 2) to induce inflow after ejection of plasmoids. In this model, fast reconnection occur as a result of strong inflow induced by plasmoid ejections. However, in order to create plasmoids, reconnection is necessary, and even after that, plasmoids are accelerated by energy release through reconnection, whereas the plasmoid ejection induces strong inflow and reconnection, which then accelerate plasmoids even further, vice versa. Hence both plasmoid ejection and magnetic reconnection are strongly coupled and form a kind of nonlinear instability (Shibata and Tanuma 2001). It is also noted that the current sheet tend to show many plasmoids with different sizes, i.e., fractal structure. Observed fractal-like hard X-ray and microwave emissions (Benz and Aschwanden 1992)

(a,b):   giant   arcades,
LDE/impulsive   flares,
CMEs

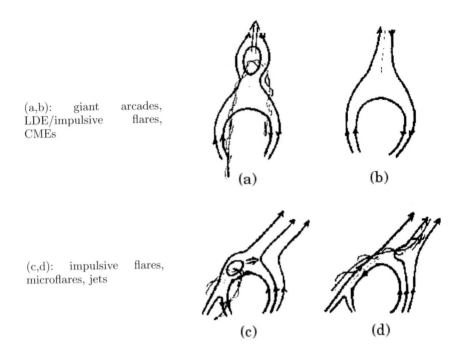

(c,d):   impulsive   flares,
microflares, jets

Figure 4. Unified model (Shibata 1999)

may be a result of such fractal-like reconnection induced by various plasmoid ejections with different size.

6.   Two-step Reconnection Model as Triggering Model

Chen and Shibata (2000) presented an MHD simulation model of eruptive flares and CMEs on the basis of idea from observational data analysis on the triggering of filament eruption by emerging flux (Feynmann and Martin 1995). In this model, small scale reconnection (cancellation) associated with emerging flux triggers large scale reconnection in the X-point high above or far from the emerging flux region. In this sense, it can be classified as two-step reconnection model (Wang and Shi 1993). The magnetic helicity annihilation model by Kusano *et al.* (2003), the break out model by Antiochos *et al.* (1999) and the tether cutting model by Moore & Roumeliotis (1992) also belongs to this model.

More recently, Shiota *et al.* (2003) extended the Chen-Shibata model and compared the model with Yohkoh observations of Y-shaped ejections above giant arcades (helmet streamer), finding the signature of slow and fast mode MHD shocks.

7. Remaining Questions

Finally, we list several key remaining questions in the following:
1) What is the condition of fast reconnection? Is it related to plasmoid ejections?

2) What is the energy storage and trigger mechanisms? Are they related to emerging flux?

3) Where are reconnection jet, inflow, and MHD shocks? Though there are several new findings on this subject (e.g., Yokoyama *et al.* 2001, Shiota *et al.* 2003), the Doppler shift observations of these phenomena are remained as important subjects for Solar B which will be launched in 2006.

Acknowledgements

The author would like to thank Prof K. Dere for his continuous encouragement and help on writing this proceedings paper.

References

Aly, J.J. 1984, Astrophys. J. 283, 349

Aly, J.J. 1991, Astrophys. J. 375, L61

Antiochos, S.K., DeVore, C.R., and Klimchuk, J.A. 1999, Astrophys. J. 510, 485

Aschwanden, M. 2002, Space Sci. Rev., 101, 1.

Barnes, C.W. and Sturrock, P.A. 1972, Astrophys. J. 174, 659

Benz, A.O. and Aschwanden, M.J. 1992, in Proc. Eruptive Solar Flares, IAU Colloq. No. 133, (eds.) Z. Svestka *et al.*, Lecture Notes in Physics, 399, Springer-Verlag, Berlin, p. 106

Biskamp, D. and Welter, H. 1989, Solar Phys., 129,49

Cargill, P.J. and Priest, E.R. 1982, Solar Phys. 76, 357

Carmichael, H. 1964, in Proc. of AAS-NASA Symp. on the Physics of Solar Flares, W.N. Hess (ed.), NASA-SP 50, p. 451

Chen, P.F. and Shibata, K. 2000, ApJ, 545, 524

Choe, G.S. and Lee, L.C. 1996, ApJ, 472, 372

Choe, G.S. and Cheng, C.Z. 2002, ApJL, 574, 17

Feynman, J. and Martin, S.F. 1995, J. Geophys. Res., 100, 3355

Forbes, T.G. and Priest, E.R. 1984, Solar Phys., 94, 315

Forbes, T.G. and Malherbe, J.M. 1986, Astrophys. J. 302, L67

Forbes, T.G. 1990, JGR, 95, 11919

Heyvaerts, J., Priest, E.R., and Rust, D.M. 1977, ApJ, 216, 123

Hirayama, T. 1974, Sol. Phys., 34, 323

Hoshino, M., Stenzel, L., Shibata, K., (ed.) 2001, special volume on "Magnetic Reconnection in Space and Laboratory Plasma", Earth, Planets, and Space, volume 53

Hu, Y.Q. 2000, Solar Phys. 200, 115

Hu, Y.Q. 2005, this volume

Kopp, R.A. and Pneuman, G.W. 1976, Solar Phys., 50, 85

Kusano, K., Suzuki, Y., and Nishikawa, K. 1995, ApJ, 441, 942

Kusano, K., Yokoyama, T., Maeshiro, T., and Sakurai, T. 2003, Adv. Sp. Res., 32,1931

Kusano, K. 2005, this volume

Lynden-Bell, D. and Boily, C. 1994, MNRAS, 267, 146

Magara, T., Shibata, K. Yokoyama, T., 1997, ApJ, 487, 437

Mikic, Z., Barnes, D.C., and Schnack, D. 1988, ApJ, 328, 830

Mikic, Z. and Linker, J.A. 1994 Astrophys. J. 430, 898

Moore, R.L. and Roumeliotis, G. 1992, in Eruptive Solar Flares (ed. Z. Svestka, B.V. Jackson and M.E. Machado), Springer-Verlag, Berlin, p69

Priest, E.R. and Forbes, T.G. 2002, A&A Rev. 10, 313

Priest, E.R., Parnell, C.E., and Martin, S.F. 1994, ApJ, 427, 459

Shibata, K., 1991, in *Proc. of "Flare Physics in Solar Activity Maximum 22"*, eds. Y. Uchida, R. Canfield, T. Watanabe and E. Hiei, in the series of Lecture Note in Physics, No. 387, Springer Verlag, pp. 205-218

Shibata, K., Nozawa, S., and Matsumoto, R. 1992, PASJ, 44, 265

Shibata, K. 1997, in Proc. 5-th SOHO workshop, (ESA SP-404) p. 103

Shibata, K. 1998, in Proc. Observational Plasma Astrophysics, Watanabe, T., Kosugi, T., and
    Sterling, A.C. (eds), Kluwer, p. 187
Shibata, K. 1999, Astrophys. Space Sci., 264, 129
Shibata, K. and Yokoyama, T. 1999, ApJ, 526, L49
Shibata, K. and Tanuma, S. 2001, Earth, Planets, Space, 53, 473
Shibata, K. and Yokoyama, T. 2002, ApJ, 577, 422
Shibata, K 2003, in Proc. IAU 8th Asian Pacific Regional Meeting, Ikeuchi, S. *et al.* (eds.), ASP,
    p.371
Shiota, D., *et al.* 2003, PASJ, 55, L35
Sturrock, P.A. 1966, Nature, 211, 695
Sturrock, P.A. 1991, Astrophys. J. 380, 655
Sturrock, P.A. 1992, in Eruptive Solar Flares, IAU Colloq. No. 133, (eds.) Z. Svestka
    et al., Lecture Notes in Physics, 399, Springer, Berlin, p. 397
Sweet, P.A. 1958, IAU Symp. 6, 123
Svestka, Z. and Cliver, E.W. 1992, in Eruptive Solar Flares, IAU Colloq. No. 133, (eds.)
    Z. Svestka, *et al.*, Lecture Notes in Physics, 399, Springer, Berlin, p. 1
Tajima, T. and Shibata, K. 1997, Plasma Astrophysics, Addison-Wesley
Tsuneta, S., *et al.* 1992, PASJ, 44, L63
Uchida, Y., Hirose, S., Cable, S., Morita, S., Torii, M., Uemura, S., and Yamaguchi, T. 1999,
    PASJ,51, 553
Wang, J. and Shi, Z. 1993, Solar Phys. 143, 119
Yang, W.H., Sturrock, P.A., and Antiochos, S.K. 1986, ApJ., 309, 383
Yokoyama, T. and Shibata, K. 1995, Nature, 375, 42
Yokoyama, T. and Shibata, K. 1996, PASJ, 48, 353
Yokoyama, T. and Shibata, K. 1997, ApJ, 474, L61
Yokoyama, T. and Shibata, K. 1998, ApJ, 494, L113
Yokoyama, T. and Shibata, K. 2001, ApJ, 549, 1160
Yokoyama, T., *et al.* 2001, ApJ, 546, L69

Discussion

SCHWENN: Please comment on non-reconnection CME model

SHIBATA: As for this, I had an interesting discussion with B. C. Low. He claims that
there is a regime where a CME can occur without reconnection, and such CMEs can
be accelerated by magnetic buoyancy after draining mass from the prominence due to
gravity. I think even in such case, reconnection is necessary to drain mass from the
prominence. Hence I do not believe that many CMEs belong to a non-reconnection model.
Only a small fraction of CMEs may belong to a non-reconnection model, if the CME is
similar to a slowly rising arch filament observed in an emerging flux region. But in this
case there is no explosive energy release nor rapid mass motion.

KAHLER: What does it mean that reconnection is fractal - size structure or energy
releases? what is the evidence that reconnection is not only structured but also fractal?

SHIBATA: Here I used the word fractal from two points of view: spatial structure and
temporal variation. As a result, the energy release shows also fractal: the occurrence
frequency vs released energy would show power law distribution. We have proposed that
the current sheet consist of a number of plasmoids (flux rope) which have difference sizes
with power law distribution (fractal) (Tajima and Shibata 1997, Shibata and Tanuma
2001). These plasmoids collide each other or expelled from the current sheet, both of
which induce intermittent reconnection. Since the released energy and time scale are
determined by the size of plasmoid, the energy spectrum and the power spectrum of time
variability would also show power law distribution (fractal). There are indirect evidence

for this, i.e., time variability of radio microwave emissions and hard X-ray emissions (e.g., Benz and Aschwanden 1992), both of which show power law spectrum for time variation of the intensities of these emissions.

GOPALSWAMY: Most filament eruptions are accompanied by mass down flows? How does this down flows fit in your eruption scenario?

SHIBATA: Once the eruption of plasmoid (flux rope) occurs, the core of flux rope (prominence) would expand like Omega shape. So gravity acts along the rising Omega (helical) loop, enabling mass draining. So such observations fit our eruption scenario.

STERLING: You said that the important point about break out is the reconnection in the corona. But isn't it also important that the coronal reconnection be slow initially, and then fast? Without the initial slow reconnection, you would not have a stress buildup between the two flux system followed by explosive eruption. Therefore I think that the important point is that the reconnection rate in the corona be slow at first, and then fast. I think that the key is for the reconnection to be inhibited initially. This is a basic point for many astrophysical circumstances.

SHIBATA: Yes I agree with you about the importance of inhibiting reconnection. Without inhibiting reconnection, it would be difficult to store energy in the corona and also difficult to get explosive energy release. At present, we do not know how we can have slow reconnection initially and then get fast reconnection later. There is no established answer about this question, which is of course a key basic question of the reconnection physics.

Coronal and Stellar Mass Ejections
Proceedings IAU Symposium No. 226, 2005
K. P. Dere, J. Wang & Y. Yan, eds.

© 2005 International Astronomical Union
doi:10.1017/S1743921305000657

The Effects of Gravity of an Loss-of-Equilibrium CME Initiation Model

Katharine K. Reeves and Terry G. Forbes

University of New Hampshire, Institute for Earth, Oceans and Space, Durham, NH 03824, USA
email: kreeves@unh.edu

Abstract. We include gravity in a loss of equilibrium model for the initiation of coronal mass ejections (CMEs). We examine equilibria for both normal and inverse polarity and neglect the effects of current sheets. Although equilibria exist for normal polarities, in the absence of current sheets, the equilibria are unstable to horizontal perturbations. For the inverse polarity configuration, we find that gravity generally has a negligible effect if the magnetic field is strong (>50 G) but that it can have a significant effect if the magnetic field is weak. Specifically, if the characteristic magnetic field is less than about 6 G, no eruption occurs if the CME mass is on the order of 2×10^{16} gm.

Keywords. Sun: coronal mass ejections, MHD

1. Introduction

Loss of equilibrium CME models are characterized by a catastrophic unbalancing of the forces on a flux rope in the solar corona. In this paper, we will examine the effect of including gravity in a system that consists of a flux rope at a height h above the corona and two line sources embedded in the photosphere a distance 2λ apart. There are photospheric currents that are accounted for by an image current submerged a distance h below the photosphere. This magnetic configuration has been examined before (Forbes & Priest (1995), Lin & Forbes (2000)), but without the inclusion of gravity in the equilibrium equation.

The effect of gravity was previously considered by Isenberg, Forbes & Démoulin (2002) for a 2D quadrupolar configuration, but only for the inverse configuration. They determined that in the asymptotic case of an infinitely small flux rope radius, there is a limiting magnetic field below which no loss of equilibrium occurs. A similar result is suggested in Lin (2004), for an inverse configuration containing a current sheet. In this paper, we will examine the effect of gravity on the equilibria of both normal and inverse configurations (see Figure 1), and we will examine the equilibrium curves with respect to footpoint motions to determine the conditions under which a loss of equilibrium is possible for configurations with no current sheet and a finite flux rope radius.

We also examine the equilibria with respect to draining mass from the flux rope. Mass loss has been suggested as a mechanism for CME initiation (*e.g.* Zhang & Low (2004)), and it becomes a factor in the balance of forces on the flux rope only when gravity is taken in to account. The other variations of this model that include the effects of gravity (Isenberg, Forbes & Démoulin (2002), Lin (2004)) do not examine the equilibrium curves that result when the mass of the flux rope is varied.

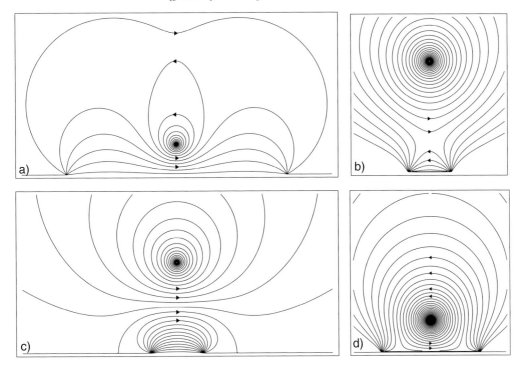

Figure 1. Magnetic configurations used in the model. Panels a) and b) are the normal and inverse polarity configurations, respectively, including an x-line. Panels c) and d) are the normal and inverse configurations, respectively, with no x-line. Arrows indicate magnetic field direction.

2. Equilibrium curves

2.1. *Governing equations*

The inverse and normal polarity magnetic configurations used in these calculations are shown in Figure 1. Some configurations do not have x-lines – the equation for the x-line location is given in Priest & Forbes (2000), and it can be shown that the x-line disappears in the normal configuration when $h/\lambda > -1/2J$, where J is the normalized current in the flux rope. For the inverse configuration, no x-point occurs if $h/\lambda < 2J$.

The equilibrium curves are curves in a parameter space where the total force on the flux rope is zero. In the absence of a current sheet, the vertical forces on the flux rope are: an upward force due to the image current, a downward force due to gravity, and a force due to the photospheric line sources which is either upwards (normal) or downwards (inverse). Normal polarity configurations can be in equilibrium only if gravity is included, since all other vertical forces on the flux rope in that configuration are upwards. In general, the equation for the vertical force per unit length on the flux rope is

$$F_y = \frac{I B_{ext_x}(x,y)}{c} - \frac{mg_0}{(1 + h/R_{sun})} \tag{2.1}$$

where I is the current in the flux rope, B_{ext} is the external field, m is the mass of the flux rope per unit length, g_0 is the value of gravity at the sun's surface and R_{sun} is the radius of the sun. The x component of the external field is given by

$$B_{ext_x}(x,y) = \frac{2A_0}{\pi} \left[\frac{J(y+h)}{x^2 + (y+h)^2} - \frac{2\lambda(\lambda^2 - x^2 + y^2)}{\lambda^4 - 2\lambda^2(x^2 - y^2) + (x^2 + y^2)^2} \right] \tag{2.2}$$

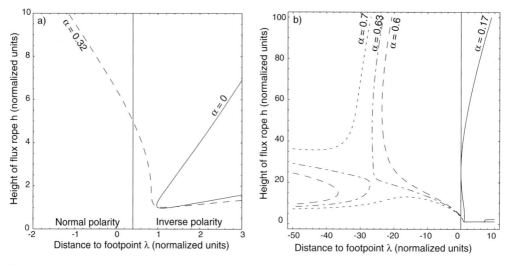

Figure 2. Equilibrium curves with respect to footpoint separation that have nose points a) in the inverse configuration and b) in the normal configuration. Negative values of λ indicate normal polarity magnetic configurations, while positive values indicate inverse polarity configurations. Distances are normalized to the height of the flux rope at the maximum current point.

where A_0 is the value of the vector potential at the footpoints, and $J = I\pi/cA_0$. The first term in B_{ext_x} is due to the image current and the second is due to the photospheric source terms.

The horizontal force on the flux rope depends only on the photospheric line sources, since the image current always lies directly under the flux rope:

$$F_x = \frac{IB_{ext_y}(x,y)}{c} = -\frac{A_0}{\pi}\left[\frac{4xy\lambda}{\lambda^4 - 2\lambda^2(x^2 - y^2) + (x^2 + y^2)^2}\right] \qquad (2.3)$$

2.2. *Footpoint motions*

Figure 2 shows select equilibrium curves with respect to footpoint motions for different values of $\alpha = mg_0h_0c^2/I_0^2$, a parameter that relates the strength of the gravitational force to the force of the characteristic magnetic field. A loss of equilibrium occurs at the point where the equilibrium curve folds back on itself, creating a "nose point" – an example can be clearly seen for the zero gravity case in Figure 2a. If the force due to the characteristic field is weak enough compared to the force of gravity, this nose point can disappear, as shown for the $\alpha = 0.32$ case in Figure 2a. In real units, if the mass per unit length, m is 2.1×10^6 gm/cm and $h_0 = 5 \times 10^9$ cm, then the characteristic field (given by I_0/ch_0) must be greater than 6 G for a nose point to exist in the inverse configuration in the resulting equilibrium curve.

For normal field configurations, there are no equilibria possible if $\alpha < 0.17$, as shown in Figure 2b. For larger values of α, equilibria do exist in the normal field configuration, and there are nose points. All nose points vanish, however, when $\alpha > 0.65$. For the same values of m and h_0 used above, the characteristic field should be between 4.2 and 8.2 G for nose points to exist in the normal configuration. It should be noted, as Jun Lin pointed out in the discussion of this paper, that the normal equilibria calculated here are not stable with respect to horizontal perturbations even though some branches are stable with respect to vertical perturbations. Stability could possibly achieved with the introduction of a current sheet at the x-line for those configurations that have x-lines (see Figure 1).

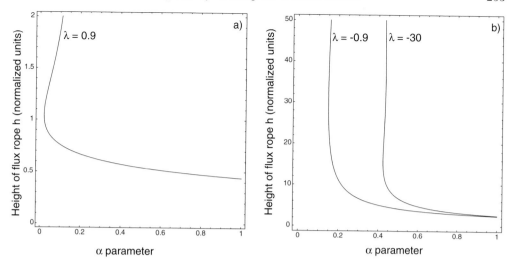

Figure 3. Equilibrium curves with respect to mass loss in the prominence showing the location of nose points for a) the inverse configuration and b) the normal configuration. Distances are normalized to an arbitrary point on the equilibrium curve, since there is no maximum current point.

2.3. *Mass loss*

Figure 3 shows equilibrium curves for inverse and normal polarity configurations with respect to a decrease in the parameter α, which is equivalent to mass loss in the flux rope since α is directly proportional to the mass. The nose point in the inverse polarity case occurs at a smaller value of α than either of the two normal polarity cases. Thus there must be more mass drained from an inverse configuration flux rope to bring it to a catastrophic loss of equilibrium than for a flux rope in the normal configuration. This finding agrees with Zhang & Low (2004), who estimated masses necessary for anchoring the flux rope in the corona and found that more mass is needed to keep the normal configuration from erupting.

Figure 4 shows the total energy plotted along the mass loss equilibrium curves for three values of λ. The x on each curve marks the position where the loss of equilibrium occurs. Zhang & Low (2004) have contended that normal magnetic configurations store more energy than inverse magnetic configurations, but these results show that it is not always the case. Specifically, normal configurations with large footpoint separations have a higher total energy at the loss of equilibrium point than inverse configurations, but normal topologies with close together footpoints have a lower energy at the loss of equilibrium point. The configuration with large footpoint separation is pushing the boundaries of our model, however, because we use a gravity term that falls off as $1/h^2$ in combination with an infinite plane cartesian geometry.

3. Conclusions

We have found that when equilibria with respect to footpoint separation are examined, the nose points in the equilibrium curves can disappear for both normal and inverse polarities if the force on the flux rope due to the characteristic magnetic field is weak compared to the force of gravity. The minimum magnetic field that allows eruptions to occur in the inverse case is approximately 6 G. This value is lower than the 17 G found in Isenberg, Forbes & Démoulin (2002), possibly because of the different magnetic

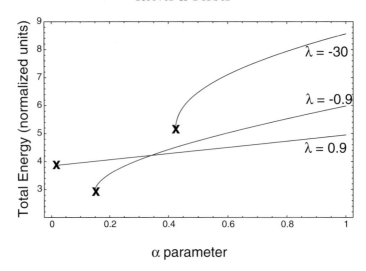

Figure 4. Plots of the total energy along the mass loss equilibrium curves for three different values of λ. Positive values of λ indicate inverse configurations and negative values indicate normal configurations. The position of the loss of equilibrium is marked with an x.

configurations used. Lin (2004) finds a minimum value for the characteristic field of 13.5 G, which is higher than our value possibly because he is concerned with CME propagation rather than the actual disappearance of the nose point. When a current sheet is included, as in Lin (2004), there could be parameters for which the CME does not propagate out in to the corona, but a loss of equilibrium still occurs.

We have also examined the equilibria with respect to mass loss in the flux rope, and we find that more mass must drain to attain a loss of equilibrium in the inverse configuration than in the normal configuration, which agrees with the conclusions of Zhang & Low (2004). Plotting the energy along the mass loss equilibrium curves for several different values of λ, we find that the total energy at the nose point can be higher in the normal configuration than in the inverse configuration if λ is large enough, however, this conclusion is weakened by the constraints of our model. For small values of λ the energy at the loss of equilibrium point is smaller in the normal configuration than in the inverse configuration, contradicting the assertion of Zhang & Low (2004) that normal configurations store more energy than inverse ones.

Acknowledgements

This work was supported by NSF grant ATM-0327512 and NASA LWS TR+T grant NNH05AA13I to UNH as well as DoD MURI grants to UC Berkeley and the University of Michigan.

References

Forbes, T.G. & Priest, E.R. 1995, *ApJ* 446, 377
Isenberg, P.A., Forbes, T.G. & Démoulin, P. 1993, *ApJ* 417, 368
Lin, J. & Forbes T.G. 2000, *J. Geophys. Res.* 105, 2375
Lin, J. 2004, *Sol. Phys.*, 602, 422
Priest, E.R. & Forbes, T.G. 2000, *Magnetic Reconnection* (Cambridge University Press), p.378
Zhang, M. & Low, B. C. 2004, *ApJ* 600, 1043

Discussion

KOUTCHMY: Could your flux rope correspond to the cavity we see rising up during a CME possibly due to buoyancy?

REEVES: No, in this model the radius of the flux rope is very small for reasons of mathematical simplicity.

JUN LIN: We might need to worry about the stability of the system to the horizontal disturbance in the normal polarity case.

REEVES: This point is an extremely good one and we will examine the stability of the normal polarity equilibria with respect to horizontal perturbations.

GOPALSWAMY: Is your stability analysis consistent with the observational fact that inverse polarity filaments are more common than the normal polarity ones?

REEVES: Yes, especially in light of Jun Lin's comment about stability with respect to perturbations in the horizontal direction. Also, there are no equilibria in the normal configuration when the fields are strong compared to the gravitational force on the flux rope, but there are inverse equilibria in this case.

Coronal and Stellar Mass Ejections
Proceedings IAU Symposium No. 226, 2005
K. P. Dere, J. Wang & Y. Yan, eds.

© 2005 International Astronomical Union
doi:10.1017/S1743921305000669

Study of the Relationship between Magnetic Helicity and Solar Coronal Activity

Kanya Kusano

Graduate School of Advanced Sciences of Matter, Hiroshima University

Abstract. Magnetic helicity, which is a measure of magnetic flux linking, is widely believed to play a crucial role for the solar coronal energetic processes, e.g. heating, flares and eruptions. In this paper, we introduce the several new findings of magnetic helicity physics in the solar corona both from the observational and theoretical points of view. The new observations based on the vector magnetograms successfully revealed that the solar coronal activity is indeed related not only to the intensity of magnetic helicity injection from the photosphere, but also to the complexity in the structure of magnetic shear. In particular, we recently found through the advanced analyses of vector magnetogram data that steep reversal of magnetic shear may efficiently activate the liberation of free energy stored in the coronal magnetic field. Motivated by the results, we developed the large scale three-dimensional simulation to investigate the causal relationship between the magnetic helicity injection and the energy liberation in the solar corona. The simulations clearly demonstrated that the reversal of the magnetic shear in a magnetic arcade can cause the sudden onset of plasmoid eruption. The mechanism is able to be explained as a self-exciting process of multiple magnetic reconnections. Finally, we propose a new scenario for the triggering mechanism of eruptive flares, which is called the reversed-shear flare model.

Keywords. Sun: coronal mass ejections (CMEs)

References

Kusano, K., *et al.*, 2002, ApJ, 577, 501
Kusano, K., *et al.*, 2004, ApJ, 610, 537
Maeshiro, T., *et al.*, 2004, ApJ, submitted

Discussion

SUBRAMANIAN: What is the measure of structural complexity you use?

KUSANO: Length of shear inversion layer where the sign of magnetic shear is changed.

LAKHINA: What is the magnetic Reynolds number used in your simulation?

KUSANO: The magnetic Reynolds number is 10^5 in our simulation except in the thin current layer, where it is reduced to 2×10^3 by anomalous resistivity.

Coronal and Stellar Mass Ejections
Proceedings IAU Symposium No. 226, 2005
K. P. Dere, J. Wang & Y. Yan, eds.

© 2005 International Astronomical Union
doi:10.1017/S1743921305000670

Magnetic Structure Equilibria and Evolutions due to Active Region Interactions

J. Lin and A. A. van Ballegooijen

Harvard-Smithsonian Center for Astrophysics, 60 Garden Street, Cambridge, MA 02138, USA
email: jlin@cfa.harvard.edu, vanballe@cfa.harvard.edu

Abstract. Equilibria and evolutions in the coronal magnetic configurations due to the interactions among active regions are investigated. The magnetic structure includes a current-carrying flux rope that is used to model the prominence or filament. We use either two dipoles or four monopoles on the boundary surface to model active regions, and the change in the boundary conditions corresponds to either the horizontal motion of magnetic sources or decaying of the active regions. Both cases show the catastrophic behavior in the system's evolutions. The results have important observational consequences: most eruptive prominences that give rise to CMEs are driven by the interactions between two or more active regions. Such eruptions may not necessarily take place in the growing phase of the active regions, instead they usually occur at the decay phase.

Keywords. Sun: coronal mass ejections (CMEs), evolution, filaments, MHD

1. Introduction

In response to slow motions in the photosphere and in the convective zone, the coronal magnetic field evolves and builds up the stress and the extra energy. In addition to the usual forms of motions, such as shearing and converging of the footpoints of the magnetic field, and the new emerging flux, the system's energy can also increase due to the interaction among two or more active regions as they approach one another. In observations, the horizontal motions of active regions (or sunspots) have long been a well-known precursor for some types of eruptions (Tanaka & Nakagawa 1973; Martres *et al.* 1986; Dezsö *et al.* 1984; Kovács & Dezsö 1986; Dezsö & Kovács 1998).

Previously, we investigated the detailed consequences of the new emerging flux, and discussed the results of the horizontal motions of the magnetic source regions on the boundary surface (Lin *et al.* 2001). In the present work, we consider the equilibria and evolutions in the magnetic configurations with background fields resulted from two dipoles or four monopoles located on the boundary surface. All the configurations include a current-carrying flux rope that is usually used to model the prominence or filament. The system evolves as a result of either changes in the strength of the background field, which is utilized to model the magnetic cancellation occurring below and on the surface, or displacement of the source regions on the surface. As we usually do, we are not fuzzy with the details of how a prominence forms. The interested readers may refer to the works by van Ballegooijen & Martens (1989), van Ballegooijen *et al.* (2000), as well as MacKay & van Ballegooijen (2001 and 2005).

2. Description of Models and Equilibria in Systems

The model consists of equilibrium solution of the two-dimensional, ideal-MHD equations in the semi-infinite x-y plane ($y \geqslant 0$) with $y = 0$ being the photospheric boundary

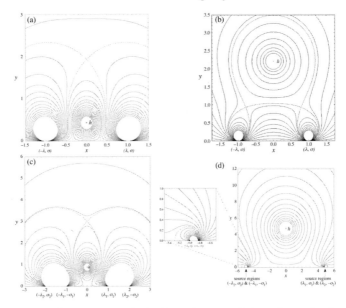

Figure 1. Magnetic configurations with the boundary condition of two dipoles located on the surface: (a) $\lambda = 1$, $\sigma = -2.02$, and $h = 0.43$, (b) for $\lambda = 1$, $\sigma = 1$, and $h = 2.12$; and magnetic configurations with the boundary condition of four monopoles located on the surface: (c) for $\lambda_1 = 1$, $\lambda_2 = 2$, $\sigma_1 = 1$, $\sigma_2 = 1$, and $h = 0.89$; and (d) for $\lambda_1 = 4.84$, $\lambda_2 = 5$, $\sigma_1 = 1.46$, $\sigma_2 = 1$, and $h = 4.75$. Dashed curves in all the panels plot the separatrices and the magnetic interfaces.

(or more properly, the base of the corona) and $y > 0$ corresponding to the corona which is perfectly conducting. So, magnetic field lines in the corona are frozen to the plasma and any reconnection which occurs is restricted to the region where a neutral point or the current sheet exists, and to the photosphere.

In the environment of the low corona, the magnetic field can be treated as force-free and its corresponding evolution prior to the eruption is considered ideal-MHD. We also assume that all the currents are confined in the flux rope and the photosphere, and that there is no current sheet present before the catastrophe takes place (e.g., see Lin *et al.* 2001; and Lin & van Ballegooijen 2002). The magnetic configurations of interest are given in Figures 1a and 1b for the case in which the background field is produced by two dipoles on the boundary surface; and in Figures 1c and 1d for the case in which the background field results from four monopoles. In each of these panels, h is the height of the flux rope, σ is the strength of each dipole, λ is the distance between each dipole and the origin, $\pm\sigma_1$ and $\pm\sigma_2$ are the strengths of monopoles, and $\pm\lambda_1$ and $\pm\lambda_2$ are the distance of each monopole to the origin respectively.

2.1. *Evolution in the System with Two-Dipole Boundary Condition*

In this part of work, we investigate the equilibrium and the evolution in the magnetic configuration with the background field produced by two dipoles located on the boundary surface (Figures 1a and 1b). The evolution in such a system is caused by the changes in either σ or λ or both. Plotting the equilibrium height of the flux rope h versus λ and σ gives two equilibrium surfaces in (λ, σ, h) space (Figure 2a), which are in the regions of $\sigma > 0$ and $\sigma < 0$, respectively. The shadows located at the bottom of the figure outlines the domains of (λ, σ) over which the equilibria in the system exists. Obviously, the two surfaces do not join one another within the domain of physical meaning ($\lambda > 0$ and $h > 0$). This implies that the smooth transition between the configurations in equilibria

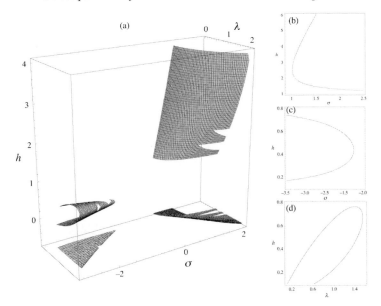

Figure 2. (a) Evolutionary features of the system manifested in the parameter space (λ, σ, h). (b) Variations of the equilibrium heights h versus σ for $\lambda = 1$ and $\sigma > 0$. (c) Variations of h versus σ for $\lambda = 1$ and $\sigma < 0$. (d) Variations of h versus λ for $\sigma = -2.5$.

governed by different surfaces shown in Figure 2a is impossible since any such tendency will directly result in the catastrophic loss of equilibrium.

To reveal the detailed catastrophic behavior implied by Figure 2a, we slice the surfaces with a plane of $\lambda = \text{const}$, say $\lambda = 1$, and thereby obtain the functional behavior of h as a function of σ. Two equilibrium curves (h versus σ) are deduced as shown in Figure 2b: the curve comes from the surface at right ($\sigma > 0$), and in Figure 2c: the curve comes from the surface at left ($\sigma < 0$). Both curves display the bifurcation and thus suggest the catastrophic characteristics of the evolution as the background field decays.

The evolution in the system may also occur in response to the change in λ, the distance between each source region and the origin. This equivalent to the case in which two active regions approach or depart from one another. Because we are just interested in the configuration with the flux rope located below the X-point (Figure 1a), we only need to focus on the evolution governed by the surface at left in Figure 2a. Slice this surface with a plane of $\sigma = -2.5$, thereby we obtain a curve of h versus λ as shown in Figure 2d. This curve consists of two branches and suggests the catastrophic behavior as the system evolves along the lower branch in response to the separation of two dipoles.

A similar situation with the point sources on the boundary surface was investigated by Forbes & Priest (1995), but the quasi-static evolution in their case eventually turns to the catastrophe with the two source regions on the surface approaching, instead of separating from, one another. Furthermore, the range of λ over which the equilibrium exists in the present case is much smaller than that of Forbes & Priest (1995). Such differences in the evolutionary behaviors are basically due to the different properties of the background field in each case.

2.2. *Evolutions in the System with Boundary Condition of Four Monopoles*

In this part of work, we model the evolution in the system with the background field produced by four monopoles on the boundary surface. Comparing with the case of Forbes & Priest (1995), two more monopoles are included. Adding two extra monopoles of opposite polarities to the magnetic system studied by Forbes & Priest (1995) results in the magnetic configurations shown in Figure 1c and 1d. The complex boundary condition leads the X-point to existing in the background field. The interactions among different magnetic units in this system can be understood to occur either between the active region consisting of $-\sigma_1$ and σ_1 and two individual spots $-\sigma_2$ and σ_2, or between the two active regions of opposite orientations located at each side of the origin. For simplicity, we arrange these four monopoles in the symmetric way as shown in Figures 1c and 1d. In reality, the interacting systems may form much more sophisticated structures without any symmetry, but the fundamental physical courses should be the same.

Without losing the generality, we assume that λ_2 is always larger than λ_1, the evolution in the system is due to the change in either σ_1 or λ_1, and the sources $\pm\sigma_2$ exist as a background in distance. Such a scenario provides us with an opportunity to investigate how a large scale structure in background impacts the evolutionary behaviors of the local magnetic units, and how the re-organization of the local magnetic components in a confined area, in turn, disturbs the global field.

In this case, the equilibrium height of the flux rope h is a function of σ_1, σ_2, λ_1, and λ_2. This is a five-dimensional function in (σ_1, λ_1, σ_2, λ_2, h) space, which can hardly be investigated directly. In the present work, due to the limited space, we will focus on the evolutions in response to the changes in σ_1 and σ_2 only. Those in response to the changes in λ_1 and λ_2 can be found in our another work (Lin & van Ballegooijen 2005). Figure 3a plots h as a function of λ_1 and σ_1 for $\lambda_2 = 5$ and $\sigma_2 = 1$. (Gaps on the surface are purposely left so that the details behind the surface, if any, can be seen.) Figure 3b displays a group of $h - \sigma_1$ curves for $\lambda_1 = 1, 2, 3$, and 4, respectively. All the curves show the typical property of bifurcation that results in the catastrophe as the system evolves in response to the change in the background field. Furthermore, the height of the turning (critical) point (σ_{1c}, h_c) at which the catastrophe occurs increases with the value of λ_1. This suggests that the eruption due to the magnetic cancellation between two approaching magnetic sources takes place more easily if the cancellation occurs as the two sources are still distant. We can understand this conclusion in another way: the equilibrium in the configuration with source regions close to one another can be kept more easily than that in the configuration of which the sources are located far apart.

Figure 3c plots h versus λ_2 and σ_2 for $\lambda_1 = \sigma_1 = 1$. We notice that when sources σ_1 and σ_2 are close to one another, the equilibrium occurs only if $\sigma_2 < \sigma_1$. This is because strong σ_2 weakens the capability of σ_1 of maintaining the system in equilibrium. Figure 3d plots the equilibrium curves, $h - \sigma_2$, for $\lambda_2 = 1.5, 3, 5$, and 10, respectively. These curves are nearly the duplicates of their counterparts in Figures 3b, except the opposite concave directions, which is indicating that the roles played by λ_1 and λ_2 (also by σ_1 and σ_2) in driving the evolutions of the system are just opposite. This suggests that the importance of sources σ_1 and σ_2 is relative, and their roles are symmetry in governing the evolution in the system.

3. Conclusions

Consequences of interactions between two magnetic systems (active regions) were investigated in the present work. The magnetic configurations include a current-carrying flux rope that is used to model the prominence or filament. The configurations in equilibrium evolve in response to the change in parameters for the background field, such as the strength and the distances between sources. Our main results are listed as follows:

1. The equilibrium positions of the flux rope can be either below or over the neutral points. In the case that the background field is produced by two dipoles, smooth transition

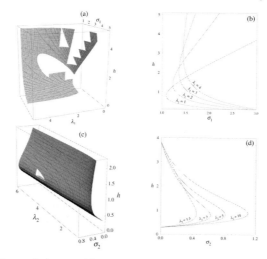

Figure 3. (a) Variations of the equilibrium heights of the flux rope h versus λ_1 and σ_1. (b) Equilibrium curves $h - \sigma_1$ for various λ_1. (c) Variations of the equilibrium heights of the flux rope h versus λ_2 and σ_2. (d) Equilibrium curves $h - \sigma_2$ for various λ_2.

between the configurations with the flux rope below and over the neutral points is impossible; in the case of four monopoles, such transition seems to be possible.

2. No matter whether the background field is produced by two dipoles or by four monopoles, evolutions of the system in response to the change in the sources may eventually perform catastrophic feature. This indicates the impact of the varying magnetic sources on driving the disruption of the magnetic structure in the corona.

3. We focus on the cases in which the flux rope is located below the neutral point. But the impact of the closed field lines over the neutral point on the consequence of the catastrophe is not clear. They may be quite possible to behave like those described by the break-out model (e.g., Antiochos *et al.* 1999; and MacNeice *et al.* 2004).

Acknowledgements

This work was supported by NASA under the grants NNG04GE84G and NAG5-12827 to the Smithsonian Astrophysical Observatory.

References

Antiochos, S.K., DeVore, C.R., & Klimchuk, J.A. 1999, *ApJ*, 510, 485.
Dezső, L., Csepura, G., Gerlei, O., Kovács, Á., & Nagy, I. 1984, *Adv. Space Rev.*, 4(7), 57.
Dezső, L. & Kovács, Á. 1998, in *The 14th Consultation on Solar Physics: Conference Proceedings*, (eds.) B. Rompolt, J. Jakimiec, & P. Heinzel, p. 17.
Forbes, T.G. & Priest, E.R. 1995, *ApJ*, 446, 377
Lin, J., Forbes, T.G., & Isenberg, P.A. 2001, *JGR*, 106, 25053
Lin, J. & van Ballegooijen, A.A. 2005, *ApJ*, submitted.
MacKay, D.H. & van Ballegooijen, A.A. 2001, *ApJ*, 560, 445.
MacKay, D.H. & van Ballegooijen, A.A. 2005, *ApJ*, 621, 77.
MacNeice, P.J., Antiochos, S.K., Phillips, A., Spicer, D.S., DeVore, C.R., & Olson, K. 2004, *ApJ*, 614, 1028.
Martres, M.J., Mouradian, Z., & Soru-Escaut, I. 1986, *A& A*, 161, 376.
Tanaka, K. & Nakagawa, Y. 1972, *Solar Phys.*, 33, 187.
van Ballegooijen, A.A. & Martens, P.C. H. 1989, *ApJ*, 343, 971.
van Ballegooijen, A.A., Priest, E.R. & MacKay, D.H. 2000, *ApJ*, 539, 983.

Discussion

P. F. CHEN: Forbes & Priest (1995), compared two solutions at the critical point, one with a current sheet, and the other without a current sheet. Have you considered this in your work?

LIN: In fact, the solution deduced by Forbes & Priest (1995) consists of two components, one for the configuration including a vertical current sheet attached to the boundary surface and another one for the without the current sheet. With the specific background field that they chose, the current sheet did not develop below the flux rope in the relevant magnetic field until the system loses the mechanical equilibrium. In the present case, on the other hand, the magnetic configuration possesses more complex structure since the extra magnetic sources are included. The complexity of such structure can be seen first from the X-type neutral point or the vertical current sheet of finite length located above the current sheet prior to the eruption, and further from the multiple patterns of the catastrophe shown by the evolutionary behaviors of the system in response to the change in the background field. In our work, we considered the situation in which a vertical current sheet of finite length is located above the flux rope during the slow evolution. But whether such a configuration makes any sense in reality is not clear. For the time being, we could not find the closed form of the solution for the evolution in the magnetic configuration that includes a vertical current sheet attached to the boundary surface due to the complexity in mathematics. Further investigation on this issue is worthwhile and will be arranged for our work in the future.

Coronal and Stellar Mass Ejections
Proceedings IAU Symposium No. 226, 2005
K. P. Dere, J. Wang & Y. Yan, eds.

© 2005 International Astronomical Union
doi:10.1017/S1743921305000682

The Catastrophe of Coronal Magnetic Flux Ropes in CMEs

Y. Q. Hu[1]†

[1]School of Earth and Space Sciences, University of Science and Technology of China, Hefei 230026, China
email: huyq@ustc.edu.cn

Abstract. A brief review is given on the progress made in the study of the catastrophe of coronal magnetic flux ropes with implication in coronal mass ejections (CMEs). Relevant studies have been so far limited to 2.5-D cases, with a flux rope levitating in the corona, either parallel to the photosphere in Cartesian geometry or encircling the Sun like a torus in spherical geometry. The equilibrium properties of the system depend on the features of the flux rope and the surrounding background state. Under certain circumstances, the flux rope exhibits a catastrophic behavior, namely, the rope loses equilibrium and erupts upward upon an infinitesimal variation of any control parameter associated with the background state or the flux rope. The magnetic energy of the system right at the catastrophic point may exceed the corresponding open field energy so that after the background field is opened up by the erupting flux rope, a certain amount of magnetic free energy is left for the heating and acceleration of coronal plasma against gravity. The flux rope model has been used to reveal the common features of CMEs and to simulate typical CME events, proving to be a promising mechanism for the initiation of CMEs. Incidentally, the Aly conjecture on the maximum magnetic energy of force-free fields places a serious constraint on 2.5-D flux models. Nevertheless, current sheets must form during a catastrophe on the Alfvén timescale, and magnetic reconnection across the newly formed current sheets may contribute to circumventing such a constraint. In this sense, the catastrophe simply plays a role of driver for the fast magnetic reconnection, and a combination of them is thus responsible for the initiation of CMEs.

Keywords. Sun: coronal mass ejections (CMEs), magnetic fields, MHD

1. Introduction

Magnetic flux rope, defined as a twisted magnetic loop anchored in the photosphere, is believed to be a typical structure in the solar corona, though there is no direct evidence based on very few and vague magnetic field observations over there. Nevertheless, Yan *et al.* (2001) found a flux-rope like structure in the corona through a reconstruction of the coronal magnetic field based on the vector magnetograms (figure 1). Theoretically, the existence of such a flux rope serves as a necessary condition for supporting prominences in equilibrium against gravity (Low & Hundhausen, 1995). Based on observations, prominences have two distinct magnetic configurations, normal and inverse, according to whether their magnetic field is consistent with or opposite to the photospheric magnetic polarity beneath them (Anzer, 1989; Leroy, 1989). Consequently, the associated coronal flux ropes also have two types of magnetic configurations (Zhang & Low, 2004), but most flux rope models so far belong to the inverse type.

If a flux rope exists in the corona, its eruption must lead to an eruption of the associated prominence below, a coronal mass ejection (CME) above, and possibly, a fast reconnection

† Present address: School of Earth and Space Sciences, University of Science and Technology of China, Hefei 230026, China

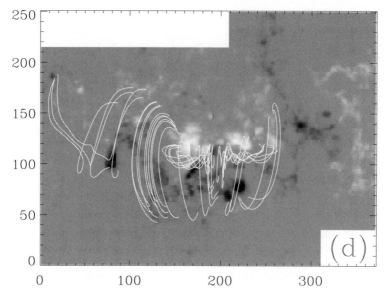

Figure 1. Reconstructed field lines projected on the photospheric magnetogram show clearly a flux rope along the neutral line embraced by overlying arcades (form figure 3d of Yan *et al.*, 2001). The tick labels are in the unit of arc seconds.

across the newly-formed vertical current sheet caused by the rope eruption, leading to a two-ribbon solar flare. In terms of this scenario, the magnetic flux rope is naturally and organically related to various solar explosive phenomena. On the other hand, the eruption of the flux rope is caused by a change of either the background state or the rope itself, and such a change is believed to be created by photospheric activities such as random motions, magnetic emergence and cancellation, and magnetic reconnection, which are generally slow compared to explosive events. Therefore, the rope eruption should take a catastrophic manner, namely, when the flux rope system evolves to a certain critical state, an infinitesimal perturbation will cause a sudden transition of the system from equilibrium into a dynamic state: the flux rope suddenly loses equilibrium and erupts upward at a fraction of the local Alfvén speed.

Both analytical studies and numerical simulations have been made to explore the equilibrium properties and catastrophic behaviors of coronal flux ropes in order to find a physically sound mechanism for solar explosive phenomena. We will review various flux rope models with emphasis on the catastrophic conditions in section 2. The catastrophic energy threshold and its relation to the Aly conjecture is discussed in section 3. We conclude with the implication of the flux rope catastrophe in CMEs in section 4.

2. MHD Coronal Flux Rope Models

To our knowledge, the earliest flux rope model is attributed to Van Tend & Kuperus (1978) who approximated the flux rope by a wire current filament (figure 2a) and concluded that a loss of equilibrium occurs if the current in the filament exceeds a critical value. However, in their model and subsequent similar ones, the field of the wire filament and the background field are freely reconnected, so the ideal magnetohydrodynamic (MHD) condition is disregarded. Soon their simple wire filament model was refined and replaced by the so-called thin-rope model, in which the ideal MHD condition is taken into account and thus electric current sheets appear in the solution (figure 2b).

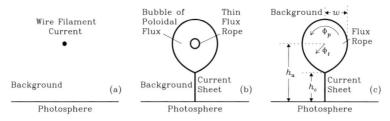

Figure 2. Schematic representation of coronal flux rope models: (a) wire filament current, (b) thin flux rope with a bubble, and (c) flux rope of finite cross section without bubble.

The flux rope is thin in the sense that its radius is far smaller than the length scale of the photospheric field, an approximation purely for analytical tractability. The thin-rope model was then extended to numerical rope models, where the rope is finite in radius (figure 2c).

2.1. *Description of the Flux Rope System*

Before moving on to discuss different types of flux rope models, it is opportune at this point to describe the fundamental features of the flux rope system and introduce several parameters which control and characterize the flux rope.

As shown schematically in figure 2c, the system consists of two topologically disconnected regions, a flux rope and a surrounding background state. The rope is characterized by its annular magnetic flux Φ_p per length in Cartesian geometry or per radian in spherical geometry, and axial flux Φ_t. If the gravity is incorporated, then the total mass M inside the rope becomes another crucial parameter. In addition, the following parameters may be introduced to represent the geometrical features of the flux rope: the height h_a of the rope axis, the length h_c of the vertical current sheet formed as the rope is detached from the photosphere, and the half-width w of the rope. For thin-rope models, the axial flux is limited inside a rope of circular section, which is surrounded by a "bubble" of purely poloidal flux (figure 2b). As a result, Φ_t is replaced by the ratio between the current intensity of the rope and the force-free factor (Lin *et al.*, 1998), w by the radius of the rope, and Φ_p by the poloidal flux in the bubble. For numerical models, the field in the bubble is also twisted so that, as the term suggests, the flux rope comes in contact with the background field, needless to introduce an additional bubble between them.

There exist various choices for the background field. A simplest one is a closed bipolar potential field, which was mostly adopted in thin-rope models. Closed multipolar potential field was also used; it gives a more dynamic evolution or jump of the flux rope than the bipolar field does. For numerical solutions, one may have more choices, for instance, a partly-open potential field with an equatorial current sheet, a quadrupolar field with a neutral point in the corona, or a helmet streamer with several bipolar fields inside. The background field may be changed by photospheric motions, magnetic reconnections, magnetic emergence and cancellation, or simply by a variation of the source strength. Moreover, a more realistic choice is a partly-open magnetic field with a quasi-magnetostatic helmet streamer in the closed field region and a steady solar wind outside.

Theoretically, one may introduce a set of parameters that control the physical properties of the background state and the flux rope. By solving the equilibrium equations of the system, we may determine the variation of the geometrical parameters caused by a change in the control parameters. If the variation is discontinuous at a certain point, then a catastrophe occurs for the system when the control parameters cross that point. For simplicity, we may let one of the control parameters changeable, denoted by λ, while keeping others fixed. On this basis, each geometrical parameter, denoted by h, can be

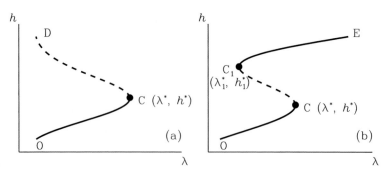

Figure 3. Schematic representation of equilibrium curves associated with a catastrophe of a system. The solid curves represent stable equilibrium states of the system whereas the dashed unstable ones. C and C_1 are the nose points at which a catastrophe takes place.

determined as a function of the single control parameter λ. Figure 3 shows schematically two typical h-λ profiles associated with a catastrophe, where the solid curve represents stable equilibrium states and the dashed unstable equilibrium states. On the fold-shaped profile OD in figure 3a, there is one nose point C, at which $\lambda = \lambda^*$ and no equilibrium is available beyond for the system. For $\lambda < \lambda^*$, there exist two equilibrium solutions, one stable and the other unstable. As λ increases beyond λ^*, the system loses equilibrium, and h approaches to infinity. In figure 3b, the profile OE is of S-type, and it has two nose points, C and C_1, with $\lambda = \lambda^*$ and λ_1^* respectively. For $\lambda_1^* < \lambda < \lambda^*$, there are three equilibrium solutions, one unstable and the other two stable. Across both C and C_1, a catastrophe occurs, and h jumps up and down respectively with a finite amplitude, namely, the system loses its original equilibrium and reaches a new one. The catastrophe associated with the nose point C is more relevant to our purpose, since we are interested in the eruption and expansion of the system with a positive jump of h.

2.2. Catastrophe Associated with the Rope Property Change

As mentioned above, the rope system evolves with the change of the rope properties, including the annular flux Φ_p, the axial flux Φ_t, and the total mass M within the rope. For thin-rope models, the gravity is ignored, Φ_t is usually fixed, and Φ_p, identified as the poloidal flux in the surrounding bubble, is to be adjusted. In the thin-rope model proposed by Forbes & Isenberg (1991) in Cartesian geometry, the background field is produced by a line dipole at a depth d below the photosphere, and a flux rope of radius $r \ll d$ is embedded in the bubble. A magnetic reconnection at the photosphere right below the flux rope leads to a gradual increase of Φ_p in the bubble. It was found that for $r/d < 10^{-3}$, the rope system exhibits a catastrophic behavior in relation to Φ_p. The rope loses equilibrium at the catastrophic point and enters a new one with a vertical current sheet below. By taking a background field created by a submerged line quadrupole instead, Isenberg *et al.* (1993) found a similar catastrophe for $r/d < 0.23$.

A 2.5-D, time-dependent ideal MHD model in Cartesian coordinates was used to find equilibrium solutions associated with a coronal flux rope of large cross section embedded in a background field (Hu & Liu, 2000; Hu, 2001, 2002; Hu & Jiang, 2001; Li & Hu, 2001, 2003; Wang & Hu, 2003). It was demonstrated that the catastrophic behavior of the system depends crucially on the pattern of the background magnetic field. For a completely closed background field, catastrophe does not exist no matter how the background field is produced, say, by two separated extended sources of opposite sign (Hu & Liu, 2000), or by a line dipole, quadrupole, or octapole at a depth below the photosphere (Wang & Hu, 2003). This result is consistent with the above-mentioned conclusion reached by thin-rope models that no catastrophe occurs for flux ropes of large radius. However, there exists a steep segment in the profiles of the geometric parameters versus Φ_p or Φ_t, and the faster the background field decays with height, the larger both the gradient and the growth amplitude within the segment will be (Wang & Hu, 2003).

On the other hand, a flux rope underneath a partly-open background field exhibits a catastrophic behavior during the slow change of the magnetic fluxes of the rope, namely, there exists a certain point, at which an infinitesimal enhancement of these fluxes causes a jump of the geometrical parameters of the rope (Hu, 2001, 2002). In addition, the amplitude of the jump depends on the extent to which the background magnetic field is open, approaching infinity for fully-open background field (Li & Hu, 2001, 2002), and zero for closed one. Incidentally, a drainage of mass from the rope may also lead to the expansion of the rope and thus a catastrophe of the system, as expected and demonstrated by tentative simulations.

In an extension of the flux rope model to spherical geometry, Hu *et al.* (2003) found that a flux rope embedded in a bipolar background field also exhibits a catastrophic behavior with respect to Φ_p or Φ_t, which does not depend on whether the background field is partly-open or closed. The rope sticks to the solar surface in equilibrium below the catastrophic point, and escapes to infinity above. The catastrophic amplitude is infinite in this case. However, under some special circumstances when a downward force on the rope dominates, a flux rope in spherical geometry may also levitate stably with a vertical current sheet below in the frame of ideal MHD, and catastrophe may either be non-existent or finite in amplitude. For instance, in a quadrupolar background field with a neutral point in the corona between the central and ambient bipolar fields, a flux rope underneath the central bipolar field may lead to the formation of a transverse current sheet right above the flux rope with a current opposite to the rope current. The two currents are repellent so that the rope is subject to a downward force (Zhang *et al.*, 2004). Besides, the gravitational force on the rope is also downward, whatever the background field is. In the analytical solutions of magnetic flux ropes of both inverse and normal configurations obtained by Zhang & Low (2004), the gravity associated with the total mass inside the rope is crucial to hold the system in equilibrium especially for ropes of normal configuration. If these downward forces happen to dominate, the flux rope may levitate stably high in the corona, and a catastrophe, if any, must be finite in amplitude, or even no catastrophe occurs for the system. This issue is left for future numerical studies.

2.3. *Catastrophe Associated with Photospheric Motions*

The background field is subject to change under the action of photospheric motions, which have a subtle influence on the behavior of the flux rope. In terms of a thin-rope model, Forbes & Priest (1995) found that the converging motion of the two magnetic sources of opposite polarity leads to a catastrophe of the rope if the radius of the rope is smaller than a certain critical value. Lin *et al.* (2002) studied the evolution of a semi-circular flux rope with two ends anchored in the photosphere with varying strength and distance of the background field sources, and concluded that the evolution due to the change in source strength shows the likelihood of catastrophic loss of equilibrium. In a model developed by Hu & Jiang (2001) and Hu (2002), the two magnetic fluxes of the rope, Φ_p and Φ_t, were fixed, and the background field was partly open, produced by two separated magnetic sources of opposite sign at the photosphere and changed by three types of photospheric motions: a reduction of the interval between the two sources, a contraction of each source, and a shear of the closed part of the field respectively. The system exhibits a catastrophe such that there exists a certain critical point for each type of photospheric motions, across which an infinitesimal displacement on the photosphere causes a finite jump of the geometrical parameters of the rope.

2.4. *Catastrophe with Both Background Field and Rope Properties Changed*

There were flux rope models in which both the background field and the flux rope are changed in property. For instance, Isenberg *et al.* (1993) let the background field decrease with time in strength and the decreased flux transfer to the bubble around the rope. As such, they found a catastrophe with respect to the background field strength. The same approach was taken by Lin *et al.* (1998) but for a dipole background field and a torus-shaped flux rope in spherical geometry, leading to a similar conclusion. In these models the flux transfer between the bubble and background is implemented via a reconnection at the photosphere right below the rope. A reconnection across the newly formed current sheet in the corona during the rope ascending might be also invoked to achieve the transfer, provided that the evolution of the system is sufficiently slow. Lin & van Ballegooijen (2002) assumed that the vertical current sheet appearing in the thin-rope model obtained by Forbes & Isenberg (1991) is suppressed by magnetic reconnection and replaced by an X-type neutral point, and found that the new configuration exhibits a catastrophic behavior that is no longer constrained by the radius of the rope. In the study of the interaction between an existing flux rope system and a new emerging flux, Lin *et al.* (2001) allowed the background field, the field in the bubble around the rope, and the new emerging field to reconnect freely such that the resultant field is potential everywhere except in the thin flux rope. As a result, no current sheet appears in the solution. Then, in terms of the profiles of the positions of the flux rope axis versus various control parameters, they found quite a lot of cases with catastrophe.

2.5. *Flux Rope Catastrophe in a Background Solar Wind*

Most flux rope models available in literature assumed that the background coronal plasma is in magnetostatic equilibrium. However, solar wind does exist and certainly exerts a critical influence on the flux rope equilibrium and the associated catastrophe. Wu *et al.* (1997a) presented a numerical model with a helmet streamer surrounded by a steady solar wind and a flux rope embedded in the streamer. They found that the flux rope either sticks to the solar surface in equilibrium or erupts upward depending on the magnetic energy level of the system. For the former, a photospheric shear motion applied to the streamer base (Wu & Guo, 1997a) or an enhancement of the axial flux (Φ_t) of the rope (Wu & Guo, 1997b) destabilizes the system, leading to an eruption of the rope. This implies an existence of the catastrophe of the flux rope in the solar wind background, though the catastrophic point was not identified yet. Sun & Hu (2004) used a similar model to Wu and Guo's and found the catastrophic point with respect to the variation of Φ_t, Φ_p, and M, respectively. In comparison with magnetostatic models in the absence of solar wind, the coronal plasma helps open up the external part of the helmet streamer that is expanding with the increase of magnetic fluxes of or the drainage of mass from the flux rope. This reduces the tension force of the helmet streamer on the flux rope and makes it easier for the catastrophe to occur.

3. Catastrophic Energy Threshold of the Flux Rope System

An important issue in the study of flux rope catastrophe is the magnetic energy threshold, referred to as W_c hereafter, across which a catastrophe takes place. One hopes that W_c would exceed the energy of the corresponding open field, referred to as W_{open}, so that after the field is opened up by a catastrophe, there is a certain amount of magnetic free energy left for the heating and acceleration of coronal plasma. However, such an expectation is faced with a serious challenge since Aly (1984) put forward a conjecture saying that in an infinite domain and for a given distribution of normal field at the lower boundary, the maximum energy of force-free fields with at least one end of each field line anchored in the lower boundary is W_{open}. The Aly conjecture was supported by numerical (Yang *et al.*, 1986; Mikić & Linker, 1994; Roumeliotis *et al.*, 1994; Amari *et al.*, 1996) and analytical (Lynden-Bell & Boily, 1994; Aly, 1994; Wolfson, 1995) examples. Meanwhile, Aly (1991) and Sturrock (1991) addressed proofs of the conjecture, respectively, based on some intuitive assumptions. Recently, the Aly conjecture was extended by Hu (2004) in such a way that it is impossible to store more magnetic energy in the corona by photospheric shear motions at the base of any part of the closed flux of a force-free field than that of the field in which the sheared closed flux opens but the rest remains closed. Several authors argued that counter-evidence was found to deny the Aly conjecture (e.g., Choe & Cheng, 2002; Wolfson & Low, 1992). However, these authors took a boundary condition at infinity, which forced all allowable force-free fields to be completely closed and might thus change the nature of the solutions in a dramatic way (Hu, 2004), so their conclusions seem to be questionable.

Under 2-D approximations, one has to select a spherical geometry so as to make a physically reasonable analysis in energetics. For force-free fields in the exterior of a sphere of radius R_0 with a given radial field at the sphere, the magnetic energy satisfies the following inequality (Aly, 1984)

$$W_p \leqslant W \leqslant W_{\text{max}} \equiv \frac{R_0^3}{2\mu} \int_0^{2\pi} d\phi \int_0^\pi B_r^2|_{r=R_0} \sin\theta d\theta,$$

where W_p is the energy of the corresponding potential field. Therefore, we have $W_{\text{open}} \leqslant W_{\text{max}}$. For 2-D flux rope configurations, there exist field lines which are detached from the solar surface, so the force-free field energy may exceed W_{open} (Priest & Forbes, 1990). The percentage of this energy in excess of W_{open} has an upper bound $(W_{\text{max}}/W_{\text{open}} - 1)$. Taking the dipole field as an example, we have $W_p = 4\pi B_0^2 R_0^3/(3\mu)$ (where B_0 is the field strength at the equator and R_0 the solar radius), $W_{\text{open}} = 1.662 W_p$ (Low & Smith, 1993; Mikić & Linker, 1994), and $W_{\text{max}} = 2W_p$, so the maximum energy in excess of W_{open} is 20.3%.

In terms of a 2.5-D flux rope model in spherical coordinates, Hu *et al.* (2003) concluded that W_c is slightly larger than W_{open}, and the gravity associated with the prominence supported by the flux rope raises W_c by an amount that is approximately equal to the magnitude of the excess gravitational energy associated with the prominence. Such a conclusion was further quantified by Li & Hu (2003): W_c exceeds W_{open} by 8%. Flyer *et al.* (2004) found a 2-D force-free field solution with detached field lines which should be below the catastrophic point but has an energy larger than W_{open} by 3.5%. Using a cold plasma approximation, Zhang & Low (1004) found magnetostatic equilibrium solutions of both inverse and normal prominence fields which have magnetic energy larger than W_{open}. The flux rope is kept in equilibrium by gravity, and more mass is needed for trapping flux ropes of normal type. Therefore, a drainage of plasma out of a prominence will lead to a rope eruption.

In the presence of a background solar wind, W_c may be also larger than W_{open}. Using a polytropic solar wind model, Guo & Wu (1998) found quasi-static helmet streamer solutions that contain a flux rope with or without cavity, and calculated the magnetic energy in the computational domain $(1\text{-}6R_0)$. The open field energy in the same domain was also calculated for comparison. They found that the solution associated with a cavity flux rope has more energy than the open field. Sun & Hu (2004) used a similar model to study the catastrophe of the flux rope system and to determine the value of W_c. The threshold was found to be larger than W_{open} by about 8% that is the same as obtained for magnetostatic equilibrium solutions (Li & Hu, 2003). Also, with increasing mass inside the flux rope, W_c increases by an amount that is approximately equal to the magnitude of the excess gravitational energy associated with the enhanced mass in the flux rope (Sun & Hu, 2004).

In general, a realistic coronal flux rope should have its ends anchored in the photosphere. If we believe that the Aly conjecture is correct, a system with a flux rope anchored in the photosphere can never have an energy in excess of W_{open}. If a catastrophe exists for such a system, W_c must be smaller than W_{open}, the rope ascends and expands after catastrophe, and the background field around the rope remains to be invariant in topology. To make the catastrophe develop into a plausible eruption, one has to invoke magnetic reconnection across the newly formed current sheet below the ascending rope, which leads to a transfer of magnetic flux from the background field into the rope and the newly formed helmet arcade below the rope. W_c needs only to be larger than the energy of the corresponding partly-open field instead of W_{open}. It is the catastrophe that creates a current sheet at the Alfvén timescale as an ideal MHD process and provides a favorable site for fast magnetic reconnection. A combination of the catastrophe and the follow-up fast reconnection is thus responsible for the rope eruption and the initiation of CMEs. On the other hand, if the Aly conjecture becomes invalid for magnetic configurations with a catastrophic behavior, W_c may be still larger than W_{open}, as inferred by Li & Hu (2003). It is interesting to check such a possibility by 3-D numerical calculations.

4. Flux Rope Catastrophe and CMEs

The coronal flux rope structure and its catastrophe were used to explain the observed features of CMEs, mostly in qualitative levels. As mentioned above, present flux rope models have been limited to 2-D cases in either Cartesian or spherical geometry. The two types of models differ from each other while applied to CMEs.

For models in 2-D Cartesian geometry it is energetically impossible for the flux rope to open the background field in the frame of ideal MHD. When a catastrophe occurs, the flux rope jumps at the Alfvén timescale but by a finite height, so the catastrophe plays a trigger of CMEs at most. In order for a real eruption of the rope, one has to rely on a magnetic reconnection across the newly formed current sheet below the rope. The motion pattern is determined entirely by the reconnection rate, so it is of no importance whether a catastrophe really takes place. In the absence of catastrophe, Lin & van Ballegooijen (2002) concluded that a slow reconnection at the photosphere leads to a continuous upward motion of the flux rope, which may account for slow CMEs. On the other hand, Lin & Forbes (2000), Lin (2002), and Lin et al. (2004) introduced a reconnection somewhere in the middle of the current sheet, allowing the system to evolve into a configuration consisting of an expanding helmet arcade below and an ascending bubble containing the rope above, connected by the vertical current sheet. For a given constant reconnection rate M_A, the Alfvén number of the inflow into the reconnection site, the motion patterns of these features are determined by the coronal Alfvén speed and its variation with height. Taking $M_A = 0.1$ and a realistic density profile with height, these authors explained the peculiar motion of giant X-ray arches and anomalous post-flare loops and the observed features of three-component (bright dome, dark cavity, and dense core) CMEs at heights of a few solar radii.

On the other hand, the flux rope may erupt to infinity after catastrophe in 2-D spherical geometry, at least for a bipolar background field. Therefore, by itself, a catastrophe is sufficient to implement the eruption process associated with CMEs. Hu *et al.* (2003) took an isothermal static corona as the background and obtained profiles of the geometrical parameters of the rope with height after catastrophe. The flux rope erupts at the Alfvén timescale that increases with height, leading to an initial sharp acceleration followed by a gradual deceleration. The eventual speed of the rope axis is less than 100 km/s. However, if the effects of the solar wind and the magnetic reconnection are considered, the motion pattern of the erupting rope will be different. In a solar wind background, the speed of the flux rope axis approaches the solar wind speed at large distance from the Sun (Wu *et al.*, 1997a). Wu *et al.* (1997b, 1999) applied their flux rope model in the presence of solar wind to simulate two CME events observed by the Large-Angle and Spectrometric Coronagraph Experiment (LASCO) in July 1996 and January 1997. The eruption was presumed to be caused by an enhancement of Φ_t (azimuthal flux) of the rope for the second event, and a simultaneous increase of Φ_t and decrease of M (mass) of the rope for the first. The simulation results were compared with LASCO observations in the near Sun region and Wind observations at 1 AU, showing a reasonable agreement in shape with LASCO images and a qualitative resemblance to Wind observations. Note that in the model by Sun & Hu (2004), a special measure is taken to completely suppress any reconnection across the vertical current sheet below the rope, whereas in Wu *et al.*'s model, a numerical reconnection exists across the sheet and a helmet arcade forms during the eruption of the rope. The numerical reconnection helps the rope erupt, but ignores the Joule heating that would have occurred and caused a coronal heating and a further acceleration of the rope eruption.

So far we were limited to the flux rope of inverse type. Based on a qualitative analysis, Low & Zhang (2002) and Zhang & Low (2004) argued that flux ropes of normal type are apt to produce fast CMEs in which magnetic reconnection plays a crucial role, whereas the expulsion of flux ropes of inverse type seems likely to involve a gradual acceleration without magnetic reconnection necessarily playing a principle role. Some observational evidence was found in support of such an argument (Zhang *et al.*, 2002).

Although the available flux rope catastrophe models are still too simplified and idealized to really explain the exact triggering mechanisms and quantitative dynamical behaviors of CMEs, they may after all be accepted as a promising mechanism at least for the initiation of three-component CMEs. Further work needs to be done to refine these models so as to examine the interplay between ideal and nonideal MHD processes and to consider three-dimensional effects. configurations.

Acknowledgements

This work was supported by grants NNSFC 40274049 and 10233050, and grant NKBRSF G2000078404 in China.

References

Aly, J. J. 1984, *ApJ* 283, 349
Aly, J. J. 1991, *ApJ* 375, L61
Aly, J. J. 1994, *A&A* 288, 1012
Amari, T., Luciani, J. F., Aly, J. J., & Tagger M. 1996, *ApJ* 466, L39
Anzer, U. 1989, in: E. R. Priest (ed.), *Dynamics and Structure of Quiescent Solar Prominences*, (Kluwer Academic Publishers, Dordrecht, Holland), p. 143
Choe, G. S. & Cheng, C. Z. 2002, *ApJ* (Letters) 574, L179
Flyer, N., Fornberg, B., Thomas, S., & Low, B. C. 2004, *ApJ* 606, 1210
Forbes, T. G. & Isenberg, P. A. 1991, *ApJ* 373, 294
Forbes, T. G. & Priest, E. R. 1995, *ApJ* 446, 377

Guo, W. P. & Wu, S. T. 1998, *ApJ* 494, 419

Hu, Y. Q. 2001, *Solar Phys.* 200, 115

Hu, Y. Q. 2002, in: H. N. Wang & R. L. Xu (eds.), *Solar-Terrestrial Magnetic Activity and Space Environment*, COSPAR Colloquia Series (Pergamon), Vol. 14, p. 117

Hu, Y. Q. 2004, *ApJ* 607, 1032

Hu, Y. Q. & Jiang, Y. W. 2001, *Solar Phys.* 203, 309

Hu, Y. Q. & Liu, W. 2000, *ApJ* 540, 1119

Hu, Y. Q., Li, G. Q., & Xing, X. Y. 2003, *J. Geophys. Res.* 108(A2), 1072, doi:10.1029/2002JA009419

Isenberg, P. A., Forbes, T. G., & Démoulin, P. 1993, *ApJ* 417, 368

Leroy, J. L. 1989, in: E. R. Priest (ed.), *Dynamics and Structure of Quiescent Solar Prominences* (Kluwer Academic Publishers, Dordrecht, Holland), p. 77

Li, G. Q. & Hu, Y. Q. 2001, *Chinese Science (Series A* (Supplement) 31, 53

Li, G. Q. & Hu, Y. Q. 2003, *Chin. J. Astron. Astrophys.* 3, 555

Lin, J. 2002, *Chin. J. Astron. Astrophys.* 2, 539

Lin, J. & Forbes, T. G. 2000, *J. Geophys. Res.* 105, 2375

Lin, J. & van Ballegooijen, A. A. 2002, *ApJ* 576, 485

Lin, J., Forbes, T. G., & Démoulin, P. 1998, *ApJ* 504, 1006

Lin, J., Forbes, T. G., & Isenberg, P. A. 2001, *J. Geophys. Res.* 106, 25053

Lin, J., Raymond, J. C., & van Ballegooijen, A. A. 2004, *ApJ* 602, 422

Lin, J., van Ballegooijen, A. A., & Forbes, T. G. 2002, *J. Geophys. Res.* 107, 1438

Low, B. C. & Hundhausen, J. R. 1995, *ApJ* 443, 818

Low, B. C. & Smith, D. F. 1993, *ApJ* 410, 412

Low, B. C. & Zhang, M. 2002, *ApJ* 564, L53

Lynden-Bell, D. & Boily C. 1994, *MNRAS* 267, 146

Mikić, Z. & Linker, J. A. 1994, *ApJ* 430, 898

Priest, E. R. & Forbes, T. G. 1990, *Solar Phys.* 126, 319.

Roumeliotis, G., Sturrock, P. A., & Antiochos, S. K. 1994, *ApJ* 423, 847

Sturrock, P. A. 1991, *ApJ* 380, 655

Sun, S. J. & Hu, Y. Q. 2004, Coronal flux rope catastrophe in a background solar wind, to be submitted.

Van Tend, W. & Kuperus, M. 1978, *Solar Phys.* 59, 115

Wang, Z. & Hu, Y. Q. 2003, *Chin. J. Astron. Astrophys.* 3, 241

Wolfson, R. 1995, *ApJ* 443, 810

Wolfson, R. & Low, B. C. 1992, *ApJ* 391, 353

Wu, S. T. & Guo, W. P. 1997a, *Adv. Space Sci.* 20, 2313

Wu, S. T. & Guo, W. P. 1997b, in: N. Crooker, J. Joselyn & J. Feynman (eds.), *Coronal Mass Ejections*, Geophysical Monograph 99 (AGU, Washington, D. C.), p. 83

Wu, S. T., Guo, W. P., & Dryer, M. 1997a, *Solar Phys.* 170, 265

Wu, S. T., Guo, W. P., Andrews, M. D., *et al.* 1997b, *Solar Phys.* 175, 719

Wu, S. T., Guo, W. P., Michels, D. J., & Burlaga, L. F. 1999, *J. Geophys. Res.* 104, 14789

Yan, Y. H., Deng, Y. Y., Karlicky, M., *et al.* 2001, *ApJ* (Letters) 551, L115

Yang, W. H., Sturrock, P. A., & Antiochos, S. K. 1986, *ApJ* 309, 383

Zhang, Y. Z., Hu, Y. Q., & Wang, J. X. 2004, in this proceeding

Zhang, M. & Low, B. C. 2004, *ApJ* 600, 1043

Zhang, M., Golub, L., Deluca, E., & Burkepile, J. 2002, *ApJ* 574, L97

Discussion

KOUTCHMY: As an observer, I would like to better understand what would be a good candidate for seeing a rising flux rope which eventually drives an eruption or a flare. Is it a filament or is it the cavity which is around a filament, or something else?

HU: Theoretically, the filament is situated right below the axis of a flux rope, so the cavity around the filament might be a good candidate for either static or rising flux rope.

ZHUKOV: The Aly-sturrock conjecture (or theorem) deals with force-free fields. Why do you mention it in connection with CMEs, when the magnetic force is almost certainly playing a role and the field is thus not force-free?

HU: There are two issues in connection with CMEs, one about the energetics and the other the dynamical process. Force-free field is a good approximation for the first issue, whereas the magnetic force plays a crucial role for the second, and thus the field cannot be considered as force-free.

FORBES: Comment: I agree with your remark that the Aly-Sturrock conjecture has not been rigorously proved for all possible configurations. However, even if it is true, it is also possible to get an eruption by just partly opening the field.

HU: Yes, but then non-ideal MHD effect such as magnetic reconnection must be invoked in order to make part of the field lines close back to the photosphere.

Coronal and Stellar Mass Ejections
Proceedings IAU Symposium No. 226, 2005
K. P. Dere, J. Wang & Y. Yan, eds.

© 2005 International Astronomical Union
doi:10.1017/S1743921305000694

Exact Solution of Jump Relations at Discontinuities in a Two-And-Half-Dimensional Compressible Reconnection Model

Marina Skender[1], Bojan Vršnak[2] and Mladen Martinis[1]

[1]Department for Theoretical Physics, Rudjer Bošković Institute, Bijenička 54,
Zagreb, Croatia HR-10001
email: marina@rudjer.irb.hr

[2]Hvar Observatory, Faculty of Geodesy, Kačićeva 26, Zagreb, Croatia, HR-10000

Abstract. Two–and–half–dimensional reconnection is examined for a compressible plasma: Exact solution of jump relations in the system of discontinuities is used to investigate how the outflowing jet and the conditions in the intermediate region depend on the characteristics of the inflow. The most significant implications concerning large-scale eruptive phenomena of solar atmosphere are presented.

Keywords. MHD, shock waves, turbulence, Sun: magnetic fields

When merging magnetic fields are skewed one to the other, there is a component of the magnetic field perpendicular to the plane of reconnection. Petschek & Thorne (1967) extended the fast reconnection of Petschek (1964) by considering this two–and–half–dimensional ($2\frac{1}{2}$ D) case, and introduced two pairs of rotational discontinuities (RDs) in front of the two pairs of slow-mode shocks (SMSs). The $2\frac{1}{2}$ D compressible reconnection was put on the firm mathematical foundation by Soward (1982). Yet, the jump conditions on the RD/SMS discontinuity system were greatly simplified by assuming that the inflow is slow and perpendicular to the outflowing jet.

The analytical solution of a full set of MHD jump relations on the RD/SMS discontinuity system of the symmetrical $2\frac{1}{2}$ D reconnection model has recently been developed by Skender, Vršnak & Martinis (2003). The jump relations are derived from the continuity equation, equation of motion under conditions of electrical neutrality and no influence from gravity and viscosity, energy conservation equation for fully ionized H-plasma, which has the ratio of specific heats $\hat{\gamma} = \frac{5}{3}$, magnetic divergence relation, and magnetic flux conservation equation. The general forms of these equations are, respectively,

$$\frac{\partial \rho}{\partial t} + \rho \vec{\nabla} \cdot \vec{v} = 0, \tag{0.1}$$

$$\rho \frac{D\vec{v}}{Dt} = -\vec{\nabla}p + \vec{j} \times \vec{B}, \tag{0.2}$$

$$\vec{\nabla} \cdot [(\frac{5}{2}p + \frac{1}{2}\rho v^2)\vec{v}] + (\vec{v} \times \vec{B}) \cdot \vec{j} = 0, \tag{0.3}$$

$$\vec{\nabla} \cdot \vec{B} = 0, \tag{0.4}$$

$$\vec{\nabla} \times \vec{E} = 0. \tag{0.5}$$

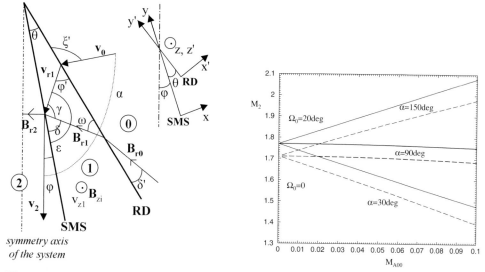

Figure 1. Left – One quadrant of the geometry of the discontinuity system in the $2\frac{1}{2}$ D symmetrical reconnection problem. The symmetry axis of the system (dash–dotted line) is defined by the outflow jet. The magnetic field generally has x-, y-, and z- components in all regions, as does the velocity in region "1", while the velocities in regions "0" and "2" have the xy-plane components only. All quantities in a $2\frac{1}{2}$ D model are independent of the z-axis.

Figure 2. Right – The dependence of the outflow magnetosonic Mach number M_2 on the reconnection rate M_{A00} is presented. The results are presented for $\Omega_0 = 0, 20°$, and $\alpha = 30°, 90°, 150°$, at the inflow plasma–to–magnetic pressure ratio $\beta_0 = 0.01$. We see that the inflow with a component in the direction of the outflowing jet increases M_2, while the inflow with an oppositely directed component decreases M_2.

The exact solution of the jump conditions enables us to follow changes in the current sheet characteristics which are due to the faster and non-perpendicular inflow, and in the limiting case when the transversal component of the inflowing magnetic field approaches zero.

Figure 1 shows one quadrant of the geometry of the discontinuity system in the $2\frac{1}{2}$ D symmetrical reconnection problem. The system of discontinuities separates the outflow from the inflow region. The inflow region is marked by "0", the region between RD and SMS (the intermediate region) by "1", and the outflow region by "2". In the inflow region plasma of the density ρ_0 and the pressure p_0 flows into the RD with velocity $\vec{v}_0$, carrying the magnetic field $\vec{B}_0$. Rotated and accelerated plasma proceeds towards the SMS through the intermediate region. At the SMS plasma is heated, compressed, and further deflected and accelerated.

The exact solution reveals how the outflow magnetosonic Mach number M_2 depends on the reconnection rate M_{A00} (see figure 2). The reconnection rate M_{A00} is defined as the ratio of the inflow velocity component perpendicular to the symmetry axis of the system to the Alfvén speed based on the component of the magnetic field parallel to the symmetry axis of the system. The angle between the inflow and the outflow velocity is α (figure 1), while the angle Ω_0 describes the influence of the transversal component of the inflowing magnetic field ($\tan \Omega_0 = B_{z0}/B_{r0}$). Figure 2 exposes that M_2 depends significantly on the incidence angle α, even more than on the angle Ω_0, however, only at comparatively large inclinations of the inflow. Around $\alpha = 90°$ the outflow Mach number depends very weakly on M_{A00}, which indicates that the decrease/increase of M_2 for incidence angles smaller/larger than $\alpha = 90°$ is primarily associated with the inflow

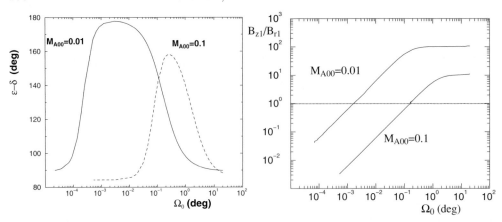

Figure 3. Properties of the system in the transition to the 2 D configuration. The results are presented for the perpendicular inflow $\alpha = 90°$ and two reconnection rates M_{A00}, at the inflow plasma–to–magnetic pressure ratio $\beta_0 = 0.01$. The angle Ω_0 defines the ratio of the transversal to the reconnection–plane component of the inflowing magnetic field. The rapid change of the flow/field geometry in the intermediate region at the transition $\Omega_0 \rightarrow 0$ is shown as dependence of the reconnection–plane inclinations of the velocity and magnetic field (left figure) and dimensionless ratio of transversal and reconnection–plane components of the magnetic field (right figure).

velocity component parallel to the symmetry axis of the system. The slight decrease of $M_2(M_{A00})$ at the perpendicular inflow $\alpha = 90°$ is related to a slight increase of the compression with the increasing reconnection rate (for more details please see Skender, Vršnak & Martinis (2003)).

The ratio of the transversal and reconnection–plane component of the inflowing magnetic field approaches zero when, e.g., the erupting prominence stretches the initially sheared magnetic arcade. The exact solution provides a possibility for tracking changes of the system in transition to the two–dimensional (2 D) case. It is found (see figure 3) that when $\Omega_0 \rightarrow 0$, a dramatic change takes place in the intermediate region: The inclination of the velocity ϵ and the magnetic field δ to the RD in the reconnection–plane (figure 1) change rapidly, as well as the ratio of the transversal to the reconnection–plane component of the magnetic field. In the transition from $2\frac{1}{2}$ D to 2 D a very small change in the inflowing transversal magnetic field leads to a dramatic change of the field and flow in the intermediate region, suggesting that turbulence might take place.

References

Petschek, H.E. 1964 in: W.N. Hess (ed.), *AAS-NASA Symp. on Physics of Solar Flares* (NASA, Greenbelt, Maryland), pp. 425-439
Petschek, H.E. & Thorne, R.M. 1967, *Astrophys. J.* 147, 1157
Soward, A.M. 1982, *J. Plasma Phys.* 28, 415
Skender, M., Vršnak, B. & Martinis, M. 2003, *Phys. Rev. E* 68, 46405

Coronal and Stellar Mass Ejections
Proceedings IAU Symposium No. 226, 2005
K. P. Dere, J. Wang & Y. Yan, eds.

© 2005 International Astronomical Union
doi:10.1017/S1743921305000700

On the Coronal Current-Free Global Field Configuration

Yihua Yan

National Astronomical Observatories, Chinese Academy of Sciences, A20 Datun Road,
Chaoyang District, Beijing 100012, China
email:yyh@bao.ac.cn

Abstract. A coronal current-free field model, which applies the asymptotic condition of no field at infinity and the boundary condition on the solar surface specified, is presented. Some Applications of the method to practical solar events indicate that the extrapolated global magnetic field structures effectively demonstrate the case for the disk signature of the radio CME and the evolution of the radio sources during the CME/flare processes.

Keywords. Sun: corona, Sun: coronal mass ejections (CMEs), Sun: magnetic fields

1. Introduction

The coronal magnetic field configuration is important for understanding the energy storage and release processes that account for CMEs. The source-surface is generally employed to reconstruct the coronal field configuration by fitting the observed coronal structures and a parameter is needed in order to obtain closed field region versus open field lines for the source-surface method or its variants (Hoeksema & Scherrer (1986),Zhao & Hoeksema (1994)). Here we present a model which is based on the work for potential magnetic field problems that only applies the condition at infinity with the boundary condition on the solar surface specified (Yan, Yu & Shi (1993), or recently in Wang, Yan, Wang, *et al.* (2002)). For some event analyses, we have employed MDI/SOHO longitudinal magnetogram inserted into the synoptic magnetogram to obtain whole boundary condition over the solar surface. Due to the projection effect, we employed MDI magnetogram data in the central region of $\pm 50°$ longitude and $\pm 50°$ latitude and the MDI synoptic magnetogram data in the regions of $-180°$ to $-50°$ and $50°$ to $180°$ longitude, and $\pm 60°$ latitude. The extrapolated global magnetic field structures effectively demonstrate the case for the disk signature of the radio CME and the evolution of the radio sources during the CME/flare processes.

2. Method

The extrapolation code is based on the work for potential magnetic field problems (Yan, Yu & Shi (1993); e.g., recently in Wang, Yan, Wang, *et al.* (2002)). For the potential condition, or current-free field, which is governed by $\nabla \times \mathbf{B} = 0$ and $\nabla \cdot \mathbf{B} = 0$, the magnetic field can be represented by a scalar potential Ψ with $\mathbf{B} = -\nabla \Psi$ Then we have the Laplacian equation.

$$\nabla^2 \Psi = 0. \tag{2.1}$$

On the solar surface S we have line-of-sight field component or its normal component B_n specified, i.e.,

$$B_n = -\frac{\partial \Psi}{\partial n}. \tag{2.2}$$

Therefor we can obtain a boundary value problem of (2.1-2.2), e.g., as described in Sakurai (1989). In general, the potential Ψ at any position $\mathbf{r}_i$, in space V can be determined from (Courant & Hilbert 1962):

$$\Psi(\mathbf{r}_i) = \oint_S [G(\mathbf{r}_i; \mathbf{r}) \frac{\partial \Psi(\mathbf{r})}{\partial n} - \frac{\partial G(\mathbf{r}_i; \mathbf{r})}{\partial n} \Psi(\mathbf{r})] dS \qquad (2.3)$$

where $G = 1/4\pi |\mathbf{r}_i - \mathbf{r}|$ is Green's function of Laplacian equation in free space. The Green's function solution of the above boundary value problem was applied to practical solar magnetic field observations in early 1960s (Sakurai (1989)). Here, the magnetic field is, however, obtained from:

$$\mathbf{B} = \oint_S \left[\Psi \frac{\partial}{\partial r} \left(\frac{\partial G}{\partial n} \right) - \frac{\partial \Psi}{\partial n} \frac{\partial G}{\partial r} \right] ds \qquad (2.4)$$

with $\partial \Psi / \partial n$ known over the boundary and Ψ solved numerically by the boundary element method (Yan, Yu & Shi (1993)), which is a well-established method for science and technology applications (Brebbia, Telles & Wrobel (1984)).

3. Results

For the event analysis, we have employed MDI/SOHO longitudinal magnetogram inserted into the synoptic magnetogram to obtain whole boundary condition over the solar surface. Due to the projection effect, we employed MDI magnetogram data in the central region of $\pm 50°$ longitude and $\pm 50°$ latitude and the MDI synoptic magnetogram data in the regions of $-180°$ to $-50°$ and $50°$ to $180°$ longitude, and $\pm 60°$ latitude. The method has been applied in Wang, Yan, Wang, *et al.* (2002) for a trans-equatorial CME event in May 1998.

For the 17-Mar-2002 event, the extrapolated global magnetic field structures effectively demonstrate the case for the disk signature of the radio CME and the evolution of the radio sources during the CME/flare processes. It is obtained that the radio counterpart of the CME as well as source of type III burst was propagating along the open field lines. The result shows that the extrapolated field structures are very helpful to understand the flare/CME process. The detailed analysis is to be presented elsewhere separately.

Acknowledgements

The work is supported by CAS, NSFC grants 10225313, 10333030, and MOST grant G2000078403.

References

Brebbia, C. A., Telles, J. C. F., & Wrobel, L. C. 1984, *Boundary Element Techniques* (Berlin: Springer)
Courant, R. & Hilbert, D. 1962, *Methods of Mathematical Physics* (New York: Wiley), Vol. 2
Hoeksema, J. T. & Scherrer, P. H. 1986, *Sol. Phys.* 105, 205
Sakurai, T. 1989, *Space Sci. Rev.* 51, 11
Wang, T. J, Yan, Y. H., Wang, J. L., Kurokawa, H., & Shibata, K. 2002, *ApJ* 572, 280
Yan, Y. H. & Sakurai, T. 2000, *Sol. Phys.* 195, 89
Yan, Y. H., Yu, Q., & Shi, H. L. 1993, in: J.H. Kane, G. Maier, N. Tosaka & S. N. Atluri (eds.), *Advance in Boundary Element Techniques* (New York: Springer), p. 447
Zhao, X. P. & Hoeksema, J. T. 1994, *Sol. Phys.* 151, 91

Coronal and Stellar Mass Ejections
Proceedings IAU Symposium No. 226, 2005
K. P. Dere, J. Wang & Y. Yan, eds.

© 2005 International Astronomical Union
doi:10.1017/S1743921305000712

Neoclassical Effects on Solar Plasma Loops

B.L. Tan^{1,2}† and G.L. Huang^{1}

^{1}Purple Mountain Observatory CAS, Nanjing 210008, China
email: bltan@pmo.ac.cn

^{2}Graduate School of the CAS, Beijing, China

Abstract. Tokamak physics shows that there will be a neoclassical effect in current-carrying plasma loops. We apply the theory to solar coronal loops and hope to find a fast magnetic reconnection mechanism for understanding solar flares and CMEs.

Keywords. Sun: coronal mass ejections (CMEs), flares, magnetic fields, MHD

1. Introduction

Studies on Tokamak show, in current-carrying plasma loop, the transport coefficient of energy and particles is much greater than that of the loop without current, and will trigger a new kind of MHD instabilities-neoclassical tearing mode (NTM). These phenomena can be called neoclassical effect. It includes: (A) Neoclassical resistance (Wesson 1997), $\eta^{nc} = \eta^{sp}/(1 - \sqrt{\epsilon})^2$, $\epsilon = a/R$, η^{sp} is classical resistance. (B) Bootstrap current (Bickerton et al. 1971): $\jmath_b = -2\sqrt{2}\frac{\epsilon^{1/2}T}{B_\theta}\frac{dn}{dr}$. Since 1988, in large Tokamaks, bootstrap current is observed up to 0.85MA, and the fraction f_b is up to 77 % (Wesson 1997).(C) Neoclassical Tearing mode (NTM):When f_b is much enough, NTM will be triggered(Qu and Callen, 1985). Its growth rate is: $\gamma_{NTM} = 6.61(\tau_h)^{-1}\Delta^{-\frac{2}{3}}(j_b\alpha\dot{B}_\theta)^{\frac{2}{3}}S^{-\frac{1}{3}}$. S is magnetic Reynolds number, Δ is tearing mode instability factor, α is mode number. As a contrast, the classical tearing mode(CTM) is: $\gamma_{CTM} = 0.755(\tau_h)^{-1}\Delta^{\frac{4}{5}}(\alpha\dot{B}_\theta)^{\frac{2}{5}}S^{-\frac{3}{5}}$. For solar plasma, S $\sim 10^8 - 10^{11}$, $\gamma_{NTM} \sim 10^2 - 10^3\gamma_{CTM}$.

2. Analysis on solar coronal loops

Typical coronal loops: Similarly, coronal loop are always current-carrying plasma loops, It is reasonable to suppose that neoclassical effect will play a role in the evolution of such loops. The first & second column in table 1 is the parameters in typical coronal loops (Zhang,2000) and calculating results. Loop's current is typical value (Zaitsev et al. 1998, Khodachenko et al. 2003). We can find, in typical cool loops $f_b < 0.3$ %, and in hot loops $f_b < 2$ % . Tokamak experiences show neoclassical effect isn't obvious when $f_b < 20\%$, we can't detect any measurable information about it. This shows in typical coronal loops the neoclassical effect is so faint that we may neglect it.

A special solar coronal loop: In fact, the range of coronal loops' parameters is always very large. It is reasonable to suppose that in some special loops the neoclassical effect may become measurable. The third column in table 1 is a M3.6 GOES flare event on 25 August 1999 from 01:32 UT to 01:40 UT in AR8674 (S28E21). (Huang et al. 2003). Figure 1(a) is the image, from it we may estimate the geometrical parameters. Figure 1(b) shows the time profile of GOES 8 X-ray. It reflects the thermal emission feature. The duration of rising phase is about 4 minutes. The bootstrap current is $f_b = 29.1\%$, this

---

† Present address: No.2, West Beijing road, Nanjing 210008, China.

Table 1. (Loop's parameters and bootstrap current calculating results)

Parameters	Typical cool loop	Typical hot loop	Special loop
Loop radius(m)	1.0×10^7	5.0×10^7	7.0×10^6
Section radius(m)	1.0×10^6	5.0×10^6	2.8×10^6
Plasma temperature(eV)	10	200	900
Loop magnetic field	0.1	0.1	0.1
Plasma mean density	1.0×10^{16}	2.0×10^{15}	1.3×10^{16}
Electric current(A)	4.0×10^{10}	1.0×10^{11}	1.0×10^{11}
e mean free path(m)	1.02×10^4	2.05×10^7	6.4×10^7
Bootstrap current, $j_b(A/m^{-2})$	3.79×10^{-5}	2.02×10^{-5}	1.17×10^{-3}
Fraction, $f_b(\%)$	0.299	1.59	29.1

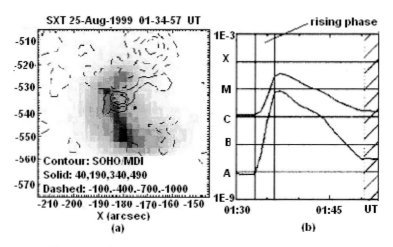

Figure 1. A special coronal loop and its emission feature

value is higher than that of JET in 1989's experiments. As the neoclassical effect is proportion to f_b, the neoclassical effect in above loop is very obvious. Similar to tokamak experience, NTM will be triggered. Then we may get: $\gamma_{NTM} = 2.18 \times 10^{-3}$, $t_{NTM} = 4.5 \times 10^2(s)$, $\gamma_{CTM} = 1.38 \times 10^{-6}$, $t_{CTM} = 7.2 \times 10^5(s)$. We find that NTM is more consistent with the observations than that of CTM.

3. Conclusions

(1) In typical coronal loops $f_b < 2\ \%$, it is too small to consider the neoclassical effect.

(2) For special coronal loops, bootstrap current exists well and truly, f_b may be over 29%, the neoclassical effect will play an important role in the loop's evolution. NTM will be triggered and provide a mechanism of fast magnetic reconnection.

References

Bickerton, R.J., Connor, J.W. & Taylor, J.B. 1975, *Nature* 229, 110

Huang, G.L., Wu, H.A., Grechnev, V.V., Sych, R.A. & Altyntsev, A.T. 2003, *Sol.Phys.* 213, 341

Khodachenko, M., Haerendel, G. & Rucker, H.O. 2003, *Astron.Astrophys.* 401, 721

Qu, W.X. & Callen, J.D. 1985, UWPR-85-5, University of Wisconsin, Wisconsin.

Wesson, J. 1997, Tokamak, Claarendon Press, Oxford.

Zaitsev, V.V., Stepanov, A.V., Urpo, S. & Pohjolainen, S. 1998, *Astron.Astrophys.* 337, 887

Zhang, Z.D. 2000, Coronal Physics(in Chinese), Science Press, Beijing.

Coronal and Stellar Mass Ejections
Proceedings IAU Symposium No. 226, 2005
K. P. Dere, J. Wang & Y. Yan, eds.

© 2005 International Astronomical Union
doi:10.1017/S1743921305000724

Coronal Magnetic Flux Ropes in Quadrupolar Magnetic Fields

Yingzhi Zhang[1], Youqiu Hu[2] and Jingxiu Wang[1]

1.National Astronomical Observatories, Chinese Academy of Sciences, Beijing 100012, China
email:zhangyingzhi@ourstar.bao.ac.cn

2.School of Earth and Space Sciences, University of Science and Technology of China, Hefei 230026, China

Abstract. Using a 2.5-D, time-dependent ideal MHD model in spherical coordinates, we carry out a numerical study of the equilibrium properties of coronal magnetic flux ropes in a quadrupolar background magnetic field. For such a flux rope system, a catastrophic occurs: the flux rope is detached from the photosphere and jumps to a finite altitude with a vertical current sheet below. There is a transversal current sheet formed above the rope, and the whole system stays in quasi-equilibrium. We argue that the additional Lorentz force provided by the transversal current sheet on the flux rope plays an important role in keeping the system in quasi-equilibrium in the corona.

Keywords. Sun: corona, magnetic fields, MHD

1. Introduction

A common conclusion of many numerical studies is that catastrophe exists under certain conditions. In spherical geometry, however, catastrophe exists for the flux rope system with a bipolar background field that may be either partly open or completely closed (Hu *et al.*, 2003), and the catastrophe amplitude is infinite.

2. Numerical Results

We work out the 2.5-dimensional ideal MHD equations in spherical coordinates(r, θ, φ). The initial magnetic field is of Antiochos type (Antiochos et al, 1999). The multistep implicit scheme (Hu, 1989) is used to solve the 2.5-dimensional ideal MHD equations.

In this study, calculations were carried out for different values of the axial magnetic flux ϕ_z for a fixed annular magnetic flux $\phi_p = 0.5$. Figure 1 shows the height of the rope axis h_a and the length of the vertical current sheet h_c as a function of ϕ_z for the quasi-equilibrium states thus obtained. As seen from this figure, for $\phi_z \leqslant 0.0437$, the flux rope remains attached to the solar surface but expands with increasing ϕ_z. As a result, h_c is zero and h_a increases slightly with increasing ϕ_z. For $\phi_z \approx 0.0438$, the two geometric parameters jump to about 1.7 and 2.6, respectively, implying a catastrophe of finite amplitude. We show the magnetic field lines at several separate times in Figure 2 for ϕ_z=0.0437 (Figure 2(A-E))and ϕ_z =0.0438 (Figure 2(F-J)). For both cases, the original neutral point changes to a transversal current sheet under the action of the flux rope, and it provides a downward Lorentz force on the rope. After a temporal evolution of 150 τ_A(in Figure 2 τ represents τ_A), the system as a whole approaches quasi-equilibrium for both cases. The magnetic rope sticks to the photosphere for ϕ_z =0.0437, and breaks away from the photosphere, leaving a vertical current sheet stretching from the photosphere

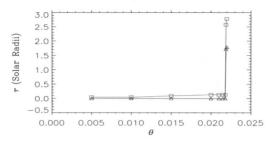

Figure 1. Geometrical properties of the coronal magnetic flux rope in equilibrium versus the axial magnetic flux ϕ_z. The solid squares and the triangles are symbols of h_a and h_c.

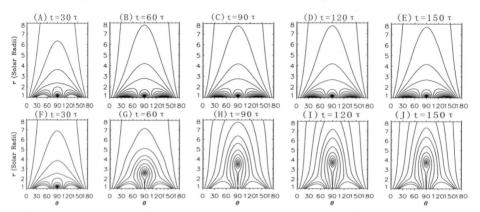

Figure 2. Magnetic field lines at several times for $\phi_z = 0.0437$(A-E) and $\phi_z=0.0438$(F-J)

to the bottom of the rope for $\phi_z=0.0438$. Besides, the flux rope stands in the corona for a long time in the second cases, as seen from Figure 2(F-J).

3. Concluding Remarks

This study shows that the system exhibits a catastrophic behavior in response to the enhancement of the axial magnetic fluxes of the rope, but the catastrophic amplitude is finite. The flux rope may levitate stably in the corona instead of escaping to infinity. This forms a striking contrast to the bipolar background field case as discussed by Hu *et al.* (2003). A transversal current sheet forms above the flux rope because of the presence of the neutral point and the subsequent action of the expanding flux rope.We argue that the sheet provides a downward Lorentz force on the flux rope and thus has made the flux rope levitate stably in the corona and has made the associated catastrophe be finite in amplitude.

Acknowledgements

The work is supported by the National Natural Science Foundation of China(10233050) and the National Key Basic Science Foundation(TG2000078404).

References

Antiochos, S.K., Devore, C.R. & Klimchuk, J.A. 1999, *ApJ* 510, 485
Hu, Y.Q. 1989, *J. Comput. Phys.* 84, 441
Hu, Y.Q., Li, G.Q. & Xing, X.Y. 2003, *J. Geophys. Res.* 108, 1072, DOI: 10.1029/2002JA009419

Coronal and Stellar Mass Ejections
Proceedings IAU Symposium No. 226, 2005
K. P. Dere, J. Wang & Y. Yan, eds.

© 2005 International Astronomical Union
doi:10.1017/S1743921305000736

The Effect of Viscosity on Magnetic Generation in Solar Plasmas

Zhiliang Yang[1] and Hairong Jing[1]

[1]Department of Astronomy, Beijing Normal University, Beijing 100875, P. R. China
email: zlyang@bnu.edu.cn

Abstract. Based on the dynamics of charged particles in plasma with the two-fluid description, the equation for magnetic field generation and maintenance is derived. The nature of magnetic field in cosmic plasma and the generation can be understood from the equation. The ions and electrons are considered as two independent fluids with the collision connected. Due to the different viscosity between ions and electrons, there will be a velocity difference in ions and electrons. This causes a current in the plasma and generates magnetic field. IN the plasma system, the differential velocities and the viscous forces of the electrons are the source of magnetic field.

Keywords. Sun: magnetic fields, Plasmas

1. The nature of magnetic field generated in plasma

In the universe magnetic fields are detected everywhere (see some of the reviews (Han, et al. 2002; Schekochihin, et al. 2003). The magnetic fields play a curious role in astrophysics for its commonplace and poorly understood. Dynamo theory is the theory that describe the existence of magnetic fields in conducting fluid masses. Dynamo problems are of two kinds, the kinematic dynamos as the fluid velocities are regarded as given and the hydromagnetic dynamos with the fluid velocities are determined by equations of motion including forces of magnetic origin. The reviews on dynamo theories could be seen in a lot of literatures (Cowling, 1981; Schmitt, 1993).

Plasma is the collection of charged particles with a neutralized charge as a whole. Considering plasma as two-fluid of ions and electrons (Huba & Fedder, 1993), We can get the magnetic field at r in laboratory system

$$\mathbf{B} = \Sigma \frac{mu_0(n_i e \mathbf{v_i} - n_e e \mathbf{v_e}) \times \mathbf{r}}{4\pi r^3} \tag{1.1}$$

$n_i e \mathbf{v_i} - n_e e \mathbf{v_e}$ is the current in the plasma. We can see that the condition for the generation of magnetic field is $n_i \mathbf{v_i} - n_c \mathbf{v_e} \neq 0$. we can get:

$$\mathbf{B} = \Sigma \frac{mu_0 n e (\mathbf{v_i} - \mathbf{v_e}) \times \mathbf{r}}{4\pi r^3} \tag{1.2}$$

2. Assumptions and Equations

Suppose the plasma consists of protons and electrons only. For the case that higher ions existed in plasma, we can easily get the result following the way we discussed. The plasma fluid equations are given by:

$$\frac{d\mathbf{v}_\alpha}{dt} = \frac{q_\alpha}{m_\alpha} \left(\mathbf{E} + \frac{1}{c} \mathbf{v}_\alpha \times \mathbf{B} \right) - v_{\alpha\beta}(\mathbf{v}_\alpha - \mathbf{v}_\beta) - \frac{\nabla P_\alpha}{n_\alpha m_\alpha} + F_\alpha \tag{2.1}$$

283

where $\mathbf{v}_\alpha$, $\mathbf{v}_\beta$ refers to electrons and/or protons (e, i), $\mathbf{F}_\alpha$ is the external force acted on protons or electrons except the electromagnetic force.

The electromagnetic induction equation:

$$\frac{\partial \mathbf{B}}{\partial t} = \nabla \times (\mathbf{v}_i \times \mathbf{B}) - \frac{1}{ne} \nabla \times (\mathbf{J} \times \mathbf{B}) + \eta \nabla^2 \mathbf{B} + \frac{c\nabla \times \nabla P_e}{ne} - \frac{c}{e} \nabla \times \mathbf{F}_e - \frac{c}{e} \nabla \times [(\mathbf{v}_e \cdot \nabla)\mathbf{v}_e]$$

$$(2.2)$$

The first term of equation (2.2) $\nabla \times (\mathbf{v}_i \times \mathbf{B})$ is the convective term, the second term $\frac{1}{ne}\nabla \times (\mathbf{J} \times \mathbf{B})$ is the Hall term, the third term $-\frac{cm_i\nu_{ie}}{ne^2}\nabla \times \mathbf{J}$ is magnetic diffusion (the resistivity is $\eta = \nu_{ic}c^2/\omega_{ie}^2$, where $\omega_{ie}^2 = 4\pi ne^2/m_e$, and the terms $-\frac{c}{e}\nabla \times \mathbf{F}_e$ and $\frac{c\nabla \times \nabla P_e}{ne}$ is the source term for the magnetic field. In regular flow, $-\frac{c}{e}\nabla \times [(\mathbf{v}_e \cdot \nabla)\mathbf{v}_e$ is ignored in the assumption of incompressible fluid.

3. The Viscosity on the Dynamo Effect in Differential Rotated Plasma

Eventually, the famous Biermanns 'battery is one of the results of two-fluid model of a fully ionized plasma. If there is no magnetic field in the plasma, then $\mathbf{E} + \frac{\nabla P_e}{n_e e} = 0$. In general stars, $\nabla \times \nabla P_e = 0$. The Biermanns Battery is not important. The magnetic field should be generated from the different velocities of ions and electrons.

In the two-fluid model of fully ionized plasma, the viscosity difference between ions and electrons is the possible mechanism to cause the different velocities of ions and electrons. Since the motion of plasma is assumed as conditions, the motion of ions is a fixed condition, only the viscous force of electrons should be included in equation (2.2). The viscous force of electron can be expressed as:

$$\mathbf{f}_{vis-e} = \nu\nabla^2\mathbf{v}_i \tag{3.1}$$

The magnetic induction equation (2.2) can be written as the following simple form with the Hall term ignored:

$$\frac{\partial \mathbf{B}}{\partial t} = \nabla \times (\mathbf{V}_i \times \mathbf{B}) + \eta\nabla^2\mathbf{B} + -\frac{cm_e}{e}\nabla \times (\nu\nabla^2\mathbf{v}_i) \tag{3.2}$$

In the spherical coordinate, we can get the magnetic field in the $\mathbf{r}$ and θ direction with the assumption $\mathbf{v} = v_\varphi\varphi$.

$$\mathbf{B}_{\mathbf{r}\theta} = \frac{c\nu}{\eta e}\nabla \times v_{i-\varphi} = \frac{cm_e}{e}P_m\nabla \times v_{i-\varphi} \tag{3.3}$$

The parameter P_m is the magnetic Prandtl number defined as the ratio of viscous to magnetic diffusivity, $P_m = \frac{\nu}{\eta}$.

we can see that the magnetic generation is from $\nabla \times \mathbf{v}$ and depends on the Prandtl number.

As a example, we can estimate the poloidal component of magnetic field inside the sun. From the differential rotation inside the sun (Dikpati & Charbonneau, 1999), we can estimate $\nabla \times \mathbf{v} \sim$ inside the sun. Using the Spitzer values for the viscosity and magnetic diffusivity of a fully ionized plasma, one finds

$$P_m \sim 2.6 \times 10^{-5}T^4/n \tag{3.4}$$

Where T is temperature. In the case of a partially ionized gas with neutral dominated viscosity, the formula for P_m is

$$P_m \sim 1.7 \times 10^{-7}T^2/n \tag{3.5}$$

In solar plasma, the Prandtl number is $P_m \sim 10^{10} - 10^{14}$. With small differential velocity $\nabla \times v$ we can get strong magnetic field.

4. Summary and Discussions

In this poster, we stress that the current in plasma is the result of different velocities of ions and electrons. The difference is due to the difference of viscosity for ions and electrons. In the two-fluid model, the ions and electrons are two fluids. Each has own viscosity. The kinematic viscosity v is anti-proportional to the density of fluid. The viscosity for ions is smaller then that for electrons due to their different mass. In the frame of ions or plasma, the viscous force of electrons is the main mechanism for magnetic generation. The reverse process is the dissipation of current, which is the collision between ions and electrons and decreases the velocity difference between ions and electrons. This is the Ohms dissipation.

Acknowledgements

The research is supported by NFSC under grant number 10273003 and the 973 Project of China under grant number G200078400.

References

Han, J.-L. & Wieleninsk, R. 2002, Chin. J. Astron. Astrophys. 2, 293

Schekochihin, A. A., Cowley, S. C., Taylor, S. F., Maron, J. L., & McWilliams, J. C. 2003, Astrophys. J., Astro-ph/0312046

Cowling, T. G. 1981, Ann. Rev. Astron. Astrophys., 19, 115

Schmitt, D. 1993, The Cosmic Dynamo, IAU, Printed in the Netherlands, p1

Huba, J. D. & Fedder, J. A. 1993, Phys. Fluids B., Vol. 5, No. 10, 3779

Dikpati, M. & Charbonneau, P. 1999, Astrophys. J., 518, 508

Session 5
Comparisons of CME models and observations

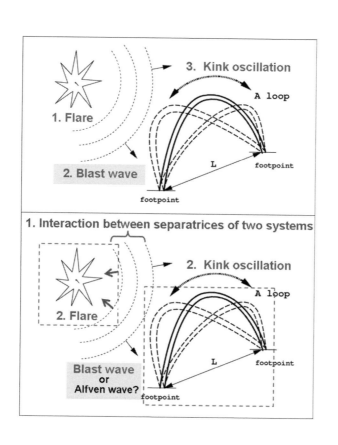

Coronal and Stellar Mass Ejections
Proceedings IAU Symposium No. 226, 2005
K. P. Dere, J. Wang & Y. Yan, eds.

The Connections Between CME Models and Observations

Terry Forbes

University of New Hampshire, Durham, New Hampshire, USA

Keywords. Sun: coronal mass ejections (CMEs)

Discussion

SCHWENN: About the spiral pattern in the U.Michigan animation: We saw such unwinding spirals. Question: what are they?

FORBES: They are difficult to interpret because what is plotted in the animation are really flux surfaces rather than field lines. In the regions where there is a strong component of the field out of the plane of the figure, the flux surfaces do not correspond well to the true field line structure. Thus, one needs to be cautious in interpreting the unwinding motion. The expansion of the flux surface corresponds to the decrease of current in the flux rope, but the rotation could be an illusion.

KOUTCHMY: You correctly pointed out that theoretical models are dealing with the behavior of the magnetic field which is not observed on the corona. However observations of CME show very important feature like (1) filament eruption and (2) the heating as well as the (3) mass loading process and finally during the evolution of the CME we see also downflowing gas which mean that (4) gravity could be important for models. Why modelers are not trying to deal with some of these parameters?

FORBES: The reason I emphasize observations which give information about the magnetic field is because the magnetic energy is the dominate factor controlling the dynamics of a CME. In CME models, it is the magnetic field and the current density which control the evolution of the plasma and which distinguish one model from another. Other factors, such as gravity, can play a significant role, but for the modelers, they do not have the same importance as the magnetic field.

JIE ZHANG: My question is about the escape of slow, gradually accelerated CMEs. For fast/impulsive CME, one may introduce magnetic reconnection to allow fast energy release and remove overlying field. But for slow CMEs without strong energy release, how do they remove the overlying field and escape? By the tendency of flux rope system expansion, as you mentioned for basic principles? Or by slow wind dragging?

FORBES: I don't know the answer. It may be that slow CMEs are directly driven by a combination of emerging flux and the force exerted by the solar wind. Reconnection and the decrease of the flux rope current may not be so important for the slowly accelerating CMEs.

ZHUKOV: Do you agree that it should be a concern for models to include more realistic photospheric motions involved in trigging the eruptions?

FORBES: Yes. The motion of the photospheric plasma, whether it is due to flux emergence or simple surface flows, is closely associated with the build up of currents in the corona. From the modelers point of view, these currents are the key factor in determining the equilibrium and stability properties of the coronal magnetic field.

STERLING: 1) Comment: can you publish your quantitative criticisms of the flux injection model, in order to assist in a productive dialog? I understand that you would want to do this tactfully.
2) Soft X-rays often precede HXRs in eruptions: which model best matches this observation?
3) Regarding to Serge Koutchmy's question: some CMEs do not involve filament eruptions; doesn't this rule out mass loading as a mechanism?

FORBES: 1) I've been asked this before, but I am reluctant to just publish a negative criticism of someone else's work.
2)I think this fact best matches the tether cutting model, where the soft X-rays and also the filament activation period, corresponds to the reconnection process which destabilizes the magnetic field.
3) For the eruptions without filaments, Low and [?] suggest that mass in the corona can account for the mass loading. This could explain some slow CMEs. However, even some fast CMEs do not involve filament eruptions.

Proceedings Coronal and Stellar Mass Ejections IAU Symposium
Proceedings IAU Symposium No. 226, 2004
A.C. Editor, B.D. Editor & C.E. Editor, eds.

A Three-Dimensional Magnetohydrodynamic (MHD) Model of Active Region Evolution

S. T. Wu[1,2], A. H. Wang[1], and D. A. Falconer[3,4]

[1]Center for Space Plasma and Aeronomic Research, [2]Department of Mechanical and Aerospace Engineering, [3] Department of Physics, University of Alabama in Huntsville, Huntsville, AL 35899 USA email: wus@cspar.uah.edu, wanga@cspar.uah.edu, David.Falconer@msfc.nasa.gov

[4]XD12, NASA/Marshall Space Flight Center, Huntsville, AL 35812 USA

Abstract. A three-dimensional, time-dependent, magnetohydrodynamic (MHD) model is constructed for the study of active region (AR) evolution. The new physics included in this model is differential rotation, meridional flow, effective diffusion and cyclonic turbulence effects, which means, that the photospheric shear is automatically generated instead of prescribed as is usually done for modeling. To benchmark this newly developed model, we have used observed active region NOAA/AR-8100 (October 29 - November 3, 1997) to verify the model by computation of the total magnetic flux and magnetic field maps of that active region. Then, we apply this model to compute the non-potentiality magnetic field parameters for possible coronal mass ejection production. These parameters are: (i) magnetic flux content (Φ), (ii) the length of strong shear, strong-field main neutral line, (L_{ss}), (iii) the net electric current (I_N) and (iv) the flux normalized measure of the field twist ($\alpha = \mu \frac{I_N}{\Phi}$). These parameters are compared with the measured values which showed remarkable agreement.

Keywords. Sun:Active Region, Sun:MHD, Sun:Coronal Mass Ejections (CMEs), Sun:Flares, Sun:Magnetic Fields

1. Introduction

Understanding the sources of solar eruptive phenomena requires knowledge of the evolution of the active region. By looking at the full disk of the photospheric magnetogram, it is immediately recognized that the evolution of sunspots and sunspot groups are the sources of the most powerful solar eruptions (Wang *et al.* 2002, Wang *et al.* 2004). In an early study, Leighton (1964) modeled the sunspots and solar cycle in relation to the expansion and migration of unipolar (UM) and bipolar (BM) magnetic regions. Since then, a number of investigators (DeVore, *et al.* 1984; McIntosh & Wilson, 1985; Sheeley, *et al.* 1985; Sheeley & Devore 1986; Wilson & McIntosh, 1991; Wang & Sheeley, 1991; McKay 2003) have extensively investigated the magnetic flux transport in relation to the solar cycle by means of a modified Leighton model with additional physics. Wang and Sheeley (1991) have presented a numerical simulation including differential rotation, supergranular diffusion, and a poleward surface flow (i.e. meridional flow) of the redistribution of magnetic flux erupting in the form of bipolar magnetic regions (BMRs). They reproduced many of the observed features of the Sun's large scale field not encompassed by the Leighton (1964) model. Wilson and McIntosh (1991) compared observed evolution of large-scale magnetic fields with simulated evolution based on the kinematic model of Devore & Sheeley (1987). They concluded that there must be significant contributions to the evolving patterns by non-random flux eruptions within the network structure, independent of active regions. McKay (2003) presented a magnetic flux transport simulation of the Sun's surface distribution of magnetic fields during Maunder minimum. All these works are focused on the large-scale field and long-time-scale (i.e. solar cycle) evolution. In the case of small-scale field and short-time-scale (i.e. hours and days), the basic flux transport model was also applied with additional physical features (Schrijver, 2001; Schrijver & Title, 2001). However, all these investigations have not invoked full magnetohydrodynamic (MHD) theory; that means the nonlinear dynamic interactions among the plasma flow field and magnetic field are ignored. In order to include this nonlinear dynamic interaction, Wu, *et al.* (1993) have constructed a quasi-three-dimensional, time-dependent incompressible MHD model with differential rotation, meridional flow and effective diffusion as well as cyclonic turbulence to study evolution of BMRs. In their limited quasi-three-dimensional theoretical study, they have demonstrated that the observed complexity pattern could arise on the Sun's surface due to the dynamic interaction between the flow fields and magnetic field (i.e. MHD effect) and growth and decay of a BMR.

In this paper we will present a full three-dimensional, time-dependent, compressible MHD model with differential rotation, meridional flow, effective diffusion due to random motion of the granules or the super-granules and cyclonic turbulence effect to study the active region evolution to deduce the non-potential magnetic field parameters for possible initiation of solar eruptive events using observed magnetic field data as the initial conditions. The mathematical model, initial and boundary conditions are presented in Section 2, numerical results and concluding remarks are given in Section 3 and 4, respectively.

2. Mathematical Model, Initial and Boundary Conditions

2.1. *Mathematical Model*

On the basis of magnetohydrodynamic (MHD) theory, the mathematical model appropriate for the physical scenario we described in the previous section can be expressed by a set of compressible MHD equations consisting of conservation of mass, momentum, energy and the induction equation resulting from Maxwell's equations. These equations account for non-linear dynamic interactions of plasma flow and magnetic field. These governing equations are:

$$\frac{\partial \rho}{\partial t} + \nabla \cdot (\rho \vec{u}) = 0 \tag{2.1}$$

$$\rho \left[\frac{\partial \vec{u}}{\partial t} + \vec{u} \cdot \nabla \vec{u} \right] = -\nabla p + \frac{1}{4\pi}(\nabla \times \vec{B}) \times \vec{B} + \vec{F}_g - 2\pi\vec{\omega}_o \times \vec{u} - \rho\vec{\omega}_o \times (\vec{\omega}_o \times \vec{r}) + \Psi \tag{2.2}$$

where, $\Psi = -\frac{2}{3}\nabla(\mu\nabla \cdot \vec{u}) + \mu\left[\nabla^2\vec{u} + \nabla(\nabla \cdot \vec{u})\right] + 2\left[(\nabla\mu) \cdot \nabla\right]\vec{u} + \left[(\nabla\mu) \times (\nabla \times \vec{u})\right]$

$$\frac{\partial p}{\partial t} + \vec{u} \cdot \nabla p + \gamma p \nabla \cdot \vec{u} = (\gamma - 1)\nabla \cdot \vec{Q} + (\gamma - 1)\left[\eta J^2 + \frac{\mu}{2}(\nabla \cdot \vec{u})^2\right] \tag{2.3}$$

$$\frac{\partial \vec{B}}{\partial t} = \nabla \times (\vec{u} \times \vec{B}) + \lambda(\nabla \times \vec{B}) + \eta\nabla^2\vec{B} + \vec{S} \tag{2.4}$$

where ρ is the plasma mass density, $\vec{u}$ the plasma flow velocity vector, p the plasma thermal pressure, $\vec{B}$ the magnetic induction vector, $\vec{J}$ the electric current and $\vec{S}$ the energy source term, respectively. The other quantities are defined as follows: $\vec{\omega}_o$ is the angular velocity of solar differential rotation referring to the center of the solar coordinate system, that is given by Snodgrass (1983) and the meridional flow profile used here is given by Hathaway, (1996). $\vec{F}g$ is the gravitational force, $\nabla \cdot \vec{Q}$ represents the heat conduction, γ, μ, λ, and η are the specific heats ratio (1.05), viscosity, coefficients of the cyclonic turbulence and effective diffusion. Finally, the Ψ represents the viscous dissipation.

This set of MHD equations differs from first principle MHD theory because of the inclusion of additional physics. For example, the additional terms in Eq(2.2) represent the inertial centrifugal force (i.e. $2\rho\vec{\omega}_o \times \vec{u}$) and the coriolis force ($2\rho\vec{\omega}_o \times \vec{\omega}_o \times \vec{r}$) due to the Sun's differential rotation. The terms $\eta\nabla^2\vec{B}$ and $\lambda(\nabla \times \vec{B})$ in Eq. (2.4) represent the effective diffusion due to random motion of granules or super-granules and cyclonic turbulence effect, respectively.

2.2. *Initial and Boundary Conditions*

To simulate the active region evolution, we have cast the set of governing equations in a rectangular coordinate system. The computational domain includes six planes (i.e. four side planes, top and bottom). The boundary conditions used for the four sides are linear extrapolation, top boundary is non-reflective boundary and the bottom boundary is derived from the method of characteristics (Wu and Wang, 1987; Wang 1992) which is given in the Appendix. This set of boundary conditions is the time-dependent boundary conditions. In such a way, we are able to model the emerging and submerging magnetic flux in a self-consistent manner.

To implement this evolutionary simulation of the active region, we apply the following steps:

2.2.1. *Initializing the Simulation of the Active Region*

(*a*) Use the magnetic field data from photospheric magnetogram together with potential field model to construct a three-dimensional field configuration.

(*b*) Since there is no density measurement on the photosphere, we simply assume that the density distribution at the photospheric level is directly proportional to the absolute value of the magnitude of the transverse field and decreases exponentially with the scale height, thus $\rho(x,y,z,0) = \sqrt{\frac{B_x^2+B_y^2}{B_o^2}}e^{-\frac{z}{H_g}}$ where ρ_o and B_o are the constant reference values with H_g as the scale height, and

(*c*) Input the results of (*a*) and (*b*) into the MHD model described in Section 2 to allow its relaxing to a quasi-equilibrium state. This will be our initial state for the study of the evolution.

2.2.2. *Evolutionary Simulation of the Active Region*

To evolve the corona, we apply differential rotation and meridian flow to the lower boundary (photospheric magnetic field) in 5-second time steps. After each time step, we allow the corona to respond to the changes in the lower boundary condition. Once, every ~96 minutes, when a new MDI magnetogram is available, we add or subtract magnetic flux at the lower boundary according to the expression given by Eq. (A6) to mimic the emergence and submergence of the magnetic flux where the first term represents the general increasing and decreasing of the magnetic flux in the whole region and the second term is represented by a "delta" function which takes account the pop-up flux at a specific location. The coefficient "a" in Eq. (A6) is chosen according to the cadence of the observation and the computation time step. In this calculation a is 10^{-4}.

With these two procedures, we obtain the time and spatial evolution of the active region represented by this initial state.

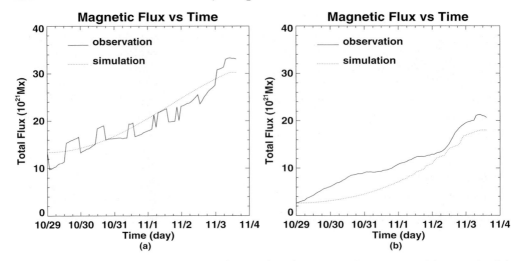

Figure 1. The simulated and measured (SOHO/MDI) magnetic flux content of AR8100 for (a) all fields and (b) strong fields (i.e. $|\vec{B}| \geqslant 100$ G) during the period October 29 - November 3, 1997.

3. Numerical Results

To carry out this simulation study, we have chosen the SOHO/MDI magnetic field measurements of NOAA/AR 8100 from October 29, 11:15 UT to Nov 3, 15:59UT, 1997 for this study. The SOHO/MDI field measurements of the active region have a resolution of ~ 2 arc sec with 198×198 pixels with a candence of ~ 96 min. In order to assure the computational grids are compatible with the measurements, the computational domain is set as a rectangular region with $99 \times 99 \times 99$ grid points in Carrington longitude (x), latitudinal direction (y) and height (z), respectively. To match the data with the grids, we have taken four point average of the pixels inside the domain. On the boundary we have taken a two point average from the measurements. At the four corners, the measurements are used. Before we can carry out the simulation study, we need to know two important coefficients; effective diffusivity (η) and cyclonic turbulence (λ). There are no precise theory and observations and laboratory experiments to determine these coefficients. However, there are some previous works which have discussed the choice of these two coefficients. For example, $\eta = 160$ - 300 km^2 s^{-1} given by Parker, (1979); Leighton's value of η is 800 - 1600 km^2 s^{-1} (1964); DeVore, et $al.$ (1985) selected $\eta = 300$ km^2 s^{-1} for their study. Wang (1988) derived a value of η being 100 - 150 km^2 s^{-1} on the basis of observation of sunspot's decay. We noticed that there is a wide range of values for the effective diffusivity. The value of cyclonic turbulence is chosen according to the scale law $(\lambda \sim \eta/L)$, given by Parker, (1979) where L is the characteristic length of the sunspot, it is chosen to be $6{,}000$ km for this study and η is 200 km^2 s^{-1}.

3.1. *Model Verification*

Since there is no analytical solution to test this three-dimensional time-dependent MHD model of active region evolution, the only test which could be accomplished is to use observations. To carry out this test, we compare the simulated total flux content and the contours of the line-of-sight component of the magnetic field of the active region with the observation. Figure 1 shows the simulated and measured total flux content of NOAA/AR8100 during the period of October 29 - November 3, 1997 for strong fields (Fig1(b)) and all fields (Fig 1(a)), respectively. Simulated total flux content is obtained

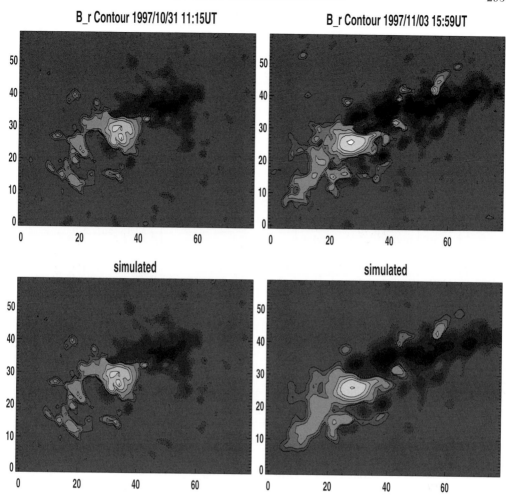

Figure 2. Comparison between the SOHO/MDI measured (upper panel) and simulated (lower panel) contours of the line-of-sight magnetic field maps at 1997 Oct 31, 11:15 UT and 1997 Nov 3, 15:50 UT, respectively for AR8100.

by input of the SOHO/MDI measured line-of-sight magnetic field together with potential field model into the three-dimensional, MHD AR evolution model described in Section 2. Then, the total flux content of the region is computed (i.e. $\Phi = \int_A \vec{B} \cdot d\vec{A}$; where $\vec{A}$ is the area of AR). The measured total flux content is obtained by computing $\Phi = \int_A \vec{B} \cdot d\vec{A}$ from the measured magnetic field at various time. Let us examine these results shown in Figure 1; it is clearly shown that the model simulated and measured total flux content (Φ) agree well. We also notice that the weak field does contribute significantly to the amount of total flux content.

Figure 2 shows the simulated and measured (SOHO/MDI) contours of the line-of-sight component of magnetic field at 1991 Oct 31, 11:15 UT and 1997 Nov 3, 15:59UT, respectively. Again the model results have mimicked the observed features well.

3.2. *Simulation of Non-potential Magnetic Field Parameters*

Using the definitions given by Falconer *et al.* (2002), the non-potential magnetic field parameters are computed based on the model outputs. These non-potential magnetic field

Table 1. Non-Potential Magnetic Field Parameters of AR8100
during the Period of October 31 - November 3, 1997

Time (Date)	(UT)	$L_{ss}(10^3\text{km})$ (Obs)	(Sim)	$I_N(10^{11}\text{A})$ (Obs)	(Sim)	$\alpha(10^{-8}/\text{m})$ (Obs)	(Sim)	$\Phi(10^{21}\text{Mx})$ (Obs)	(Sim)
10/31	11:15	—	0	—	0	—	0	—	5.8
10/31	12:51	—	12.6	—	0.4	—	0.050	—	5.9
10/31	14:27	—	20.0	—	0.7	—	0.080	—	6.0
10/31	15:07	28±8*	22.5	1±1*	0.8	0.13±0.1*	0.092	6±1*	6.2
10/31	16:03	—	24.0	—	0.9	—	0.105	—	6.3
10/31	17:39	—	27.0	—	1.1	—	0.120	—	6.4
10/31	20:48	—	29.4	—	1.3	—	0.140	—	6.8
—	—	—	—	—	—	—	—	—	—
11/01	01:39	—	33.1	—	1.6	—	0.148	—	7.5
11/01	06:27	—	36.2	—	1.8	—	0.152	—	8.0
11/01	11:25	—	37.5	—	1.9	—	0.155	—	8.8
11/01	16:03	—	39.4	—	2.2	—	0.158	—	9.5
11/01	20:48	—	42.5	—	2.4	—	0.159	—	10.6
—	—	—	—	—	—	—	—	—	—
11/02	01:39	—	41.5	—	2.5	—	0.161	—	11.7
11/02	06:27	—	48.0	—	2.7	—	0.165	—	12.6
11/02	11:12	—	51.0	—	3.0	—	0.172	—	14.0
11/02	16:00	—	54.1	—	3.2	—	0.181	—	14.7
11/02	20:48	—	57.2	—	3.6	—	0.192	—	15.8
—	—	—	—	—	—	—	—	—	—
11/03	01:39	—	61.6	—	4.1	—	0.218	—	16.8
11/03	06:27	—	67.5	—	4.8	—	0.260	—	17.7
11/03	09:37	—	71.4	—	5.4	—	0.295	—	18.0
11/03	12:51	—	77.0	—	6.5	—	0.354	—	18.1
11/03	14:24	—	81.6	—	7.5	—	0.392	—	18.0
11/03	14:58	85± 17*	83.3	7±2*	7.8	0.44±0.1*	0.413	14±3*	17.9
11/03	15:59	—	86.4	—	8.8	—	0.460	—	17.7

*These two data points are obtained from NASA/MSFC vector magnetogram (Falconer, *et al* 2002); Obs = Observed and Sim = simulated

parameters are; (i) total magnetic flux context, (Φ); (ii) the length of strong magnetic shear ($\geqslant 45°$) and strong transverse field ($\geqslant 150$ gauss) of the main neutral line, (L_{ss}); (iii) the net electric current (I_N); and (iv) the flux normalized measure of the field twist ($\alpha = \mu \frac{I_N}{\Phi}$). Table 1 shows these parameters as a function of time and these values are compared with the values given by Falconer *et al.* (2002) at two specific times. Their results are obtained by using the MSFC vector magnetograph data. There is no time series of observations with MSFC's vector magnetograms, thus, we can only make comparison with these two specific times. This is the reason why we have chosen SOHO/MDI data for this study because the SOHO/MDI has made continuous measurements. Examination of the tabulated values of non-potential parameters further shows that at the initial state (October 31, 11:15 UT), three of four non-potential parameters simulated are null because the initial state is approximated by potential model. As time progresses, all non-potential parameters increase due to the Sun's rotation and meridional flow as well as the flux emergence and submergence according to the effects of the lower boundary (i.e. photosphere shown in the Appendix). As such, the MHD effects occur due to the nonlinear dynamical interactions between the magnetic fields and plasma flow, in which, the magnetic shear is generated where the magnetic field is stressed, it leads to field expansion. Subsequently the mass is lifted up by the movement of magnetic field shown in Figure 3 and then partially opens up, which looks similar to the "break-out" model

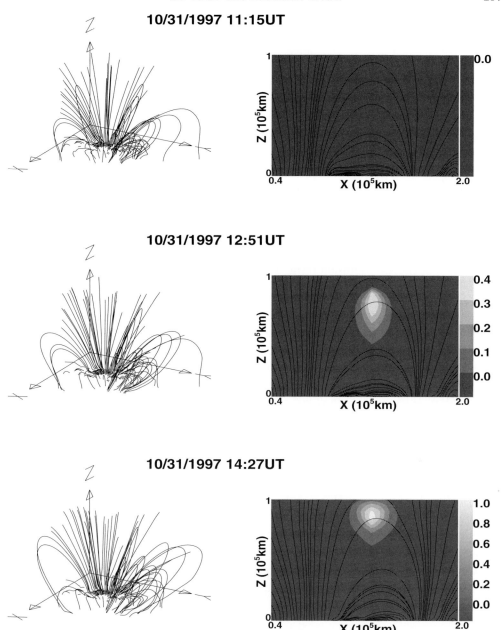

Figure 3. The computed three-dimensional magnetic field evolution (left column) and corresponding field line projection on the x-z plane with density enhancement contours $\frac{(\rho - \rho_o)}{\rho_o}$ (right column).

suggested by Antiochos, *et al.* (1999). As such, a coronal mass ejection could be initiated. These features can be seen in Figure 3.

Figure 3 shows the three-dimensional magnetic field evolution in three specific times (left column) and the corresponding projection of these field lines on the x-z plane together with density contours $\frac{(\rho - \rho_o)}{\rho_o}$ in gray scale (right column) to show the movement of the plasma blob. By looking at the field lines and density contours on the x-z plane together with the L_{ss} shown in Table 1, it is easily recognized that when L_{ss} increases, the field lines rise up and begin to open. Consequently, the mass is lifted up by the field movement. As soon as the field opens, then, the mass will release to form a CME. It should be noted that the initial density shown at 11:15 UT is null, because the background density has been subtracted (i.e. $\frac{\rho - \rho_o}{\rho_o}$).

4. Summary and Concluding Remarks

A three-dimensional, time-dependent magnetohydrodynamic (MHD) model for the Active Region (AR) evolution is presented. This newly developed MHD model includes new physics which has usually been neglected. These new physical terms in the governing equations are caused by the introduction of the Sun's differential rotation and meridional flow. These new terms are the inertial centrifugal force and coriolis force in the momentum equation and the cyclonic turbulence effect in the induction equation (Eq. 2.4). In addition, the effect due to random motion of the granules or super-granules is included in the form of an effective diffusion term. This newly developed model is tested by using a specific data set (i.e. SOHO/MDI magnetogram recorded NOAA/AR8100). Results show that the model reproduces observations well (see Fig. 1 and 2). Then, we have employed this model to simulate the non-potential parameters developed by Falconer *et al.* (2002). Again, the model performed reasonably well (see Table 1).

In summary, some of the important properties of this model can be described as follows:

(*a*) the new formulation with inclusion of differential rotation and meridional flow give rise to the magnetic shear. It leads to the complexity of the magnetic field features which give understanding of the growth and decay of an active region.

(*b*) this model has the potential to be utilized to quantify the critical parameters for the initiation of solar eruptive phenomena (i.e. flare/CME), such that a predictive model for solar eruptive events could be achieved.

Appendix A. Bottom Boundary conditions

This set of boundary conditions are obtained by method of characteristics (Wu & Wang 1987, Wang 1992) with the assumption of positive vertical velocity and less than characteristic speeds (i.e. Alfven, slow and fast wave speed). These expressions which describe the physical parameters of pressure, density, the components of velocity, and magnetic field vary with time on the boundary are:

$$\frac{\partial p}{\partial t} = \frac{V_s^2 B + V_f^2 C}{2V_A^2(V_f^2 - V_s^2)} \tag{A 1}$$

$$\frac{p}{\rho^\gamma} = const. \tag{A 2}$$

$$\frac{\partial u_x}{\partial t} = \frac{B_y \left(V_s(V_A^2 - V_s^2)B - V_f(V_f^2 - V_A^2)C \right)}{2B_z(B_x^2 + B_y^2)V_s V_f(V_f^2 - V_s^2)} + \frac{B_x A}{2B_z(B_x^2 + B_y^2)} \tag{A 3}$$

$$\frac{\partial u_y}{\partial t} = \frac{B_x \left(V_s(V_A^2 - V_s^2)B - V_f(V_f^2 - V_A^2)C \right)}{2B_z(B_x^2 + B_y^2)V_s V_f(V_f^2 - V_s^2)} - \frac{B_y A}{2B_z(B_x^2 + B_y^2)} \tag{A 4}$$

$$\frac{\partial u_z}{\partial t} = \frac{(-V_s B + V_f C)}{2\rho V_s V_f (V_f^2 - V_s^2)} \tag{A 5}$$

$$B_z = B_{zo}(x, y, t)(1 + a\Delta t) + \sum_{x,y,t} \delta B_z(x, y, t) \tag{A 6}$$

$$\frac{\partial B_x}{\partial t} = \frac{B_x \left((V_A^2 - V_s^2)B - (V_f^2 - V_A^2)C \right)}{2V_A^2 (V_f^2 - V_s^2)(B_x^2 + B_y^2)} + \frac{B_y A}{2V_A (B_x^2 + B_y^2)} \tag{A 7}$$

$$\frac{\partial B_y}{\partial t} = \frac{B_y \left((V_A^2 - V_s^2)B - (V_f^2 - V_A^2)C \right)}{2V_A^2 (V_f^2 - V_s^2)(B_x^2 + B_y^2)} - \frac{B_x A}{2V_A (B_x^2 + B_y^2)} \tag{A 8}$$

where the coefficients A, B, and C are given below.

$$A = -(u_z - V_A) \left[B_y B_z \frac{\partial u_z}{\partial z} - B_x B_z \frac{\partial u_y}{\partial z} + B_y V_A \frac{\partial B_x}{\partial z} - B_x V_A \frac{\partial B_y}{\partial z} \right]$$
$$+ (B_x B_y V_A - u_x B_y B_z)\frac{\partial u_x}{\partial x} - (B_x^2 V_A - B_x B_z u_x)\frac{\partial u_y}{\partial x} - \frac{B_y B_z}{\rho}\frac{\partial p}{\partial x}$$
$$- \frac{B_z}{\rho}(B_x^2 + B_y^2)\frac{\partial B_y}{\partial x} + B_x V_A u_x \frac{\partial B_y}{\partial x} - u_x B_y V_A \frac{\partial B_x}{\partial x} - \frac{B_y B_z^2}{\rho}\frac{\partial B_z}{\partial x}$$
$$+ (B_y^2 V_A - u_y B_y B_z)\frac{\partial u_x}{\partial y} - (B_x B_y V_A - B_x B_z u_y)\frac{\partial u_y}{\partial y} + \frac{B_x B_z}{\rho}\frac{\partial p}{\partial y}$$
$$+ \frac{B_z}{\rho}(B_x^2 + B_y^2)\frac{\partial B_x}{\partial y} - u_y B_y V_A \frac{\partial B_x}{\partial y} + u_y B_x V_A \frac{\partial B_y}{\partial y} + \frac{B_x B_z^2}{\rho}\frac{\partial B_z}{\partial y}, \tag{A 9}$$

$$B = -(u_z - V_f) \left[(B_x B_z V_f \frac{\partial u_x}{\partial z} + B_y B_z V_f \frac{\partial u_y}{\partial z} + \rho V_f (V_f^2 - V_A^2)\frac{\partial u_z}{\partial z} \right.$$
$$\left. + (V_f^2 - V_A^2)\frac{\partial p}{\partial z} + B_x V_f^2 \frac{\partial B_x}{\partial z} + B_y V_f^2 \frac{\partial B_y}{\partial z} \right]$$
$$- \left(B_x B_z V_f u_x + a^2\rho(V_f^2 - V_A^2) + B_y^2 V_f^2 \right)\frac{\partial u_x}{\partial x} - B_y V_f(u_x B_z + B_x V_f)\frac{\partial u_y}{\partial x}$$
$$+ \rho u_x V_f(V_f^2 - V_A^2)\frac{\partial u_z}{\partial x} - \left(\frac{B_x B_z V_f}{\rho} + u_x(V_f^2 - V_A^2) \right)\frac{\partial p}{\partial x} - u_x B_x V_f^2 \frac{\partial B_x}{\partial x}$$
$$- u_x B_y V_f^2 \frac{\partial B_y}{\partial x} - B_x V_f^3 \frac{\partial B_z}{\partial x} + (B_x B_y V_f^2 - u_y B_x B_z V_f)\frac{\partial u_x}{\partial y}$$
$$- \left(u_y B_y B_z V_f + a^2\rho(V_f^2 - V_A^2) + B_x^2 V_f^2 \right)\frac{\partial u_y}{\partial y} + \rho V_f(V_f^2 - V_A^2)u_y \frac{\partial u_z}{\partial y}$$
$$- \left(\frac{B_y B_z V_f}{\rho} - u_y(V_f^2 - V_A^2) \right)\frac{\partial p}{\partial y} - u_y B_x V_f^2 \frac{\partial B_x}{\partial y} - u_y B_y V_f^2 \frac{\partial B_y}{\partial y}$$
$$- B_y V_f^3 \frac{\partial B_z}{\partial y} + \rho g V_f(V_f^2 - V_A^2), \tag{A 10}$$

$$C = -(u_z - V_s) \left[(B_x B_z V_s \frac{\partial u_x}{\partial z} + B_y B_z V_s \frac{\partial u_y}{\partial z} + \rho V_s (V_s^2 - V_A^2) \frac{\partial u_z}{\partial z} \right.$$
$$\left. + (V_s^2 - V_A^2) \frac{\partial p}{\partial z} + V_s^2 \left(B_x \frac{\partial B_x}{\partial z} + B_y \frac{\partial B_y}{\partial z} \right) \right]$$
$$+ \left[a^2 \rho (V_s^2 - V_A^2) + B_y^2 V_s^2 + B_x B_z V_s u_x \right] \frac{\partial u_x}{\partial x} + B_y V_s (u_x B_z + B_x V_s) \frac{\partial u_y}{\partial x}$$
$$- \rho u_x V_s (V_s^2 - V_A^2) \frac{\partial u_z}{\partial x} + \left[u_x (V_s^2 - V_A^2) + \frac{B_x B_z V_s}{\rho} \right] \frac{\partial p}{\partial x} + u_x B_x V_s^2 \frac{\partial B_x}{\partial x}$$
$$+ u_x B_y V_s^2 \frac{\partial B_y}{\partial x} + B_x V_s^3 \frac{\partial B_z}{\partial x} - (B_x B_y V_s^2 - u_y B_x B_z V_s) \frac{\partial u_x}{\partial y}$$
$$+ \left[a^2 \rho (V_s^2 - V_A^2) + B_x^2 V_s^2 + u_y B_y B_z V_s \right] \frac{\partial u_y}{\partial y} - \rho V_s (V_s^2 - V_A^2) u_y \frac{\partial u_z}{\partial y}$$
$$+ \left[\frac{B_y B_z V_s}{\rho} + u_y (V_s^2 - V_A^2) \right] \frac{\partial p}{\partial y} - u_y B_x V_s^2 \frac{\partial B_x}{\partial y} + u_y B_y V_s^2 \frac{\partial B_y}{\partial y}$$
$$+ B_y V_s^3 \frac{\partial B_z}{\partial y} + \rho g V_s (V_s^2 - V_A^2) \quad \text{(A 11)}$$

Alfvén Speed

$$V_A = \frac{B_z}{\sqrt{4\pi\rho}}, \quad \text{(A 12)}$$

Fast MHD Wave Speed

$$V_f^2 = \frac{1}{2} \left((a^2 + b^2) + \left((a^2 + b^2)^2 - 4a^2 V_A^2 \right)^{1/2} \right), \quad \text{(A 13)}$$

Slow MHD Wave Speed

$$V_s^2 = \frac{1}{2} \left((a^2 + b^2) - \left((a^2 + b^2)^2 - 4a^2 V_A^2 \right)^{1/2} \right), \quad \text{(A 14)}$$

with

$$b = \sqrt{\frac{B_x^2 + B_y^2 + B_z^2}{4\pi\rho}} \quad \text{(A 15)}$$

Sound Speed

$$a = \sqrt{\gamma R T} \quad \text{(A 16)}$$

Acknowledgements

The work performed by STW and AHW is supported by NASA grant NAG5-12843, NSF grant ATM0316115, and AAMU subcontract under NNG04GD59G. DAF is supported by NASA Cooperative Agreement NCC8-200 and NSF grants ATM-0352834 and ATM-0203098. The authors wish to thank Dr. C. D. Fry for reading the manuscript and making useful comments.

References

Antiochos, S.K., DeVore, C.R., & Klumchuk, J.A. 1999, *ApJ* 510, 258

Devore, C.R., Sheeley, N.R., Jr., Boris, J.P., Young, R.T., Jr., & Harvey, K.L. 1984, *Solar Phys.* 92, 1

Devore, C. R., Sheeley, N.R., Jr., Boris, J.P., Young, R.T., Jr., & Harvey, K.L. 1985, *Solar Phys* 102, 41

Devore, C. R., & Sheeley, N. R., 1987, *Solar Phys* 108, 47

Falconer, D.A., Moore, R.L. & Gary, G.A., 2002, *ApJ* 569, 1016.

Leighton, R.B. 1964, *ApJ* 140, 1547

McKay, D.H. 2003, *Solar Phys* 213, 173

Hathaway, D.H 1996 *ApJ* 460, 1027

McIntosh, P.S. & Wilson, P.R. 1985 *Solar Phys* 99, 59

Parker, E.N. 1979 in: *Cosmic Magnetic Fields* (Oxford University Press, England), p. 509

Schrijver, C.J. 2001 *ApJ* 547, 475

Schrijver, C.J. & Title, A.M. 2002 *ApJ* 551, 1099

Sheeley, N.R., Jr., Devore, C.R., & Boris, J. P. 1985 *Solar Phys* 98, 219

Sheeley, N.R., Jr., & Devore, C. R. 1986 *Solar Phys* 103, 203

Wang, A.H. 1992, PhD thesis, University of Alabama in Huntsville.

Wang, H.M. 1988 *Solar Phys* 116, 1

Wang, H.M., Spirock, T.J., Qiu, J., Ji, H., Yurchyshyn, V., Moon, Y-J., Denker, C. & Goode, P.R. 2002 *ApJ* 576, 497

Wang, H.M., Qiu, J., Jing, J. Spirock T. J., Yurchyshyn, V., Abramenko, V., Ji, H. & Goode, P.R. 2004 *ApJ* 605, 931

Wang, Y.-M. & Sheeley, N. R., Jr., 1991 *ApJ* 375, 761

Wilson, P.R. & McIntosh, P.S. 1991 *Solar Phys* 136, 221

Wu, S.T., Wang, J.F. 1987 *Comp. Methods in Appl. Mech.* 64, 267

Wu, S.T., Yin, C.L., McIntosh, P.S. & Hildner, E. 1993 in: H.Zirin, G. Ai & H.M. Wang (eds.), *The Magnetic and Velocity Fields of Solar Active Regions*, ASP Conference Series (New York: AIP), vol. 46, p. 98

Discussion

SCHMIEDER: I appreciate your model. The total flux follows the observed total flux. The shear parameter α however is increasing during all of your time sequences

WU: In the example of Falconer, the CME arrives 5 days later. The simulation does not reach the phase of eruption.

KUTCHMY: Usually, when we observe an increase of the total magnetic flux we think about a flux emergence process of fluxes produced in sub-photospheric layers. In your case, you produce an increase of the flux using surface phenomena. How can one decide which phenomenon is more important and do they have the same sign of variation?

WU: In the model, we are able to accommodate sub-photospheric effect through the time-dependent characteristic boundary included in the Appendix. In the present calculation, we prescribe the flux emergence according to a receipt given in Eq. (A6) on the basis of the MDI observation.

GRECHNEV: If your first plot of the total magnetic flux versus time, the difference of the observed and calculated values is a quasi-periodic function, which can be of observational origin. If this quasi-periodic difference can be eliminated, the coincidence of the calculated and observed values would still be there.

WU: Yes, you are correct.

Coronal and Stellar Mass Ejections
Proceedings IAU Symposium No. 226, 2005
K. P. Dere, J. Wang & Y. Yan, eds.

The Importance of Topology and Reconnection in Active Region CMEs

Robert J. Leamon†

L-3 Communications at NASA Goddard Space Flight Center,
Code 612.5, Greenbelt, MD 20771, USA
email: leamon@grace.nascom.nasa.gov

Abstract. A distinctive characteristic of interplanetary magnetic clouds is their rope-like magnetic structure, *i.e.*, their smoothly-varying helical field lines whose pitch increases from their core to their boundary. Because this regular structure helps to make MCs particularly geoeffective, it is important to understand how it arises.

We discuss recent work which relates the magnetic and topological parameters of MCs to associated solar active regions. This work strongly supports the notion that MCs associated with active region eruptions are formed by magnetic reconnection between these regions and their larger-scale surroundings, rather than simple eruption or entrainment of pre-existing structures in the corona or chromosphere. We discuss our findings in the context of other recent works on both the solar and interplanetary sides, including ion composition and various MHD models of magnetic cloud formation.

Keywords. Sun: coronal mass ejections (CMEs) — Sun: solar-terrestrial relations

1. Introduction

Coronal mass ejections (CMEs) are an almost daily occurrence on the sun. Interplanetary manifestations of CMEs often have a very distinct magnetic structure, namely a large-scale helix. Burlaga *et al.* (1981) described this magnetic structure as a "Magnetic Cloud." According to Burlaga's definition, three properties are required to identify a structure as a magnetic cloud: (*i*) a very low proton temperature; (*ii*) a large, smooth, monotonic rotation of the field direction; and (*iii*) enhanced magnetic field strength. The general case of a smooth, monotonic rotation of the field direction is called a magnetic flux rope. At least one-third (Gosling 1990), and, according to some researchers (*e.g.*, Webb 2000) considerably more, of the interplanetary manifestations of CMEs observed *in situ* have magnetic flux rope (large-scale helix) structure.

When one can associate a solar source with interplanetary ejecta, the source is almost always an active region or a quiescent solar filament. At first inspection, those magnetic clouds spawned from ARs and filaments seem to be similar, but as we shall demonstrate, source-related differences exist.

1.1. Magnetic Clouds Associated With Filaments

The relationship between magnetic clouds and their solar progenitors may be much more straightforward for filaments (Bothmer & Rust 1997; Crooker 2000) than active regions (Pevtsov & Canfield 2001; Leamon *et al.* 2002), as we shall now discuss.

Left-handed (Right-handed) MCs associated with erupting filaments tend to come from the Northern (Southern) hemisphere—only about 1 in 30 filament eruptions violate this

---

† Work performed while at Department of Physics, Montana State University, Bozeman, MT 59717, USA.

rule (Bothmer & Rust 1997). The tilts of their axes with respect to the ecliptic plane are correlated with tilt angles (relative to the heliomagnetic equator) of the associated filaments. The north-south component of the magnetic fields inside the MC tends to reverse at or around cycle maximum; the field of MCs is made up of the large-scale dipole fields. In short, we can say that such MCs carry the imprint of the large-scale solar magnetic field out into the heliosphere.

1.2. *Magnetic Clouds Associated With Active Regions*

The same hemispheric preference for handedness holds for AR-associated CMEs, but only at about a 2:1 ratio, which is about the same as the handedness of active regions themselves (Joy's Law). However, the tilts of MC axes with respect to the ecliptic plane are not correlated with tilt angles (relative to the heliomagnetic equator) of associated ARs. The major difference between the two north-south component of MC fields tends to flip at around cycle minimum, with the polarity of sunspots. So unlike the global nature of filament-based eruptions, the field of active region MCs is made up both active-region and large-scale dipole fields, and such MCs carry the imprint of both active regions and the large-scale solar magnetic field out into the heliosphere.

1.3. *Outline*

In the next section we shall see that, for AR-associated eruptions at least, magnetic clouds are formed by magnetic reconnection between these regions and their larger-scale surroundings, rather than simple eruption of pre-existing structures in the corona or chromosphere. We shall then investigate what the plasma composition data can tell us about the nature and dynamics of that reconnection, and that composition highlights further differences between active region-based events and those associated with filaments.

2. Magnetic Structure: Solar-Interplanetary Comparisons

Following Lepping *et al.* (1990), we model magnetic clouds using the constant-α force-free field solution of Lundquist (1950), which assumes cylindrical symmetry. In a cloud-centered frame of reference, the axial and tangential fields are modeled as zeroth- and first-order Bessel functions, respectively, and there is no radial component. Full details of our magnetic cloud model and code are described in Leamon *et al.* (2004), along with sample fits. Once a satisfactory fit is found, it is a simple matter to deduce current density from the induction equation, and calculate the integrated current and flux in the cloud. There are some (sometimes gross) approximations to be made in modeling interplanetary ejecta as an infinite right cylinder, as well as constant-α; however, since we are comparing the helicity of magnetic clouds to that of solar active regions, which are represented by constant-α force-free field models (Pevtsov *et al.* 1995) the model and assumptions used are a valid first step.

Table 1, reproduced from Leamon *et al.* (2004), contains a comparison of flux, current and twist between 12 magnetic clouds and their identified progenitor active regions. We draw the reader's attention to four significant results: (*i*) the flux ratios Φ_{MC}/Φ_{AR} tend to be large; (*ii*) the current ratios I_{MC}/I_{AR} tend to be orders of magnitude less than the flux ratios Φ_{MC}/Φ_{AR}; (*iii*) there is a statistically significant proportionality between the flux and current ratios; and (*iv*) in four of the 12 events (Nos. 3, 4, 11, and 12) the cloud and active region have opposing senses of twist.

These features can be discussed in the framework of three highly simplified, but conceptually useful, models of the solar genesis of magnetic clouds. These models are that the flux in the magnetic cloud originates in: (1) the active region alone (the AR model); (2)

Table 1. Comparison of flux, current, and twist values in ARs and MCs.

No.	Year	Event Times: Eruption	Magnetogram	AR	I_{MC}/I_{AR} $\times 10^{-3}$	Φ_{MC}/Φ_{AR} %	$(\alpha L)_{AR}$	$(\alpha L)_{MC}$
1	1995	Feb 04 15:56	Feb 04 01:51	7834	0.15–0.53	20–38	−0.34	−53.2
2	–	Feb 28 08:46	Feb 27 07:17	7846	0.05–0.30	8–18	−0.01	−62.2
–	–	–	Feb 28 19:45	–	0.12–0.28	8–12	−0.91	–
3	–	Dec 11 03:31	Dec 10 17:22	7930	0.13–1.25	115–279	1.94	−25.8
–	–	–	Dec 11 17:15	–	0.58–4.02	122–314	1.30	–
4	1998	Feb 14 02:29	Feb 12 17:33	8156	0.30–0.66	22-30	−0.85	+43.1
5	–	Apr 29 16:58	Apr 28 16:38	8210	0.09–0.83	35–69	1.64	+26.3
–	–	–	Apr 29 16:39	–	0.36–1.19	40–72	1.27	–
6	1999	Feb 14 11:16	Feb 11 19:26	8457	0.36–1.07	43–89	−0.46	−25.0
7	–	Aug 04 04:11	Aug 02 16:36	8651	0.06–0.13	5–7	−1.06	−37.6
8	–	Sep 17 22:28	Sep 20 17:00	8700	0–0.53	0–13	−0.59	−124.8
9	2000	Jul 14 09:27	Jul 11 16:52	9077	0.58–0.71	61–69	−7.79	−33.5
–	–	–	Jul 14 16:39	–	0.55–0.68	52–112	−5.45	–
–	–	–	Jul 17 16:32	–	1.78–6.12	138–274	−1.60	–
10	–	Jul 25 02:48	Jul 21 20:28	9097	0.13–0.14	6.0–6.4	−1.42	−53.9
11	–	Jul 27 22:18	Jul 26 16:33	9097	0.19–0.49	40–106	1.42	−21.6
–	–	–	Jul 28 16:40	–	0.83–4.00	121–317	−1.14	–
12	–	Aug 09 16:30	Aug 08 16:37	9114	0.59–1.35	56–79	−1.70	+35.2
–	–	–	Aug 11 16:58	–	0.60–1.14	46–62	−1.19	–

the overlying large-scale dipole alone (the LSD model); and (3) a region of reconnection of the active region and the large-scale dipole (the AR-LSD reconnection model).

First, the large flux ratios are inconsistent with the AR model (If 90% of the flux from an AR were ejected, there wouldn't be much of an active region left—this is certainly not observed). Large flux ratios are at least consistent with the LSD model and the AR-LSD reconnection model.

That the current ratios tend to be orders of magnitude less than the flux ratios makes sense only in the context of the AR and AR-LSD reconnection models, if there is no significant current present outside the cores of active regions. Although this view is reasonable, it is hard to defend, since as a practical matter vector magnetographs lack sufficient sensitivity to measure currents beyond the strong-field areas of active regions. It remains possible that there exist on the Sun very extended regions outside ARs with current density that is low, but nevertheless not so low that their contribution to the total current of MCs is negligible.

Figure 1 illustrates in graphical form of the proportionality between flux and current. Whether or not the relationship is linear, it is demonstrably real, with Spearman rank-order correlation coefficient of 0.811 at a 99.8% confidence level. This trend can most easily be explained in terms of the AR-LSD reconnection model. The combination of I_{MC}/I_{AR} and Φ_{MC}/Φ_{AR} argues against the AR model.

The quantity αL, shown in the last two columns of Table 1 represents the total twist in length L. For active regions, we determine L from magnetograms, and for magnetic clouds we take the only directly measured value, 2.5 AU (Larson *et al.* 1997). It is immediately obvious that $(\alpha L)_{MC} \approx 10(\alpha L)_{AR}$ and that there is no systematic sign or amplitude relationship between $(\alpha L)_{MC}$ and $(\alpha L)_{AR}$. As we have already noted, one-third of the MC-AR pairs have mismatched senses of twist.

$(\alpha L)_{MC} \gg 1$ rules out the AR and LSD models, and only the AR-LSD reconnection model can produce so many turns within the resulting magnetic cloud. Further, only the AR-LSD reconnection model can explain the lack of relationship between $(\alpha L)_{MC}$ and

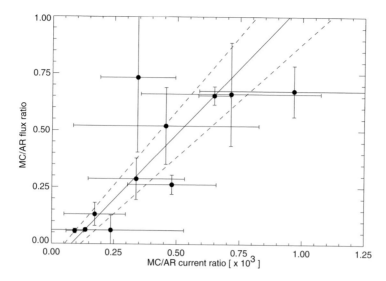

Figure 1. Flux ratio versus current ratio. Reproduced from Leamon *et al.* (2004, their Figure 3). Whether or not the true relationship is linear, the best fit straight line, with errors, is over-plotted.

$(\alpha L)_{AR}$ (Zhang & Low 2003). Hence, we are compelled to believe that only models that invoke magnetic reconnection between active regions and their large-scale overlying fields in the eruptive process can explain the relationships between total flux, total current, and associated twist of magnetic fields of magnetic clouds and active regions revealed in this study.

3. Plasma Structure and Composition

Charge compositional signatures are very sensitive to identifying ejecta *in situ*. The best signatures are oxygen and iron charge states (Cane & Richardson 2003; Lepri & Zurbuchen 2004):

$$O^{7+}/O^{6+} > 1 \qquad \langle Q_{Fe} \rangle > 12 \qquad Fe^{16+}/Fe_{total} > 0.01$$

An enhanced alpha/proton ratio is also indicative of ejecta, but at a weaker confidence level and is not unique (for example, it is also indicative of coronal holes).

Higher charge states are indication of a hotter corona at time of freezing-in. They are a better measure of thermal history than the proton temperature observed at 1AU, which has undergone changes due to ejecta evolution, possible *in situ* heating, and of course depends on the pre-existing solar wind conditions into which the CME was launched. O^{7+}/O^{6+} freezes in at $\sim 1R_\odot$, whereas $\langle Q_{Fe} \rangle$ is fixed at 3–$4R_\odot$ (i.e., at lower temperatures) Enhanced O^{7+}/O^{6+} (i.e., hotter) is observed in large flare events and correlates with BDEs (Reinard & Fisk 2004).

The 'Breakout' model of Antiochos (1998) predicts that if the flux rope forms as result of flare reconnection, one would expect hot ejecta in the center of the cloud. On the other hand, if the flux rope forms before flare onset (e.g., Lin & Forbes 2000) one would expect a relatively cool flux rope with hot sheaths before and after. The only test of this hypothesis to date (Lynch *et al.* 2003, their Figure 9) is inconclusive. They observe cooler ends to a relatively uniform Q_{Fe} profile, but there is no systematic profile in oxygen.

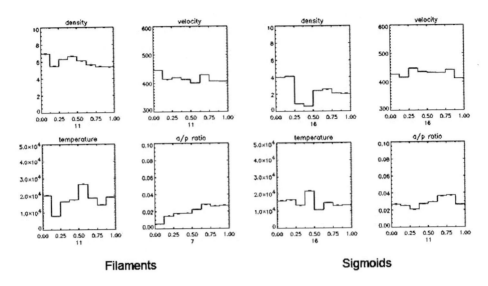

Figure 2. Epoch superposition and weighted averaging of the plasma structure of magnetic clouds associated with active region flares and filament eruptions. There are 11 filament-sourced clouds (of which only 7 have composition data) and 16 AR-sourced clouds (11 with composition data).

Finally, Figure 2 shows tentative first results from a recent study (Leamon, McKenzie and Wilson, "Geoeffective CMEs, Filaments, and Sigmoids", in preparation) suggesting that these tests are indicative of more differences between sigmoids and filaments. Specifically, compare the radically different density and α/p ratio profiles, which suggest different creation mechanisms. (We are in the process of extending these results by including iron and oxygen tests.)

4. Summary

Considering 3 simple models—that magnetic cloud flux originates in: (1) active regions alone; (2) the overlying large-scale dipole alone; and (3) a region that has experienced reconnection of the active region and the large-scale dipole, the observational results of the ratios of flux, current and twist only allow the AR-LSD reconnection model to escape unscathed.

Compositional tests, such as those based on iron and oxygen charge states, are sensitive in identifying *in situ* ejecta. However, there is insufficient statistically significant data to prove or disprove either Antiochos' or Lin & Forbes' reconnection models.

Filament-related MCs have identifiably different compositional signatures. More study is needed to categorically disprove whether such MCs might be formed either by the LSD model or from pre-existing chromospheric structures.

Acknowledgements

Funding for this work was provided by AFOSR under grant number F49620-00-1-0128 and the MURI grant to the University of California, Berkeley, and NASA under SR&T grant NAG5-6110. I would like to express thanks to Dick Canfield, Alexei Pevtsov and Dave McKenzie. More thanks are due to students Keith Lambkin, Brian Lundberg, Sarah Jones and Karen Wilson, all of whom participated in this project through the Research

Experience for Undergraduates (REU) Program at Montana State University, which is supported by the National Science Foundation under Grant No. 0243923.

Travel assistance to the Beijing IAU symposium from the NSF is also gratefully acknowledged.

References

Antiochos, S.K. 1998, *ApJ* 502, L181

Bothmer, V. & Rust, D.M. 1997, in Geophysical Monograph Series, Vol. 99, Coronal Mass Ejections, ed. N.U. Crooker, J.-A. Joselyn, & J. Feynman (Washington, D.C.: American Geophysical Union), 139

Burlaga, L.F., Sittler, E.C., Mariani, F., & Schwenn, R. 1981, J. Geophys. Res., 86, 6673

Cane, H.V. & Richardson, I.G. 2003, *J. Geophys. Res.* 108, 6

Crooker, N.U. 2000, *J. Atmos. and Solar-Terr. Phys.* 62, 1071

Gosling, J.T. 1990, in Geophysical Monograph Series, Vol. 58, Physics of Magnetic Flux Ropes, ed. C.T. Russell, E.R. Priest, & L.C. Lee (Washington, D.C.: American Geophysical Union), 343

Larson, D.E., Lin, R.P., McTiernan, J.M., McFadden, J.P., Ergun, R.E., McCarthy, M., Rème, H., Sanderson, T.R., Kaiser, M., Lepping, R.P., & Mazur, J. 1997, *Geophys. Res. Lett.* 24, 1911

Leamon, R.J., Canfield, R.C., Jones, S.L., Lambkin, K., Lundberg, B.J., & Pevtsov, A.A. 2004, *J. Geophys. Res.* 109, 5106, doi:10.1029/2003JA010324

Leamon, R.J., Canfield, R.C., & Pevtsov, A.A. 2002, *J. Geophys. Res.* 107, 1234, doi:10.1029/2001JA000313

Lepping, R.P., Jones, J.A., & Burlaga, L.F. 1990, *J. Geophys. Res.* 95, 11 957

Lepri, S.T. & Zurbuchen, T.H. 2004, *J. Geophys. Res.* 109, 1112

Lin, J. & Forbes, T.G. 2000, *J. Geophys. Res.* 105, 2375

Lundquist, S. 1950, *Ark. Fys.* 2, 361

Lynch, B.J., Zurbuchen, T.H., Fisk, L.A., & Antiochos, S.K. 2003, *J. Geophys. Res.* 108, 1239, doi:10.1029/2002JA009591

Pevtsov, A.A. & Canfield, R.C. 2001, *J. Geophys. Res.* 106, 25 191

Pevtsov, A.A., Canfield, R.C., & Metcalf, T.R. 1995, *ApJ* 440, L109

Reinard, A.A. & Fisk, L.A. 2004, *ApJ* 608, 533

Webb, D.F. 2000, *IEEE Trans. Plasma Sci.* 28, 1795

Zhang, M. & Low, B.C. 2003, *ApJ* 584, 479

Discussion

JINGXIU WANG: Your results are consistent with the observations at the solar surface. Although many CMEs are initiated from active regions, a favorite large-scale magnetic structure is important for the occurrence of a CME. But can you say something about the characteristics of your large-scale dipoles, have you made some studies on these large-scale structures?

LEAMON: In this work, we only argue the need for reconnection with the large-scale fields. We have made some tentative studies comparing the mutual helicity of an active region and the overlying fields from, say, a PFSS ('hairy ball') model, but they are only preliminary and not conclusive.

STERLING: Several folks, including myself, have so far used only a qualitative morphological definition of sigmoids. At some point, we should improve upon this. What definition did you use for sigmoids?

LEAMON: I, too, only used a qualitative, morphological definition. Perhaps it would have been better to say "active region events" or "flare-related events" and to differentiate them from "quiescent filament events."

P.F. CHEN: Generally CME modeling involves magnetic reconnection below the erupting flux-rope, which leads to small-scale solar flares. Could you explain magnetic reconnection between active region and large-scale surroundings in detail?

LEAMON: Like my answer to Jingxiu Wang, the conclusions of this work are that the flux-rope magnetic cloud observed at 1 AU must be formed in the course of eruption, by reconnection between the active region and overlying fields. Clearly the next step is to model that reconnection quantitatively, but we have not done that yet.

Coronal and Stellar Mass Ejections
Proceedings IAU Symposium No. 226, 2005
K. P. Dere, J. Wang & Y. Yan, eds.

Progress in the Heating of Active Region Loops

L. Feng and W.Q. Gan

Purple Mountain Observatory, Nanjing, China
email: lfeng@pmo.ac.cn, wqgan@pmo.ac.cn

Abstract. Coronal heating is an important problem in solar physics. With the development of highly qualified instruments, such as TRACE, SOHO and Yohkoh, more and more observations about coronal loops have been obtained. The coronal loops' heating, being an important ingredient of coronal heating, has been paid particular attentions recently. But there are still some key issues about the structure and mechanism of the loops' heating unresolved. In this paper, after a brief review on the latest progress in both observations and modeling of coronal loops, we emphatically discuss the heating of hydrostatic loops and hydrodynamic loops based on the 1D model. The prospect of the subject is presented.

Keywords. Sun: corona

1. Introduction

One of the main goals in observing the solar corona is to determine the coronal heating mechanism, a long-standing fundamental problem in solar and stellar physics. EUV and X-ray observations have shown that the building blocks in the solar atmosphere are loop-like structures that outline the coronal magnetic field. In this paper, we mainly discuss properties of nonflaring loops ($T > 10^6 K$) in the solar active regions, especially the heating of them. A lot of work has been done on the heating mechanism both in observation and theory, and one of the mechanisms named the nanoflare heating raised by Parker (1988) at first has been paid more and more attentions, but no definite conclusion has been made on it.

An important first step to understand the problem of coronal heating is to determine the magnitude, duration, and location of heating along a coronal loop. In §2 we present the general observations of the loops, and briefly describe the loop models. The heating of hydrostatic and hydrodynamic loops are discussed in §3. Finally, we will give some discussion about the study of coronal loops' heating.

2. Observations and models

Usually, the observations of coronal loops in EUV and soft X-ray wave bands can be classified into two types, images and spectra. EUV observations are carried out by TRACE, EIT and CDS, and among them TRACE has the highest resolution, while soft X-ray data can be acquired through SXT loaded on Yohkoh satellite.

Due to the limitations of loop observations, at present the heating structure of the loops can only be obtained by comparison between results of the 1-D models and observations. In the three hydrodynamic equations of models, the equation of energy reservation is the most important one, in which conductive loss, radiative loss and the supposed heating term are considered. The loops are evolved from the initial equilibrium state, with their boundaries located in the deep chromosphere.

3. Heating of hydrostatic and hydrodynamic loops

Early in the study of coronal loops, it is supposed that they were in hydrostatic state due to their long lifetime. So we should only consider their spatial distribution of heating. In the famous RTV model proposed by Rosner, Tucker and Vaiana (1978), coronal loops were thought to be uniformly heated. but Aschwanden *et al.* (2000) favored the footpoint heating, because such a nonuniform heating model was more consistent with the flat temperature distribution derived from TRACE data. However, it is still an open question. Different heating structures can be obtained from the same observation and different instruments often lead to different heating structures.

With the development of higher resolution instruments, more and more evidences indicate that majority of the loops are not hydrostatic but hydrodynamic. For example, in the coronal loops blobs exist moving from one footpoint to another; Pressure scale height of EUV loops is often several times of their hydrostatic scale height; And relative to hydrostatic loops, hotter and shorter loops are under-dense, while cooler and longer loops are over-dense.

When a loop is hydrodynamic, we should consider not only its spatial distribution but also its time structure. A representative diagnostic example of heating structure was the work done by Reale *et al.* (2000). (3.1) is a heating function adopted by Warren *et al.* (2003). The function g(t) is chosen to be a simple triangular pulse. What we shall do is to change the parameters in the heating functions to make the results of models more close to the observations.

$$E_H(s,t) = E_0 + g(t)E_F exp[-(s - s_0)^2/2\sigma^2] \tag{3.1}$$

To get more consistent results with the observations, the multi-thread models are included. They are useful to explain the flat temperature distribution and the long time scale of cooling. A kind of multi-thread model that must be mentioned, is the nanoflare-heated model raised by Cargill (1994), in which the main idea was one nanoflare event heated one thread and the nanoflares were distributed randomly in both time and space.

4. Summary and discussion

Recently, more and more efforts have been made to the study of coronal loops heating. But there are still some key issues unresolved both in theory and observation. Perhaps some imaging observations such as TRACE data of cooling loops do not provide adequate information to determine the heating parameters, hence observations must be made early in the evolution of a loop, or we shall depend more on the spectral data, so that to establish models more close to the observations.

References

Aschwanden M.J., Nightingale R.W. & Alexander D. 2000, *ApJ* 541, 1059
Cargill P.J. 1994, *ApJ* 422, 381
Parker E.N. 1988 *ApJ* 330, 474
Reale F., Peres G. & Serio S. *et al.* 2000, *ApJ* 535, 423
Rosner R., Tucker W.H. & Vaiana G.S. 1978, *ApJ* 220, 643
Warren H.P., Winebarger A.R. & Mariska J.T. 2003, *ApJ* 593, 1174

Coronal and Stellar Mass Ejections
Proceedings IAU Symposium No. 226, 2005
K. P. Dere, J. Wang & Y. Yan, eds.

© 2005 International Astronomical Union
doi:10.1017/S1743921305000785

Magnetic Reconnection Inflow near the CME/Flare Current Sheet

J. Lin[1], Y.-K. Ko[1], L. Sui[2], J. C. Raymond[1], G. A. Stenborg[2], Y. Jiang[3], S. Zhao[3], and S. Mancuso[1],[4]

[1] Harvard-Smithsonian Center for Astrophysics, 60 Garden Street, Cambridge, MA 02138, USA, email: jlin@cfa.harvard.edu

[2] Department of Physics, Catholic University of America, 620 Michigan Avenue, Washington, DC 20064, USA

[3] National Astronomical Observatories of China/Yunnan Observatory, Chinese Academy of Sciences, P. O. Box 110, Kunming, Yunnan 650011, China

[4] INAF/Osservatorio Astronomico di Torino, 20 Strada Osservatorio, 1-10025 Pino Torinese, Italy

Abstract. This work reports direct observations of the magnetic reconnection site during an eruptive process occurring on November 18, 2003. The event started with a rapid expansion of a few magnetic arcades located over the east limb of the Sun and developed an energetic partial halo coronal mass ejection (CME), a long current sheet and a group of bright flare loops in the wake of the CME. It was observed by several instruments both in space and on ground, including the EUV Imaging Telescope, the Ultraviolet Coronagraph Spectrometer, and the Large Angle and Spectrometric Coronagraph experiment on board the *Solar and Heliospheric Observatory,* the Reuven Ramaty High Energy Solar Spectroscopic Imager, as well as the Mauna Loa Solar Observatory Mark IV K-coronameter. We combine the data from these instruments to investigate various properties of the eruptive process, including those around the current sheet. The composite of images from different instruments and the corresponding results specify explicitly how the different objects developed by a single eruptive process are related to one another.

Keywords. Sun: coronal mass ejections, flares, MHD

---

1. Introduction

The catastrophe model of solar eruptions suggests that as the catastrophe takes place, the closed magnetic field lines in the configuration are stretched so severely that they effectively open up, a long current sheet forms separating two magnetic fields of opposite polarities, and magnetic reconnection invoked in the current sheet creates flare ribbons on the solar surface and flare loops in the corona (e.g., Lin & Forbes 2000; Priest & Forbes 2002; Lin *et al.* 2003).

As a result of the high electrical conductivity and the force-free environment in the corona, the current sheet is confined to a very local and thin region. This makes direct observation of the current sheet extremely difficult (refer to Ko *et al.* 2003 for the discussions and brief review on this issue). Therefore, information about magnetic reconnection inside the current sheet is usually deduced indirectly by observing the dynamical behaviors of the products of magnetic reconnection, such as the separating flare ribbons on the solar surface (Poletto & Kopp 1986; Qiu *et al.* 2004), the growing flare loops in the corona (Sui *et al.* 2004), and so on. The magnetic reconnection inflow around the X-type neutral point over the flare loops was also reported and used to analyze the properties of magnetic reconnection in an eruptive process (Yokoyama *et al.* 2001).

Utilizing the ultraviolet spectroscopic observations from the Ultraviolet Coronagraphic Spectrometer (UVCS) on board the *Solar and Heliospheric Observatory (SOHO)*, together with other remote-sensing data, Ciaravella *et al.* (2002) and Ko *et al.* (2003) conducted comprehensive analyses from various aspects on several eruptive processes that clearly manifested both coronal mass ejections (CMEs) and solar flares, and confirmed the existence and development of the long current sheet in the events as predicted by Lin & Forbes (2000). Webb *et al.* (2003) surveyed 59 CMEs observed by the Solar Maximum Mission from 1984 to 1989, and found that about half were followed by co-axial, bright rays suggestive of newly formed current sheets.

The event of November 18, 2003 provides us another opportunity to observe directly the current sheet and the associated magnetic reconnection process. It was observed by instruments both in space and on the ground.

2. Observations and Results

This event occurred on the east limb of the Sun. No apparent magnetic structure appeared in the nearby region in Hα filtergrams. After having looked at both the magnetograms obtained by the Michelson Doppler Imager (MDI) on board *SOHO* and the Hα images obtained on ground within the following three successive days, we find that the eruption took place right between the two active regions, AR0507 and AR0508, located north and south of it, respectively, but its relations to these two active regions are not clear. On the other hand, the initial stage and the subsequent development of the eruption in the lower corona were clearly recorded in 195 Å by the EUV Imaging Telescope (EIT) on board *SOHO,* and the Reuven Ramaty High Energy Solar Spectroscopic Imager (RHESSI, Lin *et al.* 2002). The consequences in the higher corona were observed by UVCS and the Large Angle and Spectrometric Coronagraph Experiment (LASCO) on board *SOHO,* and the Mauna Loa Solar Observatory Mark IV K-coronameter (MLSO MK4).

We were first impressed by this event with the inward motions of the magnetic structures that were severely stretched by the eruption, and we were then impressed with morphological features of the CME developed during the eruption. The initial phase of the eruption at in the low corona observed in 195 Å by EIT is manifested clearly by the movies, the LASCO movie shows the magnetic structures before and after the CME passed through the field of views of LASCO C2 and C3 †, and more pictures and detailed discussions can be easily found in our recent work (Lin *et al.* 2005). The CME velocities at maxima are 1939 km/s at the front edge and 1484 km/s at the core, respectively.

3. Discussions Conclusions

Following Ciaravella *et al.* (2002), Ko *et al.* (2003) and Webb *et al.* (2003), the current sheet developing in the wake of a CME is investigated once again. The event discussed here occurred on the east limb of the Sun on November 18, 2003. It started from 08:48 UT with the fast expansion of a group of sheared arcades. No apparent prominence structure can be recognized in either Hα filtergrams or EIT 195 Å images prior to the eruption. This event developed a long and thin current sheet behind the CME. The magnetic reconnection inflow near the current sheet following the arcade expansion was observed, and the corresponding rate of magnetic reconnection M_A ranges from 0.01 to 0.23. The value of M_A in reality may not be constant and should cover a wider range.

† Movies can be found at: http://hesperia.gsfc.nasa.gov/∼ sui/20031118.

A similar process of magnetic reconnection inflow was also reported by Yokoyama *et al.* (2001) for the event on March 18, 1999. But the inflow velocity and M_A deduced by them are both one order of magnitude smaller than what we obtained in the present case. Another difference between two events lies in the initial configurations. The magnetic arcades in the present case possess a much more compact structure than those of Yokoyama *et al.* (2001). One of the advantages of this work over that of Yokoyama *et al.* (2001) is that the reconnection outflow in the present case might be observed. This allows us to estimate V_A directly in the reconnection inflow region in a more reliable way without making any extra assumptions. Furthermore, the UVCS observations make it possible for us to avoid unnecessary confusion in measuring the inflow speed of reconnection (e.g., see Chen *et al.* 2004).

The eruptive process manifests an energetic CME and a bright flare loop system that was covered by a cusp structure. The maximum velocities of the CME leading edge and the core were 1939 km s^{-1} and 1484 km s^{-1}, respectively. This event was observed by various instruments from both space and ground. Analyzing the observational data from these instruments yields the conclusion that the CME and the flare are connected by a stretched current sheet in which magnetic reconnection occurs and converts the magnetic energy into heating and kinetic energy. The morphological features of the disrupted magnetic field involved in this event fit those of the cartoon shown in Figure 1 of Lin *et al.* (2004) very well, which implies that the cartoon depicts the common characteristics of the eruptive processes that give rise to both flare and CME. The magnetic configuration in reality may be much more complex than that shown in the cartoon, but the fundamental physical processes should be the same!

Acknowledgements

JL is grateful to T. G. Forbes for valuable discussions. This work was supported by NASA under grants NNG04GE84G and NAG5-12827 to the Smithsonian Astrophysical Observatory. The work of YJ and SZ was supported by the National Science Foundation of China under grant 10173023 to the Yunnan Astronomical Observatory. *SOHO* is a joint mission of the European Space Agency and US National Aeronautics and Space Administration.

References

Chen, P. F., Shibata, K., Brooks, D. H., & Isobe, H.: 2004, *ApJ*, 602, L61.
Ciaravella, A., *et al.* 2002, *ApJ*, 575, 1116.
Ko, Y., Raymond, J. C., Lin, J., Lawrence, G., Li, J., & Fludra, A. 2003, *ApJ*, 594, 1068.
Lin, J. 2002, *Chinese J. Astron. Astrophys.*, 2, 539.
Lin, J. & Forbes, T. G. 2000, *JGR*, 105, 2375.
Lin, J., Raymond, J. C., & van Ballegooijen, A. A. 2004, *ApJ*, 602, 422.
Lin, J., Soon, W., & Baliunas, S. L. 2003, *New Astron. Rev.*, 47, 53.
Lin, R. P., *et al.*, 2002, *Solar Phys.*, 210, 3.
Poletto, G. & Kopp, R. A.: 1986, in *The Lower Atmosphere of Solar Flares*, (ed.) D. F. Neidig, NSO, Sunspot, NM, P. 453.
Priest, E. R. & Forbes, T. G. 2002, *A&A Rev.*, 10, 313.
Qiu, J., Wang, H., Cheng, C. Z., & Gary, D. E.: 2004, *ApJ*, 604, 900.
Sui, L., Holman, G. D., & Dennis, B.: 2004, *ApJ*, 612, 546.
Webb, D. F., Burkepile, J., Forbes, T. G., & Riley, P., 2003, *JGR*, 108 (A12), 1440.
Yokoyama, T., Akita, K., Morimoto, T., Inoue, K., & Newmark, J. 2001, *ApJ*, 546, L69.

Coronal and Stellar Mass Ejections
Proceedings IAU Symposium No. 226, 2005
K. P. Dere, J. Wang & Y. Yan, eds.

© 2005 International Astronomical Union
doi:10.1017/S1743921305000797

Energetics of Coronal Mass Ejections

Prasad Subramanian[1] and Angelos Vourlidas[2]

[1]Inter-University Centre for Astronomy and Astrophysics, P.O Bag 4, Ganeshkhind, Pune - 411007, India, email: psubrama@iucaa.ernet.in

[2]Code 7663, Naval Research Lab, Washington, DC 20375, USA, email: vourlidas@nrl.navy.mil

Abstract. We examine the energetics of the best examples of flux-rope CMEs observed by LASCO in 1996-2001. We find that 69% of the CMEs in our sample experience a driving power in the LASCO field of view. For these CMEs which are driven, we examine if they might be deriving most of their driving energy by coupling to the solar wind. We do not find conclusive evidence to support this hypothesis. We adopt two different methods to estimate the energy that can possibly be released by the internal magnetic fields of the CMEs. We find that the internal magnetic fields are a viable source of driving power for these CMEs.

Keywords. Sun; corona, Sun: coronal mass ejections (CMEs), Sun: magnetic fields

1. Introduction

The energy budgets involved in Coronal mass ejection (CME) propagation at heights $\gtrsim 2R_\odot$ are indicative of the manner in which the CME interacts with streamers and the solar wind and also provide constraints on the energies required during the initiation phase. We study the evolution of potential and kinetic energies of 39 individual FR CMEs between 1997 and 2001. This comprises a complete sample of the best examples of FR CMEs observed by LASCO in 1996-2001 (out of about 4000 events). We first generate mass images of each CME and then calculate the potential and kinetic energies using a procedure similar to that described in Vourlidas *et al.* (2001). For 27/39 CMEs in our sample, the mechanical energy rises linearly with time, whereas 12/39 CMEs show no such trend. For the 27/39 CMEs for which mechanical energy rises linearly with time, there is a clear external driving power.

2. Source of driving power: solar wind, or internal magnetic energy?

If CMEs are propelled via coupling with the solar wind, larger CMEs should have larger driving powers, since they have larger areas for interacting with the solar wind and would therefore be better coupled with it. For the CMEs which have a driving power, figure 1 shows a scatterplot of size versus driving power. From figure 1, the correlation between the driving power and CME size is evidently poor, and we find little evidence to suggest that larger CMEs have more driving power. This casts doubt on the hypothesis that these CMEs are powered by momentum coupling with the ambient solar wind.

We have computed an estimate of the rate of energy released by the magnetic field advected by each CME using two different methods. We envisage that the propelling force is provided by some sort of $\vec{J} \times \vec{B}$ forces (due to misaligned magnetic fields and currents) within the flux rope. One method uses the average magnetic flux carried by near-earth magnetic clouds (e.g., Lepping *et al.* 1997) and assumes that this value is representative of the average magnetic flux that is frozen into a typical CME in our dataset. According to this method, the internal magnetic field of a CME can provide 0.744 ± 1.352 of

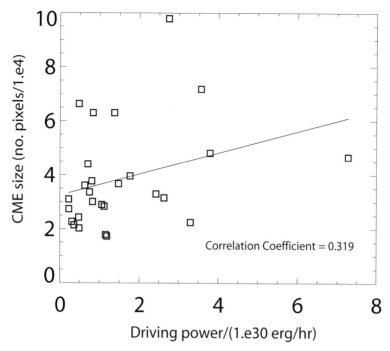

Figure 1. The mean size (in number of pixels) for CMEs plotted as a function of their driving power. The low correlation coefficient suggests that there is no evidence to claim that larger CMEs have larger driving powers.

the required driving power on the average. The other method uses direct estimates of the CME magnetic field from radio measurements (Bastian *et al.* 2001). According to this method, the internal magnetic field of a CME can provide 12.819 ± 1.677 of the required driving power on the average. The details of these calculations can be found in Subramanian & Vourlidas (2004).

3. Conclusions

We find no evidence to suggest that the driven CMEs in our sample derive their driving power primarily via coupling with the solar wind. We employ two different approaches to investigate if release of the internal magnetic energy in the CME can possibly provide the driving power. One approach suggests that, on the average, energy released by the internal magnetic field of a CME can provide upto 74% of the required driving power. Another suggests that the energy released by the internal magnetic field of a CME can be around an order of magnitude greater than what is required to drive it.

References

Bastian, T.S., Pick, M., Kerdraon, A., Maia, D., & Vourlidas, A. 2001, *ApJ (Letters)* 558, L65
Lepping, R.P., Szabo, A., DeForest, C.E., & Thompson, B.J. 1997, in *Proc. 31st ESLAB Symp.,*
   'Correlated Phenomena at the Sun, in the Heliosphere and in Geospace, ESTEC, Noordwijk,
   The Netherlands, 22-25 September 1997 (ESA SP-415, December 1997)
Subramanian, P. & Vourlidas, A. 2004, *ApJ, submitted*
Vourlidas, A., Subramanian, P., Dere, K.P., & Howard, R.A. 2000, *ApJ* 534, 456

Session 6

Coronal mass ejections and energetic particles

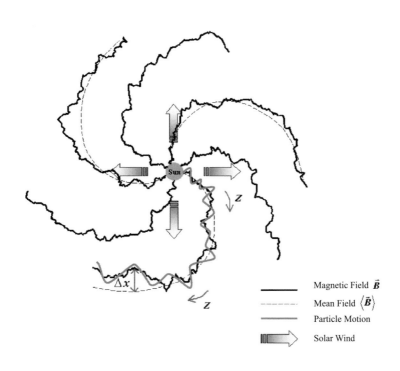

——	Magnetic Field $\vec{B}$
- - -	Mean Field $\langle\vec{B}\rangle$
——	Particle Motion
⟹	Solar Wind

Coronal and Stellar Mass Ejections
Proceedings IAU Symposium No. 226, 2005
K. P. Dere, J. Wang & Y. Yan, eds.

© 2005 International Astronomical Union
doi:10.1017/S1743921305000803

Transport and Acceleration of Solar Energetic Particles from Coronal Mass Ejection Shocks

David Ruffolo

Dept. of Physics, Faculty of Science, Mahidol Univ., Bangkok 10400 Thailand,
email: david_ruffolo@yahoo.com

Abstract. After a brief overview of solar energetic particle (SEP) emission from coronal mass ejection (CME) shocks, we turn to a discussion of their transport and acceleration. The high energy SEP are accelerated near the Sun, and because of their well-known source location, their transport can be modeled quantitatively to obtain precise information on the injection function (number of particles emitted vs. time), including a determination of the onset time to within 1 min. For certain events, transport modeling also indicates magnetic topology with mirroring or closed field loops. Important progress has also been made on the transport of low energy SEP from very strong events, which can display exhibit interesting saturation effects and compositional variations. The acceleration of SEP by CME-driven shocks in the interplanetary medium is attributed to diffusive shock acceleration, but the spectrum of SEP production is typically modeled empirically. Recent progress has largely focused on using detailed composition measurements to determine fractionation effects of shock acceleration and even to clarify the nature of the seed population. In particular, there are many indications that the seed population is suprathermal (pre-energized) and the injection problem is not relevant to acceleration at interplanetary CME-driven shocks. We argue that the finite time available for shock acceleration provides the best explanation of the high-energy rollover.

Keywords. Sun: particle emission — Sun: coronal mass ejections (CMEs) — interplanetary medium — solar-terrestrial relations

---

1. Overview of solar energetic particle transport

This presentation aims to provide a brief introduction to the basic issues and some appreciation of state of the art in solar energetic particle (SEP) transport and acceleration, for a broad audience of specialists in different aspects of coronal mass ejections (CMEs).

Figure 1 shows the first report, in 1962, of energetic particles associated with an interplanetary shock, which we now believe to be driven by a CME (Bryant, Cline, Desai, *et al.* 1962). This shows the flux of protons in different energy ranges as a function of time. There are evidently two distinct populations. The first arrives shortly after the time of the flare [which we now know to be closely related to the time of CME liftoff; see Zhang *et al.* (2004)]. While the CME and shock were still very close to the Sun, protons were accelerated to several hundred MeV. On a finer timescale, SEP of higher velocity are seen to arrive first; this is termed a dispersive onset. On the other hand, there is a delayed, non-dispersive peak that dominates at low energies, associated with shock passage by the observer (in this case near Earth, as identified by a sudden storm commencement, SC). This evidently corresponds to particles accelerated by the shock as it proceeds through the interplanetary medium. These have been termed "energetic storm particles," although in recent usage both these and prompt population are referred to collectively as solar energetic particles, because at lower energies the two populations are not cleanly separated. Finally, Figure 1 shows the response of a ground-based neutron monitor, which measures the flux of galactic cosmic rays (GCR) impacting the atmosphere from a specific direction in space (by means of secondary atmospheric neutrons). Interestingly, the flux of GCR is depressed when a shock passes the Earth and sweeps these particles away. This phenomenon is known as a Forbush decrease (Forbush 1937). However, a very strong event can produce SEP to GeV energies and register an increase in neutron monitor rates; such a event is called a ground level enhancement (GLE).

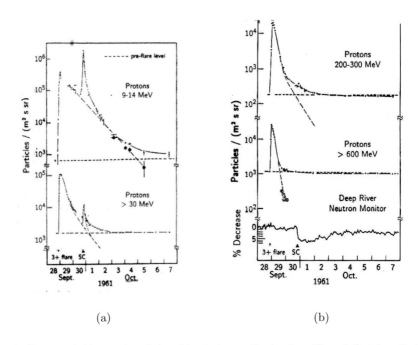

(a)   (b)

Figure 1. Representative proton intensities between September 28 and October 7, 1961; the decay of the solar proton event and the arrival of the energetic storm particles late on September 30 are shown. The Deep River neutron monitor record is shown for comparison. [Based on Fig. 18 of Bryant *et al.* (1962)]

There is now overwhelming physical evidence that for the class of "gradual" events that have a solar flare and a coronal mass ejection (including the geoeffective events with greatest SEP intensity), the escaping SEP are accelerated at the CME shock and not deep in the corona, e.g., not at the site of the flare or primary energy release (Mason, Gloeckler, & Hovestadt 1984; Lee & Ryan 1986; Reames 1990; Ruffolo 1997). Therefore, the two populations shown in Figure 1, with very different energy spectra, both correspond to shock acceleration, but under very different physical conditions while the shock is still close to the Sun and later as it moves in the interplanetary medium.

I would like to comment that discussions of geoeffectiveness typically stress the effects when a CME and its associated shock impact the Earth's magnetosphere, which is typically days after its liftoff from the Sun. For example, the largest SEP event of 2003 had a flare and CME on October 28 and the CME arrived at Earth on October 29. However, in a recent presentation, a NASA representative stated that more satellite anomalies occurred on October 28 than on October 29 (L. Barbieri, private communication, 2004). Therefore, the flare/CME at the Sun is immediately geoeffective in the sense of producing prompt space weather effects.

The main types of SEP populations are summarized in Table 1. In addition to the gradual events we have discussed so far, associated with a CME (and for major events, a flare as well), another type of event is an impulsive solar flare, with no associated CME. In this case the energetic particles are believed to result from stochastic acceleration, and there are very interesting compositional effects, such as enhancements in the isotope ^{3}He (Hsieh & Simpson 1970) and heavy ions (Hurford, Mewaldt, Stone, *et al.* 1975; Reames 2000) by factors up to 10^3 or even 10^4, an enhancement in electrons (Evenson, Meyer, Yanagita, *et al.* 1984; Cane, McGuire, & von Rosenvinge 1986), and high charge states (Klecker, Hovestadt, Gloeckler, *et al.* 1984; Luhn, Klecker, Hovestadt, *et al.* 1987).

Table 1. Populations of escaping solar energetic particles

Impulsive flares	CME shocks (gradual events)	
	Near Sun	Interplanetary
^{3}He enhanced, electron-rich, high ion Q (stochastic acceleration)	Up to high E dispersive onset	At low E non-dispersive
	(shock acceleration)	

2. Injection near the Sun: Precision modeling

According to Figure 1 and Table 1, SEP at high energy are almost always injected near the Sun. With this well-determined source, and given that the basic transport processes are well established, we are able to undertake precision modeling to determine transport parameters, the magnetic field configuration in space, and the injection vs. time near the Sun. We discuss transport of the interplanetary component in §4.

We describe the propagation of protons from a solar event by numerically solving a Fokker-Planck equation of pitch-angle transport that includes the effects of interplanetary scattering, adiabatic deceleration and solar wind convection (Roelof 1969; Ruffolo 1995; Nutaro, Riyavong, & Ruffolo 2001). We are assuming transport along the mean magnetic field, as expected when there is good magnetic connection between the source and the observer. Following Ng & Wong (1979), we define the particle distribution function F depending on time, t, pitch-angle cosine, μ, distance from the Sun along the interplanetary magnetic field, z, and momentum, p, as

$$F(t,\mu,z,p) \equiv \frac{d^3N}{dz\,d\mu\,dp}, \tag{2.1}$$

where N represents the number of particles inside a given flux tube. The derived transport equation takes the form:

$$
\begin{aligned}
\frac{\partial F(t,\mu,z,p)}{\partial t} = &-\frac{\partial}{\partial z}\mu v F(t,\mu,z,p) - \frac{\partial}{\partial z}\left(1-\mu^2\frac{v^2}{c^2}\right)v_{\rm sw}\sec\psi F(t,\mu,z,p) \\
&-\frac{\partial}{\partial\mu}\frac{v}{2L(z)}\left[1+\mu\frac{v_{\rm sw}}{v}\sec\psi - \mu\frac{v_{\rm sw}v}{c^2}\sec\psi\right](1-\mu^2)F(t,\mu,z,p) \\
&+\frac{\partial}{\partial\mu}v_{\rm sw}\left(\cos\psi\frac{d}{dr}\sec\psi\right)\mu(1-\mu^2)F(t,\mu,z,p) \\
&+\frac{\partial}{\partial\mu}\frac{\varphi(\mu)}{2}\frac{\partial}{\partial\mu}\left(1-\mu\frac{v_{\rm sw}v}{c^2}\sec\psi\right)F(t,\mu,z,p) \\
&+\frac{\partial}{\partial p}pv_{\rm sw}\left[\frac{\sec\psi}{2L(z)}(1-\mu^2)+\cos\psi\frac{d}{dr}(\sec\psi)\mu^2\right]F(t,\mu,z,p). \tag{2.2}
\end{aligned}
$$

The particle velocity is denoted by v and the solar wind velocity by v_{sw}. The angle between the field line and the radial direction is specified by the function $\psi(z)$, the focusing length by $L(z) = -B/(dB/dz)$, and the pitch-angle scattering coefficient by $\varphi(\mu)$. The simulation program to solve this equation runs in a few minutes on a personal computer.

In the next step, we can simultaneously fit observed data for the SEP intensity and anisotropy vs. time. It is computationally efficient to use least squares fitting to determine the optimal piecewise linear injection function, i.e., the rate of particle injection onto the local magnetic field line vs. time near the Sun (Ruffolo, Khumlumlert, & Youngdee 1998). We find the χ^2 values of fits for different transport assumptions to determine the optimal model. For a standard Archimedean spiral field configuration (Figure 2), typically the only parameter we vary is the interplanetary scattering mean free path. Note that anisotropy data are important for constraining the optimal scattering mean free path.

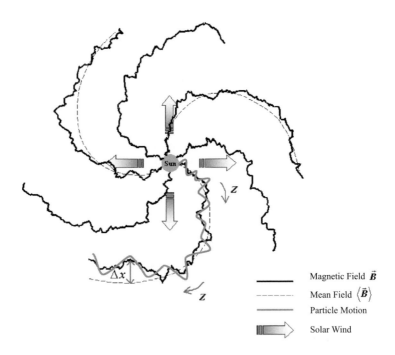

Figure 2. Typical Archimedean spiral configuration of the interplanetary magnetic field as it is dragged out of the rotating Sun by the radial solar wind.

An example of such precision modeling for the GLE of 2001 April 15 (Easter 2001) is shown in Figure 3. The intensity and anisotropy of relativistic solar protons (at rigidity ~1-3 GV) are derived from count rate increases in the *Spaceship Earth* network of polar neutron monitors, which provide high count rates and excellent directional sensitivity, and the data are then fit by the above procedure. The injection function is interpreted as the time profile of relativistic particle acceleration. Table 2 compares the injection timing with electromagnetic emissions converted into "solar time," ST, or UT minus 8 minutes to account for the propagation time. It is of particular interest that the start time of relativistic particle acceleration is coincident with the soft X-ray peak, which

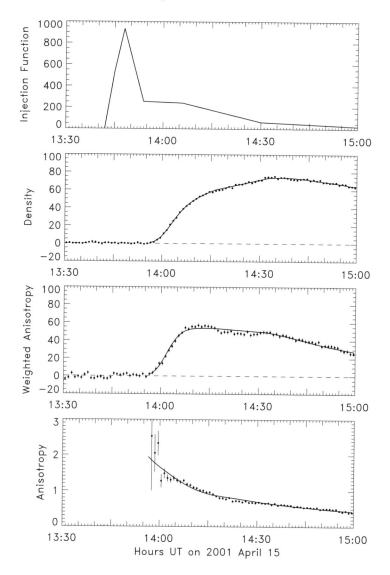

Figure 3. Precision modeling of relativistic solar proton data from neutron monitors on Easter, 2001 [Bieber, Evenson, Dröge, *et al.* (2004)].

actually marks the end of energy input in the flare. It is also later than the extrapolated CME liftoff time. Our interpretation is that the CME shock takes some time to develop and accelerate relativistic particles. Nevertheless, it does occur quite quickly; the time of relativistic particle injection corresponds to a CME altitude of only a few solar radii (Cliver, Kahler, & Reames 2004).

In some cases, such detailed fitting allows us to infer a non-standard magnetic field configuration. For example, in the GLE of 2000 July 14 (Bastille Day 2000), Bieber, Dröge, Evenson, *et al.* (2002) inferred a magnetic bottleneck configuration as in Figure 4. This corresponds to distortion of interplanetary magnetic field lines beyond the Earth by a preceding CME from the same active region a few days earlier. This is not as unusual as you might think, because major flare/CME events typically occur in sequences a few

Table 2. Timing of flare, CME, and particle emission on Easter, 2001 [Bieber *et al.* (2004)], in "solar time" (see text).

Emission	2001 April 15		
	Start	Peak	End
Relativistic protons	**13:42**	**13:48**	
Soft X-rays	13:11	13:42	13:47
Hα	13:28	13:41	15:27
Type III radio burst	13:36		13:38
CME liftoff	13:24-31		
Type II radio burst	13:40		13:47
Type IV radio burst	13:44		14:57

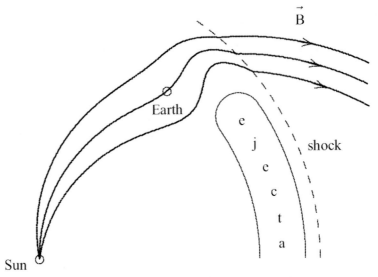

Figure 4. Magnetic bottleneck configuration inferred at the time of the Bastille Day, 2000 GLE [Bieber *et al.* (2002)].

days apart from the same active region. Indeed for two other GLEs we infer from the angular distributions that relativistic solar protons were propagating inside a magnetic loop configuration (Ruffolo, Tooprakai, Rujiwarodom, *et al.* 2004; Bieber, Clem, Evenson, *et al.* 2005). This important information about particle transport again relies on accurate measurements of directional distributions of SEP, such as those from the worldwide neutron monitor network or from rotating spacecraft with multiple sensor heads.

3. Transport perpendicular to the mean magnetic field

So far we have discussed SEP transport parallel to the mean magnetic field, commonly called "parallel transport." Another important issue is perpendicular transport, i.e., perpendicular to the mean magnetic field, which governs the latitudinal and longitudinal transport of SEP. In the classic work of Jokipii (1966), such transport is considered to be dominated by the field line random walk. As illustrated in Figure 2, the interplanetary magnetic field fluctuates strongly due to solar wind turbulence, and individual field lines can undergo a random walk that deviates quite far from the mean magnetic field. This was classically viewed as a diffusive random walk, and one can define a field line diffusion coefficient in terms of the lateral deviation Δx compared with the distance along the mean field, Δz:

$$\text{field line diffusion} \quad \rightarrow \quad D = \frac{\langle \Delta x^2 \rangle}{2 \Delta z}. \tag{3.1}$$

The field line diffusion is related to the particle diffusion coefficient:

$$\text{particle diffusion} \quad \rightarrow \quad \kappa = \frac{\langle \Delta x^2 \rangle}{2 \Delta t}, \tag{3.2}$$

and in the limit that particles exactly follow the field lines, one obtains $\kappa = Dv/4$.

For a realistic model of solar wind turbulence, Matthaeus, Gray, Pontius, *et al.* (1995) derived an expression for D, and Bieber & Matthaeus (1997) developed a theory for κ based on concepts of dynamical turbulence. Giacalone & Jokipii (1999) used Monte Carlo simulations to derive κ values intermediate to those for the classic field line random walk model and for dynamical turbulence. Recently, Qin, Matthaeus, & Bieber (2002) and Matthaeus, Qin, Bieber *et al.* employed numerical simulations and a nonlinear guiding center theory to show that in the ensemble average, particles undergo diffusion, then subdiffusion (also known as compound diffusion), and finally a second régime of diffusion at a slower rate.

Interestingly, Mazur, Mason, Dwyer *et al.* (2000) presented observations of SEP from impulsive flares (which are particle sources of narrow lateral extent) with "dropouts" or sudden disappearance and reappearance of flux as a function of time, which is interpreted as due to the spacecraft's motion through a filamentary distribution of magnetic flux tubes, of typical width 0.03 AU, that are filled with particles because they connect back to the source region. This shows that the lateral transport of SEP, and presumably the field lines themselves, is highly non-diffusive over a distance scale of 1 AU. To address this, Ruffolo, Matthaeus, & Chuychai (2003) replaced ensemble statistics with conditional statistics dependent on the starting point. For a standard description of solar wind turbulence, with no free parameters, they were able to simultaneously reproduce dropout structures of field lines at 1 AU connected to a small initial region and also explain the high rate of lateral diffusion κ inferred from observations by the Ulysses spacecraft (McKibben, Lopate, & Zhang 2001). The resulting picture (Figure 5) is that field lines starting near O-points in the turbulence are topologically trapped for some distance beyond 1 AU, whereas other field lines escape very rapidly. This accounts for a "core" region of outgoing SEP, with dropouts, and the SEP missing from the interstitial core regions are instead in a "halo" of low SEP density over a wide lateral region. At long radial distances all the field lines are found to escape and undergo diffusive random walks, so that particles undergo parallel and perpendicular diffusion throughout the inner heliosphere at later times.

4. Particle acceleration by coronal mass ejection shocks in the interplanetary medium

Referring to Table 1, we now turn to SEP accelerated by CME shocks traveling through the interplanetary medium (also referred to as energetic storm particles). We are fortunate to have other presenters who will show detailed results about these particles, so I will present only a broad-brush overview to help orient the non-specialist reader. While such SEP are certainly important, and also relevant to space weather effects, the underlying processes of acceleration and transport are poorly understood because they are both time-dependent and difficult to separate. (We saw in Section 1 that when acceleration and transport can be considered individually, the observations can clearly address each

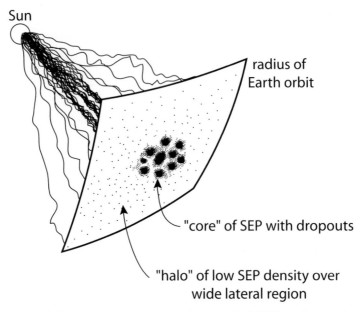

Figure 5. Illustration of the temporary trapping of magnetic field lines due to the small scale topology of solar wind turbulence [Ruffolo *et al.* 2003]. Trapped field lines form a core region with high SEP density and dropouts, while escaping field lines form a wider halo of lower SEP density. At long distances the field lines and particles ultimately escape to participate in diffusive random walks.

of them.) Modeling the simultaneous acceleration and transport of particles in the time-dependent system of a CME, shock, magnetic field topology, and magnetic fluctuations (the last of which are also affected by the particles) is necessarily difficult and involves many simplifying assumptions that are not well constrained by observations. Because of the complicated time dependence, recent research has concentrated on variations in ionic composition to probe the underlying physical processes.

That said, there have been important improvements in understanding. In a series of papers, Ng and others have examined saturation effects in very intense SEP events, based on the idea that the particles generate waves that in turn enhance interplanetary scattering and inhibit their transport. Ng, Reames, & Tylka (1999) provided a remarkable explanation of observed changes in element ratios as a function of time, confirming that wave generation probably plays a major role in these very intense events. However, other predictions of the theory, such as very intense waves and extremely low scattering mean free paths (below 10^{-3} AU) have not been confirmed by observations.

Before proceeding further, let me present a simple introduction to the process of diffusive shock acceleration. Figure 6 is a schematic of a shock, i.e., a discontinuity in fluid properties caused by a collision between fluids (or a fluid and an obstacle) with a relative speed greater the speed of sound. In general, the magnetic field (slanted lines in Figure 6) also has a different direction on the upstream and downstream sides. Usually we can enter a reference frame in which the fluid flow $\vec{u}$ is along $\vec{B}$ both upstream and downstream, called the de Hoffmann-Teller frame (de Hoffmann & Teller 1950). As the microscopic particle scatters off the ubiquitous macroscopic magnetic irregularities flowing with speed u in the space plasma, it is analogous to a game our Chinese audience knows and loves: ping-pong. If you hit the ping-pong ball with your paddle moving forward, the ball is accelerated. This is what we see occurring on the upstream side, after the head-on collision. However, if you imagine moving your paddle backwards (not that a good Chinese ping-pong player would ever do this), the ball loses energy. This is analogous to the particle's deceleration on the downstream side. However, a shock always has $u_1 > u_2$, so there is always a net gain in energy after a complete cycle. The particle has some probability of crossing and recrossing the shock plane, and a small number of particles can achieve a very high energy. Indeed, the number of particles per unit momentum, which we call the spectrum, is a power law for standard theories of diffusive shock acceleration (Drury 1983).

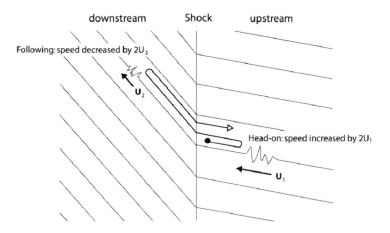

Figure 6. Illustration of diffusive shock acceleration as a particle scatters off magnetic irregularities, crossing and recrossing a shock discontinuity.

Figure 6 shows how an energetic seed particle can gain further energy at a shock. Until recently, interplanetary shocks were generally believed to accelerate particles out of the thermal solar wind population. However, there is a theoretical difficulty in understanding how thermal particles can join the shock acceleration process, the so-called "injection problem." Quite recently, Desai, Mason, Dwyer, *et al.* (2003) have analyzed the elemental composition of energetic storm particles to infer that those SEP were accelerated from a seed population of suprathermal (pre-energized) particles that happen to be upstream of the shock. Perhaps these were remnants from previous SEP events. Therefore, the injection problem is not relevant to acceleration at interplanetary CME-driven shocks.

Desai, Mason, Wiedenbeck, *et al.* (2004) have also examined the energy spectra of both the energetic storm particles and their upstream seed populations. The spectra are examined at the time of shock passage, which is one way to isolate the issue of acceleration from that of transport. They confirm a well-known rollover in the spectrum at 0.1-10 MeV nucleon^{-1} (see also Gosling, Asbridge, Bame, *et al.* 1981; van Nes, Reinhard, Sanderson, *et al.* 1985), where the power-law spectrum in particle energy changes to decline more rapidly above a critical energy, T_c. Such spectra are typically modeled empirically using the spectral form of Ellison & Ramaty (1985) for T_c as a fit parameter. However, this limit to the acceleration process is an important component of our understanding of SEP acceleration, and should be understood physically. Indeed, the critical energy must be higher near the Sun as the more energetic SEP originate there (see Table 1).

Of the possible rollover mechanisms listed by Ellison & Ramaty (1985) that might explain the rollover at 0.1-10 MeV nucleon^{-1} in the spectrum of particles accelerated by a CME-driven shock in the interplanetary medium, Ruffolo & Channok (2003) argue that the rollover is due to the finite time available for shock acceleration (see also Klecker, Scholer, Hovestadt, *et al.* 1981; Lee 1983). This allows one to derive spectra for various ionic species in terms of the physical quantities that underlie the acceleration time. For a rollover energy well above the initial energy of the seed particle, and if the mean free path λ is proportional to rigidity to the power α, one expects a rollover energy per nucleon of

$$T_c/A \propto t^{2/(\alpha+1)} (Q/A)^{2\alpha/(\alpha+1)} \tag{4.1}$$

as a function of t, the time duration of shock acceleration.

References

Bieber, J. W., Clem, J., Evenson, P., Pyle, R., Ruffolo, D., & Sáiz, A. 2005, *Geophys. Res. Lett.* (submitted)

Bieber, J.W., Dröge, W., Evenson, P., Pyle, R., Ruffolo, D., Pinsook, U., Tooprakai, P., Rujiwarodom, M., Khumlumlert, T., & Krucker, S. 2002, *ApJ* 567, 622

Bieber, J.W., Evenson, P., Dröge, W., Pyle, R., Ruffolo, D., Rujiwarodom, M., Tooprakai, P., & Khumlumlert, T. 2004, *ApJ (Letters)* 601, L103

Bieber, J.W., & Matthaeus, W.H., 1997, *ApJ* 485, 897

Bryant, D.A., Cline, T.L., Desai, U.M., McDonald, F.B. 1962, *J. Geophys. Res.* 67, 4983

Cane, H.V., McGuire, R.E., & von Rosenvinge, T.T. 1986, *ApJ* 301, 448

Cliver, E.W., Kahler, S.W., & Reames, D.V. 2003, *ApJ* 605, 902

de Hoffmann, F., & Teller, E. 1950, *Phys. Rev.* 80, 692

Desai M.I., Mason, G.M., Dwyer, J.R., Mazur, J.E., Gold, R.E., Krimigis, S.M., Smith, C.W., & Skoug, R.M. 2003, *ApJ* 558, 1149

Desai, M.I., Mason, G.M., Wiedenbeck, M.E., Cohen, C.M.S., Mazur, J.E., Dwyer, J.R., Gold, R.E., Krimigis, S.M., Hu, Q., Smith, C.W., & Skoug, R.M. 2004, *ApJ* 611, 1156

Drury, L.O'C. 1983, *Rep. Prog. Phys.* 46, 973

Ellison, D.C., Ramaty, R. 1985, *ApJ* 298, 400

Evenson, P., Meyer, P., Yanagita, S., & Forrest, D.J. 1984, *ApJ* 283, 439

Forbush, S.E. 1937, *Phys. Rev.* 51, 1108

Giacalone, J., & Jokipii, J.R., 1999, *ApJ* 520, 204

Gosling, J.T., Asbridge, J.R., Bame, S.J., Feldman, W.C., Zwickl, R.D., Paschmann, G., Sckopke, N., & Hynds, R.J. 1981, *J. Geophys. Res.* 86, 547

Hsieh, K.C., & Simpson, J.A. 1970, *ApJ (Letters)* 162, L191

Hurford, G.J., Mewaldt, R.A., Stone, E.C. & Vogt, R.E. 1975, *ApJ (Letters)* 201, L95

Jokipii, J.R. 1966, *ApJ* 146, 480

Klecker, B., Hovestadt, D., Gloeckler, G., Ipavich, F.M., Scholer, M., Fan, C.Y., & Fisk, L.A. 1984, *ApJ* 281, 458

Klecker, B., Scholer, M., Hovestadt, D., Gloeckler, G., & Ipavich, F.M. 1981, *ApJ* 251, 393

Lee, M.A. 1983, *J. Geophys. Res.* 88, 6109

Lee, M.A., & Ryan, J.M. 1986, *ApJ* 303, 829

Luhn, A., Klecker, B., Hovestadt, D., & Möbius, E. 1987, *ApJ* 317, 951

Mason, G.M., Gloeckler, G., & Hovestadt, D. 1984, *ApJ* 280, 902

Matthaeus, W.H., Gray, P.C., Pontius, D.H., Jr., & Bieber, J.W. 1995, *Phys. Rev. Lett.* 75, 2136

Matthaeus, W.H., Qin, G., Biebier, J.W., & Zank, G.P. 2003, *ApJ (Letters)* 590, L53

Mazur, J.E., Mason, G.M., Dwyer, J.R., Giacalone, J., Jokipii, J.R., & Stone, E.C. 2000, *ApJ (Letters)* 532, L79

McKibben, R.B., Lopate, C., & Zhang, M. 2001, *Space Sci. Rev.* 97, 257

Ng, C.K., Reames, D.V., Tylka, A.J. 1999, *Geophys. Res. Lett.* 26, 2145

Ng, C.K., & Wong, K.-Y. 1979, *Proc. 16th Internat. Cosmic Ray Conf.* 5, 252

Nutaro, T., Riyavong, S., & Ruffolo, D. 2001, *Comp. Phys. Comm.* 134, 209

Qin, G., Matthaeus, W.M., & Bieber, J.W. 2002, *ApJ (Letters)* 578, L117

Reames, D.V. 1990, *ApJ (Letters)* 358, L63

Reames, D.V. 2000, *ApJ (Letters)* 540, L111

Roelof, E.C. 1969, in: H. Ögelmann & J.R. Wayland (eds.), *Lectures in High Energy Astrophysics*, NASA SP-199 (Washington, DC: NASA), 111

Ruffolo, D. 1995, *ApJ* 442, 861

Ruffolo, D. 1997, *ApJ* (Letters) 481, L119

Ruffolo, D., & Channok, C. 2003, *Proc. 28th Internat. Cosmic Ray Conf.* 6, 3681

Ruffolo, D., Khumlumlert, T., Youngdee, W. 1998, *J. Geophys. Res.* 103, 20591

Ruffolo, D., Matthaeus, W.H., & Chuychai, P. 2003, *ApJ* (Letters) 597, L169

Ruffolo, D., Tooprakai, P., Rujiwarodom, M., Khumlumlert, T., Bieber, J.W., Evenson, P., & Pyle, R. 2004, *Eos Trans. AGU* 85, Jt. Assem. Suppl., Abstract SH31A-04

van Nes, P., Reinhard, R., Sanderson, T.R., Wenzel, K.-P., & Roelof, E.C. 1985, *J. Geophys. Res.* 90, 398

Zhang, J., Dere, K.P., Howard, R.A., & Vourlidas, A. 2004, *ApJ* 604, 420

Discussion

UNKNOWN: Comment: At the time of the 2001 April 15 event halo observations show the CME at $\sim$0.3 $R_\odot$. So if it is shock acceleration it is operating from very low heights.

RUFFOLO: Yes, I agree.

JIE ZHANG: For the two events you studied, you showed that the proton onset time is close to soft X-ray peak time. This implies that you start to see SEP at the end of CME acceleration based on my observation of CME flare relation. My question is whether the coincidence (proton onset - soft X-ray peak), is true for many other events? Any statistics on this?

RUFFOLO: We would certainly like to study more events! The analysis I showed was for data from the Spaceship Earth network of polar neutron monitors. This has only been operational, with one-minute resolution, since 2001. Thus we have only been able to analyze 3 events, the two shown here and also a small GLE on Aug. 24, For these, the proton onset is consistent with the soft X-ray peak. 2002.

SCHWENN: GeV particles accelerated near the Sun early on and MeV particles accelerated in IP space – What evidence do you have that they all came from one identical shock? There are people who think in a 2-shock scenario: 1) CME shock(driven) and 2) flare shock (blast wave), associated with Type II radio burst.

RUFFOLO: If the flare shock is delayed from the primary energy release, then we cannot rule this out based on timing alone. But let me point out that at lower ion energies, tens of MeV/n, there is strong physical evidence that the ions accelerated near the Sun came from the CME-driven shock (Mason et al, 1984, Lee & Ryan 1986, Reames 1990, Ruffolo 1997) and not from a localized source, or a source deep in the corona.

BOTHMER: There is a problem of importance: (a) Transport vs Position of the source; (b) Species: Electrons, Protons. Just a comment, not really a question.

RUFFOLO: Yes, I had actually prepared a review of these issues but I had to drop them due to a lack of time.

Coronal and Stellar Mass Ejections
Proceedings IAU Symposium No. 226, 2005
K. P. Dere, J. Wang & Y. Yan, eds.

Energetic Particle from CME-Driven Shocks: Spectra, Composition, and Timing

Allan J. Tylka

Naval Research Laboratory, Washington DC, USA

Abstract. Very large solar energetic particle (SEP) events occur at the rate of about 10 per year during solar maximum. The primary accelerators in these events are shocks driven by fast coronal mass ejections, in which speeds are greater than about 1000 km/s. Solar Cycle 23 probably produced the largest fluence of >10 MeV/nucleon SEPs seen since the start of the Space Age. New instruments on Wind, ACE, SOHO, and other satellites have provided unprecedented detail on the energy spectra, elemental and isotopic composition, ionic charge states, and temporal evolution of these SEP events, as well as their associated CMEs and flares. In this talk, I will review some of the new insights provided by these data. A particular challenge in SEP studies has been the very large event-to-event variability in composition and spectral characteristics, particularly at energies above a few tens of MeV per nucleon. I will discuss recent efforts to understand this variability in terms of seed populations and shock geometry. I will also review recent studies of the time at which SEPs first appear on the Sun-Earth magnetic field line and the implications of these studies for the conditions under which SEP production was initiated.

Keywords. Sun: coronal mass ejections (CMEs), Sun: particle emission

Discussion

YOUSEF: 1) You have shown that Fe/O increases with energy, would that imply preferential acceleration of heavy particles at this domain of energy?

2) Do all relativistic nuclei (particularly) the heavy one arrive with the same speed. Please do comment on Pb observation.

3) Do you also find or find indication of relativistic electron accelerated at the same time with same velocities? Please look at my acceleration paper in this symposium.

TYLKA: 1) No. I would not make that interpretation. Instead, the increase with energy reflects a change in the accessible seed population, coupled with a steepening in the shock-accelerated spectra, as the CME moves outward from the Sun.

2) No, there is velocity dispersion in the particle arrivals. All species have a spectrum of energies. The ultra-heavy ions $(Z > 30)$ have been observed from 100 keV/nuc to ∼10 MeV/nuc. But this is a limitation of presently-available instrumentation. For more on the ultra heavies, see Reames (2000, ApJLett); Reames & Ng (2004, ApJ); Mason et al. (2004, ApJ).

3) A spectrum of electrons is also produced along with the heavy ions.

KAHLER: Is it not the case that we can assume a simple shock geometry based on the CME longitude, in which a CME near central meridian is associated with a quasi-perpendicular shock and a CME around W80 with a quasi-parallel shock?

TYLKA: These high energy SEPs are generally being produced at very low altitude, perhaps just a few solar radii according to the timing studies of Bieber et al. (2004). At these altitudes, the open magnetic field is nominally radial, and the shock geometry will be determined by the details of the CME and the ambient medium. Intuition based on the large scale Parker spiral is unlikely to be very helpful at these altitudes.

KOUTCHMY: What about more "geometric" parameters determining SEP composition/ energy, like center-limb effect and E/W asymmetry of the solar source of SEP?

TYLKA: The gradual events with enhanced Fe/O at high energies and Fe with a harder power-law than O are found from sources all across the Sun. One such event (1999, January 20) had its source region region at E95 according to Cane et at 2001. This behavior contrasts to classic, impulsive events in which ^{3}He/^{4}He $> 10\%$ at a few MeV/nuc. In those events, 85% are found at longitudes between W30 and W80, according to the longitude distribution provided by Reames (1999). The spread in the distribution is primarily determined by variation in solar-wind speed. By comparison only $\sim 40\%$ the events with enhanced Fe/O at >30 MeV/nuc are found at W30-W80.

GANG LI: In the spectrum show in the bow shock acceleration diagram, does the seed population extend to high energies and the power law is a simple increase of intensity? Or does the seed population is more a power-law with a roll-over, but the acceleration process produces the power law at high energies?

TYLKA: To my recollection, the authors (Meziane et al., JGR 2001) of the bow shock papers did not discuss the spectrum of the incident seed particles. However, given the small size of the bow shock, the slope of the power law may very well reflect the spectra of the seed particles. Never- theless, it is significant that the power-law spectra are only seen at quasi-perpendicular configurations.

Coronal and Stellar Mass Ejections
Proceedings IAU Symposium No. 226, 2005
K. P. Dere, J. Wang & Y. Yan, eds.

Some aspects of particle acceleration and transport at CME-driven shocks

G. Li and G. P. Zank [1]

[1]IGPP, University of California at Riverside, CA 92026, USA
email: gang.li@ucr.edu, zank@ucr.edu

Abstract. Gradual solar energetic particle (SEP) events are now believed to be associated with CME-driven shocks. As the shock propagates out from the Sun, particles are accelerated diffusively at the shock front and some will escape upstream and downstream into the interplanetary medium. This is in contrast with "impulsive" events, which are believed to be due to solar flares. However, recent observations have found that in some gradual SEP events, the time intensity profile show a two peak feature, suggesting a mixture of particles from solar flares with particles from CME-driven shock. Furthermore, the observed spectra of large SEP events show tremendous variability. The Fe/C (Fe/O) ratio behave oppositely in events which have similar solar progenitors. In this work, we use a numerical model to follow particle acceleration and transport at CME-driven shocks. We investigate a possible scenario for the re-acceleration of flare particles by CME-driven shocks and calculate the Fe/O ratio for two exemple shocks. These simulations are helpful in interpreting observations of particle data obtained in situ at 1 AU by spacecraft such as ACE and WIND.

Keywords. shock waves, Sun: coronal mass ejections (CMEs), particle emission

1. Introduction

Solar flares and Coronal Mass Ejections (CMEs) are catastrophic solar events. Energies up to 10^{31-32} erg are released in typical CMEs and large flares. In both cases, particles can be accelerated up to $\sim GeV/$Nuc. The underlying mechanism of each process, however, is still presently unknown, although magnetic reconnection is widely believed to be the cause of both phenomena. Observationally, solar flares usually happen in active regions and are not spatially extended. CMEs, on the other hand are spatially extended and have an average angular width of 50 degree (see for example Burkepile (2004)). Furthermore, large CMEs are likely to have flares associated with them (Kahler (1984) and Cliver *et al.* (2004)).

Evidence of electron acceleration in both cases can be inferred from the accompanying X-rays (Johns & Lin (1992)), which result from the interaction of accelerated electrons and ambient solar atmosphere. One important question for understanding the x-ray spectrum is how much X-ray emission is due to non-thermal electrons. Since non-thermal electrons lose their energy mainly through Coulomb scattering and X-rays only amount up to 10^{-5} of the total released energy, thus an X-ray due to non-thermal electrons would impose a strong relationship between the energy release process and the particle acceleration process (Lin (2002)). Similarly, the acceleration of ions can be inferred from various nuclear Gamma-ray lines. These nuclear Gamma-ray lines are produced when energetic ions collide with ambient protons and/or alpha particles in the solar atmosphere (corresponding to broad gamma-ray lines) or when energetic protons collide with ambient heavy ions (corresponding to narrow gamma-ray lines).

Besides various observations aimed directly at the solar surface, in-situ observations of Solar Energetic Particles (SEPs) provide another way for studying CMEs and flares. Historically, SEP events have been classified into two categories, "impulsive" and "gradual". The nomenclature was used initially to refer to the duration of the associated soft X-ray signals, and later on was refined as a classification of "flare accelerated" (the "impulsive") and CME-driven shock accelerated (the "gradual") events (see for example, Reames 1997 and Cliver and Cane (2002)).

Earlier studies of energetic particle abundances (see for example, Mason *et al.* (1984)) have found that for "impulsive" flare events, the measured states of Si and Fe indicate a hot temperature of $\approx 2 \times 10^7$ K, close to that of the hot flare plasma. On the other hand, the ionic charge states for "gradual" CME-driven shock events tend to correspond to a temperature of $\approx 2 \times 10^6$ K, suggesting a source of a cooler coronal material (Mason *et al.* (1995) and Tylka *et al.* (1995)). However, as more data were obtained, especially after the launch of spacecraft ACE in the past decade, one sometimes finds examples of SEP events which possess impulsive characteristics but are associated with CME events. Recently, Cane *et al.* (2003), studied 29 intense SEP events and found four mixed events. These events have a time intensity profile that looks like those due to CME-driven shock acceleration, but have a flare-like composition. It is possible that these are events in which CMEs and accompanying flares occur temporally close to each other and some of the solar flare material will undergo shock re-acceleration (see Li and Zank (2004a)), thus having a time intensity profile similar to shock acceleration particles but having a flare-like composition.

Another noteworthy phenomenon in large SEP events is the spectral variability. In a recent survey, Tylka *et al.* (2004) found that about 1/3 of the large SEP events observed by ACE and Wind in Solar Cycle 23 exhibit explicit energy dependence in Fe/C. However, the energy dependence of Fe/C vary significantly between events that have similar solar progenitors. For example, the CME ejections of two large SEP events, 21 April 2002 event and the 24 August 2002 event, are very similar, † but the Fe/C ratio above 10 MeV/Nuc.behaves very differently. The Fe/C ratio decreases with energy for the 21 April 2002 and increases with energy for the 24 August 2002 event. Tylka *et al.* (2004) has invoked particle acceleration at a perpendicular shock to explain the opposite energy dependence of Fe/C in these two events. They argued that the 24 August 2002 event may correspond to a perpendicular shock, which has higher injection energies. Since more remnant flare material are populated at higher energies, and there are relatively more seed Fe ions than seed CNO ions, we thus expect to see a different energy dependence of Fe/C in the 24 August 2002 event from that of the 21 April 2002 event.

In this paper, we use the model developed initially by Zank *et al.* (2000) and later improved by Rice *et al.* (2003) and Li *et al.* (2003) to follow the acceleration and transport of protons and heavy ions at a CME-driven shock. In the following, we discuss the possible re-acceleration of flare particles and perform a simulation that calculates the Fe/O ratio at two exemple shocks. For details of the model, readers are refereed to the above mentioned papers.

2. Re-acceleration of flare particles

Large CMEs are usually associated with solar flares, and to understand the corresponding SEP observations, it is crucial to understand the temporal relationship between CMEs

---

† The initial CME speed and transit time to Earth are 2400 km/s, 51 hours and 1900 km/s, 58 hours for the 21 April 2002 and the 24 August 2002 events respectively. The associated solar flares for these two events are also comparable with the 21 April 2002 event being X1.5/1F at $S14°W84°$ and 24 August 2002 event being X3.1/1F at $S02°W81°$

and associated solar flares. Zhang *et al.* (2001), using C1, C2 and C3 of LASCO and X-ray data from GOES studied four CME events and found that the main CME acceleration and the main energy release of associated flare occur almost simultaneously. If CMEs and associated solar flares occur at approximately the same time, then we would expect to see some interaction between processes associated with these two events. Li and Zank (2004a) studied the re-acceleration of flare particles by assuming particles are accelerated at both solar flares and CME-driven shocks. The spectrum of the flare particle is assumed to be p^{-4} and the acceleration last 1000 seconds. The spectra of CME-driven shock particles are followed numerically. When particles escape upstream of the shock, they propagate along the interplanetary magnetic field (IMF) with occasional pitch angle scatterings. Because of the presence of the CME-driven shock, large portion of the flare particles are subject to absorption by the shock and are re-accelerated. Figure 1 is a plot of the time-intensity profile adapted from Li and Zank (2004a). In the plot, the upper panel assumes particle acceleration due to a pure flare source. The second panel assumes particle acceleration due to a pure shock source and the third panel assumes a mixed source with a ratio of the initial number of CME-driven particles to the initial number of flare particles to be 2 : 1. The particle mean free path in front of the shock due to pitch angle scattering is taken to be $\lambda = 0.4\mathrm{AU}(\frac{pc}{1\mathrm{GeV}})^{1/3}(\frac{r}{1\mathrm{AU}})^{2/3}$ in all three panels. From figure 1 we can see that the time intensity profile for the "flare-only" case (the upper panels) shows an abrupt rise followed by a decay. This is because the point source lasts a short time only. The middle panel of Figure 1 corresponds to the "shock-only" case. Compared to the "flare-only" case, it is clear that the traveling shock, which is neither spatially nor temporally confined, leads to a much slower rise of the intensity profile. The bottom panel of Figure 1 corresponds to the "flare + shock" case. At early times, the intensity profile resembles the "flare-only" case. These are particles that were accelerated at the flare site, and then propagate to 1 AU and not absorbed by the shock. At later times, the contribution from the shock becomes more pronounced. The contribution comes from two populations, these being particles that were injected from the ambient solar wind and accelerated at the traveling shock, and particles that were accelerated at the flare but have been absorbed and re-accelerated by the shock. The latter particles, once absorbed, will behave like those particles that originate at the shock. Thus, at late times, we expect the intensity profile of the "flare + shock" case to behave like that of the "shock-only" case. Indeed, the profile shown in the bottom panel resembles that observed by Cane *et al.* (2003).

3. The Fe/O ratio

Understanding how heavy ions are accelerated at CME-driven shocks and then transported in the interplanetary medium is essential to understand the observational data obtained by spacecraft such as ACE and WIND. Unlike protons, which comprise $\sim 98\%$ material in the solar system and when accelerated at the CME-driven shocks, will generate Alfvén waves, heavy ions such as C, N, O and Fe are "rare" elements and can be treated as test particles. The differences between various heavy ions are the charges and masses. Even though the diffusive shock acceleration mechanism does not differentiate between ions of different charges and masses and predicts power laws for different ions with the same spectral index, the maximum attainable energy for each element, however, depends on the combination of Q/A. Here Q is the charge of the heavy ion and A is the mass number. Furthermore, the mean free paths of ions also depend on Q/A. Thus the spectra observed at 1 AU of different heavy ions can be quite different. One useful observational quantity of an SEP event is the Fe/O ratio as a function of time at

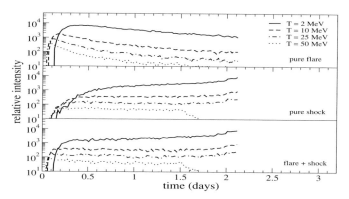

Figure 1. The re-acceleration of flare particle by a CME-driven shock. From Li and Zank (2004a). The upper panel assumes particle acceleration due to a pure flare source. The second panel assumes particle acceleration due to a pure shock source. The third panel assumes a mixed source. See text for details.

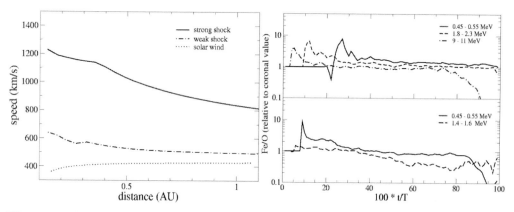

Figure 2. The shock profile for a strong and a weak shock and the corresponding Fe/O ratio (from Li and Zank (2004b)). The left part plot the shock speed as the function of heliospheric distance for a strong and a weak shock. The right part plot the Fe/O ratio as a function of time.

different energies. Li and Zank (2004b) performed simulations of heavy ion acceleration and transport at a strong and a weak shock and obtained the corresponding Fe/O ratio, shown in Figure 2. The left part of Figure 2 is the shock profile. The solid line is for the strong shock and the dashed line is for the weak shock. The solar wind speed is also plotted in the figure for reference. The right part of Figure 2 contains the Fe/O ratio. The upper panel is for the strong shock and the lower panel is for the weak shock. In the upper panel, three energy bins: 0.45 - 0.55 MeV/Nuc., 1.8 - 2.3 MeV/Nuc. and 9 - 11 MeV/Nuc. are shown. In the lower panel, two energy ranges, $0.45 - 0.55$ MeV/Nuc. and $1.4 - 1.6$ MeV/Nuc. are shown. From the figure we see that the Fe/CNO ratios rise above 1 at early times and then slowly decrease, approaching 1 or even going below 1 at later times. The curve corresponding to higher energies rises faster, but has a smaller peak value. The early rise of the Fe/CNO ratio is because iron, at the same energy per nucleon, has a longer mean free path than CNO particles, thus arriving at 1 AU sooner. The ratios shown in Figure 2 are consistent with observations. (Tylka *et al.* (1999)).

4. Conclusion

With the launch of spacecraft such as ACE, WIND and RHESSI in the past decades, much data for SEP events has been obtained. These include X-ray, gamma ray, radio signals at or near the surface of the Sun and in-situ particle and magnetic field data at 1 AU. We now know that large gradual SEP events are due to CMEs and large CMEs are often accompanied by solar flares. However, the exact relationship between CMEs and solar flares is still unknown. Equally intriguing questions are, for example, how particles are accelerated at CME-driven shocks; what is the seed population and can perpendicular shock acceleration explain the spectral variability? To answer these questions, it is essential to develop a sophisticated numerical model which captures the essence of the underlying physics. In this paper, using a model we developed earlier, we discuss the possibility of a mixture of flare accelerated particles with CME-driven shock accelerated particles and its possible implications for the time intensity profiles of SEP events. We also considered the Fe/O ratio as a function of time for a parallel shock configuration and find qualitative agreements with observations. We believe this work is helpful on interpreting the observations of particle data obtained in situ at 1 AU by spacecraft such as ACE and WIND.

Acknowledgements

This work has been supported in part by a NASA grant NAG5-10932 and an NSF grant ATM-0296113. GL would like to thank Drs Y.H. Yan and J. Wang for their hospitality during his stay in Beijing.

References

Burkepile, J. T. *et al.* , 2004, *J. Geophys. Res.* 109, A03103 doi: 10.1029/2003JA010149.
Cane, H.V., *et al.*, 2003, *Geophy. Res. Lett.* 30, 10.1029/2002GL016580.
Cliver, E.W., Cane, H.V., 2002, *EOS Vol. 83, number 7.*
Cliver, E. *et al.* , 2004, *ApJ*, 605, 902-910.
Johns, C. and Lin, R. P., 1992, *Sol. Phys.* 137, 121.
Kahler, S.W. *et al.*, 1984, *J. Geophys. Res.* 89, 9683.
Lin, R. P. *et al.*, 2002, *Sol. Phys.* 210, 3.
Li, G, *et al.*, 2003, *J. Geophys. Res.*,108(A2), 1082, doi:10.1029/2002JA009666.
Li, G. and Zank, G. P., 2004a, *Geophys. Res. Lett.*, in press.
Li, G. and Zank, G. P., 2004b, *J. Geophys. Res.*, submitted.
Mason, G.M., Gloeckler, G., Hovestadt, D., *Astrophys. J., 280*, 902-916.
Mason, G.M., *et al.*, 1995, *Astrophys. J., 452*, 901-911.
Reames, D.V., Kahler, S.W., and Ng, C.K., 1997 *Astrophys. J., 491*, 414-420.
Rice, W.K.M., *et al.*, 2003, *J. Geophys. Res.*, 108, A10, pp. SSH 5-1, doi:10.1029/2002JA009756.
Tylka, A.J., *et al.*, 1995, *Astrophys. J., 444*, L109-L113.
Tylka, A.J. *et al.*, 1999. *Geophys. Res. Lett.*, 26, 2141-2144
Tylka, A. *et al.* , 2004, *submitted to ApJ.*
Zank, G.P. *et al.*, 2000, *J. Geophys. Res.*, 105, 25079-25095.
Zhang, J., *et al.*, 2001, *Astrophys. J., 559*, 452.

Discussion

P.F. CHEN: Which parameters govern the final spectral index of the energetic particles in your model?

LI: The spectrum at the shock front is decided by the instantaneous shock parameters such as compression ratio, etc. The observed particles at 1AU is a combination of particle acceleration at the shock front and particle transport in the interplanetary medium.

Coronal and Stellar Mass Ejections
Proceedings IAU Symposium No. 226, 2005
K. P. Dere, J. Wang & Y. Yan, eds.

© 2005 International Astronomical Union
doi:10.1017/S1743921305000839

The Production of Near-Relativistic Electrons by CME-Driven Shocks

S.W. Kahler[1], H. Aurass[2], G. Mann[2], and A. Klassen[3]

[1]Space Vehicles Directorate, Air Force Research Laboratory, Hanscom AFB, MA 01731, USA
email: stephen.kahler@hanscom.af.mil

[2]Astrophysikalisches Institut Potsdam, An der Sternwarte 16, D-14482 Potsdam, Germany

[3]Institut für Experimentelle und Angewandte Physik, University of Kiel, Leibnizstrasse 11, D-24118 Kiel, Germany

Abstract. The solar sources of near-relativistic ($E > 30$ keV) electron events observed at 1 AU are poorly understood. In general, the solar injection times deduced from the observed 1 AU onset times and assumed 1.2 AU travel distances yield injection times about 10 minutes after the associated flare impulsive phases and type III radio burst times. One interpretation is that the apparent delays occur in the interplanetary medium, probably due to scattering of the electrons. If the injection times are delayed from the impulsive phases, the electron acceleration might take place in CME-driven shocks. Here a large number of electron events observed with the UC/Berkeley 3DP detector on the Wind spacecraft are compared with CMEs observed by the Lasco coronagraph on SOHO and with type II bursts observed by the 40 to 800 MHz radio receiver at the Astrophysikalisches Institut Potsdam (AIP) and by the 20 kHz to 14 MHz WAVES instrument on the Wind spacecraft. The acceleration of at least some of the electron events is not consistent with the shock hypothesis.

Keywords. Sun: coronal mass ejections (CMEs), Sun: particle emission, Sun: radio radiation, shock waves

---

1. Introduction

1.1. *Delayed Solar Injections of Near-Relativistic Electrons*

It has long been understood that bursts of 2 to $\geqslant 100$ keV electrons accelerated near the Sun and observed at 1 AU are nearly always accompanied by solar type III radio bursts (Lin (1985)). The velocity dispersion of the electron fluxes produces an instability which results in Langmuir waves that are converted into electromagnetic waves at the local electron plasma frequency and its second harmonic (Melrose (1985)). Observations of 326 $E > 2$ keV electron events in space during a 15-month period in 1978-79 on the ISEE-3 satellite seemed to confirm the result that the type III radio bursts are produced by 2 to $\geqslant 10$ keV electrons (Lin (1985)).

The early analysis of solar electron events observed near solar minimum during 1994-95 with the 3-D Plasma and Energetic Particle (3DP) experiment on the Wind spacecraft showed surprising differences from the previous ISEE-3 results. The predominately higher energies ($E > 30$ keV) of the Wind 3DP electron events allowed a more accurate determination of solar electron injection times based on plots of electron onset times at 1 AU versus v^{-1}, where v is the electron speed, and on an assumption that the first electrons arrive scatter-free. That analysis led to the unexpected result that many impulsive electron injections at the Sun occurred up to half an hour later than the onset of the accompanying type III burst. Krucker *et al.* (1999) found that injection times of 41 of 58 $E > 25$ keV 3DP electron events were delayed beyond the timing onset uncertainties of each electron event and coronal and/or interplanetary type III burst. Twelve low energy ($E \leqslant 25$ keV) electron events, on the other hand, were much better associated with type III burst times, and there were at least 2 cases of hybrid events with type III-related $E \leqslant 25$ keV electrons and $E \geqslant 25$ keV electrons showing the respective simultaneous and delayed release times.

The existence of delayed solar electron events was then confirmed by Haggerty & Roelof (2001, 2002) using observations of 79 impulsive $38 < E < 315$ keV electron events with the Electron, Proton, and Alpha Monitor (EPAM) on the ACE spacecraft. The electron injections of the 45 EPAM electron events with associated metric type III bursts were characterized by a median electron injection delay of 9.5 minutes. A similar result was found for the electron injection delays relative to the starts of other kinds of flare electromagnetic emission. The EPAM study was recently extended through March 2002 to include 113 electron events by Haggerty *et al.* (2003), who found a median delay of 13 minutes between the metric type III bursts and the electron injection times. Haggerty & Roelof (2002) argued that the correlations of $r \sim 0.5$ between the $38 < E < 62$ keV peak electron intensities and several flare radio and X-ray parameters provided evidence of only a loose relationship between the electron events and the accompanying flares.

1.2. *The Shock-Acceleration Interpretation*

The first attempts to explain the origins of the delayed electron events invoked shock acceleration. Krucker *et al.* (1999) considered the association of electron events with coronal waves observed in the Extreme Ultraviolet Imaging Telescope (EIT) on board the SOHO spacecraft (Thompson *et al.* (1998)). They suggested that fast moving wave fronts at high altitudes allowed a wave front to link the flare site to the magnetic connection point to the Earth. Haggerty & Roelof (2001, 2002) suggested that the ~ 10 minute injection delay corresponds to the time for a shock forming in the low corona and propagating at 1000 km s^{-1} to reach 1 R$_\odot$, at which point electrons could be accelerated to near-relativistic energies by the shock and reach open field lines. Mann *et al.* (2003) found a local minimum of the Alfvén speed in the middle corona, i.e., 1.2–1.8 R$_\odot$, and a broad maximum around 2–6 R$_\odot$ (see Fig. 6 in Mann *et al.* (2003)). Since shocks form more easily in regions with low Alfvén speed, a disturbance would need a period to travel from the flare or CME source site up to the local minimum Alfvén speed, where shock acceleration of electrons could occur. That would lead to a delay between the flare peak time and the shock acceleration.

Simnett *et al.* (2002) tested this interpretation of delayed shock acceleration by comparing the injection times of impulsive electrons with the launch times of associated coronal mass ejections (CMEs). For 47 EPAM electron events associated with CMEs the electron injections were typically delayed by ~ 20 minutes from the CME launch times. In addition, the peak electron intensities and energy spectra spanning ~ 40 to 300 keV were correlated with associated CME speeds. Their results were confirmed by the extended analysis of EPAM electron events by Haggerty *et al.* (2003). Simnett *et al.* (2002) suggested that most of the impulsive near-relativistic electrons were produced in shocks driven by CMEs and released into space when the CMEs reached heights of ~ 2 to 3 R$_\odot$ from Sun center. Note that Mann *et al.* (1999, 2003) argued for a global maximum of the Alfvén speed in just this region of the high corona.

A similar shock interpretation for the near-relativistic electron events was proposed by Klassen *et al.* (2002), who investigated four near-relativistic impulsive electron events. In each case herring-bone (HB) and shock-associated (SA) emission soon after the start of the type II burst indicated the release of $E < 30$ keV electrons, but the near-relativistic electrons were released later, at least 11 minutes after the start of the type II burst and with no corresponding radio signatures. In their view both groups of electrons were accelerated by the shock, but the near-relativistic population was injected when the associated type II shock and CME were 2 to 5 $R_\odot$ from Sun center.

1.3. *Alternatives to the Shock Interpretation for the Delayed Electron Onsets*

While the case for shocks as producers of the near-relativistic electrons has been made as described above, alternative interpretations for the delayed injections have been proposed. Pick *et al.* (2003) and Maia & Pick (2004) studied Nancay decametric array and radio heliograph (NRH) images associated with EPAM near-relativistic electron events with delays > 5 min from type III bursts and argued that coronal radio brightenings, resulting from post-eruptive magnetic reconnection or from interactions between coronal structures and passing coronal waves or CME bow shocks, are the sources of the delayed electron injections. Classen *et al.* (2003) gave a similar interpretation to an electron event observed on 2002 June 2 by the 3DP instrument.

Further evidence against the shock interpretation comes from a statistical survey comparing 57 3DP near-relativistic electron events with Nancay radio data. Klein *et al.* (2003a) found that two thirds of those events showed some burst or enhancement of decimetric or metric radio emission in the western hemisphere and within the electron injection windows. Type II bursts were found in only a minority of the events with enhancements, and in those cases the brightenings occurred at coronal heights lower than the shock. In a different approach to the shock interpretation, Klein *et al.* (2003b) used X-ray observations of occulted flares with type II bursts to put upper limits on interplanetary electron fluxes produced in coronal shocks. They found that the numbers of near-relativistic electrons in large events are close to or exceed the upper limits of their analysis.

A very different interpretation of the delays was suggested by Cane (2003), who found that the inferred injection delay times correlated directly with the times for the radio-generating electrons to transit to 1 AU. In addition, a correlation of the delays was also found with the 1 AU ambient solar wind densities, supporting her conclusion that interaction effects in the interplanetary medium were the cause of the anomalous delays of the electron onsets.

1.4. *Open Questions about Impulsive Near-Relativistic Electron Injection*

Although inferred near-relativistic electron injection delays after the type III burst onsets are clearly established by two independent studies (Krucker *et al.* (1999), Haggerty & Roelof (2002)), the statistical studies have shown neither a bimodal distribution of events, with one group clearly temporally associated with the type III bursts and the other clearly delayed, nor a single group of event onsets distinctly delayed beyond the type III burst onsets. Krucker *et al.* (1999) distinguished two groups simply on the basis of the timing uncertainties of the type III bursts and of the electron injection onsets (their Figure 3), and Haggerty & Roelof (2002) only determined median values of the delays (their Figure 5). Maia & Pick (2004) selected 21 EPAM near-relativistic electron events for study and found two groups – 10 weak, short events with essentially no delays and 11 events with variable delays associated with solar radio-complex bursts. This may suggest two physically separate groups of near-relativistic electron events, but their sample was too small for a firm conclusion.

It is further puzzling that in all the impulsive near-relativistic electron events we find an associated type III burst (Haggerty & Roelof (2002)) which, at least for the delayed events, is presumed to have no direct relevance to the near-relativistic electron events. Since all the impulsive near-relativistic electron events are associated with decametric-hectometric (DH) type III bursts observed in the Wind WAVES experiment (Haggerty & Roelof (2002)), another major question is why the delayed electron injections do not also produce type III bursts.

The case for the shock origin for the near-relativistic impulsive electron events is not without problems, such as the association of the electron events with CME-driven shocks. Although Simnett *et al.* (2002) found that 47 of the 52 electron events of their sample could be associated with Lasco CMEs, a direct comparison of their electron event list with the CMEs listed at the web site of the SOHO Lasco CME catalog (http://cdaw.gsfc.nasa.gov/CME_list/) shows that 17 of the 52 (33%) electron events do not have associated CMEs within a reasonable ($\sim$ 1 hour) time preceding the event onset. In 14 cases Simnett *et al.* (2002) associated features they termed blobs and jets with the electron events, but the solar corona observed in Lasco movies allows many more dynamic features than only the CMEs to be identified. Some such features may indeed be associated with shocks, but Gopalswamy *et al.* (2001) found that only 6 of 101 fast ($v > 900$ km s^{-1}) CMEs associated with DH type II bursts were narrower than 60°. Thus the association of the blobs and jets with coronal shocks would appear doubtful. Further, Simnett *et al.* (2002) point out that about a third of their event-associated CMEs have speeds below 400 km s^{-1}, presumed to be generally too low to drive shocks.

A related question is the determination of the association of the impulsive electron events with metric and DH type II bursts as another way to test the validity of the shock acceleration hypothesis discussed in Section 1.2. The association of either metric or DH type II bursts with either the EPAM or the 3DP electron events has not been systematically examined. A low rate of type II burst associations would call into question the presence of the presumed basic acceleration mechanism. A high correlation with both shocks and CMEs would lend support to the shock hypothesis. Here we undertake a more comprehensive examination of the associations of near-relativistic electron events with type II bursts and CMEs.

2. Data Analysis

2.1. *Selection of the Impulsive Near-Relativistic Electron Events*

We will work with two combined sets of near-relativistic impulsive electron events. The first set consists of the list of event onset times observed in the EPAM instrument, given in Table 2 of Haggerty & Roelof (2002) and extended through the end of 2001 (D. Haggerty, private communication). The event selection criteria and a brief description of the instrument are given in Haggerty & Roelof (2002). Each of the two EPAM detectors measures electrons in the $\sim$ 40 to $\sim$ 315 keV range in four channels. From their list we select for analysis only those events observed in their third (103 to 175 keV) or fourth (175 to 315 keV) energy channels.

The second set of events are the 3DP impulsive electron events from the solid state telescope (SST) given at the web site http://sprg.ssl.berkeley.edu/~bezerkly/all_events.htm. The event selection criteria and a brief description of the 3DP instrument and data reduction are given in Ergun *et al.* (1998) and in Krucker *et al.* (1999). We selected events from two previously posted lists: the period 15 November 1994 to 22 June 2001, and a second period coinciding with the observations by the Ramaty High Energy Solar Spectroscopic Imager (RHESSI) from February 2002 to October 2002. The lists gave the estimated injection times, qualitative assessments of the event data, and links to SST intensity plots and pitch-angle distributions (PADs). From those lists we eliminated the "poor" events and used the "mediocre" and "good" events for the analysis.

Table 1. Associations with EPAM and 3DP Events

Electron event	AIP Type II	WAVES II/IV	CMEs
EPAM	16 of 41	5 of 40	27 of 34
3DP	35 of 90	17 of 90	57 of 69
Total	37 of 100	17 of 99	63 of 79

For this study we also required that metric radiospectrographic observations from the Tremsdorf station (Mann *et al.* (1992)) of the AIP be available from about 1 hour before the electron injection time through the injection time. These observations are normally made in the 800 to 40 MHz range from about 0700 to 1500 UT. Finally, we require that Wind/WAVES 20 kHz to 14 MHz observations be available through the duration of the electron event, which eliminates a single EPAM event, although we include that event in the statistics of Section 2.2. With these requirements we have a total of 41 EPAM events and 90 3DP events, of which 31 events are common to both lists, leaving a total of 100 different events.

2.2. *The Associations Statistics*

We now proceed to compare the electron events with three different solar signatures of shock acceleration - metric type II bursts, DH type II bursts, and CMEs. We first consider the associations between the AIP metric type II bursts reported at the web site http://www.aip.de./People/AKlassen/ and the EPAM and 3DP electron events. For an association the type II burst times had to be within $\sim$ 15 minutes of the inferred electron injection time. We found that 16 of the 41 (39%) EPAM events and 35 of the 90 (39%) 3DP events were associated with reported AIP type II bursts. For the combined 100 different events, 37 were associated with metric type II bursts, as shown in Table 1. We then compared the electron events with listings of possible DH type II bursts observed by the Wind/WAVES instrument (Bougeret *et al.* (1995)), requiring the DH burst onset to be within about 1 hour of the electron injection time. There are only 17 DH type II bursts associated with the combined 99 electron events with simultaneous Wind/WAVES data coverage.

The last comparison is with Lasco CMEs listed at the Lasco CME catalog website at http://cdaw.gsfc.nasa.gov/CME_list/. Sixty three of the combined 79 electron events (80%) with Lasco observations were associated with CMEs with onsets within $\sim$ 20 minutes of the inferred injection times. This is somewhat larger than the 67% CME association rate for the Simnett *et al.* (2002) list of EPAM events we discussed above. We further ask how many of the 63 associated CMEs had observed widths $> 60°$ or speeds > 400 km s^{-1}, and hence were likely to be associated with interplanetary shocks. The result is that only 43 of the 79 electron events (54%) were associated with wide ($> 60°$) CMEs and only 52 of the 79 events (66%) with fast (> 400 km s^{-1}) CMEs.

Finally, we did a reverse association, starting with AIP type II bursts and the complete list of all beamed EPAM events and all listed 3DP electron events, and found that 17 of 245 (7%) type II bursts could be associated with an EPAM event and 41 of 214 (19%) type II bursts could be associated with a 3DP event. We attribute this lower number of EPAM associations to the fact that only about 30% of all EPAM electron events were selected as beamed events (Haggerty & Roelof (2002)).

3. Discussion

We have compared EPAM and 3DP near-relativistic electron events with metric type II bursts, DH type II bursts, and CMEs in an effort to determine whether the resulting associations support the conclusion of Simnett *et al.* (2002) that most of the near-relativistic electrons observed in impulsive events are accelerated at CME-driven shocks. The associations of the electron events with metric (37%) and DH (17%) type II burst shocks are not high, suggesting that most of the electron events do not originate in shocks. The 80% CME association is much higher, but our understanding that shocks are driven only by wide and fast CMEs suggests a more realistic result of $\sim 50\%$ association of electron events with fast and wide CMEs. In particular, we find that 11 of the 63 associated CMEs have speeds < 400 km s^{-1}, a fraction lower than that of Simnett *et al.* (2002), who found 17 of their 47 CMEs to have such low speeds. The difference is probably due to the fact that Simnett *et al.* (2002) selected their events from the Haggerty & Roelof (2002) list of all EPAM events, which extended to the EPAM energy channels 2, 3, or 4, while we have limited our event selection to only those extending to the higher EPAM energy channels 3 or 4. Thus a higher CME association might be expected for our more energetic electron events. In both studies, however, the significant number of low-speed CME associations raises doubt about the CME-driven shock hypothesis for electron acceleration in those events.

We pointed out that only 37% of the selected near-relativistic electron events were associated with metric type II bursts. As in the case with CMEs, this association is higher than that reported by Simnett *et al.* (2002) (3 of their 52, or 6%) probably because we restricted our study to the most energetic electron events. The low inverse association of metric type II bursts with all observed electron events ($\sim 20\%$) shows that most type II bursts are not associated with observed electron events. However, Figure 2 of Haggerty & Roelof (2002) indicates that the electron events are observed only from a broad longitude range of $\sim 60°$ on the visible disk. Let us assume a uniform type II burst visibility over a 200° longitude range (Kahler *et al.* (1985), Cliver *et al.* (2004)). Then $\sim 200/60 \times 20\% = 67\%$ of all type II bursts can be associated with near-relativistic electron events. We can compare this result with the case of solar energetic proton (SEP) events, for which shock acceleration is presumed to be the dominant acceleration mechanism. Cliver *et al.* (2004) compared SEP ($E \sim 20$ MeV) events and metric type II bursts favorably located in the western hemisphere and over a similar time interval. In that study 56% of the western hemisphere metric type II bursts were not associated with SEP events, but 82% of the western hemisphere SEP events were associated with metric type II bursts. Thus most SEP events (82%) *are* and most electron events (63%) *are not* associated with metric type II bursts, supporting a possible but weak connection between coronal shocks and near-relativistic electrons. The inverse case of the higher association of type II bursts with electron events ($\sim 67\%$) than with SEP events (56%) suggests a higher occurrence of electron events than of SEP events.

Pick *et al.* (2003) and Maia & Pick (2004) have divided EPAM near-relativistic electron events into two groups, one of which show essentially only type III radio bursts and no inferred injection delays from the type III bursts. The second group were associated with major changes in complex radio bursts which coincided with the inferred injection times that are delayed from the type III bursts; 6 of those events had metric type II bursts, and 5 did not. Maia & Pick (2004) argued that even when type II bursts were present, the shocks may have worked as restructuring agents rather than as direct accelerators of electrons. Our results are consistent with the conclusion that at least some, and perhaps a majority, of the near-relativistic electron events do not originate in CME-driven shocks.

However, a role for shock-accelerated electrons in broad, fast CMEs remains a viable concept.

Acknowledgements

We acknowledge use of the Lasco CME catalog, which is generated and maintained by NASA and The Catholic University of America in cooperation with the Naval Research Laboratory. SOHO is a project of international cooperation between ESA and NASA. SK thanks the AIP for their hospitality and the Air Force Office of Scientific Research for funding his Window on Europe visit to the AIP. The AIP acknowledges the European Office for Aerospace Research and Development for its support in maintaining the solar radio spectral observations at Potsdam.

References

Bougeret, J.-L., *et al.* 1995, *Space Sci. Rev.* 71, 5

Cane, H.V. 2003, *ApJ* 598, 1403

Classen, H.T., Mann, G., Klassen, A., & Aurass, H. 2003, *A & A* 409, 309

Cliver, E.W., Kahler, S.W., & Reames, D.V. 2004, *ApJ* 605, 902

Ergun, R. E., *et al.* 1998, *ApJ* 503, 435

Gopalswamy, N., Yashiro, S., Kaiser, M.L., Howard, R.A., & Bougeret, J.-L. 2001, *J. Geophys. Res.* 106, 29219

Haggerty, D.K. & Roelof, E.C. 2001, *Proc. 25th Int. Cosmic-Ray Conf. (Hamburg)* 3238

Haggerty, D.K. & Roelof, E.C. 2002, *ApJ* 579, 841

Haggerty, D.K., Roelof, E.G., & Simnett, G.M. 2003, *Adv. Space Sci.* 32(12), 2673

Kahler, S.W., Cliver, E.W., Sheeley, N.R., Jr., Howard, R.A., Koomen, M.J., & Michels, D.J. 1985, *J. Geophys. Res.* 90, 177

Klassen, A., Bothmer, V., Mann, G., Reiner, M.J., Krucker, S., Vourlidas, A., & Kunow, H. 2002, *A & A* 385, 1078

Klein, K.-L., Krucker, S., & Trottet, G. 2003a, *Adv. Space Sci.* 32(12), 2521

Klein, K.-L., Schwartz, R.A., McTiernan, J.M., Trottet, G., Klassen, A., & Lecacheux, A. 2003b, *A & A* 409, 317

Krucker, S., Larson, D.E., Lin, R.P., & Thompson, B.J. 1999, *ApJ* 519, 864

Lin, R.P. 1985, *Sol. Phys.* 100, 537

Maia, D.J.F. & Pick, M. 2004, *ApJ* 609, 1082

Mann, G., Aurass, H., Voigt, W., & Paschke, J. 1992, *in Coronal Streamers, Coronal Loops, and Coronal and Solar Wind Composition: The First SOHO Workshop, ed. C. Mattok (ESA SP-348; Noordwijk: ESA)* 129

Mann, G., Aurass, H., Klassen, A., & Estel, C. 1999, *in Proc. 8th SOHO Workshop, ESA-SP446* 447

Mann, G., Klassen, A., Aurass, H., & Classen, H.-T. 2003, *A & A* 400, 329

Melrose, D. 1985, *in Solar Radiophysics, ed. D.J. McLean & N.R. Labrum, Cambridge Univ. Press* 177

Pick, M., Maia, D., Wang, S.J., Lecacheux, A., Haggerty, D., & Hawkins, S.E., III 2003, *Adv. Space Sci.* 32(12), 2527.

Simnett, G.M., Roelof, E.C., & Haggerty, D.K. 2002, *ApJ* 579, 854

Thompson, B.J., *et al.* 1998, *Geophys. Res. Lett.* 25, 2461

Discussion

TYLKA: This is really a question for David Ruffolo, who showed this morning that the time vs $1/\beta$ analysis may not be reliable for determining onsets, at least for ions. Might similar effects contribute to the delay here?

KAHLER: The delays are comparable to the travel time. Ruffolo says he does not expect such large systematic errors in the velocity dispersion analysis.

RUFFOLO: (to question of Tylka after talk of Kahler) Was the delay on the order of tens of minutes? (Kahler: Yes.) Then, no, I don't think a systematic error in the onset time vs. 1/v fit could account for that delay.

SCHWENN: You told us what these relativistic electrons are not due to. So what are they due to, in your opinion, and why?

KAHLER: At this time I favor the interpretation that the delays are due to propagation, but we can not rule out any of the several suggested scenarios.

GRECHNEV: What for the velocity did you use in your study? If it is the velocity from the SOHO/LASCO CME catalog, then the velocity of the shock can be a little bit different than the velocity of the frontal structures. This can change the selection which you had.

KAHLER: I used the speed of the CME leading edge. You are right that the shock speed should be higher than the CME speed, but the shock speed should increase with the CME speed.

GOPALSWAMY: In a recent study (Gopalswamy *et al.* 2004, JGR) we found that the electron intensity in the 108 KeV channel correlated with flare peak intensity in X-rays better than it did with CME speed. Does this have any bearing on your result?

KAHLER: That result would suggest that acceleration occurs in the flares rather than in shocks.

Coronal and Stellar Mass Ejections
Proceedings IAU Symposium No. 226, 2005
K. P. Dere, J. Wang & Y. Yan, eds.

Analysis of the Acceleration Process of SEPs by an Interplanetary Shock for Bastille Day Event

G.M.Le[1,2]†, and Y.B.Han[1]

[1]The National Astronomical Observatories, Chinese Academy of Sciences, Beijing 100012

[2]National Center for Spaceweather Monitor and Forecast, Chinese Meteorological Administration, Beijing 100081

Abstract. Based on the solar energetic particle (SEP) data from ACE and GOES satellites, the acceleration of SEP by CME-driven shock in interplanetary space was investigated. The results showed that the acceleration process of SEP by the Bastille CME-driven shock ran through the whole space from the sun to the magnetosphere. The highest energy of SEP accelerated by the shock was greater than 100MeV. A magnetic bottle associated with the CME captured a lot of high energy particles with some of them having energy greater than 100MeV. Based on magnetic field data of solar wind observed by ACE data, we found that the the magnetic bottle associated with the Bastille CME was the sheath caused by the CME in fact.

Keywords. shock waves, Sun: coronal mass ejections (CMEs), particle emission

1. Introduction

Reames (1999) introduced SEP events' properties in great detail. He think that there are two kinds of SEP events, impulsive type related to flare and gradual type related to fast shock driven by a CME. An important point of view of Reames(1999,2002) is that large SEP events are only related to CME-driven shock. Gopalswamy's (2002) view of point is that the interaction between two CMEs' came from the same region within 24h play an important role for large SEP events. All these work mainly emphasize CME's role and dismissed flare's function. Recently Richardson (2003) argued that the CME interaction may be not important for accelerating major solar energetic events. Some results (Cohen et al., 1999; Mazur et al., 1999) showed that flare may provide some seed particles for CME-driven shock. Torsti et al. (2001) pointed out that the SEP event on 1999 May 9 was a hybrid event, the flare first accelerated particles and then CME accelerated the particles produced by the flare. Mewaldt et al. (2003) pointed out that remnant of interplanetary material from earlier impulsive can't provide sufficient Fe in hybrid events, which also implied that flare may provide seed particles for CME-driven shock. Kallenrode (2003) suggested SEP events should consist of flare accelerate particles (FAPs) and CME-driven shock accelerate particles (CSAPs), with the only FAPs or only CSAPs being the limiting case.

The solar active region AR9077 produced a X5.7 flare and a intensive coronal mass ejection(CME) causing very strong disturbance in Sun-earth connection space on 2000 July 14. This day was called Bastille day and the solar event on that day was called Bastille Day Event, CME on 2000 July 14 was called Bastille CME in this paper. The event has been paid attention greatly in the world. Many paper have been devoted to

† Present address: 20A Datun Road, Chaoyang District, Beijing, China

the event study. Reames et al. (2001) analyzed the heavy ion abundance and spectra of the event. Bieber et al. (2002) investigated the intensity-time and anisotropy-time profiles and pitch-angle distributions of energetic protons near Earth; Tylka et al. (2001) explained the energy spectra of Fe by shock acceleration from solar wind suprathermals and small (∼5%) admixture of remnants flare particles. Maia et al (2001) study the energetic electron on Bastille Day Event by analyzing the radio data observed by Nancay radio heliograph. Le et al (2004) ever studied the moving direction of the Bastille CME by using galactic cosmic ray data. MÄKELÄ et al. (2001) analyzed the time variation of 2.7-3.3MeV nucl^{-1} and 10-13MeV nucl^{-1} protons and some heavy ions at energy 8.5-15MeV. They thought that the magnetic tubes were filled with flare-related particles. Tang et al. (2003) suggested that the Bastille Day event was a hybrid event. Very recently Tang (2004 Sep.) found that the protons were firstly accelerated by electric field in the magnetic reconnection region by analyzing the data of gama ray and EIT/SOHO. So it seems that the event is still not completely understood.

Here we report that the data of high energy protons with energy greater than 10MeV and 30MeV in interplanetary space observed by Advanced Composition Explorer (ACE). We also analyze the energetic particles with energy greater than 10MeV, 30MeV, 50MeV and 100Mev observed by Goes satellite. We study the CME-driven shock's role in acceleration of Bastille Day SEP Event. Finally we investigate which part associated with the Bastille CME captured a lot of higher energy particles with some of them having energy greater than 100MeV.

2. Data Analysis

Because only CME or flare may produce or accelerate solar energetic particle, so we firstly present the flare and CME's information related to Bastille Day SEP event. On 2000 July 14, a big flare flare X5.7 began at 10:03UT, peaked at 10:24UT and ended at 10:43, shown in Fig.1. There were no other important flare except the X5.7 during 14-16, July 2000. Based on the flare classification given by Pallavicini et al. (1977), the flare X5.7 was a gradual flare. While the whole duration of the flare is shorter than 1 hour, so the flare belonged to impulsive flare based on the definition of a flare by its time structure.

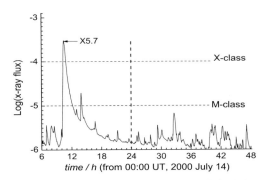

Figure 1. solar flare during 14-16, Jul 2000

The CMEs' came from the AR9077 on 2000 July 13 and Bastille day are listed below (copy from Gopalswamy's (2002) paper). CME1's onset was 14hour and 24m earlier than CME2's onset.

The data from the SIS instrument on ACE during 14∼15, July 2000 are shown in Fig.2. From Fig.2 we can see that the proton flux with energy greater than 10MeV

Table 1. 2000 Jul. 13 and 14's CME information

Date	CME	Onset time	Width	Speed(km/s)
2000 Jul 13	CME1	20:30UT	62	839
2000 Jul 14	CME2	10:54UT	360	1674

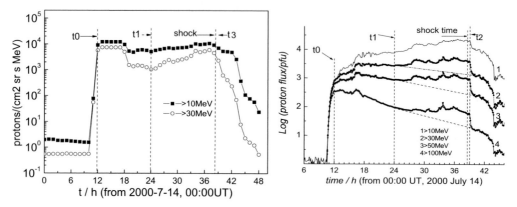

Figure 2. the SEP data observed by ACE   **Figure 3.** the SEP data observed by GOES 8

and 30 MeV reached saturation no long after the onset of the SEP event. At about 2000 July 14 18:00UT, the proton flux with energy greater than 10MeV and 30 MeV decreased which means that the shock driven by the CME2 accelerated particles very intense during 11:00UT~18:00UT, July 14, 2000. With the deceleration of the CME2 as it traveling outward, the ability of CME2-driven shock acceleration of SEP decreased. So the enhancement of SEP flux from t1 to shock time must not be caused by shock driven by the Bastille CME. Because there was no important flare around the 24:00UT and also no CME around that time, so the enhancement must be caused by the CME itself. The SEP flux observed by GOES satellite increased also from t1 for four energy channels including the channel of greater than 100MeV shown in Fig.3. After shock passed ACE, the SEP flux was still high during the period from shock time to t3 shown in Fig. 2, so there were large number of energetic particles captured by the magnetic bottle. Based on the ACE data, the location of the magnetic bottle relative to the CME-driven shock and CME itself was shown in Fig.4.

3. Summary and Discussion

The Bastille Day CME-driven shock had been accelerating SEP with energy greater than 100MeV during the period from the onset time to the moment that shock reached the the magnetosphere. The magnetic bottle associated the Bastille CME captured a lot of high energy particles with some of them having energy greater than 100MeV. The location of the magnetic bottle associated with the Bastille CME was behind the shock but in front of the magnetic cloud.

In fact the magnetic bottle associated with Bastille CME was the sheath caused by the CME. The Bastille CME-driven shock accelerated SEP very intensively causing the SIS SEP instrument saturation during the period from the onset time of SEP to about 18:00UT, 2000 July 14. After that the SEP flux observed by ACE decreased for a while shown in Fig.2. The SEP flux with energy greater than 10MeV, 30MeV and 50MeV observed by GOES satellite didn't decline obviously around 18:00UT, 2000 July 14. This

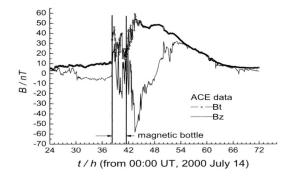

Figure 4. The location of magnetic bottle associate the Bastille CME

means that the charged particles with energy greater than 10MeV, 30MeV and 50MeV can stay in geosynchronous orbit for some time. There was big difference between the SEP data in geosynchronous orbit observed by GOES satellite and the SEP data in interplanetary space observed by ACE satellite for the particles with energy greater than 10MeV and 30MeV.

Acknowledgements

We thank Dr. Allan J. Tylka and Professor Ruffolo David for their helpful suggestions and we have made some revision for the paper. The data used in this paper are downloaded from http://cdaweb.gsfc.nasa.gov/cdaweb/sp_phys/ and http://spidr.ngdc.noaa.gov/. This work is supported by National Nature Science Foundation of China (through Grant No. 10073013) and National Center for Spaceweather Monitor and Forecast, Chinese Meteorological Administration.

References

Bieber, J.W., Droge, W., Evenson, P.A., *et al.*, 2002, *ApJ* 567, 622B
Cohen,C.M.S., Cummings, A.C., Leske, R.A., 1999, , *Geophys. Res.Lett.* 26, 149
Duldig, M.L., Bombardieri, D.J., Humble, J.E., 2003, *28th ICRC* 6, 3389
Gopalswamy, N., Yashiro, S., Michalek, G., 2002, *ApJ* 572, L103
Kallenrode, M.B., 2003, *J Phys. Nucl. Part. Phys.* 29, 965
Kocharov, L., Torsti, J., 2002, *Solar Physics* 207, 149
Le, G.M., Ye, Z.H., 2004, *Chin. J. Space Sci.* 24, 15 (in Chinese)
Maia, D., Pick, M., Hawkins, III S.E., *et al.*, *Solar Phys.* 204, 199
Makela Pertti, Torsti Jarmo., *Solar Phys.* 204, 215
Mewaldt, R.A., Cohen, C.M.S., Mason, G.M., *et al.*, 2003, *28th ICRC* 6, 3229
Mazure, J.E., Mason, G.M., Looper, M.D., *et al.*, 1999, *Geophys. Res.Lett.* 26, 173
Pallavicini, R., Serio, S., Vaiana, G.S., 1977, *ApJ* 216, 108
Raeder, J., Wang, Y.L., 2001, *Solar Phys.* 204, 325
Reames, D.V., 2001, *Astrophys. J.* 548, L233
Reames, D.V., 1999, *Space Science Review* 90, 413
Reames, D.V., 2002, *ApJ* 571, L63
Richardardson, I.G., Lawrence, G.R., Haggerty, D.K., *et al.*, 2003, *GRL.* 30(12), 8014
Shinichi, W., Manabu, K, Takashi, W., 2001, *Solar Phys.* 204, 425
Tang, Y.H., Dai, Y., 2003, *Adv. Space Res.*, 12, 2609
Tang, Y.H., private communication, 2004 Sep.
Torsti, J., Kocharov, L., Innes, D.E., *et al.*, 2001, *A&A* 365, 198
Tylka, A.J., Cohen, C.M.S., Dietrich, W.F., 2001, *ApJ* 558, L59

Coronal and Stellar Mass Ejections
Proceedings IAU Symposium No. 226, 2005
K. P. Dere, J. Wang & Y. Yan, eds.

Propagation of Energetic Particles to High Heliographic Latitudes

Trevor R. Sanderson

Research and Scientific Support Division of ESA, ESTEC, Noordwijk, The Netherlands
email: trevor.sanderson@esa.int

Abstract. The Ulysses spacecraft has now completed its second orbit over the poles of the Sun. Energetic particles associated with CMEs were observed at the highest latitudes over both poles, quite unlike the first polar pass when virtually no CME or CIR accelerated particles were observed at the very highest latitudes. We present observations of solar energetic particle events observed in the energy range ∼1 MeV to ∼100 MeV made by the COSPIN instrument, when the spacecraft was at high heliographic latitude over the northern pole, above the current sheet, and immersed in high-speed solar-wind flow coming from the northern polar coronal hole. We discuss the rise to maximum, the onset time and the anisotropy of the energetic particles. We find that, unlike the events observed at mid and low latitudes, the particle angular distributions were almost isotropic, but with a net outward flow along the magnetic field lines. We compare these events with other events observed at lower latitudes.

Keywords. Sun: coronal mass ejections (CMEs), particle emission, solar wind

Keywords. Ulysses, energetic particles, propagation, coronal mass ejection, anisotropy

1. Introduction

In February 1992 the Ulysses spacecraft used an encounter with the planet Jupiter to begin its first out-of-the-ecliptic orbit around the Sun, completing it in April 1998. This orbit began as the level of solar activity was falling. Most of the time when the spacecraft was over the south and then over the north pole of the Sun was close to the time of minimum of solar activity of solar cycle 22.

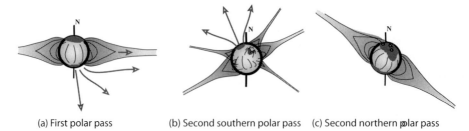

(a) First polar pass   (b) Second southern polar pass   (c) Second northern polar pass

Figure 1. Cartoon showing the coronal magnetic field observed during, from left to right, (a) the first polar passes, (b) the second southern polar pass. and (c) the second northern polar pass (Adapted from Suess et al., 1998).

At this time the axis of the dipolar component of the coronal magnetic field was nearly parallel to the Sun's spin axis, and the current sheet was almost flat, giving rise to a streamer belt which was more or less in the plane of the ecliptic. Large coronal holes were observed over the poles, giving rise to high-speed solar-wind flow in the high latitude polar regions of the heliosphere. Figure 1a, after Suess et al. (1998) shows this configuration.

As this was a period close to solar minimum, very few particle increases due to either Coronal Mass Ejections (CME) or Co-Rotating Interaction Regions (CIR) were observed in the polar regions during these passes. At mid-latitudes, significant increases due to CIRs were observed for a substantial fraction of the time that the spacecraft was above the current sheet, but again at the very highest latitudes very few, if any, were observed.

The second orbit around the sun began in April 1998 as the level of solar activity of solar cycle 23 was increasing. The second orbit was considerably different from the first. During most of the orbit, the heliosphere was dominated by the presence of CMEs as the level of solar activity increased, the spacecraft passing over the south and then the north poles of the Sun during the time around the maximum of solar activity of solar cycle 23. A summary of Ulysses observations during the maximum of cycle 23 can be found elsewhere in this volume (Marsden, 2004).

Conditions on the Sun were very different for the southern and the northern polar passes (Sanderson et al., 2003a, Sanderson, 2004). During the southern pass (Figure 1b), the dipole axis was oriented at around 135 degrees to the spin axis and the field had a significant quadrupole component. The current sheet reached up to very high latitudes, and there was no polar coronal hole over the southern pole, typical of that expected around solar maximum. So, Ulysses remained in slow solar wind flow as it passed over the pole.

Over the northern pole (Figure 1c) the situation was quite different. The tilt of the dipole was similar. The dipole strength had increased, and the quadrupole term of the coronal field had diminished. The field was therefore much more dipolar, and so the current sheet only reached up to mid latitudes. A polar coronal hole had started to develop, and by the time Ulysses reached the highest northern latitudes it was immersed in fast solar wind. Although the northern pass was still close to solar maximum, the configuration of the Sun was beginning to look more like that close to solar minimum.

Solar activity was still high, and so perhaps not surprisingly, CMEs and substantial particle increases were observed at the highest latitudes as the spacecraft passed over the northern pole of the Sun. Several large SEP events were observed at high latitude and in the fast solar wind. These events were unusual in that they were the only CME-related particle increases observed so far by Ulysses at high latitudes and in the fast solar wind. They were also unusual in that the onsets at high energies were delayed considerably, and the angular distributions at onset were almost isotropic.

In this paper, we discuss the propagation of energetic particles to high heliographic latitudes as observed when Ulysses was over the northern pole and immersed in the fast solar wind, and compare them with events observed in the slow solar wind over the southern pole.

2. The Second Northern Polar Pass

We begin by discussing the northern polar pass. In Figure 2 we show summary plots of the events number 2, 3 and 4 of Sanderson et al. (2003b). These events have also been studied in detail by Dalla et al. (2003a, 2003b), Lario et al. (2004) and McKibben at al. (2003). An additional event, Event 1, is included here as a typical event observed close to Earth. Event 1 was observed shortly after Ulysses had been launched, 11 November (day 315) 1990.

Each plot covers 16 days. Each is split into three panels. In the bottom panel we show particle intensities from two electron channels ranging from 1 to 10 MeV, in the next,

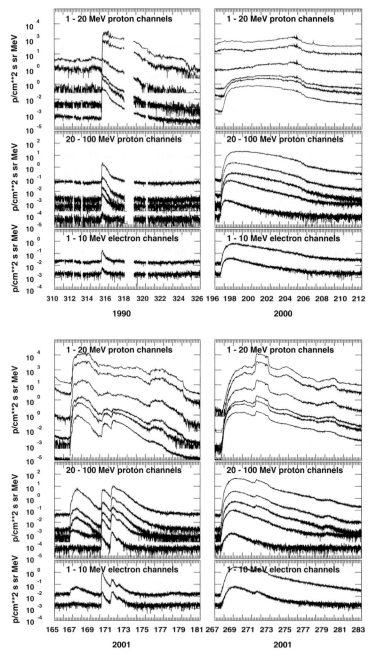

Figure 2. 16-day summary plots for events 1, 2, 3, and 4. The bottom panel of each plot shows particle intensities from 2 electron channels ranging from 1 to 10 MeV, in the next, from 5 proton channels ranging from 100 MeV to 20 MeV, and in the top panel, from 6 proton channels ranging from 20 MeV to 1MeV.

from 5 proton channels ranging from 20 MeV to 100 MeV, and in the top panel, from 6 proton channels ranging from 1 MeV to 20 MeV.

Event 1 was observed shortly after Ulysses had been launched and was on its way to Jupiter, being at a radial distance of 1.07 AU and heliographic latitude of just 4° and is included here as an example of a 'typical' 1 AU event for our instrument. Note the rapid onset, typical of events like this at 1 AU.

Event 2 is the Bastille-day event, observed at mid-latitudes and in the slow solar-wind. The onset and decay profile in the electron channels and at high proton energies was relatively smooth (bottom and middle panel). A high background masked the onset of the event at low proton energies (top panel).

Event 3 is typical of the many low-latitude events, where the event occurs at the same time as a structure in the magnetic field such as a CIR or SIR passes the spacecraft during the onset. The time intensity profile during the first onset was considerably disturbed by the presence of the structure, parts of which acted as channels to allow the lower energy protons rapid access to the position of the spacecraft, and other parts of which acted as sources of local acceleration. This gave rise to complicated intensity versus time and anisotropy parameters versus time profiles. This is the most often seen type of event at Ulysses.

Event 4 is one of the few large high-latitude events observed in the fast solar wind during the Ulysses mission, as observed during the second high-latitude pass. This event had a relatively slow onset (compared with 1 AU in-ecliptic events) and during the first few days, a smooth time-intensity profile. The CME which followed a few days after the onset showed up as an increase lasting around 1 - 2 days in the low energy particle intensity.

In Figures 3 and 4 we show particle intensity profiles and anisotropy parameters for two events, taken from Sanderson et al. (2003b). Here we show, from top to bottom, the following: low energy proton intensities, 1.8 - 3.8 MeV first order perpendicular anisotropy, first order parallel anisotropy, high energy proton intensity, 34 - 68 MeV first order perpendicular anisotropy, first order parallel anisotropy, magnetic field azimuth, elevation (in spacecraft coordinates), and magnitude. The particle instruments scan in a plane perpendicular to the spacecraft spin axis (z-axis), where the z-axis always points to the earth. After rotating the x-axis to the direction of the magnetic field projected onto the scan plane, the sectored count rates in this frame of reference are Fourier analysed. In this way, the 2-dimensional parallel and perpendicular anisotropy amplitudes in the scan plane are derived. Anisotropies presented here are the ratios of the amplitude of the first order component in the scan plane (either parallel or perpendicular to the projected field direction) to the amplitude of the zero order component.

Figure 3 is a 3-day plot showing the onset of event 1, observed on 11 November (day 315), 1990, the 1 AU baseline event. This event has a profile similar to profiles described in many models of propagation, but in fact was one of the rare occasions when we observed an onset without the disturbing effect of the presence of a magnetic field structure such as a CIR, SIR, or CME at the spacecraft.

The 34 - 68 MeV parallel anisotropy suddenly increased at the time the 34 - 68 MeV intensity started to rise, and dropped to zero within a few hours of the onset. Similarly, the 1.8 - 3.8 MeV anisotropy rose and fell, starting one or two hours after that of the 34 - 68 MeV particles. In both cases, during the rise to maximum, and for a few hours thereafter, there was a finite parallel anisotropy and a perpendicular component which was essentially zero, signifying that the particles are always field-aligned. The small fluctuations were mainly due to the limited counting statistics.

Compare this with the high-latitude, fast solar wind event in Figure 4. This shows in detail the September 2001 event. This is a 6-day plot, starting on 24 September (day 267), 2001. At high energies, the increase was about one order of magnitude less than the low latitude events. The event was most likely initiated by an X2.6 flare at 09:36 on 24 September, day 267, at S16 E23. This event has a moderately rapid increase at ~50 MeV.

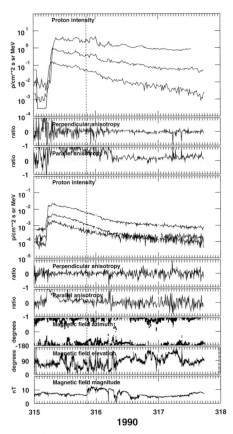

Figure 3. November 11 1990 event, from top to bottom, proton intensity from the 1.8 - 3.8, 3.8 - 8.0 and 8.0 - 19.0 MeV channels, 1.8 - 3.8 MeV first order anisotropy amplitude resolved perpendicular to the component of the magnetic field in the scan plane of the instrument, first order anisotropy amplitude resolved parallel to the component of the magnetic field in the scan plane of the instrument, proton intensity from the 24 - 31, 34 - 68 and 68 - 92 MeV channels, 34 - 68 MeV first order anisotropy amplitude resolved perpendicular to the component of the magnetic field in the scan plane of the instrument, first order anisotropy amplitude resolved parallel to the component of the magnetic field in the scan plane of the instrument, magnetic field azimuth, elevation (in spacecraft coordinates), and magnitude.

Three and one half days after the onset, an over-expanding CME (start and stop times shown by the long-dashed lines) (Reisenfeld et al., 2003), preceded by a forward shock (shown by the short-dashed line), passed the spacecraft, causing an additional increase of around one order of magnitude in the particle intensity at around ∼1 MeV. The event slowly decayed to background level after about 15 days.

In general, the duration of the events observed in the slow solar-wind events were shorter than the fast solar wind events, a typical slow solar-wind event lasting 7 - 10 days, and a typical fast solar-wind event lasting 15 days. Onsets at high latitude and in the fast solar-wind tended to be smooth and rapid, lasting typically one day. Surprisingly the anisotropies associated with the onset were very small.

Comparing event 1, the baseline 1 AU event with event 4, the high-latitude, high-speed flow event, we see immediately the large difference between the two. The 1 AU event rose to a maximum in around 3 hours. The intensity during the high-latitude event rose in

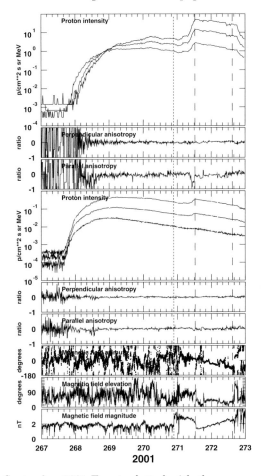

Figure 4. The September 2001 Event, plotted with the same parameters as in Figure 3.

around 2 days, a factor ∼16 slower than the 1 AU event, which was located only a factor ∼2 closer to the Sun (measured along the field line).

Larger than expected delays in onset times were observed for most of these high-latitude events (Dalla et al., 2003a, 2003b), corresponding typically to 120 to 350 minutes from the flare onset. These delay times and the path lengths were correlated by Dalla et al. (2003a, 2003b) against several variables. The best correlation was found with difference in latitude between the flare site and the latitude of Ulysses, this correlation being surprisingly better than the correlation with the angular separation between the site and Ulysses. This implies a very effective longitudinal transport of the particles, but a very inefficient transport latitudinally, which the authors concluded meant that cross field diffusion was the fundamental mechanism in getting the particles to high latitudes, in agreement with the suggestion by Zhang et al. (2003) for the Bastille Day event. However, they did not rule out the possibility that the delay was due to the time taken for the CME to reach the field lines connected to the spacecraft.

The anisotropy in the high-latitude event was very small. Note that the scales of the anisotropy panels are the same. In the 34 - 68 MeV channel, the component of the anisotropy perpendicular to the field fluctuated back and forth around zero. There was a small but finite anisotropy which persisted for around one day after the onset in the 34 - 68 MeV proton channel, signifying that the particles were propagating outwards from the

Sun (the field direction at this time was inward). This anisotropy persisted until around the time of maximum of the 34 - 68 MeV channel, and then remained around zero for the next couple of days. All along, the perpendicular component remained around zero, indicating that there was no net flow across the field.

In the 1.8 - 8.0 MeV channel, the component of the anisotropy perpendicular to the field again fluctuated back and forth around zero. There was a larger negative parallel anisotropy which persisted for around half a day after the onset in the 1.8 - 3.8 MeV channel, again signifying that the particles were propagating outwards from the Sun. A small field aligned anisotropy persisted during the next couple of days. Again, all along, the perpendicular component remained around zero, indicating that the net flow was field aligned.

The measured anisotropies were very small. Within the limits of accuracy of measurement of the anisotropy, when there was a net flow, the anisotropy directions were coincident with the magnetic field direction, projected onto the plane within which the anisotropy measurements are made. This meant that the flow was field aligned, and that particles reached high latitudes traveling along the field lines, and not by crossing over them. The particles were scattered significantly as they propagate outwards, which explains their relatively slow onset and their small anisotropy, but despite this, any net flow direction was still along the local magnetic field line direction.

3. Discussion

At high latitudes and in the fast solar wind (70 - 80°N) we found no evidence for any substantial net flow across the field lines, whereas at moderately high latitudes in the slow solar wind (62°S) Zhang et al. (2003) found evidence for cross-field flow.

In Figure 5 we show data from the Bastille-day event, using the same format as in Figure 3. This event has some similarities to the high-speed, high-latitude events, Events 4 to 7. However, the Bastille event was observed in the slow solar wind and is more like a low latitude event (such as shown in Figure 3) than a high latitude one. At Ulysses, the Bastille Day event was an unexceptional event. The increase discussed here was a secondary event sometime after the main event. It was observed at 3.17 AU, and at 62°S, which was around the same southern latitude as the southern-most extent of the current sheet, whereas Events 4, 5, 6 and 7 were observed at 2 AU and latitudes between 70° and 80°N, which at the time was ∼20° higher than the northern-most extent of the current sheet.

The main event began with an onset at high energies at Ulysses on day 193. The second, more substantial onset occurred at around 1600 UT on day 197. The anisotropy of the high energy particles suddenly increased at the time of onset, the particles streaming outwards along the field past the spacecraft. The anisotropy amplitude then started to decrease. At the beginning of day 197, a CME arrived at the position of Ulysses. The magnetic field direction in the spacecraft frame of reference changed as the spacecraft entered the CME, reversing the sign of the anisotropy amplitude. The anisotropy amplitude continued to decrease as the spacecraft entered the CME, whilst at the same time a comparable, but small, perpendicular component of the anisotropy (measured in the scan plane) was observed. Half way through the passage of the CME, the anisotropy had dropped essentially to zero. Although at a moderately high latitude, this event was observed in the slow solar wind, and has some similarities to the 1 AU event shown in Figure 3. Upstream of the CME was a region lasting around 6 hours within which the high-energy anisotropy was high. The perpendicular component was considerably smaller, which implies that the flow was essentially field aligned. Immediately after the onset,

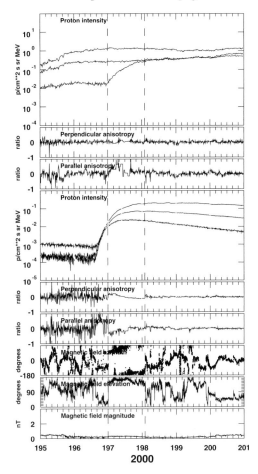

Figure 5. The Bastille day event, plotted with the same parameters as in Figure 3.

particles traveling along the field lines with a moderately substantial parallel anisotropy and essentially no perpendicular anisotropy were observed. The anisotropy amplitude then started to decay slowly, just like the 1 AU in-ecliptic event, except the time and magnitude scales were considerably different.

Approximately 10 hours after the onset, the spacecraft entered the previously-existing CME, quite unrelated to the CME which was responsible for this event. The parallel anisotropy amplitude continued to decrease, but from this time on a perpendicular component of similar magnitude was observed. Both the perpendicular and parallel component amplitudes continued to decrease until both were zero at a time half way through the passage of the CME. This could imply that particles were crossing field lines within the CME, as suggested by Zhang et al. (2003) although this is probably unlikely, as the field is usually very quiet inside a CME, so scattering across the field lines is not to be expected. A more likely explanation is that inside the previously existing CME there was a substantial un-measurable field aligned component, as at this time the magnetic field in the spacecraft frame of reference was almost perpendicular to the scan plane. This signature could also possibly be due to the existence of a gradient, but the duration of this anisotropy is probably too long for this to be true.

4. Summary and conclusions

At solar minimum, most of the high latitude region is filled with fast solar wind. At solar maximum, only a few small low-latitude fast solar wind streams exist. For the study of particle propagation at high latitudes, it is important to differentiate between propagation in slow and fast solar wind, as the characteristics of the propagation differ considerably between the two.

During the southern polar pass, the spacecraft was continually in the slow solar wind. Most of the particle events observed at this time occurred at the same time as some other pre-existing and unrelated structure, such as a CME from a previous solar flare, or a CIR, passed over the spacecraft. Particle propagation was dominated by the presence of these structures, the frequent occurrence of which meant that it was quite rare to find an event where the event was unaffected by one. The forward and reverse shocks and the stream interfaces of the CIRs, and interplanetary shocks and magnetic clouds of the CMEs all affected the particle propagation, sometimes even accelerating the lower energy particles locally. These structures tended to be full of discontinuities, which again affected the propagation, the effect depending on the size and thickness of the discontinuities and the energy of the particles.

During the southern polar pass, where structures were present, we observed events with irregular time-intensity profiles, onset times and velocity dispersion modified by the presence or the lack of structures and discontinuities and field aligned flow. In between the shocks and discontinuities, the field tended to be relatively quiet and channeling could be observed. Occasionally, close to a boundary or interface, we observe a short period of non field-aligned flow. On the rare occasions in low-speed flow when no structures were present, we observed a smoother time-intensity profile, a rapid onset with velocity dispersion and moderately high field aligned anisotropies at the onset, diminishing with time.

During the highest latitude parts of the second polar pass, Ulysses was immersed in the fast solar wind. The fast solar wind tends mainly to be homogeneous, and devoid of large scale discontinuities, but is much more turbulent than the slow solar wind, and so particles propagate to high latitudes in high-speed solar wind with some difficulty. Energetic particle events observed during the part of the northern polar pass where Ulysses was at its highest latitude and in the fast solar wind had smooth time intensity profiles, near-isotropic particle angular distributions at all energies at the onset, flow directions during the rising phase of the events along the field, and no evidence for any net flow across the field lines. These particles propagated to the highest heliographic latitudes traveling along magnetic field lines and not across them.

Our observations do not allow us to draw conclusions about propagation closer to the Sun, but most likely, to reach the high latitudes, particles must either diffuse across field lines, most likely closer to the Sun, or else there was some large scale distortion of the magnetic field lines, again, closer to the Sun.

Finally we conclude with a summary table showing the energetic particle and the plasma characteristics of the two different high-latitude regions, the slow solar wind region observed during the second southern polar pass, and the fast solar wind region observed during second northern polar pass.

Acknowledgements

The author gratefully acknowledges A. Balogh, D. McComas, and R. B. McKibben for permission to use Ulysses FGM, SWOOPS and COSPIN/HET data respectively.

Table 1. Summary of observations during the second polar passes

Characteristic	Southern Polar Pass	Northern Polar Pass
Solar wind	Slow	Fast
Time intensity profiles	Irregular	Smooth
Structures at onset	Frequent	Rare
Event onset times	Rapid	Delayed
High latitude Propagation	Modified by draping	Direct along field
Anisotropy at onsets	Large	Nearly isotropic
Particles inside CME	Intensities depressed	Intensities elevated
Particle flow directions	Field aligned	Field aligned (non-field aligned near structures)

References

Dalla, S., A. Balogh, S. Krucker, A. Posner, R. Mueller-Mellin, J. D. Anglin, M. Y. Hofer, R. G. Marsden, T. R. Sanderson, B. Heber, M. Zhang, and R. B. McKibben, Delay in solar energetic particle onsets at high heliographic latitudes, Annales Geophysicae, 21, 1367 - 1375, 2003a.

Dalla, S., A. Balogh, S. Krucker, A. Posner, R. Mueller-Mellin, J. D. Anglin, M. Y. Hofer, R. G. Marsden, T. R. Sanderson, C. Tranquille, B. Heber, M. Zhang, and R. B. McKibben, Properties of high heliolatitude solar energetic particle events and constraints on models of acceleration and propagation, Geophys. Res. Lett., 30, No. 19, 8035, 2003b.

Lario, D., R. B. Decker, E. C. Roelof, D. B. Reisenfeld, T. R. Sanderson, Low-energy particle response to CMEs during the Ulysses solar maximum northern polar passage, JGR, 109, A01108, 2004.

Marsden, R. G., Ulysses at Solar Maximum, in The Sun and the Heliosphere as an integrated system, IAU monograph, ed. G. Poletto and S.T. Suess, Kluwer, p113-146, 2004.

McKibben, R. B., J. J. Connell, C. Lopate, M. Zhang, J. D. Anglin, A. Balogh, S. Dalla, T. R. Sanderson, R. G. Marsden, M. Y. Hofer, H. Kunow, A. Posner, and B. Heber, Ulysses COSPIN Observations of Cosmic Rays and Solar Energetic Particles from the South Pole to the North Pole of the Sun during Solar Maximum, Ann. Geophys., 21, 1217, 2003.

Reisenfeld, D. B., J. T. Gosling, R. J. Forsyth, P. Riley, and O. C. St. Cyr, Properties of high-latitude CME-driven disturbances during Ulysses second northern polar passage, Geophys. Res. Lett., 30, No 19, 8031, 2003.

Sanderson, T. R., T. Appourchaux, J. T. Hoeksema, and K. L. Harvey, Observations of the Sun's Magnetic Field during the Recent Solar Maximum, J. Geophys. Res., No. A1, 1035, doi:10.1029/2002JA009388, 2003a.

Sanderson, T. R., R. G. Marsden, C. Tranquille, S. Dalla, R. J. Forsyth, J. T. Gosling, and R. B. McKibben, Propagation of energetic particles in high-latitude high-speed solar wind, Geophys. Res. Lett., 30(19), doi:10.1029/2003GL017306, 2003b.

Sanderson, T. R., Propagation of energetic particles to high latitudes, in The Sun and the Heliosphere as an integrated system, IAU monograph, ed. G. Poletto and S.T. Suess, Kluwer, 2004.

Suess, S. T., J. L. Phillips, D. J. McComas, B. E. Goldstein, M. Neugebauer, and S. Nerney, The Solar Wind - inner heliosphere. Space Science Reviews, 83, 75-86, 1998.

Zhang, M., R. B. McKibben, C. Lopate, J. R. Jokipii, J. Giacalone, M.-B. Kallenrode, and H. K. Rassoul, Ulysses observations of solar energetic particles from the July 14, 2000 event at high heliographic latitudes, J. Geophys. Res., 108, No. A4, 1154, doi: 0.1029/2002JA009531, 2003.

Discussion

KAHLER: If it is possible to organize the observations by heliomagnetic coordinates, would we find that the SEP intensities vs time scales are also organized by heliographic latitudes?

SANDERSON: Yes, I think there is one of the next steps for us with this data set. I would expect we will find some interesting correlations. I expect it will be relatively easy to do for solar minimum, but much more difficult for the most interesting solar maximum period.

SCHWENN: What is the present wisdom about energetic particles (CIR associated) at very high latitudes at activity minimum, although far away from the actual CIRs? Fisk or other?

SANDERSON: I think it is now time to re-examine this subject. During the second Ulysses orbit, a considerably greater number of CIRs with considerably higher intensity, were observed.

BOTHMER: 1) Comment on terminology: Dipolar /Quadrupolar /Multipolar structure 2) Drops in intensity of SEPs inside ICMEs just caused by background levels (higher in ecliptic than at higher latitude)?

SANDERSON: 1) The WSO data probably does not have enough resolution to show the multipolar structure. For this we should use some of the SoHo observations. 2) Background levels are so low that they do not influence the observations.

RUFFOLO: If I may ask you some basic solar physics in textbooks we read about the Babcock model of the 11-year solar cycle and magnetic reversals. However, this model still maintains basically a dipole field at all times, now you show us quadrupolar fields and how coronal holes evolve with the solar cycle. Do we need to revise the textbook explanation of magnetic reversals?

SANDERSON: The WSO observations cover more than a full 22year solar cycle. A comparison of the analysis with the Babcock and other models is beyond the scope of our analysis, but I would welcome some further interpretations by suitable theoreticians.

GOPALSWAMY: IP shocks observed at 1AU typically originate from anywhere on the disk. However, if they are accompanied by ICMEs, the source region is clustered around the disk center (± 30° in lat. and long.). This makes me wonder if located beyond N75 can observe ejecta. Are the Ulysses plasma measurements consistent with CME? If so, may be the weak dipole field after reversal has allowed the CMEs to expand unusually to have a width > 140°.

SANDERSON: Surprisingly, CMEs have been observed up to these high latitudes. The CME presented here was identified on the basis of plasma observations by the Los Alamos group (Re. Serfeld et al. SW10, 2003). It was associated with a S16 flare. So far, 4 or perhaps 5 small CMEs have been identified, propagating in the fast solar wind. I would be interested to know if any of the models can reproduce these results.

Coronal and Stellar Mass Ejections
Proceedings IAU Symposium No. 226, 2005
K. P. Dere, J. Wang & Y. Yan, eds.

© 2005 International Astronomical Union
doi:10.1017/S1743921305000864

Energetic Particle Tracing of Interplanetary CMEs: ULYSSES/HI-SCALE and ACE/EPAM Results

Olga E. Malandraki[1]†, D. Lario[2], T.E. Sarris[1], N. Tsaggas[1] and E.T. Sarris [1]

[1]Democritus University of Thrace, Space Research Laboratory, Xanthi, 67100 Greece,
email: omaland@xan.duth.gr, sarris@ee.duth.gr

[2]Applied Physics Laboratory, the Johns Hopkins University, Laurel, Maryland, USA
email: david.lario@jhuapl.edu

Abstract. Solar energetic particle (SEP) fluxes measured by the ULYSSES (ULS)/HI-SCALE experiment during its second polar orbit as well as by the identical ACE/EPAM experiment at 1 AU are utilized as diagnostics of the large-scale structure and topology of the Interplanetary Magnetic Field (IMF) embedded within Interplanetary Coronal Mass Ejections (ICMEs). Survey results are also reported.

Keywords. Sun: coronal mass ejections (CMEs), interplanetary medium, magnetic fields

1. Introduction

First estimates of the spatial extent of magnetic loops to distances ~ 3.5 AU from the Sun were obtained by Sarris & Krimigis 1982. In this paper, we use SEP measurements by the ULS/HI-SCALE (Lanzerotti, Gold, Anderson *et al.* (1992)) and ACE/EPAM experiments (Gold, Krimigis, Hawkins *et al.* (1998)) in order to trace the topology of ICMEs. We present the first observations of a high-latitude ICME that involves complex intertwined structures including regions both connected to and disconnected from the Sun. Results of an ICME survey are also presented.

2. Observations & Data Analysis

In figure 1, ion, electron (e^-), solar wind and magnetic field observations by the EPAM, SWEPAM and MAG experiments onboard ACE are presented from October 29-31 (DOY 302-304), 2003. Two of the fastest ICMEs ever measured in the solar wind were detected during this period and in early November 2003. These very fast ICMEs drove shocks identified in the ACE magnetic field data (solid lines S1 and S2 in figure 1). B was particularly enhanced following the S1 shock, briefly reaching 68 nT. The maximum speed on Oct 29th was 1900 km/sec; the largest speed associated with the second ejection, late Oct 30th, was 1940 km/sec. The first ICME is associated with an X17.2/4B (S16 E08) solar flare which started at 0951 on Oct 28 and an associated halo CME at 1130 on Oct 28 whereas the 2nd ICME is associated with an X10.0/2B (S15 W02) solar flare which started at 2037 UT on Oct 29 and an associated halo CME which first appeared in the C2 coronagraph of the SOHO/LASCO experiment at 2054 UT on Oct 29 (http://cdaw.gsfc.nasa.gov/).

We argue that the two dashed lines in figure 1 consist a more accurate determination of the boundaries of the 1st ICME. Smooth rotations of the B direction start to be observed

† Also at National Observatory of Athens, Inst. for Space Applications, Athens, 15236, Greece

at the 1st dashed line. Furthermore, at that time, there is a change in B from high to low variance in both magnitude and direction, which is a typical B signature of ejecta, and a decrease in the ~ MeV ion intensity. An abrupt increase at the low-energy ion intensity is observed at the 2nd dashed line. In ecliptic observations at 1 AU have shown that during SEP events, the particle intensity level is usually higher outside the ejecta. Low-energy ion intensities usually peak at the arrival of interplanetary (IP) shocks driven by fast CMEs (Cane *et al.* (1988)). When the spacecraft enters into the ejecta, there is often a decrease in the ion intensities with respect to the intensities observed around the arrival of the shock (Richardson 1997), often followed by a recovery of the particle intensity at/near the exit from the ejecta. At the 2nd dashed line a change in the particle Pitch Angle Distributions (PADs) from bi-directional (BD) to isotropic is also observed (see below) along with an abrupt change in the azimuthal angle of the field.

A large gradual e^- event is observed within this ICME (figure 1) superposed upon the decay phase of a previous e^- event associated with the October 28 solar events. Large gradual SEP events at 1 AU are associated with particle acceleration at CME-driven coronal and IP shocks (Kahler, Cane, Hudson *et al.* (1998), Reames 1999). The e^- fluxes are observed to rise simultaneously at all energies at 2145 UT on Oct 29. The high pre-event ambient intensities mask the onset of this e^- event. PADs with a much stronger BD character, start to be detected at the time of the e^- enhancement. BD PADs with variable anisotropy magnitudes were observed till the trailing edge of this ICME was convected over the s/c when the e^- PADs switch to isotropic.

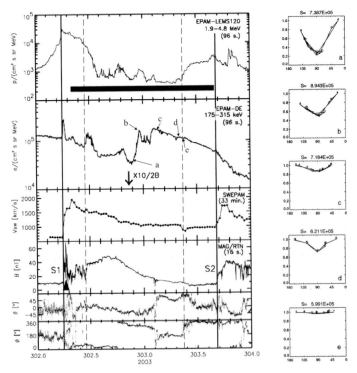

Figure 1. Ion, electron, solar wind and magnetic field observations by the EPAM, SWEPAM and MAG experiments onboard ACE from October 29-31 (DOY 302-304), 2003. The black horizontal bar indicates the time interval identified by Skoug, Gosling, Steinberg *et al.* (2004) as the 1st ICME.

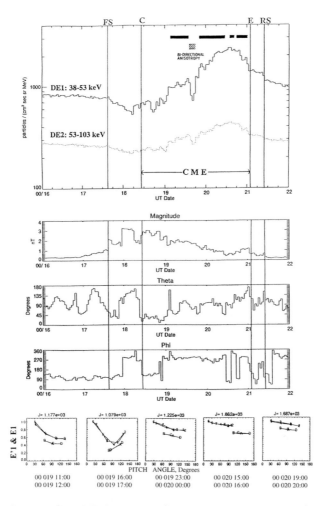

Figure 2. Vertical traces C and E denote the Commencement and End of the high latitude ICME observed by ULS. Electron PAD snapshots are presented at the bottom.

In figure 2, an overview of the hourly and spin-averaged differential intensities of e^- in two energy channels as measured by the ULS/HI-SCALE experiment from 16-22 January, 2000 is presented. IMF parameters in the heliocentric RTN system are also shown. An ICME bounded by the vertical traces C and E was detected by ULS on its way towards the Sun's south pole (Gosling & Forsyth 2001). At this time, ULS was located at 43° S, at 4.1 AU helioradius and 48° west of the Earth. Using the solar wind velocity of 400 km/sec measured by the SWOOPS experiment on Jan 19, the Parker spiral magnetic field line connecting ULS to the Sun, was computed to be 7.84 AU long. The ULS nominal magnetic connection longitude was at E60 on the visible hemisphere of the Sun.

A distinct enhancement is observed at energies $> 38 keV$ at 0105 UT on Jan 19. Since the transit time for 38-53 keV e^- with a 0° pitch angle along the spiral is 2.5 hours, an initial injection at the Sun at 2235 UT on Jan 18 would be implied. There were no significant Hα or X-ray flares reported in SGD, 2000. However, numerous IP type III radio bursts were captured by the WIND/WAVES instrument around the anticipated injection time and during Jan 19. The event has a 1.5 day risetime to maximum. The slow risetime of the event, the absence of a velocity dispersion and the non-impulsive nature of the rising phase indicate the e^- detected after their injection must have undergone drifts across the highly inhomogeneous magnetic fields near the Sun before reaching the ULS footpoint and escaping in space, populating this high latitude ICME.

PADs reveal that after the onset of the event, ULS enters a region where strong unidirectional flow is observed (figure 2). The anisotropy is directed parallel to the IMF, indicating the detected e^- comprise an e^- beam streaming away from the Sun. Unidirectional PADs persist till 1500 UT on Jan 19 (horizontal solid bars in figure 2) when PADs dramatically change from unidirectional to strongly BD (hatched horizontal bar in figure 2). This transition coincides with an abrupt decrease observed in the flux profile of the event. ULS remained in this region for 3 hrs and upon exiting the fluxes recovered their values. After a 2-hr interval of isotropy, ULS encounters a regime where e^- are again streaming antisunward for 15 hrs. PAD snapshots are also shown for the regions observed near the rear boundary of the ICME with e^- streaming away from the Sun.

3. Discussion

The observation of an SEP event within the the October 29-31, 2003 ICME implies that the field lines threading through this structure are still rooted at the Sun allowing direct access of e^-. The BD anisotropies within the ICME are signatures of strong trapping of the observed e^- population. Assuming the ICME is an open structure, when ACE is inside the ICME it connects directly to the S1 shock (beyond the observer) and it also connects to the S2 shock which is coming from the Sun. The enhanced B region reaching $\sim 68 \ nT$ and lying downstream from the preceding S1 shock passes ACE at ~ 0600 on Oct 29 and is travelling radially away the Sun with 1900 km/sec (figure 1). At the time of the acceleration and injection of the energetic e^- close to the Sun i.e. 2037 UT this compressed B region is located at a 0.6 AU radial distance upstream from ACE. In such an open magnetic field configuration, observed energetic e^- streaming in an antisunward direction, after having been reflected at this magnetic constriction in space are detected at the time of the e^- enhancement observed at 2145 UT. Subsequently, reflection from the following S2 shock of these e^- propagating sunward results to particle trapping within the region between the two shocks. Another configuration consistent with the observations is a closed magnetic field topology. When the observer is inside the ICME there is no direct connection between the observer and S1 (now beyond the observer). Hence, it is easy to explain the depression of the energetic ions observed at the leading edge of the ICME. The BD PADs in this scenario could result from e^- injected from both sides of connection of a loop-like structure to S2. Particles are trapped within this configuration by reflection from S2, bouncing back and forth between the legs of the looped B lines. A fraction of particles moving from the weak field side into the strong field are reflected by the shock magnetic field discontinuities. E^- within the loss cone can leak from the trapping region. The evolution of the particle flux is determined by the balance between the continuous particle injection from the following shock into the storage region and the loss of particles passing through the shock(s).

Within the high latitude ICME we have observed energetic e^- to be streaming for long intervals in a collimated beam along IMF lines threading through the ICME. Therefore, a large portion of the ICME most likely consisted of open IMF filaments connected to the Sun at only one end that allowed the escape of e^- to 4.1 AU and 43S. However, the 3 hr long abrupt depression observed in the flux profiles that interrupted the event risephase indicates that injected particles at the Sun could not enter this ICME portion.

We conclude that the IMF within this portion of the ICME is most likely disconnected from the Sun (Malandraki, Sarris, Lanzerotti *et al.* (2001). Furthermore, the observation of counterstreaming e^- throughout this region implies that the e^- are propagating within closed looped IMF lines in space. The e^- observed within this detached looped region may have been imitially injected at the Sun before disconnection occurred. If e^- are injected from the Sun while a plasmoid (a closed magnetically isolated structure) is in space, these e^- would be excluded from the plasmoid since they would have to diffuse across IMF lines to enter it. Such regions moving over ULS would be accompanied by substantial reductions in e^- fluxes, in agreement with what is observed.

We have performed a survey of ICMEs observed by ULS and ACE in different regions of the heliosphere and found that the large majority of the ICMEs ($\sim 92\%$) are still magnetically anchored to the Sun when they arrive at the s/c whereas for only $\sim 5\%$ of the ICMEs there is strong observational evidence that closed magnetic loop-like structures detached from the Sun are present (Malandraki (2002)). Our particle observations suggest that in rare cases ICMEs involve more complex intertwined structures including regions both connected to and disconnected from the Sun. An open magnetic field line topology with open magnetic field lines rooted to the Sun at only one end threading through the ICMEs tends to be the prevalent magnetic structure of ICMEs (e.g. Malandraki, Sarris, Lanzerotti *et al.* (2002), Malandraki, Sarris, Trochoutsos *et al.* (2004)). However, the detection of trapped energetic e^- within a number of ICMEs ($\sim 9\%$) is consistent with the presence of closed loop-like magnetic structures anchored to the Sun at both ends.

Acknowledgements

We are thankful to our HI-SCALE team colleagues for their support and encouragement. Olga Malandraki acknowledges support from the State Scholarships Foundation (I.K.Y.) through a Post-Doctoral Fellowship.

References

Cane, H.V., Reames, D.V. & von Rosenvinge, T.T. 1988, *J. Geophys. Res.* 93, 9555

Gold, R.E., Krimigis, S.M., Hawkins S.E, Haggerty, D.K., Lohr, D.A., Fiore, E., Armstrong, T.P., Holland, G., Lanzerotti, L.J. 1998, *Space Sci. Rev.* 86, 541

Gosling, J.T. & Forsyth, R.J. 2001, *Space Sci. Rev.* 97, 263

Kahler, S.W., Cane, H.V., Hudson, H.S., Kurt, V.G., Gotselyuk, Y.V., MacDowall, R.J., Bothmer, V., 1998, *J. Geophys. Res.* 103, 12069

Lanzerotti, L.J., Gold. R.E., Anderson, K.A., Armstrong, T.P., Lin, R.P., Krimigis, S.M., Pick, M., Roelof, E.C., Sarris, E.T., Simnett, G.M., Frain, W.E. 1992, *Astron. Astrophys. Suppl. Ser.* 92, 349

Malandraki, O.E., Sarris, E.T., Lanzerotti, L.J., Maclennan, C.G., Pick, M., Tsiropoula, G. 2001, *Space Sci. Rev.* 97, 263

Malandraki, O.E. 2002, Ph.D. thesis, Democritus University of Thrace, Xanthi, Greece

Malandraki, O.E., Sarris, E.T., Lanzerotti, L.J., Trochoutsos, P., Tsiropoula, G., Pick, M. 2002 *J. Atmos. Sol.-Ter. Phys.* 64/5-6, 517

Malandraki, O.E., Sarris, E.T., Trochoutsos, P., Tsiropoula, G., 2004 in: P.G. Laskarides (ed.), *Proceedings of the 6th Hellenic Astronomical Society Conference* (Athens, Greece) p. 51

Reames, D.V. 1999, *Space Sci. Rev.* 90, 413

Richardson, I.G. 1997 in: N. Crooker, J.A. Joselyn, and J. Feynman (eds.), *Coronal Mass Ejections, Geophys. Monogr. Ser.* (AGU, Washington D.C.), vol. 99, p. 189

Sarris, E.T. & Krimigis, S.M. 1982, .Geophys. Res. Lett. 9, 167

Skoug, R.M., Gosling, J.T., Steinberg, J.T., McComas, D.J., Smith, C.W., Ness, N.F., Hu, Q., Burlaga, L.F. 2004 *J. Geophys. Res.* 109, A09102, doi:10.1029/2004JA010494

Discussion

KAHLER: I am confused by seeing the interpretation of a plasmoid imbedded inside a unidirectional field, which you gave (I think) to your last event.

MALANDRAKI: We think it's the most plausible interpretation of the observations. The flux depression observed for 3 hrs signifies the electrons cannot enter this portion of the ICME, implying it is probably disconnected from the sun. Moreover, the observations of a dramatic change from unidirectional to bidirectional PADS throughout this ICME region is consistent with the electrons propagating within closed looped IMF lines in space.

KOUTCHMY: Regarding those beams of energetic electrons related to type III bursts of very short duration, what are their typical durations?

MALANDRAKI: By stating that we have observed type III burst activity associated with the electron beams detected by Ulysses/HI-SCALE, I meant a type III burst storm was observed by WIND/WAVES ie., an almost continuous succession of type III bursts for ∼ 1.5 day.

BOTHMER: The interpretation of suprathermal electrons and energetic particles is opposite previous interpretations where BDEs are signatures of closed fields throughout CME. Energetic articles may be indicative of connection to the Sun rather than of closed/open structure.

MALANDRAKI: In our work, we use Near-Relativistic electrons (38-315kev) measured by the Ulysses/HI-SCALE and ACE/EPAM experiments to trace the large-scale structure of Interplanetary Coronal Mass Ejections (ICMEs). Observations of bi-directional Pitch Angle Distributions (PADs) are consistent with a loop-like structure connected to the Sun at both ends. Furthermore, open field lines forming a magnetic constriction in space upstream from the S/C can reflect the outgoing electrons and produce the observed counterstreaming electrons. In some cases, from the time delay between the first outgoing and the first back-streaming electrons (provided by the measurement capacities of HI-SCALE and EPAM) it is possible to distinguish between the 2 topologies (see eg. Malandraki *et al.*, J. Atmos. Sol-Terr. Phys., 64/5-6, 517, 2002).

Coronal and Stellar Mass Ejections
Proceedings IAU Symposium No. 226, 2005
K. P. Dere, J. Wang & Y. Yan, eds.

© 2005 International Astronomical Union
doi:10.1017/S1743921305000876

CME Interaction and the Intensity of Solar Energetic Particle Events

N. Gopalswamy[1], S. Yashiro[1,2] S. Krucker[3], and R. A. Howard[4]

[1]Code 695.0, NASA Goddard Space Flight Center, Greenbelt, MD 20771, USA
email: gopals@fugee.gsfc.nasa.gov

[2]Department of Physics, The Catholic University of America, Washington, DC 20064, USA

[3]Space Sciences Laboratory, University of California at Berkeley, Berkeley, CA 94720, USA

[4]Solar Physics Branch, Naval Research Laboratory, Washington, DC, 20375, USA

Abstract.
Large Solar Energetic Particles (SEPs) are closely associated with coronal mass ejections (CMEs). The significant correlation observed between SEP intensity and CME speed has been considered as the evidence for such a close connection. The recent finding that SEP events with preceding wide CMEs are likely to have higher intensities compared to those without was attributed to the interaction of the CME-driven shocks with the preceding CMEs or with their aftermath. It is also possible that the intensity of SEPs may also be affected by the properties of the solar source region. In this study, we found that the active region area has no relation with the SEP intensity and CME speed, thus supporting the importance of CME interaction. However, there is a significant correlation between flare size and the active region area, which probably reflects the spatial scale of the flare phenomenon as compared to that of the CME-driven shock.

Keywords. shock waves, Sun: coronal mass ejections (CMEs), Sun: flares, Sun: X-rays, gamma rays, Sun: particle emission, solar-terrestrial relations

1. Introduction

Solar energetic particle (SEP) events with high intensity ($\geqslant$10 pfu in the > 10 MeV channel as measured by GOES) and duration exceeding a few hours are closely associated with coronal mass ejections (CMEs). This association was first pointed out by Kahler, Hildner, van Hollebeke, *et al.* (1978), leading to the idea that SEPs are accelerated by CME-driven shocks (Reames (1999)). One of the strongest evidences for shock acceleration is the observation of energetic storm particle (ESP) events (see e.g. Rao, McCracken, Bukata, (1967). ESPs are particles accelerated locally at the shock front and detected when the shock blows past the observing spacecraft. In the case of SEPs, the particles arrive at 1 AU in an hour or so, while the shock takes much longer ($\geqslant$ half a day). However, the CME that drives the shock can be detected when it is still near the Sun. The speed of the CMEs correlates reasonably well with the intensity of the associated SEP events (Kahler, Sheeley, Howard, *et al.* (1984), Kahler (2001), Gopalswamy, Yashiro, Lara, *et al.* (2003)), as one would expect if the particles are accelerated by the CME-driven shocks. Kahler, Sheeley, Howard, *et al.* (1984) showed that the SEP intensity is also well correlated with the apparent width of CMEs, suggesting that the kinetic energy of CMEs is an important factor deciding the intensity of SEP events. Since there is a reasonable correlation between CME speed and angular width, we consider just the correlation between SEP intensity and CME speed. One of the major problems with this correlation has been that the scatter is very large: for a given CME speed the SEP intensity can vary over 3 orders of magnitude. Finding an explanation for this large scatter is an important part of understanding the origin of SEPs. This is a complex problem because the particle acceleration depends on both the source (shock strength, free energy in the active region) and medium (presence of seed particles, presence of turbulence in the ambient medium, orientation of the ambient magnetic field with respect to shock normal, connectivity of the acceleration region to the observing spacecraft) properties. Kahler (2001) pointed out the presence of seed particles in the ambient medium and the spectral variation between SEP events can account for one to two orders of magnitude scatter in the CME speed vs. SEP intensity plots.

Recently, Gopalswamy, Yashiro, Krucker, *et al.* (2004) found that the presence of preceding CMEs may also affect the intensity of SEP events: while the CME properties such as speed, width, mass, kinetic energy and source longitude were similar, SEP events with preceding CMEs had higher intensity. Another possibility is to attribute the SEP intensity variation to the properties of the active region. For example, the high intensity events may be from active regions that have large free energy and hence erupt more frequently. When eruptions happen frequently, one expects CME interaction. In order to see this connection, we examine how the active region properties affect the SEP intensity.

2. Data and Analysis

The data set used for this study is the same as that of Gopalswamy, Yashiro, Krucker, *et al.* (2004) consisting of all the distinct large (intensity of protons $\geqslant 10$ pfu in the > 10 MeV channel) SEP events from 1997-2002 that had overlapping observations from the Solar and Heliospheric Observatory (SOHO) mission's Large Angle and Spectrometric Coronagraphs (LASCO). There were 57 such events. Some of these events were backsided, so we do not know their flare and source properties. We also excluded events that did not originate from active regions. We consider the remaining 41 events for this study. For each one of these events, the primary CME, the associated flare, and the associated active region are known. We use all the information compiled in Table 1 of Gopalswamy, Yashiro, Krucker, *et al.* (2004). In addition, we compiled the area of the associated active regions, which we analyze in this paper. We examine how the SEP intensity is affected by the active region area.

Following Gopalswamy, Yashiro, Krucker, *et al.* (2004), we divide the SEP events into three groups: those with preceding CMEs (P events), (ii) those with no preceding CMEs (NP events), and (iii) those with other types of interaction (O events). The preceding CMEs were required to be wide (width $\geqslant 60°$) and originate within a day from the same active region as the primary CME. In O events the primary CMEs either interacted with a streamer or with a preceding CME before it emerged above the occulting disk of the coronagraph. The P events were the largest in number (19), with roughly equal number of NP (10) and O (12) events. Figure 1 shows a P event that occurred on 2000 November 24 along with other eruptions from the same region (AR 9236). Flares and CMEs erupted in quick succession from this region over a three-day period and the SEP intensity remained high for several days (see the distributions of flare and CME recurrence times in Figure 1). The SEP event was associated with a fast halo CME (1245 km/s) from AR 9236 located at N22W07 at the time (15:30 UT) of the eruption. About 10 h earlier, another fast (994 km/s) halo CME had erupted from the same region. Clearly, what was ahead of the 15:30 UT CME was not the normal solar wind, but the aftermath of the preceding CME. We suggest that if a shock acceleration mechanism is at work, then what enters into the shock is not the normal solar wind but the disturbed plasma behind the first CME or the first CME itself.

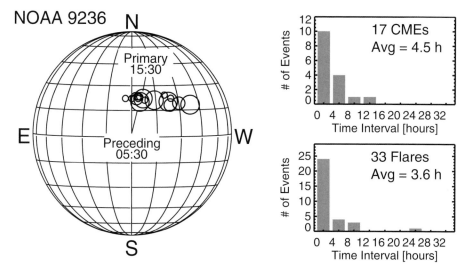

Figure 1. (left) Location of AR 9236 at the times of CME-associated flares during the period 2000 November 22-27, including one of the primary CMEs on November 24 (15:30 UT) with the preceding CME at 05:30 UT. The small, medium and large circles represent C, M, and X class flares, respectively. All the M and X class flares were associated with CMEs. (right) Histograms of CME and flare recurrence times from AR 9236. There were 33 soft x-ray flares (C, M, and X class) and 17 CMEs. The average flare and CME recurrence times are 3.6 h and 4.5 h, respectively.

2.1. *SEP Intensity and Active Region Area*

Figure 2(a-d) shows the distribution of the SEP intensities for all the events and for the three subgroups (P, NP, and O). We see that the P events have generally larger intensity compared to the NP events and the O events are have intermediate behavior. The P events are three times more likely to have higher intensity than the NP events, probably because of the presence of preceding CMEs that boost the acceleration efficiency of the primary shock via shock strengthening (Gopalswamy, Yashiro, Kaiser, *et al.* (2001)), and/or the presence of seed particles from the preceding eruption. Is it possible that the higher intensity has its origin in the active region itself? To check this, we have shown the distribution of active region area for all the SEP events as well as for the P, NP, and O events in Figure 2(e-f). The median value of the active region area is the largest for P events (880 millionths) and the smallest for NP events (500 millionths) and the O events had an intermediate value (620 millionths). The active regions of NP events have a slightly smaller area on the average. Does this mean higher intensity results from larger active regions?

The scatter plot between the SEP intensity and active region area in Figure 3(a) shows that there is little correlation between the two quantities (correlation coefficient r= 0.21). When we consider the individual subgroups, the correlation coefficient is worse for P (r= −0.02) and O (r= 0.04) events, while it seems better for the NP events (r= 0.46). It must be pointed out that there were only 10 NP events and the correlation is almost lost (r= 0.26) if we exclude the single outlier NP event. For a given active region area, the intensities of P and O events vary over three orders of magnitude. There is considerable overlap between NP and O events on the plot, but the P events clearly have higher intensity. Thus the area of the active region does not seem to significantly order the SEP intensity.

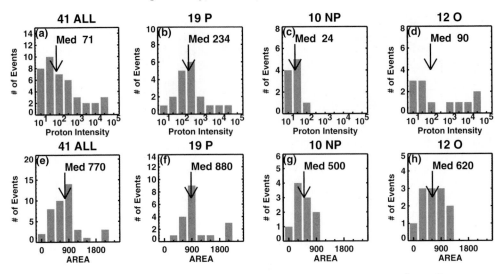

Figure 2. (a-d) Distributions of SEP intensities (pfu; 1 pfu = 1 proton $(cm^2 ssr)^{-1}$) for all the events and for the P, NP, and O events. (e-h) Distribution of active region areas (in millionths of solar hemisphere) for all events and for P, NP, and O events. The median value of the distributions is shown on the histograms.

2.2. *CME speed, Flare Size and Active Region Area*

Figure 3(b) is a scatter plot between the CME speed and the active region area. For the combined data set, the two quantities seem to be poorly correlated (r= 0.16). For the sub groups, the correlation is equally poor (r= 0.22 for P events, 0.15 for NP events and 0.01 for O events). The NP events have similar speed range as the other two, although the P events clearly occupy the higher-area side of the plot. The X-ray flare size, however, is better correlated with the active region area (r= 0.60), as can be seen in Figure 3(c). For the P and O events, the correlation is significantly smaller (r= 0.36 and 0.33, respectively). The apparent high correlation for the NP events (r= 0.74) is completely destroyed (r= 0.08) when the single outlier is dropped. Nevertheless, the flare size seems to be closely tied to the active region than the CME speed is.

3. Discussion and Conclusions

In this work, we have shown that the intensity of SEP events has no specific dependence on the area of the active region from which the associated CMEs and flares originate. The CME speed also has no significant correlation with the active region area, thus confirming the previous result that the presence of preceding CMEs makes a shock more efficient in accelerating SEPs (Gopalswamy, Yashiro, Michalek, *et al.* (2002), Gopalswamy, Yashiro, Lara, *et al.* (2003), Gopalswamy, Yashiro, Krucker, *et al.* (2004)).

We now discuss the relevance of the significant correlation that the soft X-ray flare size has with the active region area (see fig. 3c). The size of a soft X-ray flare implies intense soft X-ray emission from the flare plasma. The flare plasma is supposed to be the hot post-eruption loops containing plasma evaporated from the chromosphere (see e.g., Antonucci, Dennis, Gabriel, *et al.* (1985)). The evaporation itself is thought to be caused by electron beams precipitating into, and losing energy to, the chromosphere. Higher flare size therefore implies a higher density of heated flare plasma over a larger volume (i.e. higher emission measure). The flaring loops are generally confined to the active region, so one can understand the good correlation between flare size and active region area.

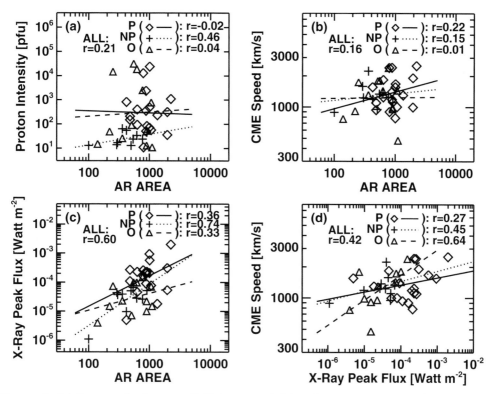

Figure 3. (a) Scatter plot between SEP intensity (pfu) and active region area (millionths of solar hemisphere). P, NP and O events are represented by diamond, triangle, and + symbols, respectively. The overall correlation coefficient, r= 0.21. The correlations coefficients for the P, NP and O events are also shown on the plots. When the single outlier NP event is excluded, the overall correlation coefficient drops to 0.15 and that for the NP events drop to 0.26. (b) Scatter plot between the CME speed and the active region area (r= 0.16). The subgroups also have similar correlation coefficients. (c) Scatter plot between X-ray flare size and active region area (r= 0.60). When the NP outlier is dropped, the overall and NP correlation coefficients drop to 0.51 and 0.08, respectively. (d) Scatter plot between the CME speed and the X-ray flare size (r= 0.42). When the NP outlier is dropped, the overall and NP correlation coefficients drop to 0.38 and 0.05, respectively.

This result may also be relevant to the previous result that the intensity of energetic (108 keV) electrons are well correlated with flare size (Gopalswamy, Yashiro, Krucker, *et al.* (2004)). If these electrons propagating away from the sun are accelerated by the same process as that of the precipitating electrons, one might expect a good correlation with the flare size.

If the SEPs are accelerated by the CME-driven shock, the SEP intensity need not have a specific relationship with the active region area. Even though the CME is rooted in the active region, the three-dimensional shock front ahead of the CME when it is within a few solar radii from the Sun is much larger than the active region area. This might explain why there is no relationship between the SEP intensity and active region area. However, a weak but significant correlation exists between CME speed and Flare size (fig. 3(d); see also Gopalswamy, Yashiro, Lara, *et al.* (2003)). This can be explained as more energetic eruptions resulting in bigger flares. It must be pointed out that the area of the active region may not be a good proxy for the free energy in an active region. The free energy is determined by the currents flowing in the corona representing the deviation from the potential field configuration. Estimates show that the free energy is of the order of the potential field energy (see e.g., Forbes (2000). One has to estimate at least the potential field energy of the active regions for a better comparison.

Acknowledgements

This research was also supported by NASA/LWS and NSF/SHINE (ATM 0204588) programs. SOHO is a project of international cooperation between ESA and NASA.

References

Antonucci, E., Dennis, B. R., Gabriel, A. H., Simnett, G. M. 1985, *Solar Phys.* 96, 129
Forbes, T.G. 1991, *J. Geophys. Res.* 105, 23153
Gopalswamy, N., Yashiro, S., Kaiser, M.L., Howard, R.A., Bougeret, J.-L. 2001, *ApJ* Letters 548, L91
Gopalswamy, N., Yashiro, S., Michalek, G. Kaiser, M.L., Reames, D.V. Leske, R. von Rosenvinge, T. 2002, *ApJ* Lett. 572, L103
Gopalswamy, N., Yashiro, S., Lara, A., Kaiser, M.L., Thompson, B.J., Gallagher, P.T., Howard, R.A. 2003, *Geophys. Res. Lett.* 30, SEP 3-1
Gopalswamy, N., Yashiro, S., Krucker, S., Stenborg, G., & Howard, R.A. 2004 *J. Geophys. Res.* in press
Kahler, S.W., Hildner, E. & van Hollebeke, M.A.I. 1978, *Solar Phys.* 57, 429
Kahler, S.W., Sheeley, N.R. Jr., Howard, R.A., Koomen, M.J., Michels, D.J., McGuire, R.E., von Rosenvinge, T.T., & Reames, D.V. 1984, *J. Geophys. Res.* 89, 9683
Kahler, S.W. 2001, *J. Geophys. Res.* 106, 20947
Rao, U. R., McCracken, K. G., Bukata, R. P. 1967, *J. Geophys. Res.* 72, 4325
Reames, D.V. 1999, *Space Sci. Rev.* 90, 413

Discussion

YOUSEF: You have showed that the time between the ejection of successive CME, is 4.30h, is that enough for building up energy? Between the successive magnetic reconnection leading to the lift off of CMEs?

GOPALSWAMY: It is likely that the active region does not release all the free energy in a single eruption. It should take several eruptions before all the built up energy is released.

P. F. CHEN: In the P-type events, did you always or often observe two type II radio burst sources?

GOPALSWAMY: Some of the preceding CMEs of the primary CMEs were associated with type II bursts. But the primary CME itself had only one type II burst.

JIE ZHANG: 'O' events also have high SEP intensity vs 'P' events. Can you explain the enhancement with the same mechanism for both 'O' events and 'P' events? As I understand, 'P' events may have seed particles. But there are no seed particles for 'O' events, since streamer is just sitting there.

GOPALSWAMY: 'O' events have two groups: (i) the preceding CME below LASCO occulting disk; (ii) streamer interaction. (i) is similar to 'P' events, except the preceding CME is not observed. The high intensity may result not only because of seed particles, but also due to other mechanisms, such as shock strengthening. When the shock of the primary CME passes through the streamer, the shock becomes stronger and accelerates more particles. This is again a speculation, but consistent with 'P' events having higher intensity.

GANG LI: It is conceivable that 'P' event can have enhanced turbulence preceding the primary CMEs. Thus the spectrum observed at 1 AU for 'P' events should differ from 'NP' events where the turbulence must be generated by the streaming protons for 'NP' events.

GOPALSWAMY: I agree with your comment. This can easily be checked once the spectral type of all the events are obtained.

KOUTCHMY: I would object against your statement "Streamers are zero-velocity CMEs" because streamers show obvious flows from movies and models of streamers confirm that there are flows inside. Both large outflow and weaken outflow are present, like in the case of CMEs.

GOPALSWAMY: You have misunderstood my statement. I say "CME with zero speed" only to indicate that the streamer sits there prior to the primary CME and is blown off. Often streamers observed at limb show the familiar three-part structure (prominence, cavity, overlying loops). I do not mean the wind-like outflows often observed in streamers.

RILEY: Statistically do you have an idea how many CMEs observed at earth are "P events", i.e., interactive CMEs, and how does the fraction vary as a function of solar cycle?

GOPALSWAMY: I have not done the statistics. But I know one of the 'P' events (2000/11/24) resulted in a complex ejecta extending over a long period (Burlaga *et al.* 2001). Also, the poster by Xie *et al.* presented during this meeting deals with extended ejecta at 1 AU.

Coronal and Stellar Mass Ejections
Proceedings IAU Symposium No. 226, 2005
K. P. Dere, J. Wang & Y. Yan, eds.

Prompt Solar Energetic Particles with Large-Scale Cross-Disk Coronal Disturbance

Y. Dai, Y.H. Tang and K.P. Qiu

Department of Astronomy, Nanjing University, Nanjing 210093, China
email: ydai@nju.edu.cn

Abstract. By using the observations of the Extreme-UV Imaging Telescope (EIT), we studied three major solar eruptions from the same super active region NOAA 10486 during the interval of October 26 to November 04, 2003. The three eruptions took place when the active region located on the eastern hemisphere, near the central meridian and on the western limb, respectively. In the first event (Oct 26 event), the coronal disturbance (indicated as an EIT wave) in EUV images was limited east to the central meridian, and there were no solar energetic particles (SEPs) detected. The second event (Oct 28 event) accompanied a nearly entire disk disturbance and very large and prompt SEP enhancements. For the last event (Nov 04 event), there was no obvious coronal disturbance on the disk, and the SEP enhancements were much more gradual. From these observational features, we suggest that different coronal disturbances correspond to different acceleration and propagation histories. A large-scale, cross-disk coronal disturbance may open quite a lot of magnetic field lines in the low corona, facilitating the direct access of flare accelerated SEPs to the Sun-Earth connected interplanetary magnetic field lines. Subsequently the SEP intensity will exhibit a very prompt enhancement.

Keywords. Sun: corona, coronal mass ejections (CMEs), magnetic fields, particle emission

1. Introduction

It is generally accepted that solar energetic particles (SEPs) can be produced on different timescales and at different site. They can be rapidly accelerated during a flare, or by coronal mass ejection-driven shocks traveling in the interplanetary space in several days (Lee & Fisk 1982; Jones & Ellison 1991; Reames, Kahler & Ng 1997). So two distinct classes of SEP events, impulsive events and gradual ones, have been defined (Reames 1999; Tylka 2001). The former ones have the origin in flares, while the latter ones are associated with CMEs.

However, there are more and more observations that impulsive and gradual component can both exist in one single SEP event. Kahler, Reames & Sheeley (2001) reported a clear connection between a narrow CME and an impulsive event. They attributed the narrow CME to flare ejecta rather than the traditionally considered CME. Mason, Mazur & Dwyer (1999) measured ^{3}He abundance from $\sim$0.5 to 2 MeV nucleon^{-1} in 12 large SEP events and found in some of these events the ^{3}He/^{4}He ratios are substantially larger than the solar wind value. In studying the very large 2000 July 14 event, Tylka *et al.* (2001) found the energy spectra of Fe are strikingly different from those of lighter species but can be well modeled with a small ($\sim$5%) admixture of flare component. These authors suggested that remnant suprathermal ions which result from previous impulsive events can be a source population available for further acceleration by interplanetary CME-driven shocks that accompany large SEP events, leading to some characteristics of impulsive events.

Recently, there are arguments that mixed acceleration processes can exist in one single flare/CME event, and the impulsive characteristic SEPs may originate from the current flare, not the suprathermal ions from the previous events. This leads to the concept of hybrid SEP events (Clive 1996). In such cases, more attention should be paid to the early phase of the solar eruptions, because the disturbances in the corona would link different SEP populations to form a very complex pattern. Torsti *et al.*(1999) found the fast enhancement of $\geqslant 10$ MeV protons in association with an EIT wave. In their opinion, the EIT wave was the signature of a quasi-perpendicular shock that gave rise to the $\geqslant 10$ MeV protons, and the transform from the quasi-perpendicular shock acceleration to the quasi-parallel shock acceleration explained the spectral softening at the maximum intensity time. However, when comparing to the perfect double power law spectrum at the maximum time, we find at the rapid rise phase, the spectrum has a considerable deviation from a power law pattern. This clearly indicates the existence of multiple acceleration processes during the coronal disturbance, and shock should not be the only acceleration agent.

In October to November 2003, the Sun became wild again. Three super active regions appeared and a lot of major solar eruptions took place from these active regions. In this paper, we choose three eruptions with distinctly different SEP behaviors to see the relationship between coronal disturbance and SEP acceleration process. In §2 we describe the observations. Discussion and summary are given in §3.

2. Observation

In this study, we use the observations of the Extreme-UV Imaging Telescope (EIT; Delaboudinière *et al.* 1995) on board the *SOHO* spacecraft to track the EUV evolutions of the solar eruptions. SEP data are obtained from *GOES* which gives the intensities of energetic protons in several energy channels.

The three solar eruptions originated from the same active region NOAA 10486 south to the equator. They occur on 2003 October 26, October 28 and November 04, respectively. The corresponding locations of the active region are on the eastern hemisphere, near the central meridian and on the western limb. All of these three solar eruptions accompany X class flares and fast CMEs. But their associated SEP behaviors and coronal disturbances are quite different.

On 2003 October 26, the position of the active region was roughly S15E44. At 05:57 UT, a flare from this active region started, and it peaked at 06:54 UT, attaining the class of X1.2. A fast CME associated with the flare first appeared in LASCO C2 at 06:54 UT, nearly the same time as the flare maximum. Linear fit for the CME's trajectory reveals a speed of 1375 km/s. The EUV evolution of the event is shown in the right panel of Figure 1. This eruption induced an EIT wave. The EIT wave traveled northwestward from the active region. But before it faded out, the wave was still limited on the eastern hemisphere. Although this flare/CME event was quite large, no SEPs associated with this event were detected. The rapid rise of SEPs shown in the left panel of Figure 1 was due to another eruption from another active region.

The second event occurred on 2003 October 28 when the active region rotated to the position of S19E15. LASCO observed a full halo CME . The CME was first observed in C2 at 10:54 UT as a bright loop front over the W limb; by 11:30 UT the front had developed into a full halo CME, very bright all around the occulting disk. The CME was associated with an X17.2 flare. GOES records this flare from AR 0486 between 09:51 - 11:24 UT with peak emission at 11:10 UT. An extremely large EUV disturbance was observed (shown in Figure 2): the large-scale cross-disk coronal brightening developed nearly simultaneously followed by a full-disk dimming. This event caused very strong geoeffects. In Figure 2, SEPs in all channels exhibited prompt enhancements during the

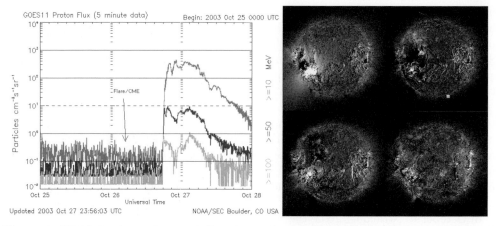

Figure 1. SEP behavior and coronal disturbance for the 2003 October 26 event. Left panel: GOES proton Flux for three energy channels. Right panel: running difference images of the EIT at 195 Å.

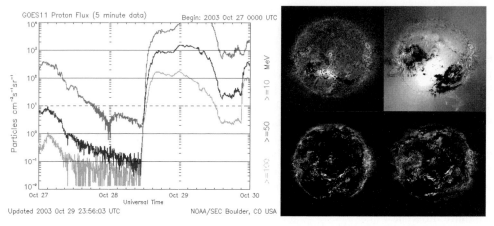

Figure 2. SEP behavior and coronal disturbance for the 2003 October 28 event.

solar eruption. After the rapid enhancement phase, the SEPs began to rise more gradually when the CMEs propagated into the interplanetary space.

The 2003 November 04 event was a limb event. Qiu *et al.* (in this proceeding) have studied the low corona signature of the event and found the disappearance of two loops which initially extended out from the active region. This interesting findings in the EIT fixed difference images (in Qiu *et al.*) are two dimmings corresponding to the footpoints of the two loops after the eruption. And in the running difference images shown in Figure 3, it seems that there was no obvious coronal disturbance on the disk. The X28 flare between 19:29 - 20:06 UT holds the record of the largest flare. The associated CME also has an unusually high speed of about 2300 km/s. But in comparison with the October 28 event, this event was in association with more gradual enhancements of the SEPs. Furthermore the peak flux of this SEP event was much smaller than that of the former one.

3. Discussion and Summary

The distinctly different SEP behaviors mentioned above imply different acceleration histories in different solar eruptions. Shock acceleration is often proposed to account for

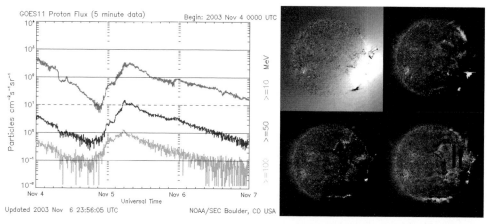

Figure 3. SEP behavior and coronal disturbance for the 2003 November 04 event.

large SEP events and power-law spectra are expected theoretically for shock acceleration below ~ 30 MeV amu^{-1} (Ellison & Ramaty 1985). In higher energies, the power-law spectra will be multiplied by exponential rollovers. This means efficiency for acceleration of high-energy ions (> 30 MeV amu^{-1}) by shocks will decrease more rapidly than those at low energies. Recently, Tsurutani *et al.* (2003) numerically studied the formation and propagation of a CME-induced shock. They found the shock formation will be sufficiently high in altitude and late in time. Even the CMEs in this paper may be fast enough to drive shocks in the low corona (Tang & Dai 2003), the low Mach numbers of the induced shocks will strongly affect the acceleration efficiency in high energies. So the prompt enhancements of the SEPs, especially for the high energy SEPs in the 2003 October 28 event, can not be mainly due to shock acceleration. Reconnection acceleration which can easily energize particles to several hundreds MeV amu^{-1} in very short timescale, must be included into the consideration.

The reconnection accelerated particles are produced quite low in the corona. So open magnetic field lines are necessary from the acceleration site to the root of interplanetary field for these SEPs to be observed quickly. In the 2003 October event, the large scale brightening and dimming in EIT images indicate the ejection of large amount of material from the low corona. At the same time, sufficient magnetic field lines are opened. This process changes the globe magnetic structure, facilitating the direct access of flare accelerated SEPs to the Sun-Earth connected interplanetary magnetic field lines.

For the other two events, the coronal disturbances are not large enough to ensure the direct access of flare accelerated SEPs. Therefore we can't see the prompt pattern of the SEPs. The 2003 November event accompanied more moderate SEP enhancements. Shock acceleration on a long timescale should be a reasonable explanation. Finally, why the 2003 October 26 event didn't produce any SEP? We notice the CMEs of 2003 October 28 and November 04 are both full halo CMEs, and the 2003 October 26 CME is the exception, whose width is 207 °. Although very board in comparison with typical CMEs, its width may be still not sufficient to produce SEPs. This is consistent with the result of Kahler & Reames (2003) that broad widths are necessary for CMEs to be associated with SEP events.

In summary, we suggest that in a large SEP event, multiple acceleration processes accelerate particles in different phases, and large-scale cross-disk coronal disturbances will link the different SEP populations to form a hybrid pattern. And in the interplanetary space, the width of CME is also an important factor in producing SEPs.

Acknowledgements

This work is supported by NKBRSF of China G2000078404 and by NNSF key project of China No.10333040, as well as NNSF under project of China No.10073005. SOHO is a project of international cooperation between ESA and NASA.

References

Cliver, E.W. 1996, *AIP Conf. Proc.* 374, 45
Delaboudinière, J.-P. *et al.* 1995, *Sol. Phys.* 162, 291
Ellison, D.C. & Ramaty, R. 1985, *Astrophys. J.* 298, 400
Kahler, S. W. & Reames, D.V. 2003, *Astrophys. J.* 584, 1063
Kahler, S. W., Reames, D. V. & Sheeley, N. R., Jr. 2001, *Astrophys. J.* 562, 558
Jones, F.C. & Ellison, D.C. 1991, *Space Sci. Rev.* 59, 259
Lee, M.A. & Fisk, L.A. 1982, *Space Sci. Rev.* 32, 205
Mason, G. M., Mazur, J. E. & Dwyer, J. R. 1999, *Astrophys. J.* 525, L133
Reames, D.V. 1999, *Space Sci. Rev.* 90, 413
Reames, D.V., Kahler, S.W. & Ng, C.K. 1997, *Astrophys. J.* 491, 414
Tang, Y.H. & Dai, Y. 2003, *Adv. Space. Res.* 32, 2609
Torsti, J., Kocharov, L.G., Teittinen, M. & Thompson, J. 1999, *Astrophys. J.* 510, 460
Tsurutani, B., Wu, S.T., Zhang, T.X. & Dryer, M. 2003, *Astron. Astrophys.* 412, 293
Tylka, A.J. 2001, *J. Geophys. Res.* 106, 25333
Tylka, A. J., Cohen, M. S., Dietrich, W. F., Maclennan, C. G., McGuire, R. E., Ng, C. K. & Reames, D. V. 2001, *Astrophys. J.* 558, L59

Discussion

BOTHMER: Paper by Bothmer *et al.* 1998 for April 7, 1997 event shows that particles arrived before EIT wave had reached the W-hemisphere. Send me e-mail so I can send you the paper.

DAI: You can e-mail to yhtang@nju.edu.cn. I will read the paper carefully.

RUFFOLO: You have interpreted your data in terms of an EIT wave. Another interpretation would be in terms of magnetic connection. The solar wind speed was unusually fast, so the best connected longitude was near central meridian. This could explain why the Oct. 28 event had the fastest onset.

DAI: When the eruption takes place, the speed of the solar wind was not very high. I think the solar wind B accelerated by the CME. So we think the eruption site was still away from the well-connected regions.

GRECHNEV: One day before yesterday I warned against artifacts, both instrumental and methodical, in handling EIT images. The Bastille Day event and October 28, 2003 event are similar in that they both show large areas subjected to the scattered light. October 26, 2003 was shown by V. Slemzin on Monday, and the coronal wave in that event was in the NE section. So please be consider EIT images carefully, otherwise you can obtain incorrect results.

DAI: The results of the Oct. 26 event we showed are consistent with that of Dr. Slemzin. And in the Oct. 28 event we just want to show a large-scale coronal disturbance. And we will reexamine the EIT images carefully.

Coronal and Stellar Mass Ejections
Proceedings IAU Symposium No. 226, 2005
K. P. Dere, J. Wang & Y. Yan, eds.

Coronal Mass Ejections and the Largest Solar Energetic Particle Events

Ruiguang Wang[1,2] and Jingxiu Wang[1]

[1]National Astronomical Observatories, Chinese Academy of Sciences, Beijing 100012, China

[2]Institute of Science and Technology for Opto-Electron Information, Yantai University, Yantai 264005, China

Abstract. We studied the association between SEP events during 1977-2003 and related CMEs and found each GLE event was associated with a primary CME, which was faster (average speed $\sim$1762 $km \cdot s^{-1}$) and wider (average angle width of 317^0) than an average CME . All SEP-related CMEs distributed within solar source regions of latitude strip of $S30^0$-$N40^0$, while 11 (85%)GLE-related CMEs originated from the western hemisphere. These fast halo CMEs (75% full-halo and 25% partial-halo) were associated with type II radio bursts in the decameter hectometer (DH) wavelengths.

Keywords. Sun: coronal mass ejections (CMEs), particle emission, radio radiation

1. Observation

Large solar energetic particles (SEPs) are thought to be accelerated by CMEs-driven shocks (see, e.g., Reames, 1999; Kahler, 2001). Nevertheless, It is still not known what makes a CME an SEP accelerator, especially for those largest SEPs or GLEs. In this paper we study the CMEs associated with SEP events which are divided three classes, the 13 GLE, 30 moderate SEP(10-100 pfu,E>10 MeV) and 62 minor SEP (1-10 pfu, E>10 MeV) during 1997-2003. Using Data are from SOHO/LASCO, SGD, EIT, SXT and GOES, we identified related CMEs and flares and collected their measured properties. CMEs and flares correlated with 13 GLEs were listed in table 1. The columns from left to right represent GLE date, peak flux time of x-ray flare, optical flare class, peak flux of SEPs with energies above 10 MeV, onset time, angular width (AW) and velocity of CME, heliocentric coordinates of solar source, NOAA active region number. Whether associated or not a decameter hectometer (DH) type II burst ("y" for yes and "n" for no). Solar surface source region distribution was plotted in figure 1 and the speed distribution of three classes of CMEs was plotted in figure 2.

2. Results

Main results of this study are: (1) Each GLE event corresponds a fast halo CME, including 9 (75%) full-halo CMEs and 3 (25%) partial-halo CMEs. (2) 9 GLEs (69%) originated from the southern hemisphere and 11 GLEs (85%) originated from the western hemisphere. Latitude of the solar source region is within a strip of $S30^0$-$N40^0$. Longitude of GLE-CMEs is west of $E10^0$ with the most probable longitude of between $W60^0$ and $W70^0$. (3) The CME average speeds are $\sim$ 1762 km $\cdot s^{-1}$, $\sim$ 1077 km $\cdot s^{-1}$ and $\sim$ 887 km $\cdot s^{-1}$ respectively for the GLE-CMEs, the moderate SEP-CMEs and the minor SEP-CMEs. There are 11 (92%) GLE-CMEs whose speed exceeding 1000 km $\cdot s^{-1}$. (4) Of the 13 GLE related flares, 11 (85%) were class x level. SEP fluxes exceeded several hundreds

Table 1. Properties of CMEs and flares correlated with 13 GLEs

GLE date	X time (UT)	Bright	flux (pfu*)	Time (UT)	AW (deg.)	V (km·s⁻¹)	Location	AR	II
1997.11.06	11:55	X9/2B	490	12:10	h	1556	S18W63	8100	y
1998.05.02	13:42	X1/3B	150	14:06	h	938	S15W15	8210	y
1998.05.06	8:9	X2/1N	210	8:29	190	1099	S11W65	8210	y
1998.08.24	22:12	X1/3B	670	-	-	-	N35E09	8307	y
2000.07.14	10:24	X5.7/3B	24000	10:54	h	1647	N22W07	9077	y
2001.04.15	13:50	X14/2B	951	14:06	167	1199	S20W85	9415	y
2001.04.18	2:14	C2/2B	321	2:30	h	2465	S20Wlimb	9415	y
2001.11.04	16:20	X1/3B	31700	16:35	h	1810	N06W18	9684	y
2001.12.26	5:40	M7/1B	779	5:30	212	1446	N08W54	9742	y
2002.08.24	1:12	X3/1F	317	1:27	h	1878	S02W81	10069	y
2003.10.28	11:10	X17/4B	29500	11:30	h	2459	S16E08	10486	y
2003.10.29	20:49	X10/2B	3300	20:54	h	2029	S15W02	10486	y
2003.11.02	17:25	X8.3/2B	1570	17:30	h	2598	S14W56	1o486	y

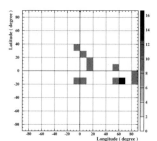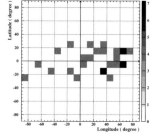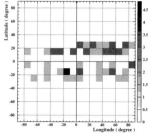

Figure 1. Distribution of heliocentric coordinates of solar surface source region of CMEs. left panel for GLE-CMEs, middle panel for moderate SEP-CMEs and right panel for minor SEP-CMEs. The numbers on the right color bar indicate appearance probability. Black represents the most probable region.

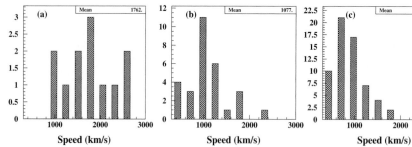

Figure 2. Distribution of CME speed. left panel for GLE-CMEs, middle panel for moderate SEP-CMEs and right panel for minor SEP-CMEs.

pfu and the highest intensity was 31,700 pfu. (5) All GLE-CMEs were associated with DH type II bursts.

Acknowledgements

The work is supported by the National Natural Science Foundation of China (10233050) and the National Key Basic Science Foundation (TG2000078404).

References

Reams, D.V. 1999, *Space Sci. Rev.*, 90, 413.
Kahler, S.W. 2001, *J. Geophys. Res.*, 106, 20947.

Coronal and Stellar Mass Ejections
Proceedings IAU Symposium No. 226, 2005
K. P. Dere, J. Wang & Y. Yan, eds.

Solar Relativistic Proton Fluxes in the Solar Flare of 14 July 2000

Ruiguang Wang[1,2] and Jingxiu Wang[1]

[1]National Astronomical Observatories, Chinese Academy of Sciences, Beijing 100012, China

[2]Institute of Science and Technology for Opto-Electron Information, Yantai University, Yantai 264005, China

Abstract. We studied the solar proton differential energy spectra with energy range of 1∼500 MeV at several time intervals during the 2000 July 14 solar flare. The results showed that before flare the spectra could be described by a power law function and after flare the power law spectra still existed above 30 MeV although spectra became softer with time. There was a spectral "knee" occurring at ∼30 MeV. We constructed a solar proton differential spectrum from 30 MeV to 3 GeV at peak flux time 10:30 UT and fitted it in the same manner. On the basis of a supposition of having the same power law spectrum in higher energy, we calculated the solar proton integrated fluxes in energy range of from 500 MeV to 20 GeV and compared them with other results obtained from experimental, modeling and theoretical calculations in other big historic SEP events.

Keywords. Sun: coronal mass ejections (CMEs), particle emission

1. Introduction

Bastille Day event is one of the most important solar energetic events in this solar cycle. A X5.7/3B flare at position N22W07 in active region NOAA 9077 was well observed with ground- and space-based instruments. X-ray flare lasted from 10:03 to 10:43 UT with a peak at 10:24 UT. SOHO/LASCO observed a full halo earth directed CME with great speed at 10:54 UT and GOES-8 observed a rapid increase of energetic proton fluxes later. On ground, more than 20 neutron monitors (NMs) observed cosmic ray intensity increases ranging from 2% to 60%. The earliest onset time of NMs increases was 10:30 UT and the highest rigidity was 6.7 GV. Modeling results from Vashenyuk *et al.* (2003) and Duldig *et al.* (2003) showed that the spectrum was soft with a power law index of between -5 and -7 during the rising phase for solar proton beam approaching the Earth. From the pitch angle distribution, it is seen that the particle arrival was anisotropic and also changed with time.

2. Energy spectra

Using GOES8 data, we constructed solar proton energy spectra at Earth in figure 1 for selected time intervals during July 14-16. The high energy range over 30 MeV of the spectra were fitted by a power law function $dJ_E/dE = A \times E^{-\delta}$. Spectral indices are showed in figure 2. Involving the flux calculated by Belov *et al.* (2001) from NM observations, the spectrum at peak time 10:30 UT was extended to 3 GeV and fitted by power law function in energy range from 200 MeV to 3 GeV, as shown in figure 3. Solar integrated spectra were plotted in figure 4, including spectra of other big historic GLE

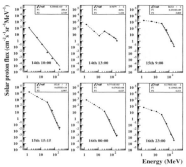

Figure 1. Solar proton energy spectra at Earth for selected time intervals during the GLE of 14 July 2000.

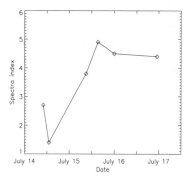

Figure 2. Spectral indices at selected time intervals during the GLE of 14 July 2000

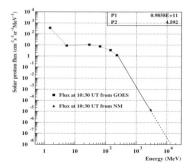

Figure 3. Solar proton energy spectrum at peak time 10:30 UT

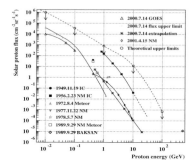

Figure 4. Integrated fluxes of some historic big GLE events.

events which were from theoretical and experimental calculations (see Miroshnichenko, 2001).

3. Results

For all time intervals shown in figure 1, a good fit could be obtained by using a power law in kinetic energy from ∼30 MeV to ∼200 MeV, which was consistent with the diffusive coronal shock acceleration mechanism. The spectral index of ∼-2.7 before flare suggested particle behavior of galactic cosmic ray during solar quite. It is evident that the spectrum varies with time in both amplitude and shape and there is a spectral "knee" occurring at ∼30 MeV. Since relativistic protons peaked in intensity earlier than the lower energy protons, the spectrum was hard at the rising phase and became soft in the declining phase. From figure 4 it is seen that our estimated fluxes were comparable to that of other big GLE events.

Acknowledgements

The work is supported by the National Natural Science Foundation of China (10233050) and the National Key Basic Science Foundation (TG2000078404).

References

Belov, A.V. et al. 2001, *Proc. 27th ICRC*, Hamburg, vol. 8, 3446.

Duldig, M.L., Bombardieri, D.J. and Humble, J.E. 2003, *Proc. 28th ICRC*, SH1.4, 3389.

Miroshnichenko, L.I. 2001, *Solar cosmic rays*, Kluwer Academic Publishers

Vashenyuk, E.V. et al. *Proc. 28th ICRC*, Tsukuba, SH1.4, 3401.

Coronal and Stellar Mass Ejections
Proceedings IAU Symposium No. 226, 2005
K. P. Dere, J. Wang & Y. Yan, eds.

On the Possibility of Acceleration of Heavy Solar and Cosmic Nuclei To GeV, TeV Energies and Beyond

Shahinaz M. Yousef

Astronomy & Meteorology Dept, Faculty of Science, Cairo University- Cairo Egypt
email: shahinazyousef@yahoo.com

Abstract. On acceleration of relativistic electrons by any means, the coherent pinch effect brings the relativistic electrons together. The inclusion of a suitable number of positive ions, preferentially heavy ions would stabilize the beam. The heavy trapped nuclei would have to go with the electrons, thus acquire relativistic speeds. The electron beam may be divided into small bunches. This mechanism can accelerate positive nuclei to GeV and TeV energies during solar flares, stellar and galactic high energy events. During the acceleration of 1000 GeV electron bunches, the energy gained by trapped positive nuclei is estimated to reach 10^{15} eV per nucleon. Higher energies can be achieved by head on collision as the energy gained by the target nucleus is proportional to γ^2, however fission reactions and very high energy gamma rays may occur. The principle of this mechanism is applied in electron-ring accelerators.

Keywords. Sun: particle emission

1. Introduction

Energetic particles from impulsive flares show that elements with $Z > 8$ are strongly enhanced relative to coronal abundance. Elements up to Si are fully ionized and Fe has charge 20. There is a strong evidence of electron beams in these events (Reames 1996). The observed number of relativistic heavy solar and cosmic ray nuclei (Schatzman 1967) can only be explained as due to an acceleration mechanism which:- 1) Preferentially accelerates heavy nuclei. 2) Simultaneously accelerates electrons and nuclei. 3) Accelerates all particles to the same velocity.

2. Coherent acceleration mechanism

At relativistic velocities, coherent effect occurs in intense electron beams in which the electrons themselves generate appreciable magnetic field (the pinch effect). The magnetic attraction force F_m is related to the repulsive electric force F_e by the relation, Kerst (1966)

$$F_m = \left(\frac{V^2}{c}\right) F_e \tag{2.1}$$

At velocities $V \sim c$, a pure beam of electrons would diverge slowly but could be rendered stable by the seeding of a suitable number of positive charges. These positive ions would be trapped electrostatically in the beam's very deep potential well and would necessarily acquire the same velocity of the electrons on acceleration. The ions are actually accelerated by the internal coherent electric field of the electrons (Veksler *et al.* 1967).

For a perfect coherent mechanism, the radius of the beam L should be $L \leqslant \lambda$ where λ is the plasma wavelength. For the bunch to be coherent in all directions, the height of the cylinder should equal $2L$.

If V is the velocity of the bunch and ω_0 is the plasma frequency, then

$$\lambda = \frac{V}{\omega_0}, \qquad where \quad \omega_0^2 = \frac{4\pi e^2 N_e}{m} \tag{2.2}$$

m and e are the mass and charge of the electron and N_e is the electron density. Assuming $V = 0.9c$, $N_e = 10^{12} \ cm^{-3}$ then $\lambda = 0.48 \ cm$ and if $N_e = 10^{11} \ cm^{-3}$ then $\lambda = 1.5 \ cm$.

3. Head on collisions

Consider the collision of a fast relativistic particle of mass M_1 with an immobile particle of mass M_2, where $M_2 \ll M_1$. If $M_1 \gg M_2\gamma$ then as a result of such head on collision, the immobile particle will get the following energy

$$W = M_2 c^2 \gamma^2 \ Where \ \gamma = \left[1 - \left[\frac{V}{c}\right]^2\right]^{\frac{-1}{2}} \tag{3.1}$$

Veksler (1956) assumed the primary relativistic particle of Mass M_1 to consist of n_1 electrons and the immobile mass M_2 to consist of n_2 particles of mass m_2, then if the condition

$$n_1 m_1 \gg n_2 m_2 \gamma \tag{3.2}$$

is fulfilled, then every ion of mass m_2 would obtain energy

$$W = m_2 c^2 \gamma^2 \tag{3.3}$$

This mechanism can give very high energies to cosmic ray particles as the energy gained by each particle is proportional to γ^2 (γ^2 increases from 1.33 for $V/c = 0.5$ to 500.5 for $V/c = 0.999$). This mechanism is capable of producing energies in the TeV and ultra relativistic range, however, the possibility of fission of heavy nuclei may exists. A target proton of rest mass 938 MeV, on moving with almost the velocity of light after the head on collision can acquire energies as large as 468.995 GeV. An iron nucleus moving with the same speed would acquire more than 26 TeV.

4. Conclusions

The steps of the proposed mechanism for the preferential acceleration of heavy solar and cosmic nuclei are:

1. The formation and acceleration of high density electron beams of radius $L \leqslant \lambda$, where λ is the plasma wavelength. Pinching will increase the density of the beam. Instabilities in the electron beam, might break it into small bunches.

2. On seeding the bunches with ions, they will be stripped of their electrons. Heavy ions get ionized more effectively than low Z ones and due to electrostatic attraction, the electron bunches preferentially collects high Z (Levy 1968).

3. The acceleration of the self focused positively seeded bunches of electrons as a whole, thus the ions are forced to acquire the same velocity of the electrons.

4. Since 1000 GeV electrons have been detected in cosmic rays, then the energy gained by a trapped nuclei can be 1.8×10^{15} eV per nucleon.

5. On head on collision, the stabilized electron bunch can give energy to target nuclei that is proportional to γ^2. A 1000 GeV electron bunch would convey ultra relativistic energies to a target nucleus and might cause fission.

6. The solar millisecond radio spikes require a coherent radiation mechanism that is also needed to explain some radio bursts on the sun and other active stars (Lang, 2001). Such spikes may be the site for coherent electron acceleration.

The general principle of this mechanism is applied in electron ring accelerators (Veksler *et al.* 1967 and Keefe 1969).

References

Kerst, D.W. 1966, The Development Of High energy Accelerators, Classics of Science (edited By M.S. Livingstone) VIII, Dover, New York, 1966.

Keefe, D. 1969, Science J., 5, 71 (1969).

Lang, K.R 2001, The Cambridge Encyclopedia of the Sun. Cambridge University Press.

Levy, R. H., in Symposium on Electron Ring Accelerators, Lawrence Radiation Laboratory Report UCRL 18103, 318 (1968).

Reames D.V., Energetic Particles from Solar Flares and Coronal Mass Ejections (1996). http://lheawww.gsfc.NASA.gov/Reames/cv.html

Schatzman, E., "High Energy Astrophysics" Lecture delivered at Les Houches during the 1966 Summer School of Theoretical Physics) edited by C. Dewitt, E. Schatzman and P. Vernon, Gordon and Breach, 1967.

Veksler, V.I., in proceedings of the CERN. symposium on High Energy Accelerators and Pion Physics, Geneva, Switzerland 1, 68, 1956.

Veksler, V.I., Sarantsev, V.P., Bonch-Osmolovsky, A.G., Dolbilov, G.V., Ivanov, G.A., Ionovitch, M.L., Kozhukhov, I,V., Kusnetsev, S.B., Makhan'kov, V.G., Perel'shtein, E.A., Rashevsky, V.P., Reshetnikova, K.A. Rubin, N.B., Ryl'stev, P.I., and Yarkovey, O.I., Proceedings of the sixth international conference on accelerators, Cambridge, Mass., U.S.A, 1967.

Session 7

ICMEs in the heliosphere

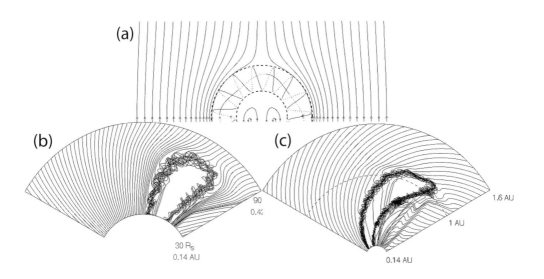

Coronal and Stellar Mass Ejections
Proceedings IAU Symposium No. 226, 2004
K. P. Dere, J. Wang & Y. Yan, eds.

Modeling Interplanetary Coronal Mass Ejections

Pete Riley

Science Applications International Corporation, San Diego, California 92121. USA.
email: pete.riley@saic.com

Abstract. Heliospheric models of Coronal Mass Ejection (CME) propagation and evolution provide an important insight into the dynamics of CMEs and are a valuable tool for interpreting interplanetary in situ observations. Moreover, they represent a virtual laboratory for exploring conditions and regions of space that are not conveniently or currently accessible by spacecraft. In this review I summarize recent advances in modeling the properties and evolution of CMEs in the solar wind. In particular, I will focus on: (1) the types of ICME models; (2) the boundary conditions that are imposed; (3) the role of the ambient solar wind; (4) predicting new phenomena; and (5) distinguishing between competing CME initiation mechanisms. I will conclude by discussing what topics will likely be important for models to address in the future.

Keywords. Sun: coronal mass ejections, magnetohydrodynamics, solar wind

1. Introduction

In this report I summarize recent advances in modeling the properties and evolution of Coronal Mass Ejections (CMEs) in the solar wind. I will describe the current state of research and suggest what topics will likely be important for models to address in the future. I will focus on the physics described by the models and not specifically on the models themselves. Given the need for brevity, references will be selective and illustrative, rather than comprehensive. Other reviews that complement the present one have been given by Linker *et al.* (2002); Cargill & Schmidt (2002); Riley (1999). While I emphasize fluid and MHD modeling in this report, it should be noted that other modeling approaches have been used with success. The extension of force-free flux rope fitting (Lepping *et al.* 1990) to include the effects of expansion (Osherovich *et al.* 1993; Marubashi 1997) and multiple spacecraft (Mulligan *et al.* 2001), for example, have allowed further classification of this important subset of CMEs. Hybrid codes have also been used to model the interaction of fast CMEs with the ambient solar wind allowing ion-kinetic effects to be explored (Riley *et al.* 1998).

Since the basic mechanism(s) by which CMEs erupt at the Sun is (are) not well known (Forbes 2000; Klimchuk 2001), it is therefore not surprising that models developed to investigate the initiation and evolution of CMEs both near the Sun and in the solar wind tend to be idealized. In fact, to make problems tractable, significant approximations must be made. For example, consider the placement of the inner radial boundary. For many years, this was chosen to be beyond the outermost critical point (e.g., Hundhausen & Gentry (1968); Dryer *et al.* (1989); Riley *et al.* (1997b); Odstrcil & Pizzo (1999b,c,a); Cargill & Schmidt (2002); Vandas *et al.* (2002)). Modeling CME propagation and evolution beyond this point is a much simpler task than including the initiation process and evolution through the lower corona. Given accurate boundary conditions at say $20-30R_S$, the physics of the medium is simpler and better understood, and the MHD equations used to describe the system are easier to solve. Further, the minimum time step required to advance the solutions are also typically much larger than would be required if the lower corona were included. Unfortunately, it is difficult to measure the plasma and magnetic field properties in this region, leading to the specification of ad hoc boundary conditions. Moreover, such an approach completely avoids the question of CME initiation.

A second, often used approximation is to neglect the magnetic field (e.g., Hundhausen & Gentry (1968)). Thus strictly speaking the simulations are valid only for high-β CMEs. The characteristic speeds at which pressure disturbances propagate in the simulation is less than in the real solar wind, and magnetic forces. Obviously such studies cannot address questions related to the magnetic structure of the CME. Nevertheless, they have proven to be extremely useful in illuminating the fundamental aspects of the processes by which both transient and corotating disturbances evolve in the solar wind (see, for example, reviews by Hundhausen (1985); Gosling (1996)).

Currently there is a trend toward "modular" modeling of space-plasma systems, where several specialized codes are integrated together, with the output from one model providing the input to the next model (e.g., http://ccmc.gsfc.nasa.gov). In some cases, such as the ionosphere-magnetosphere system, this can lead to a complex feedback loop. On the other hand, the coupling of solar coronal models with interplanetary models is considerably simpler owing to the supersonic nature of the flow at the boundary (Odstrcil et al. 2002).

Algorithms are constantly being updated to include more and more realistic physics. The methods of solution are also being improved on to take advantage of new developments in numerical techniques as well as new computing paradigms. Adaptive Mesh Refinement (AMR), for example, is a technique that allows both large- and small-scale structure to be resolved within a single simulation (e.g., Odstrcil et al. (2002); Manchester et al. (2004a)). The Message Passing Interface (MPI) is an approach that allows one to utilize a large number of processors simultaneously, leading to simulations at significantly higher resolution than previously possible.

2. Basic features of ICME models

The basic features of ICME models can be broken down into several areas, including: (1) the equations, physics, and hence approximations that are intrinsic to the model; (2) the imposed boundary conditions; (3) the ambient solar wind into which the CME propagates; and (4) the mechanism for initiating the CME.

2.1. The Physics in the Models

Generally speaking, all of the models solve some variation of: Ampere's law, Faraday's law, Ohm's law, mass continuity (conservation of mass), the equation of motion (conservation of momentum), and an energy equation (energy conservation).

In the hydrodynamic approximation the magnetic field terms are omitted and the equations are significantly simpler to solve. In some of the models the resistivity is set to zero, simplifying Ohm's law, and reducing the system to the ideal MHD limit. Some of the heliospheric models also neglect the gravity term in the equation of motion; a reasonable approximation for regions above $20 - 30R_S$.

An important distinction between the models is how the energy equation is treated. Setting the energy source term to zero reduces it to a simple polytropic relationship. This may be a reasonable first approximation for the solar wind, where the measured adiabatic index, γ, for the protons is about 1.5 (Totten et al. 1996); however, it is questionable for the solar corona, which is nearly isothermal. Moreover, models that couple the corona and heliosphere must account for the discontinuous change in γ between the two models

(Odstrcil *et al.* 2002). Several solutions to these problems have been proposed. Roussev *et al.* (2003), for example, recently implemented a γ that was a function of distance from the Sun, being ~ 1 at the Sun and approaching 5/3 further out. While this is an apparently workable practical solution, there is little physically to justify it, and, more importantly, no way to assess how the fall-off with distance should occur. Alfvén waves have also been incorporated to heat the corona and/or drive the acceleration, at least for ambient solar wind (e.g., Usmanov *et al.* (2000)). This might appear to be more physically justifiable, however, it is unlikely that Alfvén waves acting alone are sufficient, and thus, one is in effect, replacing one set of free parameters with another. The most promising long-term solution would appear to be the treatment of more complicated energy equation, including the effects radiation, thermal conduction, coronal heating, and resistive and viscous diffusion. While this is a challenging problem, progress is being made (Lionello *et al.* 2001).

2.2. *Boundary Conditions*

The heliosphere is naturally a spherical (or cylindrical in 2-D) region so most models solve the (M)HD equations in this system (note, however, that for computational reasons, some codes, e.g. BATS-R-US (Manchester *et al.* 2004a) are implemented in Cartesian coordinates). Thus at a minimum, boundary conditions at the inner spherical surface (the inner radial boundary) and outer spherical surface (the outer radial boundary) must be considered. In 1-D models, these are simply points. Some models focus on a more limited region of space, typically by reducing the coverage in latitude (e.g., Odstrcil *et al.* (2002)). The outer boundary is simple since it is always super-sonic and super-Alfvénic. Thus whatever reaches this point (plasma, fields, waves, etc) must leave the system. The same holds true for the inner boundary if it is chosen to be beyond the outermost critical point. We refer to these models as "super-critical" and those that include the lower corona as "sub-critical" models (the outer boundary of the sub-critical models may stretch out far into the solar wind). This distinction is important insofar as the physics and methods of solution are considerably simpler for the super-critical models. Moreover, the lack of any substantial observations at this heliocentric distance provides little in the way to constrain the models.

2.3. *The Ambient Solar Wind*

The medium into which CMEs propagate plays a substantial role in the subsequent evolution of ejecta and their associated disturbances. The simplest possible background wind is a single-speed wind. For 1-D models, this is really the only type of wind you can prescribe. The next level of sophistication is to set up a two-state wind consisting of slow, dense wind at low latitudes, and fast, hot, tenuous wind at high latitudes (Riley *et al.* 1997b). This mimics the declining phase and solar minimum picture that Ulysses observed during its first orbit (e.g. Riley *et al.* (1997a)), and for the purposes of running idealized, generic simulations is perfectly adequate. This wind pattern can be prescribed in two dimensions (Riley *et al.* 1997b) or as a "tilted-dipole" configuration in three dimensions (Odstrcil & Pizzo 1999b,c,a). An example of this wind pattern (at 30 R_S) is shown in Figure 1(a). However, to address the evolution of ICMEs for specific events, we need to specify the actual ambient solar wind that was present at the time of the eruption of the CME. An example of a more realistic wind is shown in Figure 1 (b). The speed was computed using the SAIC coronal MHD model (Riley *et al.* 2001b) and has been shown to reproduce the essential large-scale features of the solar wind, particularly during relatively quiet solar conditions.

(a)   (b)

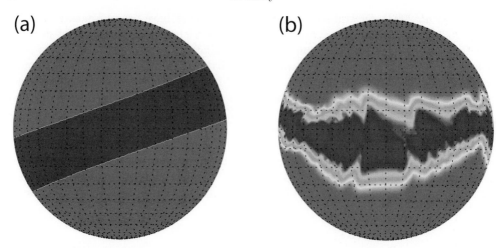

Figure 1. Ambient solar wind speed at 30 R_S for: (a) an idealized "tilted-dipole" configuration; and (b) a specific time period during solar minimum conditions (Carrington rotation 1913). Blue corresponds to the slowest speeds (400 km/s) while red corresponds to the fastest speeds (750 km/s).

2.4. *CME Initiation*

Once the ambient medium has been prescribed, the CME can be "initiated". There are a number of approaches for accomplishing this, but they can be broadly separated into those that actually model the initiation process (i.e., self-consistent models) and those that simply pass a representation of a CME through the inner boundary (i.e., ad hoc models). The oldest - but still used - example of the latter is to set up a "pulse" in one or more of the plasma and/or magnetic field parameters (e.g., Riley (1999)). This could be a strong speed change, a density and/or temperature change, or some combination (e.g., Hundhausen & Gentry (1968)) or could include a coherent magnetic structure, such as a flux rope (e.g., Vandas *et al.* (2002)). One of the appealing aspects of these ad hoc pulses is that it allows the modeler to explore a wide range of the parameter space. Input conditions can be modified to better reproduce a specific set of observations. Of course the danger is that while "tweaking" the inputs in this way can produce better model results, you do not necessarily learn much from it.

The self-consistent models attempt to model the actual eruption process of the CME. As such they have inherently more scientific value. On the other hand, it can be very difficult to explore CME evolution in the solar wind for different eruption processes, as these models currently only produce a limited range of properties. From a computational standpoint, these types of models require considerably more resources than the pulse-type simulations. Whereas a 1-D simulation incorporating a simple pulse might takes minutes on a workstation computer, a 3-D global MHD solution might require weeks of time on a multi-processor supercomputer. Thus there is a place for both types of approaches, depending on the specific problem being addressed.

3. Examples of ICME models

It would be impossible to comprehensively review all of the CME modeling work that has been undertaken in the last 40 years. Instead, I have chosen several specific examples that: (1) illustrate the main general aspects of the models that were described above; and (2) have been used to address a specific scientific problem.

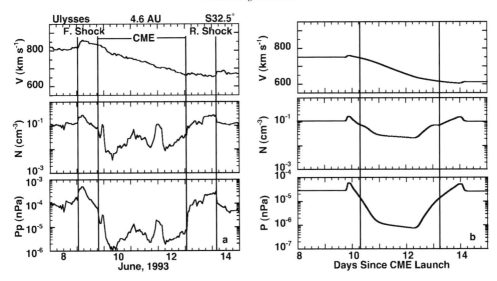

Figure 2. Comparison of Ulysses observations of a high-latitude ICME with results from a 1-D, super-critical, hydrodynamic model. Adapted from Gosling *et al.* (1998).

3.1. *One-dimensional, super-critical hydrodynamic models*

Let us start by discussing the simplest of all ICME models. These are: (1) super-critical, meaning the inner radial boundary is at 20-30 R_s; (2) idealized, single-flow solar wind; and (3) initiated by an ad hoc pulse. They are simple to implement and the results are easy to interpret. However, there is no way to include a magnetic field in a meaningful way and there will always be lingering questions about the uniqueness of the solution. On the other hand, they have been, and continue to be useful in interpreting in situ observations, providing clues about the evolution of the ejecta. As a example, in Figure 2 we compare Ulysses observations of an ICME at high heliographic latitudes, with the results of such a simulation. The ICME was observed while Ulysses was located at 4.6 AU and 32.5 degrees south latitude. Vertical lines indicate the location of the ICME. A forward and reverse shock bound the ejecta on either side. These are both fast-mode shocks. In the rest frame of the ambient plasma, the forward shock is propagating away from the Sun and the reverse shock is propagating towards the Sun. However, in the spacecraft's frame, everything is being convected away from the Sun. The general features of this event are: (1) relatively symmetric density and temperature profiles; (2) a declining speed profile (indicating that the ejecta is expanding); and (3) a decrease in the density and pressure. The inferences we draw from these profiles was that an initially high-pressure ejecta drove forward and reverse waves away from it that subsequently steepened into shocks.

The results of the 1-D, gas-dynamic simulation are shown on the right-hand side of Figure 2. After an initial equilibrium was established (with values that match what was observed at Ulysses), a pulse was initiated whereby the speed was dropped monotonically over a 10 hour period, while at the same time, the density and thus pressure were enhanced in a bell-shaped profile. How were these values chosen? Well, it is not unreasonable to expect a CME to be launched with an internal pressure higher than the ambient, although it is likely that in reality, this pressure is supplied by the magnetic field. The speed decrease merely reflects what was observed; the speed before the CME was high and lower after it.

Not surprisingly, the simulated profiles compare well with the observations. The declining speed, decrease in density and pressure are all reproduced. Even the sheath region and shocks match well. The standoff distance is not as great in the simulation because the characteristic speed is the sound speed and not the fast mode speed as it is in reality. Even the asymmetry in the locations of the shocks is reproduced. This is an evolutionary effect; the leading edge of the CME is younger than the trailing edge, when viewed at a single point in space.

So on one hand, we have successfully modeled a high-latitude ejecta from 30 Rs to 4.6 AU, and provided a plausible picture of the ejecta's evolution during this interval. And, they have been successfully applied to a number of problems, including: the evolution of CMEs are large heliocentric distances (Riley & Gosling 1998); the acceleration of CMEs near the Sun (Gosling & Riley 1996); and the relationship between density and temperature within CMEs and its implications for the polytropic index of the plasma (Riley *et al.* 2001a). However, many fundamental questions remain unanswered, such as: How was the CME initiated? What was it's early evolutionary history? What role did the magnetic field play? How does this localized picture related to the global structure of the event? To answer these questions, we must increase the dimensionality of the simulations, include the magnetic field, and consider the triggering mechanism.

3.2. 3-D, super-critical MHD models

It has long been known that the particular trajectory taken by a spacecraft through a CME and its associated disturbance can radically alter the observed profiles. In fact, the recent classifications of CMEs in to "simple" and "complex" (Burlaga *et al.* 2001) may be, at least in part, a consequence of such observational selection effects. Marubashi (1997) illustrated how a single event could be seen by one spacecraft as a non-flux-rope CME while at another it would appear as a magnetic cloud, suggesting that the delineation between magnetic clouds (or flux ropes) and CMEs may be an artificial one. CMEs simultaneously observed by Ulysses and ACE had such different profiles at the two spacecraft, that only by using global MHD modeling could we confidently infer that the events were one and the same (e.g., Riley *et al.* (2003)).

To illustrate the use of 3-D, super-critical MHD models let us consider the model developed by Vandas *et al.* (2002), who simulated the evolution of a flux-rope CME from $30R_S$ out to 1 AU within an idealized single-speed wind. The initial state of the magnetic cloud was that of a section of a torus, who's foot points were tied to the Sun. The model output is summarized at 3 times in Figure 3.

The top panel shows the flux rope embedded within the ambient solar wind field at the beginning of the simulation. The speed, density, and temperature of the torus were exactly the same as the ambient solar wind; only the magnetic field structure was different. Thus no shocks or significant disturbances would be expected to form. After emerging half of the torus, the boundary conditions were held constant, except that they were rotated in azimuth to simulate the effect of solar rotation. The two lower panels show the magnetic field in the equatorial plane at two times following the emergence of the torus. On the left, at 32 hours, the flux rope is beginning to develop some structure associated with rotation. By 120 hours, the apex of the flux rope has reached 1 AU and flattened considerably.

A particularly interesting aspect of this study was the construction of simulated time series at various longitudinal positions through the ejecta. For example, a spacecraft intercepting the apex of the flux rope would subsequently encounter the flux rope legs. The encounter with the apex displayed the usual signatures of a flux rope CME: field enhancement, rotation in the field, a modest speed decrease, indicating expansion, and a decrease in temperature and density. The encounter with the CME legs also showed characteristic signatures in the magnetic field, with no corresponding plasma signatures.

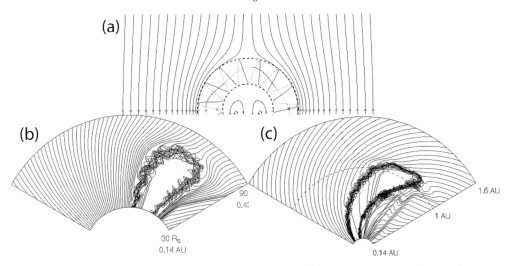

Figure 3. Results from 3-D, super-critical, MHD model. (a) Initial conditions for the flux rope and ambient solar wind; (b) Magnetic field lines at 32 hours; and (c) magnetic field lines at 120 hours.

Other more complex profiles were presented, which depended on the spacecraft's location. Some observations of "double flux ropes" (e.g., Vandas *et al.* (1993)), could be cases where a spacecraft flew through the same ICME twice.

3.3. *3-D, Sub-Critical ICME models*

As we have noted, modeling the solar environment below the critical points is more complicated because information can now travel in both directions. Nevertheless several groups are modeling the Sun's extended Corona from $1R_S$ to 1 AU, and beyond. Wu *et al.* (1999), for example, generated a CME from the eruption of a helmet streamer using an ad hoc increase in the azimuthal component of the magnetic field. The University of Michigan group (e.g., Groth *et al.* (2000b); Manchester *et al.* (2004a); Roussev *et al.* (2004)) have developed a finite-volume, AMR scheme to study CME evolution from the Sun to Earth. The CME is "initiated" in one of several ways. Groth *et al.* (2000b,a) applied a localized density enhancement at the solar surface, essentially mimicking a pressure pulse. In contrast, Manchester *et al.* (2004a,b) superimposed the magnetic and density solutions of the 3-D Gibson & Low (1998) flux rope within the coronal streamer belt; the CME being driven by the resulting force imbalance.

The University of Michigan model is summarized in Figure 4. In this simulation, a two-state wind (aligned to the rotation axis) is constructed. The Gibson-Low flux rope is superimposed onto the background field. Thus the eruption process takes place because the system begins out of equilibrium. As with the simulations initiated beyond the super-critical points, "inserting" a CME near the solar surface and allowing it to evolve from that point allows enormous freedom to adjust the initial parameters to fit the observations. It may turn out that for the purposes of space weather prediction, such an approach is the most practical. A particular set of observations, for example, may suggest an appropriate initial configuration and perturbation to produce a CME that reproduces observations near Earth.

It is unlikely, however, that such approaches will uncover the underlying eruption mechanism(s). Instead, the free parameters must be constrained by developing more self-consistent models. Towards this goal, Linker & Mikić (1997) initiated an eruption through differential rotation and followed its evolution out to 1 AU. Later, Odstrcil *et al.*

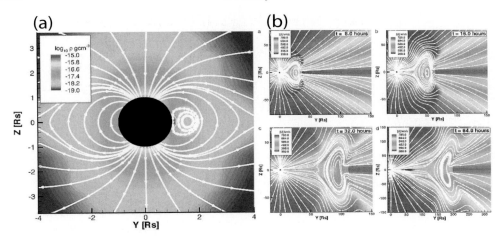

Figure 4. Simulation results from a 3-D, sub-critical, MHD model. (a) Initial conditions; and (b) Four snapshots at 8, 16, 24, and 32 hours following the launch of the CME.

(2002) coupled this coronal model with a heliospheric model to simulate the eruption of a CME at the Sun and its propagation out to 5 AU. In spite of the idealized nature of the eruption process and ambient solar wind, the solution was remarkably rich and complex. Riley *et al.* (2003), for example, used these results to interpret the plasma and magnetic field signatures of a CME observed by both ACE and Ulysses, which were aligned in longitude, but separated significantly in radial distance and latitude. Favorable comparisons with in situ observations, however, do not necessarily validate one eruption mechanism. Differentiating signatures are needed to do this (see below).

3.4. *Model Predictions*

The simulations by Odstrcil *et al.* (2002) also suggested that a jetted outflow, driven by post-eruptive reconnection underneath the flux rope occurs and may remain intact out to 1 AU and beyond (Riley *et al.* 2002). Comparison between simulations and observations of a magnetic cloud with similar signatures suggested that velocity and/or density enhancements observed trailing magnetic clouds may be the signatures of such reconnection, and not associated with prominence material, as has previously been suggested.

Figure 5 shows the relationship between this velocity enhancement and the ejecta as the flux rope approaches 1 AU. This is a view in the meridional plane from a point just below the equatorial plane. The color contours show the radial velocity, while the arrows summarize the meridional magnetic field. Superimposed are several field lines. These are the helical field lines making up the flux rope. Note how the velocity enhancement has remained intact because of the umbrella-like protection provided by the ejecta plowing into the solar wind ahead as well as the quiet, slower wind that is following the disturbance: the velocity pulse is immersed in an expansion wave (or rarefaction region). Note also that the reconnection feature has a very limited angular extent, suggesting that such a signature might be difficult to observe. A preliminary analysis of WIND observations of magnetic clouds suggests that perhaps 5% of events may contain such a feature. Gosling *et al.* (2004) have recently provided a detailed analysis of a number of cases suggestive of reconnection in the solar wind. In principle, they may be associated with this phenomenon.

If these results are further substantiated, they may provide us with a unique window into the reconnection process. Ultimately, these types of comparisons with observations may provide constraints on mechanisms for the initiation of CMEs. At present there is

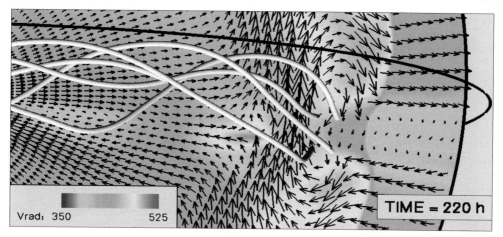

Vrad: 350   525   TIME = 220 h

Figure 5. Results from a 3-D, sub-critical MHD simulation of a flux rope erupted at the Sun via flux cancellation as it approaches 1 AU. The color contours show radial velocity and the arrows indicate the direction and magnitude of the magnetic field line the meridional plane. A selection of magnetic field lines threading the flux rope have been drawn. The solid black curve indicates the trajectory of Earth.

very little to constrain the models: Quite conveniently, the models do not yet predict any distinguishing features that we can test against.

3.5. *Differentiating between CME initiation models*

A fundamental problem in solar physics is how CMEs erupt. Many of the pieces of the puzzle have been assembled, but how they relate to one another, and what role each one plays is not known. We know, for example, that the energy for the eruption is supplied by the magnetic field. Observationally, we know that many CMEs contain flux ropes. Yet we do not know during which phase of the eruption the flux rope forms, nor do we know why some ICMEs contain flux ropes while others do not.

A number of theoretical models have been proposed (see reviews by Forbes (2000); Klimchuk (2001). Each model addresses - and is therefore consistent with - some subset of the observed properties of CMEs. Not surprisingly, few of them are in direct conflict with any of the observations. Thus it has proven difficult to distinguish between them. As an example, in Figure 6 I compare the "flux cancellation" model (Linker *et al.* 2001) with the "breakout" model (Antiochos *et al.* 1999). The models distinguish themselves primarily by the underlying structure of the region producing the eruption. In principle, the breakout model can erupt either via photospheric shear or flux cancellation. The distinguishing feature of the breakout model is the requirement of a more complex quadrupolar configuration. In addition to the differences in the large-scale magnetic field configuration, the main difference between the two models is that photospheric shear alone was used to trigger the eruption in the breakout model. The outstanding difference between the two is that a single flux rope structure is produced in the flux cancellation eruption, whereas a "double flux rope" appears to have been generated in the breakout model. This leading "flux rope" is not the same helical structure that is produced as the main ejecta since it contains essentially no azimuthal field component. It results from reconnection at the leading edge of the ejecta, together with reconnection of the streamer belt further ahead of the event. The picture is, however, quite suggestive, and if further substantiated, may provide an explanation for some of the double flux ropes that have been observed in the solar wind (e.g., Vandas *et al.* (1993)). On the other hand, these

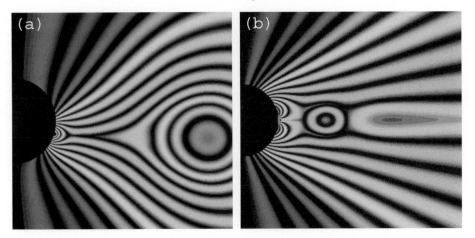

Figure 6. Comparison of (a) flux cancellation and (b) breakout eruption profiles.

are preliminary results and should not be viewed as definitive. In particular, that most ICMEs do not appear to have this double rotation should not be used to invalidate the breakout model. One feature that does appear to be robust, however, is the requirement for reconnection to proceed at the leading edge of the erupting flux rope in the breakout model. Indeed, it is this reconnection that generates the necessary force imbalance to allow the CME to be able to escape the confines of the overlying field in the first place. Thus while we expect reconnection to occur in both models underneath the erupting flux rope, we would also expect to see reconnection at the leading edge of CMEs erupting via the breakout scenario. Observationally, this might appear as a discontinuity in the sheath region between the shock and the flux rope. Discontinuities, however, are very common occurrences in the solar wind.

4. Closing Remarks and Future Research

In this review I have illustrated how models are helping us to understand the formation and evolution of CMEs in the solar wind. I would like to reiterate that these remain idealized simulations. Yet even as such, they are full of interesting phenomena. They allow us to interpret localized in situ measurements within a global perspective, and they also provide clues for interpreting remote solar observations, such as LASCO, EIT, and soft x-ray observations. Most importantly they allow us to connect these two disparate datasets. The simulations can be used to explore the role played by the ambient solar wind on the distortion of the ejecta and its associated disturbance, and they can be used to predict new phenomena that may or may not be present in the data and so eventually be used to constrain models of CME eruption.

Even a cursory glance at these results shows that the simulations are not adequately represented by cartoons typically drawn for CMEs in the solar wind. The cross section of ICMEs has been shown, for example, to be strongly distorted away from circular, a result simply of CME expansion in a spherical geometry (Riley & Crooker 2004).Given the strong interaction of CMEs with the ambient solar wind, we must view pictures of CMEs deduced from single point spacecraft measurements with caution. For example, it is unlikely that the so-called "over-expanding" CMEs are really isolated bubbles in the high speed solar wind as has been drawn (Gosling *et al.* 1994).

The models are far from perfect and it is important to read the "fine print". The availability of sophisticated visualization tools today has led to the production of amazing pictures and movies that, by their very complexity, convey the idea that everything has been solved. In reality, the important approximations and assumptions have been hidden from the audience. In many cases, these limitations may be playing a crucial role in the model. The old adage of "buyer beware" should be adhered to.

Predicting the path of future research is clearly speculation, undoubtedly driven, at least in part, by one's current interests. Nevertheless, it may be of some use to list several likely topics that may be pursued in the upcoming years. One challenge will undoubtedly involve the ability to self-consistently model CMEs with a range of properties. How do we initiate slow and fast CMEs, for example? Are they generated by the same mechanism, or are there two (or more) mechanisms that are responsible? Self-consistent models currently can only produce flux-rope CMEs. What are the underlying differences between these and CMEs that don't contain a flux rope? Is it a observational selection effect or are there intrinsically different mechanisms for producing each type?

A important new area in CME research concerns the acceleration and transport of energetic particles. Roussev *et al.* (2004), for example, have used the inferred properties of a CME-driven shock to argue that diffusive shock acceleration theory can account for solar energetic protons with energies below 10 GeV. Several groups (e.g., Sokolov *et al.* (2004)) are developing and coupling particle acceleration and transport codes with global MHD models to address which mechanisms are important for the production of SEPs.

We may be entering a new era of CME modeling. In the future, we will see models becoming increasingly capable of modeling specific events. This will require capabilities to accurately reproduce a disparate set of remote and solar observations. We may see some of these models transition from research tools to operation tools, capable of predicting the onset of geo-effective phenomena (although see Cargill & Schmidt (2002) for a more conservative opinion on this).

It is important to remember that these models are only tools that allow us to better understand CME phenomena. To close, then, we provide an illustrative selection of questions that we may be able to answer using the models described here. What are the fundamental evolutionary distinctions between CMEs and magnetic clouds? What topology is predicted for CMEs in the heliosphere by various initiation mechanisms? What is the relationship between the 3-part structure of CMEs as seen in coronagraph observations and their interplanetary counterparts? What processes control the solar connectivity of field lines embedded within CMEs? How do the properties of the ambient solar wind modify the evolution of the ejecta? How does the internal magnetic structure of a flux rope affect its distortion in the solar wind? How do in situ signatures (as would be seen by a spacecraft) change depending on where the simulated CME is sampled? What differentiating observational signatures do the models predict?

Acknowledgements

The author gratefully acknowledges the support of the National Aeronautics and Space Administration (Living with a Star Program) and the National Science Foundation (SHINE Program) in undertaking this study.

References

Antiochos, S. K., Devore, C. R., & Klimchuk, J. A. 1999, Astrophys. J., 510, 485

Burlaga, L. F., Skoug, R. M., Smith, C. W., Webb, D. F., Zurbuchen, T. H., & Reinard, A. 2001, J. Geophys. Res., 106, 20957

Cargill, P. J., & Schmidt, J. M. 2002, Ann. Geophys., 20, 879

Dryer, M., Detman, T. R., Wu, S. T., & Han, S. M. 1989, THE INTERNATIONAL HELIO-SPHERIC STUDY: COSPAR27, 75

Forbes, T. G. 2000, J. Geophys. Res., 105, 23153

Gibson, S., & Low, B. 1998, Astrophys. J., 493, 460

Gosling, J. T. 1996, Annual Review of Astronomy and Astrophysics, 34, 35

Gosling, J. T., McComas, D. J., Phillips, J. L., Weiss, L. A., Pizzo, V. J., Goldstein, B. E., & Forsyth, R. J. 1994, Geophys. Res. Lett., 21, 2271

Gosling, J. T., & Riley, P. 1996, Geophys. Res. Lett., 23, 2867

Gosling, J. T., Riley, P., McComas, D. J., & Pizzo, V. J. 1998, J. Geophys. Res., 103, 1941

Gosling, J. T., Skoug, R. M., McComas, D. J., & Smith, C. W. 2004, Accepted for publication in J. Geophys. Res.

Groth, C. P. T., de Zeeuw, D. L., Gombosi, T. I., & Powell, K. G. 2000a, J. Geophys. Res., 105, 25053

—. 2000b, Adv. Space Res., 26, 793

Hundhausen, A. J. 1985, IN: Collisionless shocks in the heliosphere: A tutorial review, 34, 37

Hundhausen, A. J., & Gentry, R. A. 1968, Astron. J., 73, 63

Klimchuk, J. A. 2001, in in Space Weather Geophys. Monogr., Vol. 125 (Washington, DC: AGU), 143

Lepping, R. P., Jones, J. A., & Burlaga, L. F. 1990, J. Geophys. Res., 95, 11957

Linker, J. A., Lionello, R., Mikić, Z., & Amari, T. 2001, J. Geophys. Res., 106, 25165

Linker, J. A., & Mikić, Z. 1997, Coronal Mass Ejections, 99, 269, edited by N. Crooker, J. Joselyn, and J. Feynmann, p. 269, AGU, Washington, D. C.

Linker, J. A., Mikić, Z., Riley, P., Lionello, R., & Odstrcil, D. 2002

Lionello, R., Linker, J. A., & Mikić, Z. 2001, Astrophys. J., 546, 542

Manchester, W., Gombosi, T., Roussev, I., De Zeeuw, D., Sokolov, I., Powell, K., Tóth, G., & Opher, M. 2004a, Journal of Geophysical Research (Space Physics), 109, 1102

Manchester, W., Gombosi, T., Roussev, I., Ridley, A., De Zeeuw, D., Sokolov, I., Powell, K., & Tóth, G. 2004b, Journal of Geophysical Research (Space Physics), 2107

Marubashi, K. 1997, in Coronal Mass Ejections (Geophys. Monogr.), ed. N. U. Crooker, J. A. Joselyn, & J. Feynman, Vol. 99 (Washington, DC: AGU), 147

Mulligan, T., Russell, C. T., Anderson, B. J., & Acuna, M. H. 2001, Geophys. Res. Lett., 29, 4417

Odstrcil, D., Linker, J. A., Lionello, R., Mikić, Z., Riley, P., Pizzo, V. J., & Luhmann, J. G. 2002, J. Geophys. Res., 107, DOI 10.1029/2002JA009334

Odstrcil, D., & Pizzo, V. J. 1999a, J. Geophys. Res., 104, 28255

—. 1999b, J. Geophys. Res., 104, 483

—. 1999c, J. Geophys. Res., 104, 493

Osherovich, V. A., Farrugia, C. J., & Burlaga, L. F. 1993, Adv. Space Res., 13, 57

Riley, P. 1999, in Solar Wind Nine Proceedings of the Ninth International Solar Wind Conference, ed. S. R. Habbal, R. Esser, V. Hollweg, & P. A. Isenberg, Vol. 471 (Nantucket, MA: The American Institute of Physics (AIP Conference Proceedings)), 131

Riley, P., Bame, S. J., Barraclough, B. L., Feldman, W. C., Gosling, J. T., Hoogeveen, G. W., McComas, D. J., Phillips, J. L., Goldstein, B. E., & Neugebauer, M. 1997a, Adv. Space Res., 20, 15

Riley, P., & Crooker, N. 2004, Astrophys. J., 600, 1035

Riley, P., & Gosling, J. T. 1998, Geophys. Res. Lett., 25, 1529

Riley, P., Gosling, J. T., & Pizzo, V. J. 1997b, J. Geophys. Res., 102, 14677

—. 2001a, J. Geophys. Res., 106, 8291

Riley, P., Linker, J., Mikić, Z., Odstrcil, D., Zurbuchen, T., Lario, D., & Lepping, R. 2003, Journal of Geophysical Research (Space Physics), 108, 2

Riley, P., Linker, J. A., & Mikić, Z. 2001b, J. Geophys. Res., 106, 15889

Riley, P., Linker, J. A., Mikić, Z., Odstrcil, D., Pizzo, V. J., & Webb, D. F. 2002, Astrophys. J., 578, 972

Riley, P., Omidi, N., & Gosling, J. T. 1998, in American Geophysical Union, Fall Meeting 1998 (San Francisco, CA: AGU), abstract

Roussev, I., Gombosi, T., Sokolov, I., Velli, M., Manchester, W., DeZeeuw, D., Liewer, P., Tóth, G., & Luhmann, J. 2003, Ap. J. Letters, 595, L57

Roussev, I., Sokolov, I., Forbes, T., Gombosi, T., Lee, M., & Sakai, J. 2004, Astrophys. J. Lett., 605, L73

Sokolov, I. V., Roussev, I. I., Gombosi, T. I., Lee, M. A., Kota, J., Forbes, T. G., Manchester, W. B., & Sakai, J. I. 2004, submitted to Ap. J. Lett.

Totten, T. L., Freeman, J. W., & Arya, S. 1996, J. Geophys. Res., 101, 15629

Usmanov, A. V., Goldstein, M. L., Besser, B. P., & Fritzer, J. M. 2000, J. Geophys. Res., 105, 12675

Vandas, M., Fischer, S., Pelant, P., & Geranios, A. 1993, J. Geophys. Res., 98, 21061

Vandas, M., Odstrčil, D., & Watari, S. 2002, J. Geophys. Res., 107, 2

Wu, S., Guo, W., Michels, D., & Burlaga, L. 1999, J. Geophys. Res., 104, 14789

Discussion

SHIBATA: What is the largest distance from the sun of the post-eruption reconnection? Did you find the signature of slow shocks in the post-eruption reconnection?

RILEY: Most of the reconnection takes place during and shortly after the eruption of the CME. But, undoubtedly, this reconnection continues for some time. I would have to re-analyze the simulation results to see how long the reconnection persisted at detectable levels. We haven't looked for the evidence of slow shocks in the post-eruption reconnection, but I would be surprised if we find such a signature.

SCHWENN: Do the models give any hint on how the 3-part (or multi-part) structure near the sun is transformed into the 2-part structure in IP space?

RILEY: The models do provide a good connection between the cavity and the flux rope in interplanetary space. Because we do not realistically model the prominence/filament, we can only infer its mapping to the trailing edge of the ICME. However, it is the loop (or bright front) that has the most ambiguity. The coronal structure suggests that much of this structure existed prior to its propagation but it is difficult to assess what contribution comes from swept-up material. This is something we are currently investigating.

RUFFOLO: How can you observationally identify when you go through the near-radial field of the legs of a CME?

RILEY: The near-radial fields are quite easy to identify in the in-situ data. To make the association with CMEs requires ancillary information. For example, you might pass through a shock and compressed plasma before encountering the radial fields. You might also see some counterstreaming suprathermal electrons. Alternatively, remote solar observations might indicate the passage of a CME past your point of observation within the right time frame.

SCHMIEDER: Is the work done by the Belgium group relevant to your work concerning the modelization of ICMEs?

RILEY: Yes it is, but I had no time to review all the simulations of ICMEs.

KOUTCHMY: Doing numerical simulations, is it not possible to include what seems important from the observational point of view: filament/prominence eruption, heating of the prominence gas, mass loading?

RILEY: These are obviously important effects, and to a limited extent, and with varying degrees of success, they have been addressed within the framework of more limited modeling. Including these effects in the global MHD models I have described here will be quite a challenge, but is something that we are in the process of addressing.

ZHUKOV: Is there any progress in modeling the north-south interplanetary magnetic field component Bz in ICMEs?

RILEY: Yes!

Coronal and Stellar Mass Ejections
Proceedings IAU Symposium No. 226, 2005
K. P. Dere, J. Wang & Y. Yan, eds.

© 2005 International Astronomical Union
doi:10.1017/S1743921305000931

A Direct Method to Estimate Magnetic Helicity in Magnetic Clouds

S. Dasso[1,2], C.H. Mandrini[1], A.M. Gulisano[1] and P. Démoulin[3]

[1] Instituto de Astronomía y Física del Espacio, IAFE (CONICET-UBA), Buenos Aires, Argentina. email: dasso@df.uba.ar

[2] Departamento de Física, Facultad de Ciencias Exactas y Naturales, Universidad de Buenos Aires, Buenos Aires, Argentina.

[3] Observatoire de Paris, LESIA, UMR 8109 (CNRS), F-92195 Meudon Cedex, France.

Abstract. Magnetic clouds are extended and magnetized plasma structures that travel from the Sun toward the outer heliosphere, carrying an important amount of magnetic helicity. The magnetic helicity quantifies several aspects of a given magnetic structure, such as the twist, kink, and the number of knots between magnetic field lines, the linking between magnetic flux tubes, etc. Since the helicity is practically conserved in the solar atmosphere and the heliosphere, it is a useful quantity to compare the physical properties of magnetic clouds to those of their solar source regions. In this work we describe a method that, assuming a cylindrical geometry for the magnetic cloud structures, allows us to calculate their helicity (per unit length) content directly from the observed magnetic field values. We apply the method to a set of 20 magnetic clouds observed by the WIND spacecraft. To test its reliability we compare our results with the helicity computed using a linear force-free field model under cylindrical geometry (i.e. Lundquist's solution).

Keywords. Sun: coronal mass ejections (CMEs), magnetic fields, solar wind, interplanetary medium

1. Introduction

Coronal mass ejections (CMEs) are huge expulsions of mass and magnetic field from the Sun. One of the most important roles of CMEs is to carry away magnetic helicity (MH) from the Sun [Low, 1996], that would accumulate incessantly in the active region oroma, since it is generated by the solar dynamo (helical turbulence and differential rotation) without changing sign with the cycle (for a recent review about chirality of magnetic features see Pevtsov and Balasubramaniam [2003]). Since magnetic helicity is well preserved even in non-ideal MHD on a time-scale less than the global diffusion time-scale [Berger, 1984], we expect to be able to trace the helicity from the time a flux tube emerges through the photosphere into the corona and is ejected into the interplanetary space, reaching the Earth in a magnetic cloud (MC).

A magnetic cloud can be distinguished [see e.g., Burlaga, 1995], from *in situ* observations in the interplanetary space, by a low proton temperature, an enhanced magnetic field strength with respect to ambient values, and a large rotation of the magnetic field vector, indicating a helical (flux rope) magnetic structure, which clearly has non-zero helicity.

In order to estimate how much helicity is transported in magnetic clouds it is necessary to determine the size and the global magnetic configuration of these astrophysical objects. The first attempt to estimate the magnetic helicity of MCs was made by DeVore [2000], who used a sample of 18 MCs analyzed by Lepping *et al.* [1990] using the classical Lundquist's (1950) model. He obtained a mean helicity value of 2×10^{42} Mx^2 (for a flux rope length of 0.5 AU) and a mean magnetic flux of 1×10^{21} Mx for these MCs. Démoulin *et al.* [2002] and Green *et al.* [2002] developed a method to measure the helicity content of active regions in the corona obtaining a typical value of $4\text{-}23 \times 10^{42} \text{Mx}^2$. Using the same method as DeVore [2000] for the estimation of the helicity content in clouds, they computed the helicity budget for two active regions (ARs) and estimated the amount of helicity carried away by the CMEs ejected from those ARs. However, they did not link the CMEs to any MC observation [see also Mandrini *et al.* 2004].

Magnetic clouds can be modeled locally using a helical cylindrical geometry as a first approximation [Farrugia *et al.* 1995]. One of the most commonly models used to describe their magnetic configuration is the linear force-free field [e.g., Lepping *et al.*, 1990]. However, several modeling and fitting methods have been used to reproduce the magnetic structure of MCs [see, e.g., Dasso *et al.*, 2005].

In this paper we present a new method to estimate the magnetic helicity of interplanetary cylindrical flux ropes. We present preliminary results of the application of this method to a set of 20 magnetic clouds. We also compare our results with the values of the helicity per unit length obtained under the assumption of a linear force-free cylindrical model (Lundquist's model, [Lundquist, 1950]) to the magnetic configuration of the cloud. In Section 2 we describe the analysis of the data and our results, while in Section 3, we present our conclusions.

2. Data Analysis and Modeling

We select all the MCs observed by the spacecraft Wind from 22-Aug-1995 to 07-Nov-1997, taking the start and the end times given in http://lepmfi.gsfc.nasa.gov/mfi/mag_cloud_pub1.html. We analyze the magnetic data measured by the Magnetic Field Instrument (MFI) aboard Wind in GSE (Geocentric Solar Ecliptic) coordinates. These observations have been downloaded with a temporal cadence of 3 seconds from http://cdaweb.gsfc.nasa.gov/cdaweb/istp_public/. Because we are only interested in the large scale magnetic structure of the clouds, and not in the magnetic fluctuations, we analyzed smoothed data, only ~ 100 averaged points per cloud.

The orientation of the axis of every cloud is obtained using a minimum variance (MV) analysis, as discussed in Bothmer & Schwenn [1998]. From this analysis we define a system of reference fixed to the cloud and we rotate the observed GSE components of the field to this frame. The cloud frame is defined such that $\hat{x}_{cloud}$ corresponds to the cylindrical radial direction ($\hat{r}$) in the ideal case of the spacecraft crossing the axis of the cloud (i.e. a null impact parameter, being the impact parameter, p, the minimum distance between the cloud axis and the spacecraft) as it leaves the structure, $\hat{z}_{cloud}$ is parallel to the axis of the cylinder (sign such that $B_{z,cloud}$ is positive at the cloud axis), and $\hat{y}_{cloud}$ completes a right handed reference system. We also determine the sign of the helicity from the global behavior of the field components. The radius (R) of the cloud is estimated from the duration of the MC and the observed solar wind speed. The list of the start and end times, radius (R), and the helicity sign, are given in Table 1 for the analyzed clouds.

2.1. *Magnetic flux and helicity from observations*

A gauge-independent relative magnetic helicity per unit length (H_R/L) can be defined for cylindrical flux ropes, independently from the reference field [Démoulin *et al.*, 2002], as:

$$H_r/L = 4\pi \int_0^R A_\varphi B_\varphi \, r dr .$$ (2.1)

Table 1. List of the studied magnetic clouds. The start and the end times, the radius (R) of the cloud, and the helicity sign, are shown.

Event	start	End	$R(10^{-2}AU)$	Helicity Sign
1	22-Aug-1995 22:00:00	23-Aug-1995 19:00:00	9.1	+
2	18-Oct-1995 19:00:00	20-Oct-1995 00:00:00	13.7	+
3	16-Dec-1995 05:00:00	16-Dec-1995 22:00:00	6.5	-
4	27-May-1996 15:00:00	29-May-1996 07:00:00	13.2	-
5	01-Jul-1996 17:00:00	02-Jul-1996 09:00:00	6.5	-
6	07-Aug-1996 13:00:00	08-Aug-1996 10:00:00	8.6	+
7	24-Dec-1996 03:00:00	25-Dec-1996 10:00:00	13.0	+
8	10-Jan-1997 05:00:00	11-Jan-1997 02:00:00	10.1	+
9	21-Apr-1997 15:00:00	23-Apr-1997 07:00:00	8.9	+
10	15-May-1997 09:00:00	16-May-1997 01:00:00	8.4	-
11	16-May-1997 07:00:00	16-May-1997 14:00:00	3.6	-
12	09-Jun-1997 02:00:00	09-Jun-1997 23:00:00	8.2	+
13	19-Jun-1997 05:06:00	19-Jun-1997 17:54:00	4.7	+
14	15-Jul-1997 06:00:00	16-Jul-1997 01:00:00	8.2	-
15	03-Aug-1997 14:00:00	04-Aug-1997 01:00:00	3.2	-
16	18-Sep-1997 00:00:00	20-Sep-1997 12:00:00	20.5	+
17	21-Sep-1997 22:00:00	22-Sep-1997 18:00:00	9.9	-
18	01-Oct-1997 16:00:00	02-Oct-1997 23:00:00	14.8	-
19	10-Oct-1997 23:00:00	12-Oct-1997 00:00:00	12.0	+
20	07-Nov-1997 05:48:00	08-Nov-1997 04:18:00	8.4	+

The azimuthal component of the potential vector, $A_\varphi(r)$, can be written in function of the partial magnetic flux, $\Phi_z(r)$, across a surface perpendicular to the cloud axis as:

$$A_\varphi(r) = \frac{1}{r}\int_0^r r' B_z(r')\, dr' = \frac{\Phi_z(r)}{2\pi r}, \qquad (2.2)$$

and thus, the relative helicity can be computed as an integral of B_φ, weighted with the accumulative flux:

$$H_r/L = 2\int_0^R B_\varphi(r)\Phi_z(r)\, dr\,. \qquad (2.3)$$

This expression to compute H_r/L is meaningful because it shows that, in this particular structure, the helicity is associated with the azimuthal field surrounding the axial flux. This expression also suggests a method to estimate H_r/L directly from the observed field.

For every cloud we construct two subseries for $B_{y,cloud}$ and $B_{z,cloud}$. The first subseries corresponds to the data in the period of time when the spacecraft is going into the cloud

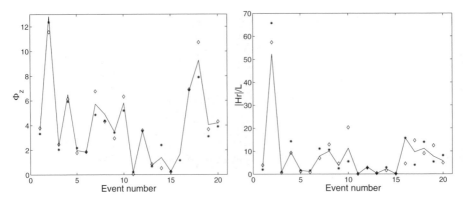

Figure 1. Left panel shows the magnetic flux Φ_z, in units of 10^{20}Mx, computed from the Lundquist's model (continuous line) and from the direct method (in-bound corresponds to diamonds and out-bound to stars). Right panel shows the absolute values of the relative magnetic helicity per unit length ($|H_r|/L$), in units of 10^{41}Mx^2/AU, with the same convention as in the left panel.

until it reaches the minimum distance to the cloud axis (in-bound), and the second one from then, during the outgoing travel (out-bound).

Thus, under the assumption of a cylindrical geometry for the cloud and $p \sim 0$, we calculate $\Phi_z(r)$ and then, using Eq. 2.3, we compute H_r/L for the set of analyzed clouds.

2.2. *Comparison of the direct method with Lundquist's model*

We model the magnetic field configuration of every cloud using the Lundquist's model in the MV coordinates. The physical parameters that fit best the observations ($B_{y,cloud}$ and $B_{z,cloud}$), and the flux and helicity, are computed following the method described in Dasso *et al.* [2003].

Figure 1 shows Φ_z and $|H_r|/L$ computed from the direct method and from the Lundquist's model (L). The left panel of this figure shows that the values of $\Phi_{z,L}$ are between the two values computed for each of the two branches (in-bound and out-bound) of the direct method in 14/20 cases. The values of $\Phi_{z,L}$ were the largest in 6/20 events, but for any of the analyzed clouds $\Phi_{z,L}$ was the lowest. The right panel of Figure 1 shows the absolute value of the relative magnetic helicity per unit length ($|H_r|/L$). The values of $|H_{r,L}/L|$ are between those obtained from the two branches of the direct method in 18/20 clouds, and there was no cloud where $|H_{r,L}/L|$ resulted the largest.

In order to estimate the in-bound/out-bound asymmetry of the clouds, we define $\Delta\Phi_z = |\Phi_{z,out} - \Phi_{z,in}|$, $< \Phi_z >= (\Phi_{z,out} + \Phi_{z,in})/2$, $\Delta H = |H_{r,out} - H_{r,in}|$, and $< H >= (H_{r,out} + H_{r,in})/2$. Figure 2 shows two histograms for $\Delta\Phi_z/ < \Phi_z >$ and $\Delta H/ < H >$, respectively. It can be seen that $\Delta\Phi_z/ < \Phi_z > (\Delta H/ < H >)$ is lower than 0.5 in 17/20 (11/20) clouds. This means that more than the 50% of the studied clouds had $\Delta\Phi_z$ lower than 0.5 times $< \Phi_z >$ and ΔH lower than 0.5 times $< H >$. Thus, we found that the global values of Φ_z and $|H_r|/L$ are affected by the in-bound/out-bound asymmetry in less than the 50% of their mean values for more than half of the studied MCs.

3. Conclusions

We have shown a method to compute the flux and the relative magnetic helicity per unit length for cylindrical flux ropes, and we have applied it to a set of 20 magnetic

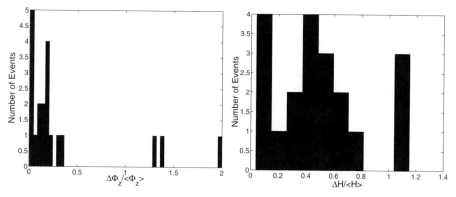

Figure 2. Histogram of the absolute value of the relative difference between in-bound/out-bound flux Φ_z (left panel) and relative helicity per unit length H_r/L (right panel).

clouds. We compare the results obtained with the new method to the values derived using the Lundquist's model.

Our results indicate that there is a relatively good agreement between the two ways to compute these quantities, and also that Lundquist's model tends to overestimate the flux and underestimate the helicity in a few cases.

From the three computed values (direct method, in-bound and out-bound, and Lundquist's model) for this particular set of clouds, we find that: $\Phi_z \sim 10^{18} - 10^{21}$Mx, with a mean value of $\Phi_z = 4\times10^{20}$Mx, and $|H_r|/L \sim 10^{39} - 10^{42}$ Mx^2/AU, with a mean value of $|H_r|/L = 8.2\times10^{41}$ Mx^2/AU.

Furthermore, the relative difference between the in-bound and out-bound estimations for the flux and helicity are lower than 0.5 for more than half of our set of clouds. Thus, from our study, Φ_z and $|H_r|/L$ resulted global magnitudes with a relative uncertainty (due to the in-bound/out-bound asymmetry) lower than 50% for the majority of the clouds.

Acknowledgements

We thank the NASA's Space Physics Data Facility (SPDF) for their data. This work was partially supported by the Argentinean grants: UBACyT X329, PIP 2693 and PIP 2388 (CONICET), and PICT 12187 (ANPCyT). C.H.M. and P.D. thank ECOS (France) and SECyT (Argentina) for their cooperative science program (A01U04). C.H.M. and S.D. are members of the Carrera del Investigador Científico, CONICET.

References

Berger, M.A. 1984. *Geophys. and Astrophys. Fluid Dyn.* 30, 79.
Bothmer, V. & Schwenn, R. 1998. *Annales of Geophys.* 16, 1.
Burlaga, L.F. 1995. *Interplanetary magnetohydrodynamics*, New York, Oxford University Press.
Farrugia, C.J., *et al.* 1995. *J. Geophys. Res.* 100, 12293.
Green, L.M., *et al.* 2002. *Solar Physics* 208, 43.
Dasso, S., *et al.* 2003. *J. Geophys. Res.* 108, 1362, doi:10.1029/2003JA009942.
Dasso S., *et al.* 2005. *Adv. in Space Res.*, in press.
DeVore, C.R., 2000. *Astrophysical Journal* 539, 944.
Démoulin P., *et al.* 2002. *Astronomy and Astrophysics* 382, 650.
Lepping, R.P., *et al.* 1990. *J. Geophys. Res.* 95, 11957.
Low, B.C. 1996. *Solar Physics* 167, 217.

Lundquist, S. 1950. *Magnetohydrostatic Fields. Ark. Fys.* 2, 361.

Mandrini, C.H., *et al.* 2004. *Astrophys. and Space Sci.* 290, 319.

Pevtsov, A.A. & Balasubramaniam, K.S. 2003. *Advances in Space Research* 32 (10), 1867.

Discussion

BOTHMER: The calculation of helicity is improper and the problems are not named! 1. Compressional effects on B. 2. Application of Bessel-functions. 3. Identification of cloud set. 4. Overall topology and improper references.

DASSO: – We presented a method to estimate the content of relative helicity per unit length in the cross section slice of MCs observed at 1AU. Our method assumes a cylindrical symmetry for that cross section, i.e., $\vec{B} = B_\phi(R)\hat{\phi} + B_z(R)\hat{z}$, and we do not consider expansion or compression effects. Under these assumptions, the relative helicity per unit length that we compute from the observations is properly defined and it is gauge and ideal MHD invariant [see, e.g., Berger & Field 1984, Demoulin *et al*, 2002, Dasso *et al*, 2003]. We compared our new method with the classical Lundquist's solution, using the Bessel's functions in the same way that they were used for modeling MCs in several previous papers [see, e.g., Burlaga, JGR, Q3, 7217, 1988; Lepping *et al.* 1990; Lepping *et al.* JGR, 102, 1404Q, 1997]. In this work we analyzed a set of 20 MCs from the Lepping's list (available at http:// Lepmfi.gsfc.nasa.gov/mfi/mag_cloud_pub1.htm). As mentioned in the last viewgraph shown (Summary and Conclusions), we propose to improve our method using expansion effects, elliptical shapes, and comparison with numerical simulations.

Coronal and Stellar Mass Ejections
Proceedings IAU Symposium No. 226, 2005
K. P. Dere, J. Wang & Y. Yan, eds.

© 2005 International Astronomical Union
doi:10.1017/S1743921305000943

Effect of Coronal Mass Ejection Interactions on the SOHO/CELIAS/MTOF Measurements

X. Wang[1], P. Wurz[1], P. Bochsler[1]
F. Ipavich[2], J. Paquette[2] and R.F. Wimmer-Schweingruber[3]

[1]Physics Institute, University of Bern, 3012 Bern, Switzerland
email: xuyu.wang@soho.unibe.ch

[2]University of Maryland, College Park, MD20742, USA
email: ipavich@umtof.umd.edu

[3]Extraterrestrische Physik, University of Kiel, 24098 Kiel, Germany
email: wimmer@physik.uni-kiel.de

Abstract. By using the plasma composition data from SOHO/CELIAS/MTOF, charge states data from ACE/SWICS, combining with the remote sensing observations from SOHO/LASCO white-light image and WIND/WAVES radio emission, we describe a coronal mass ejection (CME) observed on 2001 October 19 16:50 UT to show how the effect of CME interaction appears in the *in situ* measurements. A new narrow shock is formed while the rear CME passing through the core region of the preceding one, which moves faster than the surrounding part and has a new type II radio burst associated with it. Because of its distinguished elemental abundance and unusual low charge states, we connect a density hump observed by MTOF/PM with the preceding CME core. By comparing the relative abundances of minor ions in shock compressive region, ICME region and CME core region with respect to that in upstream slow solar wind, we indicate mass-per-charge dependence of minor thermal ions may be an important imprint of the characteristic velocity of distant acceleration region.

Keywords. shock waves, Sun: abundances, coronal mass ejections (CMEs) solar wind, radio radiation

---

1. Introduction

Our current knowledge of CME comes from two spatial domains: the near-Sun (up to 32 Rs) region remote-sensed by coronagraphs and the geospace and beyond where *in situ* observations are made by spacecraft. Comparing observations from these two domains helps us understand the propagation and evolution of CME through the interplanetary (IP) medium. The CME interaction has important implications for the space weather prediction, for instance, some of the false alarms or complex ejecta with unusual composition could be accounted for by CME interactions (Gopalswamy *et al.*, 2001). The question then arises as to how the plasma composition reflects the CME interaction in the *in situ* measurements. In other words, how does the characteristic velocity of acceleration region show up in the *in situ* measurements? How to fit the distant observations comparing with the *in situ* measurements? What can we learn from distant observations of shocks? Can the correlation study of the compressive region with upstream of shock in the *in situ* measurements be a tracer of the strength of CME interaction?

The purpose of this work is to extend our understanding of shock formation and propagation in the solar wind to the outer coronal region by separating the accelerating effects of IP shock transport and the amplified waves near the Sun on thermal ions.

Compared with energetic particles, thermal ions have advantages to address the injection efficiency of distant shock because their dispersion and transport process are not strong to wash out the initial injection signature. For thermal ions, the shock acceleration efficiency of the observed particles and resonant wave intensities both peak near the time of shock. Acceleration remains strong as shock propagates in IP medium. Therefore, thermal ions take both the information of the distant amplified waves and the IP transport. In order to separate them, we make two assumptions: first, the primary IP transport is a cascading process; second, the cascading process leads to similar features of m/q dependence to downstream particles. The distinguished features relying on the plasma conditions would reflect the injection conditions of shocks, since Mach number and plasma beta are the most important factors to determine the shock injection efficiency. They strongly reply on the local plasma conditions rather than the solar wind speed in the region before the Alfvén point where most CME interactions occur. Therefore, m/q dependence of thermal ions must take some imprints of CME interaction. In the CME interaction, the rear CME has to pass through the core, cavity and frontal of the preceding CME. We address the effects of CME interaction in these inhomogeneous regions by comparing the m/q dependence of their correlation to the upstream solar wind.

We analyze a CME event observed on 2001 October 19 16:50 UT by using the observations of SOHO/CELIAS/MTOF plasma composition, ACE/SWICS charge state, SOHO/LASCO white image, and WIND/WAVES radio emission. First, we report on a new type II spectral feature for CME interaction, then provide the evidences to connect a density hump observed by MTOF/PM with the preceding CME core and show how the effect of CME-interaction appears in the m/q dependence of the correlation of the shock compressive region and CME core region with upstream solar wind, and finally discuss the implications of varied shock strength on thermal ions.

2. Data and Analysis

The general population of CMEs observed by SOHO/LASCO is listed in the on-line catalog (http://cdaw.gsfc.nasa.gov). (For the LASCO instrument, see the review paper of Brueckner *et al.*, 1995). The halo CME onset at 2001 October 19 16:50 UT interacts with a narrow slow CME onset about 7 hours earlier with a speed 466km/s, central position-angle 241°, width 37°, accelerating rate $8.4km/s^2$. The interaction region is located in the field view of LASCO/C3 coronagraph. The image reveals that both CMEs originate from the west front side of solar disk, the preceding one is on the south (S30), the rear one is on the north (N40). So the interaction we observed in the coronagraph images should not be a superposition in the sky plane. The shock front of the rear CME shows different behavior when it passes through the core of the preceding CME at around 18:18 UT. It moves faster than the surrounding part (see the right image of figure 1). At nearly the same time, WIND/WAVES RAD1 and RAD2 radio receivers detected a sudden enhancement of type II emission at the frequency starting at 1200kHz and drifting to 250kHz (For the WAVES experiment, see the paper of Bougeret *et al.*, 1995). The left image of figure 1 shows the dynamic spectrum of the event from 16:30 UT to 21:20 UT. No other feature in the LASCO field of view can be identified as an alternative source of this radio enhancement. Therefore we connect these two phenomena to each other. A possible explanation of this type II burst is: when a fast mode shock front is refracted into a low Alfvén velocity region and the shock strength built up above a certain critical value because of a wave focusing effect. This leads to a concentration of energy favorable to the acceleration of particles in the shock front, resulting in a type II event.

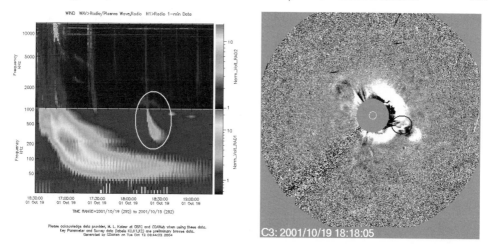

Figure 1. The WIND/WAVES dynamic spectrum on October 19, 2001 from 16:30 UT to 21:20 UT and SOHO/LASCO/C3 image on 2001/10/19 18:18:05 UT when the rear CME passing through the core of the preceding one. The circles indicate the corresponding of a type II emission to the CME interaction.

The MTOF sensor of CELIAS on board SOHO is an isochronous time-of-flight mass spectrometer with a resolution $M/\Delta M$ better than 100 and a temporal resolution of 5 min. It detects ions at solar wind bulk velocities of 300-1000 km/s, corresponding to the energy of about 0.3-3 keV/amu. A sub-sensor of MTOF, Proton Monitor, is used to select appropriate solar wind time intervals for study (Ipavich *et al.*, 1998). A detail description of the CELIAS instrument is given by Hovestadt *et al.* (1995). Another input parameter for determining the MTOF sensor response is the charge-state distribution. The charge-state distribution measured by ACE/SWICS instrument for O, Mg, Si, Fe are used to interpolate the charge state distribution of other elements. (Wurz, 1999; Wurz *et al.*, 2000; Mazzotta *et al.*, 1998).

We identified the ICME and its corresponding CME by referring to the paper of Cane and Richardson (2003). Figure 2 (left) shows the solar wind plasma parameters and mean charge states (O, Mg, Si, Fe) measured by CELIAS/MTOF/PM and ACE/SWICS, respectively. The investigated time periods are indicated in the top four panels. The CME core period with unusual low charge states corresponds to a density hump detected by MTOF/PM with a little time shift attributed to the different locations of the SOHO and ACE spacecraft. Figure 2 (right) shows the double ratio of the abundance in the shock compressive region, ICME region, and CME core region to the referenced upstream abundance of an element X relative to oxygen ($[X/O]/[X/O]_{ref}$) as a function of the mass per charge. Except the element Si (we omit this point because there is only one measured value in that core region without statistical error), the compressive region and CME core region correlated with the upstream solar wind with positive m/q dependence, and no positive m/q dependence in the ICME region.

3. Discussion

When the Mach number of the upstream flow is increased, the magnetic profile of the shock would change: a magnetic foot develops upstream of the main shock ramp and an overshoot appears at the top of the ramp. The changes in magnetic profile are closely associated with the reflected ions at the shock front, which amplify the magnitude of the waves near the shock and provide a necessary additional dissipation to the particles.

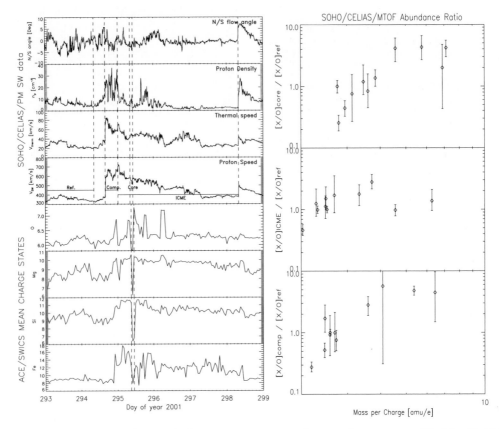

Figure 2. Left: The solar wind plasma parameters and mean charge states (O, Mg, Si, Fe) measured by CELIAS/MTOF/PM and ACE/SWICS, respectively. Right: The double ratio of the abundance for the time periods of shock compressive, ICME, and CME core to the referenced upstream abundance of an element X relative to oxygen ($[X/O]/[X/O]_{ref}$) as a function of the mass per charge. The error bars include statistical and instrumental errors.

Therefore, the strengthened shock would lower the threshold of the injection velocity and feed more energy to low-rigidity particles with an enhanced the acceleration efficiency. The distant amplified MHD waves can't be remotely sensed. They leave some observable imprints on the solar energetic particles (SEPs) in several ways, such as limit intensity of SEPs (Reames and Ng, 1998), time variations of SEP elemental abundances (Tylka, Reames and NG, 1999). To the thermal ions, turbulence effect would show up when the energy is below the threshold energy of shock acceleration by the positive m/q dependence of elemental abundances in correlation with the upstream solar wind. As we see in the event of Oct. 19, 2001, positive m/q dependences are shown in the shock compressive region and CME core region due to the CME interaction. We are looking at this point by checking more CME interaction events with different initial speed, interaction width, etc.

Once the energy of particles reach the threshold energy of shock acceleration, turbulence effect would be covered by the shock acceleration. In an ideal situation, we would expect relatively flat m/q dependence. So far the origin and the injection of the seed particles of IP shocks is not well resolved. There is considerable evidence from heliospheric shock observations and plasma simulations that shocks can directly accelerate thermal particles (Forman and Webb, 1985; Gosling, 1993), although Klecker *et al.* (1981) and

Tan *et al.* (1989) interpreted their results in terms of shock acceleration of ions left over from previous energetic particle events. Seeking the energy channel with flat m/q dependence would be significant to address the problem of the threshold energy of shock acceleration.

Acknowledgements

The CME catalog is generated and maintained by NASA and The Catholic University of America in cooperation with the Naval Research Laboratory. SOHO is a project of international cooperation between ESA and NASA. The dynamic spectrum is provided by M. L. Kalear at GSFC and CDAWeb. This work is supported by the Swiss National Science Foundation.

References

Mazzotta, P., Mazzitelli, G., Colafrancesco, S., & Vittorio, N. 1998, *Astron. Astrophys. Suppl. Ser.* 133, 403

Bougeret, J. L., Kaiser, M. L., Kellogg, P. J., Manning, R., Goetz, K., Monson, S. J., Monge, N., Friel, L., Meetre, C. A., Perche, C., Sitruk, L., & Hoang S. 1995, *Space Sci. Rev.* 71, 231

Brueckner, G. E., Howard, R. A., Koomen, M. J., Korendyke, C. M., Michels, D. J., Moses, J. D., Socker, D. G., Dere, K. P., Lamy, P. L., Llebaria, A., Bout, M. V., Schwenn, R., Simnett, G. M., Bedford, D. K., & Eyles, C. J. 1995, *Sol. Phys.* 162, 357

Forman, M. A. & Webb, G. M. 1985, *in Collisionless Shocks in the Heliosphere: Reviews of current Research, ed. B. T. Tsurutani and R. G. Stone* 91

Gopalswamy, N., Yashiro, S., Kaiser, M. L., Howard, R. A., & Bougeret, J. L. 2001, *Astrophys. J.* 548, L91

Gosling, J. T. 1993, *J. Geophys. Res.* 98, 18937

Hovestadt, D., Hilchenbach, M., Bürgi, A., Klecker, B., Laeverenz, P., Scholer, M., Grünwaldt, H., Axford, W. I., Livi, S., Marsch, E., Wilken, B., Winterhoff, H. P., Ipavich, F. M., Bedini, P., Coplan, M. A., Galvin, A. B., Gloeckler, G., Bochsler, P., Balsiger, H., Fischer, J., Geiss, J., Kallenbach, R., Wurz, P., Reiche, K.-U., Gliem, F., Judge, D. J., Ogawa, H. S., Hsieh, K. C., Moebius, E., Lee, M. A., Managadze, G. G., Verigin, M. I., & Neugebauer, M. 1995, *Sol. Phys.* 162, 441

Ipavich, F. M., Galvin, A. B., Lasley, S. E., Paquette, J. A., Hefti, S., Reiche, K.-U., Coplan, M. A., Gloeckler, G., Bochsler, P., Hovestadt, D., Grünwaldt, H., Hilchenbach, M., Gliem, F., Axford, W. I., Balsiger, H., Bürgi, A., Geiss, J., Hsieh, K. C., Kallenbach, R., Klecker, B., Lee, M. A., Mangadze, G. G., Marsch, E., Möbius, E., Neugebauer, M., Scholer, M., Verigin, M. I., Wilken, B., & Wurz P. 1993, *J. Geophys. Res.* 103, 17205

Klecker, B., Scholer, M., Hovestadt, D., Gloeckler, G., & Ipavich, F. M. 1981, *Astrophys. J.* 251, 393

Reames, D. V. & Ng, C. K. 1998, *Astrophys. J.* 504, 1002

Tan, L. C., Mason, G. M. Klecker, B., & Hovestadt, D. 1989, *Astrophys. J.* 345, 572

Tylka, A. J., Reames, D. V., & Ng, C. K. 1999, *Geophys. Res. Lett.* 26, 2141

Wurz, P. 1999, *Habilitation thesis, University of Bern, Bern, Switzerland*

Wurz, P., Wimmer-Schweingruber, R. F., Bochsler, P., Galvin, A. B., Paquette, J. A., Ipavich, F. M., & Gloeckler, G. 2001, *AIP on Solar and Galactic Composition* CP-598, 145

Coronal and Stellar Mass Ejections
Proceedings IAU Symposium No. 226, 2005
K. P. Dere, J. Wang & Y. Yan, eds.

Correlations between CME Associated Flare Magnitude and *in situ* Quantities

Alysha A. Reinard

Artep Incorporated, 2922 Excelsior Springs Court, Ellicott City, Md 21042
email: reinard@nrl.navy.mil

Abstract. We describe a study of the compositional properties of heliospheric ICME ejecta within the context of solar CME observations. In this study, we examine CME-ICME pairs with an associated flare. For each of these pairs several in situ quantities are averaged over the event and compared with flare magnitude. We find that Mg/O, He/O, and He/H are clearly correlated with flare magnitude suggesting that larger flares provide the CMEs with increased access to the low corona. We also find flare magnitude is positively correlated with velocity and negatively correlated with density, indicating that CMEs which are related to large flares are more likely to experience over-expansion during their propagation to 1 AU.

Keywords. Sun:abundances, Sun:CMEs, Sun:flares, Sun:solar wind

1. Introduction

Coronal mass ejections (CMEs) were first discovered in the 1970s (Tousey 1973) by remote sensing instruments aboard the SOLWIND spacecraft. Interplanetary CMEs (ICMEs) had been observed prior to this, but were thought to consist of flare ejecta, a misconception that prevailed for some time after CMEs were first observed (Gosling 1993). Today the detailed connection between ICMEs and CMEs and the degree to which flares affect this connection are still not well understood. Currently there are spacecraft observing the Sun by means of both remote solar imaging and in situ solar wind observations. Providing a link between these two types of data sets is essential to understanding the processes that cause CMEs.

Reinard 2005a (hereafter referred to as Paper1) investigated the relationship between heliospheric charge state and flare association for 67 CMEs. Charge states provide an in situ measurement of coronal temperatures at the solar source region (Henke *et al.* 1998) and so an association between the hot flare plasma and enhanced charge states was hypothesized. It was found that for the 43 events with flares originating in central regions (defined as 30E<x<45W) the charge state ratio had a moderate correlation with the flare magnitude (i.e. larger flares were associated with larger in situ charge states). This correlation suggests that hot ions from the flare enter and/or heat the CME plasma. In this paper we build on that framework and consider how other in situ quantities relate to flare magnitude. A description of how the events were chosen and analyzed can be found in Paper1. CME-ICME pairs were identified by Cane and Richardson (2003). CMEs were identified with the SOHO/LASCO (Large Angle and Spectrometric COronograph) instrument (Brueckner *et al.* 1995). In situ data were obtained from the ACE (Advanced Composition Explorer) spacecraft (http://www.srl.caltech.edu/ACE/ASC/). Flare identification was made using the Solar Geophysical Data reports on the website maintained by the National Geophysical Data Center, which is run by the National Oceanic and Atmosphere Administration (ftp://ftp.ngdc.noaa.gov/STP/SOLAR_DATA/).

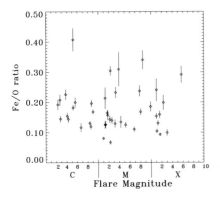

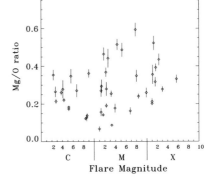

Figure 1. Fe/O is not correlated with flare magnitude

Figure 2. Mg/O is positively correlated with flare magnitude

2. Data Analysis

2.1. *Elemental composition*

In situ measurements of elemental composition reflect the coronal source region of CME ejecta. Typically, in the slow solar wind elements of low first ionization potential (FIP) such as Fe, Mg, Si are enhanced over high-FIP elements, such as Ne and He (Geiss *et al.* 1995). The solar wind is also somewhat depleted in heavy elements compared to photospheric values, because of gravitational settling in the corona. The solar wind is particularly depleted in helium (by a factor of two compared to photospheric values) because of inefficient Coulomb drag during solar wind expansion (Geiss *et al.* 1970).

To determine if elemental abundances are affected by associated flares, we take the elemental ratios Fe, Mg, Ne, and He with respect to oxygen and He/H and look at their correlations with flare magnitude. As discussed in Paper 1 we determine the statistical significance (defined as $r\sqrt{n-2}/\sqrt{1-r^2}$) of each correlation to rule out the possibility that the correlation could have occurred randomly (Larsen & Marx 1986). Conventionally, statistical significance is set at the 5% level, meaning that values above 5% are considered to be consistent with a random population. We keep in mind that this level is somewhat arbitrary. In addition, we use non-parametric, or rank-ordered, equations because of the low statistics and because the parent population may not be normally distributed. These equations reduce to parametric equations for normally distributed populations, so no information is lost by generalizing. Both parametric and nonparametric results are listed in Table 1, but in the text we only refer to the nonparametric results.

In figure 1 we plot the in situ elemental abundance ratio Fe/O versus flare magnitude for central flare events (as defined in Paper1). The vertical lines through each point are the statistical error bars, defined as $\sigma/\sqrt{n}$ where n is the number of points averaged in each event and σ is the standard deviation of the event. Clearly Fe/O is not well correlated with flare magnitude. The correlation coefficient is -0.08 and the statistical significance is 62.9%. On the other hand, Mg/O (figure 2), also a low FIP element, has a correlation coefficient of 0.33 and a statistical significance of 3.1%, which is within the 5% significance level. Ne/O (figure 3), a high FIP element, has a correlation coefficient of 0.27, and a statistical significance of 8.9%. Given the arbitrary nature of the 5% cutoff we cannot rule out the possibility that a weak correlation exists.

Turning to helium we find that He/O (figure 4) is strongly correlated with flare magnitude (r=0.40, statistical significance = 0.7%). Similarly, He/H (figure 5 has a correlation

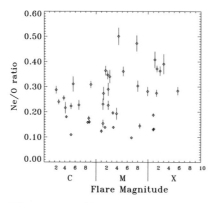

Figure 3. Ne/O has a weak positive relationship with flare magnitude

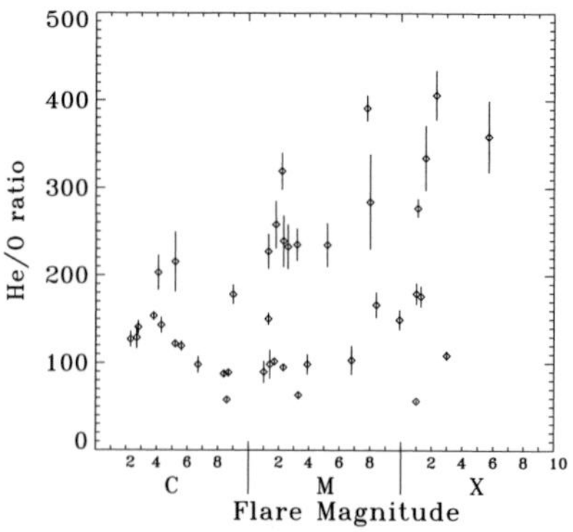

Figure 4. He/O is strongly correlated with flare magnitude

Table 1. Correlation coefficients and statistical significances for several in situ parameters as a function of flare magnitude

	# events	r_p	significance	r_{np}	significance
Fe/O	43	-0.04	82.4%	-0.08	62.9%
Mg/O	43	0.31	4.3%	0.33	3.1%
Ne/O	42	0.25	11.0%	0.27	8.9%
He/O	43	0.26	9.2%	0.40	0.7%
He/H	43	0.45	0.2%	0.49	0.07%
velocity	67	0.28	2.2%	0.24	4.9%
temperature	67	0.21	8.8%	0.135	27.3%
density	67	-0.27	2.7%	-0.35	0.3%

coefficient of 0.49, with a statistical significance of 0.07%. These results are consistent with previous studies relating helium enhancements to large flare events (Hirshberg 1971, 1972). In addition, He/H has previously been found to increase with a rising O^{+7}/O^{+6} (Reinard *et al.* 2001), so a mutual correlation with flare magnitude is not surprising.

The strong correlation between helium ratios and flare magnitude indicates that larger flares provide more access to the low corona. Heavy elements give a mixed result with Mg/O correlating well with flare magnitude, but Ne/O and Fe/O showing weak or no correlations. These results are not ordered by FIP effect, as Mg/O and Fe/O, both low FIP elements, have strikingly different correlations. Gravitational settling does not account for this effect unless Fe accumulates in deeper, inaccessible regions of the corona. However, these results are consistent with past results finding that Mg/O is well correlated with O^{+7}/O^{+6} in the solar wind (von Steiger *et al.* 1995), compared with a marginal correlation between Fe/O and O^{+7}/O^{+6} (Aellig *et al.* 1999).

2.2. *Plasma parameters*

While in situ elemental abundances are for the most part unchanged from coronal values, plasma parameters (velocity, temperature, density) are influenced during the CME expansion into the heliosphere. CME velocities tend to approach the ambient solar wind values (Lindsay *et al.* 1999), while temperatures and densities decrease due to expansion effects, particularly in large, fast CMEs (Gosling *et al.* 1994). Investigating these parameters can determine what effect flare magnitude has on CME propagation. Because plasma parameters are less localized than composition parameters (Reinard 2005b), we

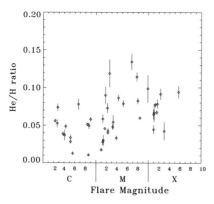

Figure 5. He/H is strongly positively correlated with flare magnitude

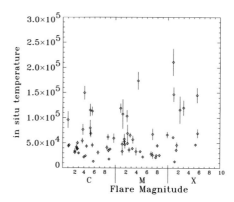

Figure 6. Temperature displays a slight positive correlation with flare magnitude.

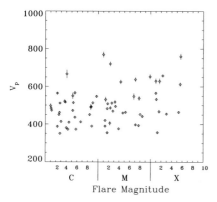

Figure 7. Velocity has a positive correlation with flare magnitude

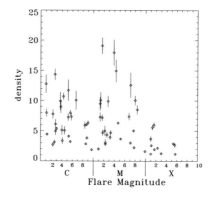

Figure 8. Density is has a strong negative correlation with flare magnitude

examine all 67 flare associated events rather than restricting our results to the central events.

Temperature (figure 6) displays a slight positive correlation, though the correlation coefficient is low (0.14) and the value is not statistically significant (27.3%). From Paper1 we would expect a higher initial temperature for larger flare events. However, Reinard *et al.* 2001 found an inverse correlation between in situ temperature and charge state. It is likely that competing effects of initial heating and subsequent expansion have an equalizing effect on the events of different initial energies.

In figure 7 we find that velocity has a positive correlation with flare magnitude (r=0.24, significance=4.9%), indicating that larger flares are associated with faster CMEs at 1 AU. This result is consistent with past studies that found that flare-associated CMEs are faster than non-flare associated CMEs (Gosling *et al.* 1974). Density (figure 8) is strongly anticorrelated with flare magnitude (r=-0.35, statistical significance=0.3%). In particular, events with a large flare association always have a low density. This result suggests that larger flares are associated with more explosive CMEs, which have a stronger overexpansion, resulting in a lower in situ density.

3. Conclusions

We present results from a study to better understand the relationship between flares, CMEs, and ICMEs. We first consider several elemental ratios. Mg/O, He/O, and He/H each correlate strongly with flare magnitude. This result indicates that large flares provide the CME with access to regions in the low corona where heavy elements are enhanced, though it is unclear why Fe/O does not follow this trend. More work needs to be done to relate these results to the conclusions from Paper1 and determine the true cause and effect nature of these correlations.

We then consider plasma parameters. We find no correlation between in situ temperature and flare magnitude. Though CMEs associated with large flares may initially have higher temperatures (given by the O^{+7}/O^{+6} ratio), these events seem more likely to overexpand and cool, which may have an equalizing effect on the in situ temperatures. Velocity is positively correlated with flare magnitude, while density is negatively correlated. We conclude that the associated flare size affects the degree to which a CME will overexpand during propagation into the heliosphere. Large flare-related CMEs are more likely to overexpand and arrive at 1 AU with higher velocities and much lower densities.

Acknowledgements

This material is based upon work supported by the National Science Foundation under Grant No. 0327715.

References

M. R. Aellig, S. Hefti, H. Grunwaldt, P. Bochsler, P. Wurz, F. M. Ipavich, & D. Hovestadt 1999, *J. Geophys. Res.* 104, 24769

G. E. Brueckner, R. A. Howard, M. J. Koomen, C. M. Korendyke, D. J. Michels, J. D. Moses, D. G. Socker, K. P. Dere, P. L. Lamy, A. Llebaria, M. V. Bout, R. Schwenn, G. M. Simnett, D. K. Bedford, & C. J. Eyles 1995, *Sol. Phys.* 162, 357

H. V. Cane & I. G. Richardson 2003, *J. Geophys. Res.* 108, A4, SSH 6-1

J. Geiss, P. Hirt, & H. Leutwyler 1970, *Sol. Phys.* 12, 458

J. Geiss, G. Gloeckler, R. von-Steiger, H. Balsiger, L. A. Fisk, A. B. Galvin, F. M. Ipavich, S. Livi, J. F. McKenzie, K. W. Ogilvie, & B. Wilken 1995, *Science* 268, 1033

J. T. Gosling, E. Hildner, R. M. MacQueen, R. H. Munro, A. I. Poland, & C. L. Ross 1974, J. Geophys. Res., 79, 4581

J. T. Gosling 1993, *J. Geophys. Res.*, 98, 18949

J. T. Gosling, S. J. Bame, D. J. McComas, J. L. Phillips, E. E. Scime, V. J. Pizzo, B. E. Goldstein, & A. Balogh 1994, *Geophys. Res. Lett.*, 21, 237

T. Henke, J. Woch, U. Mall, S. Livi, B. Wilken, R. Schwenn, G. Gloeckler, R. von Steiger, R. J. Forsyth, & A. Balogh 1998, *Geophys. Res. Lett.* 25, 3465

J. Hirshberg, J. R. Asbridge, & D. E. Robbins 1971, *Sol. Phys.*, 18, 313

J. Hirshberg, S. J. Bame, & D. E. Robbins 1972, *Sol. Phys.*, 23, 467

R. J. Larsen & L. M. Marx, 1986, An Introduction to Mathematical Statistics and its Applications, (2nd ed.; Englewood Cliffs, New Jersey: Prentice-Hall)

G. M. Lindsay, J. G. Luhmann, C. T. Russell, & J. T. Gosling, 1999, *J. Geophys. Res.*, 104, 12515

A. A. Reinard, T. H. Zurbuchen, L. A. Fisk, S. T. Lepri, G. Gloeckler, & R. M. Skoug in *Solar and Galactic Composition*, edited by R. F. Wimmer-Schweingruber, AIP conference proceedings, Woodbury, NY, 139, 2001.

A. A. Reinard 2005, *Astrophys. J.*, 618, in press.

A. A. Reinard 2005, *Astrophys. J.*, in preparation

R. Tousey 1973, *Space Res.* 13, 713

R. von Steiger, R. F. Wimmer-Schweingruber, J. Geiss, & G. Gloeckler 1995, *Adv. Space Res.* 15, (7)3

Discussion

FORBES: The oxygen abundance ratios you use cover the range from $1\text{-}3 \times 10^6$ °K, but typical flare temperatures are in the range from $10\text{-}30 \times 10^6$ °K. Do you think that one reason why your correlation with flare size is not stronger could be that the oxygen abundance ratio reflects condition during the late phase of the flare rather than during the early impulse phase as used for determine the flare size?

REINARD: That may be a factor. Other factors causing the lower temperatures include: charge states freeze in at 2-4$R_\odot$ in the slow solar wind, and by that time the 10MK plasma may have cooled to 1MK; it could be that the flare indirectly heats the CME plasma, and may produce lower temperatures for that reason; it's also possible that the upflow jet is not as hot as the downflow reconnection jet. As far as your question, I'll try doing a comparison that includes flare length or the area under the flare curve rather than just the maximum. Using the time difference from the peak until the end may indicate whether the late phase is more important. Thanks!

Coronal and Stellar Mass Ejections
Proceedings IAU Symposium No. 226, 2005
K. P. Dere, J. Wang & Y. Yan, eds.

Bidirectional Proton Flows and Comparison of Freezing-in Temperatures in ICMEs and Magnetic Clouds

L. Rodriguez[1], J. Woch[1], N. Krupp[1], M. Fränz[1], R. von Steiger[2], C. Cid[3], R. Forsyth[4] and K.-H. Glaßmeier[5]

[1]Max-Planck-Institut für Sonnensystemforschung, Katlenburg-Lindau, D 37191, Germany
email: rodriguez@mps.mpg.de
[2]International Space Science Institute, CH-3012 Bern, Switzerland
[3]Departamento de Física, Universidad de Alcalá, 28871 Alcalá de Henares, Madrid, Spain
[4]The Blackett Laboratory, Imperial College, London SW7 2BW, UK
[5]Institut für Geophysik und extraterrestrische Physik, Technische Universität Braunschweig,
D-38106 Braunschweig, Germany

Abstract. From all the transient events identified in interplanetary space by in-situ measurements, Magnetic Clouds (MCs) are among the most intriguing ones. They are a special kind of Interplanetary Coronal Mass Ejections (ICMEs), characterized by a well-defined magnetic field configuration. We use a list of 40 MCs detected by Ulysses to study bidirectional flows of protons in the ~0.5 MeV energy range. Solar wind ions are also analysed in order to compare cloud to non-cloud ICMEs.

The enhancement in freezing-in temperatures inside the clouds, obtained with data from the SWICS instrument, provides insights into processes occurring early during the ejection of the material and represents a complementary tool to differentiate cloud from non-cloud ICMEs. At higher energies, directional information for protons obtained with the EPAC instrument allows a comparison with previous results concerning bidirectional suprathermal electrons. The findings are qualitatively comparable. Apparently, the portion of bidirectional flows inside magnetic clouds is neither heavily dependent on distance from the Sun nor on parameters obtained from a flux rope model.

Keywords. Sun: corona, Sun: abundances, Sun: coronal mass ejections (CMEs), Sun: solar wind, Sun: particle emission

1. Introduction

ICMEs are the interplanetary manifestation of CMEs. As they propagate in the heliosphere, their internal properties and configuration develop to the extent that it is difficult to relate them back to what was seen at a few solar radii from the Sun. An ICME detected at several AUs from the Sun has expanded and distorted, there is no simple picture of what the global topology of these structures might look like (see e.g. Riley & Crooker (2004)). On the other hand, it is only in interplanetary space where an important subset of all the ICMEs can be detected, namely magnetic clouds (MCs). They are defined by the combination of a large-scale smooth field rotation, enhanced magnetic field magnitude, decreased plasma temperature and low plasma-beta (Burlaga (1991)). Furthermore, they represent the interplanetary manifestation of a flux rope expelled from the Sun.

From all the physical parameters which can be measured in-situ by a spacecraft located in interplanetary space, there is one which remains unaltered, not affected by the undergoing development of the solar wind parcel to which it belongs. This parameter is the ionization level of the solar wind ions, which due to the low densities prevailing already at a few solar radii from the Sun, remains unchanged as the solar wind propagates outwards. By measuring charge states of solar wind ions, thermodynamic properties present in the source region of the solar wind, can be analysed at any distance in the heliosphere.

Several previous studies (e.g. Schwenn *et al.* (1980); Galvin (1997); Henke *et al.* (2001); Lepri & Zurbuchen (2001); Rodriguez *et al.* (2004)) have explored the relation between charge states and ICMEs. The general finding is that charge states are increased inside ICMEs, with respect to values observed in quiet solar wind. Henke *et al.* (2001) adduced this increases not to all the ICMEs but mostly to those which show a flux rope structure (magnetic clouds). Rodriguez *et al.* (2004) extended Henke analysis to a set of 40 magnetic clouds detected by Ulysses, finding that the increased charge states are present at all latitudes and phases of the solar cycle. In this work, we will compare two sets of ICMEs, the first one consisting of magnetic clouds and the second one composed only of non-cloud ICMEs. The possible difference between cloud and non cloud ICMEs, regarding their ionization levels, will be studied. In this regard it is important to clarify whether charge states, in addition to the magnetic field structure, provide another in-situ tool to differentiate MCs from non-cloud ICMEs.

Early observations (Morrison (1954); Gold (1959)) prompted the possibility that the footpoints of ICMEs are still connected back to the Sun as they expand and propagate in interplanetary space. More recently an explanation for such connection has been pursued with in-situ data on bidirectional suprathermal electron flows (BDEs, e.g. Crooker *et al.* (1990)), providing that this counterstreaming particles originate in the footpoints of the ICME still anchored back to the Sun. At higher energies, bidirectional fluxes similar to the ones seen for $\sim$100 eV electrons were first reported by Rao *et al.* (1967) and have been more recently investigated by several authors (e.g. Marsden *et al.* (1987)). The explanation raised for such behavior again suggests the presence of magnetic fields loops connected to the Sun. By using a rich set of events, we are in the position to estimate the degree of bidirectionality (and corresponding connectivity) of MCs in the heliosphere. Similar studies has been carried out for suprathermal electrons, at 1 AU Shodhan *et al.* (2002) found BDE intervals within MCs, covering 0% to 100% of the total duration of the events, with 59% as average value. Less degree of counterstreaming fluxes was detected in solar minimum and the percentage decreases with decreasing cloud size. Similar result (69% average of BDE intervals) was found by Riley *et al.* (2004). We plan to compare here these results with those obtained from near relativistic particles inside MCs.

2. Data

Charge state distributions of oxygen were derived from SWICS (Gloeckler *et al.* (1992)) measurements. From charge state ratios (3-hour resolution data, described in von Steiger *et al.* (2002)) the freezing-in temperatures of oxygen (Hundhausen *et al.* (1968)) was calculated assuming the equilibrium ionization rates of Arnaud & Rothenflug (1985).

Directional information of protons (0.63-0.77 MeV) was obtained with the EPAC instrument (Keppler *et al.* (1992)). The Energetic PArticles Composition instrument EPAC was designed to provide information on the flux, anisotropy and chemical composition of energetic particles in interplanetary space. It comprises four telescopes, each of them with a geometric factor of about 0.08 cm^{2}sr and a field-of-view with a full angle of 35°. The telescopes are inclined at angles of 22.5°, 67.5°, 112.5° and 157.5°, with respect to the spacecraft spin axis (which points towards Earth). For protons, each of the four telescopes is divided in 8 sectors, providing 32 possible directions to detect incoming particles. Telescopes, sectors and the spacecraft spin, allows EPAC to sample 80% of the sphere.

The mentioned data is used in this work to analyze a list of 40 magnetic clouds detected throughout the Ulysses mission and described in Rodriguez *et al.* (2004). The non-cloud ICMEs were obtained by selecting the events present in the Ulysses ICME list, (maintained by the SWOOPS team and available at http://swoops.lanl.gov/cme_list.html) which were not defined as MCs in the first list.

3. Bidirectional protons

We have analyzed protons in the energy range 0.63 - 0.77 MeV. The information from the 32 available different incoming directions (8 sectors and 4 telescopes) have been extended using a spherical expansion method described in Fränz & Krupp (1993). With this method we obtain the harmonic coefficients up to second order which allow us to characterize the particles' directional anisotropies.

More specifically and due to the points discussed in the Introduction, we are interested in bidirectional field aligned flows. Therefore we deal mainly with the second order harmonic coefficient A_{20}. Positive values of it represent field aligned bidirectional particle fluxes with stronger bidirectionality as A_{20} increases. For values lower than zero, the particles gyrate around the magnetic field, with a pitch angle close to 90°. In this work we use the dimensionless value of A_{20}, obtained after dividing it by A_0, which is the zero order coefficient, representing the isotropic portion of the distribution. The use of this coefficient is an important aid which complements and helps to quantify the eye inspection of pitch angle plots. Figure 1 shows a colored pitch angle representation, along with the A_{20} coefficient for one event. This MC (delimited by the vertical solid lines in Figure 1) occurred in 2001, when Ulysses was located at 25° south of the ecliptic during its second fast latitude scan. During cloud passage, the pitch angle plot (top panel) shows a clear bidirectionality along the field, with high fluxes at 0° and 180° and minima close to 90°. A_{20} encompasses this description by increasing above 1, during the whole duration of the event. The bidirectional flows extent further into the trailing part of the cloud as more closed field lines seem to trail the MC. It is not the case for the frontal part in which the bidirectional flows appear suddenly and near to the commencement of the flux rope structure. This particular MC will be further analysed in a future work.

Two values of A_{20} were obtained analyzing 1-hour averages of EPAC data for each event. The first value represents the average over the positive A_{20} counts (bidirectional field aligned population); the second one was taken from the negative A_{20} values (indication for a population with ~ 90° pitch angle gyrating around the magnetic field). In Table 1, a brief summary of the results obtained is given. The energetic protons are predominantly bidirectional (in 88% of the cases) in comparison with few cases in which they were found with pitch angles close to 90° (12%). Considering their duration, relative to that of the MC, we find that (in average) positive values of A_{20} are present during 60% of the duration of the MC and negative ones during 44% (both percentages were calculated using the respective subsets to which they belong, i.e. $A_{20}>0$ or $A_{20}<0$, as 100%). Low positive values of A_{20} do not indicate a clear bidirectionality, as can be seen directly by eye inspection of the pitch angle plots. In order to assure bidirectionality, we have set a threshold for A_{20} at 0.5. In this way, we can estimate that clear bidirectionality is present in 33% of the studied cases with duration averaging 52% of the total duration of the event.

These results are qualitatively in agreement with those from Shodhan *et al.* (2002) and Riley *et al.* (2004). By inspection of the pitch angle plots it has been inferred,

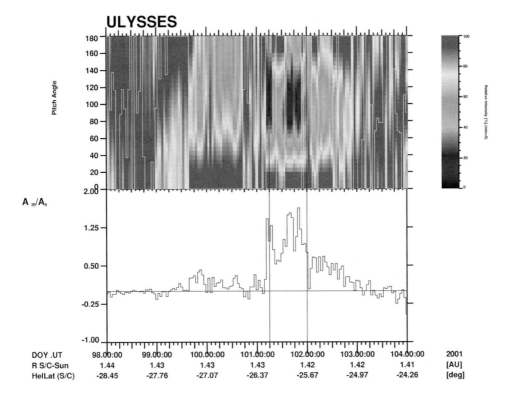

Figure 1. Pitch angle and A_{20} harmonic coefficient for the MC detected by Ulysses in April 2001. The solid vertical lines demark the MC interval as identified by Rodriguez *et al.* (2004)

	$A_{20}>0$	$A_{20}<0$	$A_{20}>0.5$
% of cases	88	12	33
% duration	60	44	52
Avg. value	0.47	-0.2	1.09

Table 1. Summary of the behavior of the A_{20} coefficient inside magnetic clouds.

nevertheless, that the bidirectional characteristics of the more energetic particles studied here show a higher degree of patchiness than its low energy counterparts. This has not been quantified here, since no difference was made on whether the values of A_{20} were contiguous or randomly spread over the event.

Based on calculations on reconnection rates, Riley *et al.* (2004) estimated a decrease of connectivity as the ICMEs propagates outwards of ∼2%/AU. A similar trend was found here using $A_{20}>0$, Figure 2. Somewhat steeper (5%/AU), the slope of the linear fit should be only carefully taken into consideration, due to the high level of scatter present in the data.

In a further approach to try to correlate these periods in which bidirectional fluxes are detected and may, therefore, represent a possible connection of the field lines back to the Sun, we have used an elliptical flux rope model described in Hidalgo *et al.* (2002). By its application to the events under study one obtains parameters such as the orientation of the flux rope axis, current densities and geometric variables describing the expected shape of the clouds. After a thoughtful comparison we can conclude that there is no

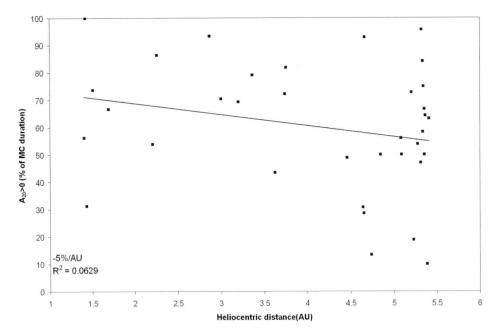

Figure 2. Development of the A_{20} coefficient with respect to distance from the Sun. A negative trend is seen, though the highly scattered points preclude further conclusions to be drawn.

striking dependence between the different parameters obtained from the model and the A_{20} coefficient. The degree of connectivity of the clouds seems to be independent of the local geometric characteristics as inferred from the model.

4. Freezing-in temperatures

Early observations of ions in the solar wind other than protons and alpha particles, such as singly ionized helium or several charge states of oxygen (e.g. Bame *et al.* (1968); Hundhausen (1968)), opened an active field in the investigation of coronal characteristics by means of in-situ interplanetary data. As the solar wind expands outward, the coronal electron density decreases to the extent that the time scale of coronal expansion is short compared to the ionization and recombination timescale. At this height in the corona (a few solar radii), the relative ionization states become constant, they 'freeze-in', reflecting the conditions at this altitude. At any further distance in space, the measurement of the charge states can be used to infer the electron temperature at the freezing-in altitude, providing thus a link between interplanetary and coronal conditions. Although this might be a simplified approximation, it constitutes a valid tool to derive coronal properties in interplanetary space. Charge states represent in this way an imprint of the solar wind source, in contrast to other plasma parameters such as density, velocity and temperature, which vary significantly between the corona and interplanetary space.

For the reasons exposed in the previous paragraph, charge state distributions of heavy ions in the solar wind are a good indicator of the solar wind type (e.g. von Steiger *et al.* (2000)), providing a robust tool for differentiating fast wind (from coronal holes), slow wind (associated with streamers) and transient-related solar wind. It is the latter the one that concern us here, as was stated in the Introduction to this paper, there have been

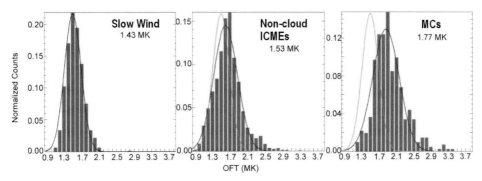

Figure 3. Oxygen freezing-in temperature distributions for slow wind, non-cloud ICMEs and MCs. Solid lines represent a Gaussian fit to the data. The slow wind Gaussian is shown for comparison in the ICME and MC distribution.

ongoing discussions in the past on whether all ICMEs or only those defined as magnetic clouds show increases in freezing-in temperature (or charge states).

Figure 3 shows a comparison of oxygen freezing-in temperatures distribution for slow wind and ICMEs with and without MC structure. The magnetic clouds are the 40 events listed in Rodriguez *et al.* (2004). Non-cloud ICMEs were obtained from the Ulysses ICME list, maintained by the SWOOPS instrument team, this list does not differentiate between cloud and non-cloud ICMEs, which were obtained simply by using those events not present in the MC list. The solar wind samples were selected from several periods in different years of the Ulysses mission, with special care taken in order to include only periods in which no transient events were present. A Gaussian curve was fitted to the three distributions and the values of the oxygen freezing-in temperature (OFT) shown in the plot are those obtained as the center of the Gaussian. Between slow wind and MCs there is a difference of 0.34 MK, whereas for the case slow wind-non-cloud ICMEs this difference reduces to 0.1 MK.

The difference with respect to the slow wind is three times higher in flux rope type ICMEs with respect to those without magnetic cloud signatures. We believe that this statistic comparison clarifies that the enhancement in freezing-in temperatures is clear for MCs, though also increased for non-cloud ICMEs to a much lesser extent.

5. Summary and conclusions

We have pursued a two-fold analysis using a set of magnetic clouds and non-cloud ICMEs relating them with energetic particles and solar wind ions.

In the first part of this work we believe to have shown that the behavior of the energetic protons of $\sim$0.5 MeV is qualitatively similar to that of the suprathermal electrons, as measured by other authors, regarding their bidirectionality. Nevertheless the energetic particles seem to be less bidirectional if one poses stricter conditions to the classifications process (i.e. $A_{20} > 0.5$), we believe that this threshold describes the bidirectional flows much clearer than simply using positive values of A_{20}. The harmonic expansion of data with sufficient directional information provides the possibility to use anisotropy coefficients to complement the analysis of pitch angle plots. There is only a weak correlation between percentage of bidirectional periods inside magnetic clouds with respect to distance from the Sun. The trend is negative as expected, but the high scatter in the data prevents us to make further conclusions. As a result of comparing the A_{20} coefficient with parameters obtained from a flux rope model we conclude that there is no apparent correlation between them.

The second part of the present paper consisted in a comparison of oxygen freezing-in temperatures between magnetic clouds and ICMEs without a cloud structure. It was established that the increases in oxygen freezing in temperature, and therefore charge states, are more than three times higher for magnetic clouds than for non-clouds ICMEs, compared to quiet solar wind values. The oxygen freezing in temperature (OFT) can be used as a further magnetic cloud identifier, complementing in this way the signatures used at present. It is clear that there is a very important relation between charge states and magnetic configuration. For example, the ∼0.3 MK difference in OFT between fast and slow wind originates from the open vs. closed (respectively) magnetic field configuration present at the source region of the solar wind. Therefore, using the same way of thinking to associate the different magnetic field topology of a MC and a non-cloud ICME similar conclusions can be drawn. It is the characteristic magnetic field configuration of a MC, most probably present at the height where the ions freeze-in, what creates a difference in temperatures that can be detected later in interplanetary space. In turn, this would mean that the magnetic field configuration in cloud and non-cloud ICMEs is present since their birth and early development, and would then not be consequence of interplanetary effects (deformation, expansion, etc).

References

Arnaud, M.& Rothenflug, R., Astron. Astrophys Suppl. Ser., 60, 425, 1985.

Bame, S. J., Hundhausen, A. J., Asbridge, J. R., & Strong, I. B., Astronomical Journal, 73, 55, 1968.

Burlaga, L. F. E., in Physics of the Inner Heliosphere II, ed. R. Schwenn & E. Marsch, Springer-Verlag, Berln, 1991.

Crooker, N. U., Gosling, J. T., Smith, E. J., & Russell, C. T., in Physics of Magnetic Flux Ropes, ed. C. T. Russell, E.R. Priest & L.C. Lee, AGU., Washington D.C., 1990.

Galvin, A. B., in Coronal Mass Ejections, ed. N. Crooker, J.A. Joselyn & J. Feynman, 1997.

Henke, T., Woch, J., Schwenn, R., Mall, U., Gloeckler, G., von Steiger, R., Forsyth, R., & Balogh, A., J. Geophys. Res., 106, 10597-10613, 2001.

Fränz, M. & Krupp, N., Internal Paper B v2, MPAe, 1993.

Hundhausen, A. J., Gilbert, H. E., & Bame, S. J., J. Geophys. Res., 73, 5485, 1968.

Gloeckler, G., Geiss, J., Balsiger, H., Bedini, P., Cain, J. C., Fischer, J., Fisk, L. A. *et al.*, Astron. Astrophys. Suppl., 922, 267, 1992.

Gold,T., J. Geophys. Res., 64, 1665, 1959.

Keppler, E., Blake, J. B., Hovestadt, D., Korth, A., Quenby, J., Umlauft, G., & Woch, J., Astron. Astrophys. Suppl., 922, 317, 1992.

Hidalgo, M. A., Nieves-Chinchilla, T., & Cid, C., Geophys. Res. Lett., 29, 13, 15-1, DOI 10.1029/2001GL013875, 2002.

Lepri, S. T., Zurbuchen, T. H., Fisk, L. A., Richardson, I. G., Cane, H. V., & Gloeckler G., J. Geophys. Res., 106, A12, 29231, 2001.

Marsden, R.G., Sanderson, T.R., Tranquille, C., Wenzel, K.-P., & Smith, E.J., J. Geophys. Res., 92, 11, 9, 1987.

Morrison, P., Phys. Rev., 95, 646, 1954.

Riley, P. & Crooker, N. U., ApJ, 600, 2, 1035, 2004.

Riley, P., Gosling, J. T., & Crooker, N. U., ApJ, 608, 2, 1100, 2004.

Rao, U.R. & McCracken, K.G., Bukata, J. Geophys. Res., 72, 4325, 1967.

Rodriguez, L., Woch, J., Krupp, N., Fr(ä)nz, M., von Steiger, R., Forsyth, R., Reisenfeld, D., & Glameier, K.-H., J. Geophys. Res., 109, A01108, doi:10.1029/2003JA010156, 2004.

Schwenn, R., Rosenbauer, H., & Muehlhaeuser, K.-H., Geophys. Res. Lett., 7, 201, 1980.

Shodhan, S., Crooker, N. U., Kahler, S. W., Fitzenreiter, R. J., Larson, D. E., Lepping, R. P., Siscoe, G. L., & Gosling, J. T, J. Geophys. Res., 105, 27, 261, 2000.

von Steiger, R., Schwadron, N. A., Fisk, L. A., Geiss, J., Gloeckler, G., Hefti, S., Wilken, B., Wimmer-Schweingruber, R. F., & Zurbuchen, T. H., J. Geophys. Res., 105 (A12), 27, 217-27, 238, 2000.

Discussion

TYLKA: Do you see these magnetic clouds at all latitudes? Does the correlation between SEP composition inside and outside the MC depend on latitude?

RODRIGUEZ: We see MCs at all latitudes. Nevertheless most of them are found below 40°. Apparently latitude is not a determinant factor for energetic particles' composition inside MCs. The important factor seems to be the presence and characteristics of shocks. This needs nevertheless, further study.

GOPALSWAMY: Have you looked at the difference composition signatures of filament and active region related ICMEs? One expects temperature difference in these two source regions.

RODRIGUEZ: It is very difficult to unambiguously find the source region of ICMEs observed by Ulysses. This is due to the orbit of Ulysses, which introduces big error margins in the back mapping procedure. Also many CMEs have originated behind the solar limb. Nevertheless this is an interesting point and we will analyze the events which source region can be identified.

Coronal and Stellar Mass Ejections
Proceedings IAU Symposium No. 226, 2005
K. P. Dere, J. Wang & Y. Yan, eds.

Magnetic Field Configuration Around Large Flux Ropes

E. Romashets[1] and M. Vandas[2]

[1]IZMIRAN, Troitsk, 149190, Russia
email: romash@izmiran.rssi.ru

[2]Astronomical Institute, 14131 Praha 4, Czech Republic
email: vandas@ig.cas.cz

Abstract. An analytical method is used to model a magnetic field distribution in the vicinity of a large interplanetary or solar flux rope. The field is a sum of the pre-existing one and an additional current-free part. An example using real data is shown.

Keywords. Sun: magnetic fields, Sun: coronal mass ejections (CMEs), methods: analytical.

1. Introduction

It is widely accepted that magnetic flux ropes reside in the solar atmosphere and that some of them erupt and propagate in the interplanetary space. In the present paper we model a magnetic field $\mathbf{B}^{\text{tot}}$ around a flux rope. We start from a pre-existing ambient (background) field $\mathbf{B}^{\text{amb}}$ and add an additional potential field $\mathbf{B}^{\text{add}}$. The rope is assumed to have a toroidal shape with a major radius R_0. Its minor radius r_0 may vary, depending on an azimuthal angle φ along the circular toroid's axis (with the radius R_0).

In toroidal coordinates, μ, η, and φ (Vandas, Romashets & Watari 2003) a boundary of the toroidal flux rope is determined by μ_0, $\cosh\mu_0(\varphi) = R_0/r_0(\varphi)$. The exterior of the flux rope has $\mu > \mu_0$.

The field $\mathbf{B}^{\text{tot}}$ must have zero normal components at the flux rope boundary, i.e.,

$$(\mathbf{B}^{\text{tot}}.\mathbf{n})|_{\mu=\mu_0} = 0 \tag{1.1}$$

must hold for all normal vectors $\mathbf{n}$ to the rope boundary. In order to fulfill the condition (1.1), an additional field $\mathbf{B}^{\text{add}}$ is added to the background field so the resulting external field is $\mathbf{B}^{\text{tot}} = \mathbf{B}^{\text{amb}} + \mathbf{B}^{\text{add}}$. The additional field is supposed to be potential and vanishing at large distances from the flux rope.

According to Romashets & Vandas (2004), the components of the resulting field are

$$B_\mu^{\text{tot}} = B_\mu^{\text{amb}} + B_0 \sinh\mu\sqrt{\cosh\mu - \cos\eta} \sum \left[\frac{1}{2}P_{n-1/2}^m(\cosh\mu)\right. \tag{1.2}$$

$$\left. + (\cosh\mu - \cos\eta)P_{n-1/2}^{m\prime}(\cosh\mu)\right]$$

$$\times \left(\alpha_n^m \cos n\eta \cos m\varphi + \beta_n^m \cos n\eta \sin m\varphi + \gamma_n^m \sin n\eta \cos m\varphi + \delta_n^m \sin n\eta \sin m\varphi\right),$$

$$B_\eta^{\text{tot}} = B_\eta^{\text{amb}} + B_0 \sqrt{\cosh\mu - \cos\eta} \sum P_{n-1/2}^m(\cosh\mu)\left[\frac{1}{2}\sin\eta\left(\alpha_n^m \cos n\eta \cos m\varphi\right.\right. \tag{1.3}$$

$$\left. + \beta_n^m \cos n\eta \sin m\varphi + \gamma_n^m \sin n\eta \cos m\varphi + \delta_n^m \sin n\eta \sin m\varphi\right) + n(\cosh\mu - \cos\eta)$$

$$\left. \times \left(-\alpha_n^m \sin n\eta \cos m\varphi - \beta_n^m \sin n\eta \sin m\varphi + \gamma_n^m \cos n\eta \cos m\varphi + \delta_n^m \cos n\eta \sin m\varphi\right)\right],$$

Figure 1. (a) Magnetic field configuration in the solar atmosphere based on real photospheric observations. An assumed loop-like flux rope is drawn only for comparison with figure (b) showing a modified field when this flux rope is taken into account.

$$B_\varphi^{\text{tot}} = B_\varphi^{\text{amb}} + B_0 \frac{(\cosh\mu - \cos\eta)^{3/2}}{\sinh\mu} \sum m P_{n-1/2}^m(\cosh\mu) \tag{1.4}$$
$$\times \left(-\alpha_n^m \cos n\eta \sin m\varphi + \beta_n^m \cos n\eta \cos m\varphi - \gamma_n^m \sin n\eta \sin m\varphi + \delta_n^m \sin n\eta \cos m\varphi\right).$$

$P_{n-1/2}^m$ is a Legendre function, its derivative by an argument is denoted by a prime. Coefficients α_n^m, β_n^m, γ_n^m, and δ_n^m are found numerically in order (1.1) to be approximately satisfied. Summations in (1.2)–(1.4) are theoretically over $n = 0, \infty$ and $m = 0, \infty$, but for numerical calculations they are cut at some suitable $n = N$ and $m = M$.

2. An example of an external field

Romashets & Vandas (2004) showed modified fields for two simple cases of ambient pre-existing fields, uniform and radial. We present a much more complex case. Figure 1a shows magnetic field lines in the solar atmosphere based on real photospheric observations. Coefficients provided by the Wilcox Solar Observatory for an ad-hoc chosen period were used to construct this ambient potential field ($\mathbf{B}^{\text{amb}}$); see, e.g., Hoeksema & Scherrer (1986). Let us introduce a flux rope into the field (thick lines). The flux rope is anchored in the photosphere. Two groups of field lines are displayed. The lines of the first group do not intersect the flux rope; these are the lines which do not enter into the silhouette of the rope in Figure 1a. The second group are remaining lines drawn in Figure 1a; all they cross near the central axial line of the rope. Figure 1b shows magnetic field lines after the modification (field $\mathbf{B}^{\text{tot}}$). No field line penetrates the flux rope. The field lines of the first group are nearly not or only slightly modified, in dependence on their distance from the flux rope. The field lines of the second group drape around the flux rope.

3. Conclusions

A new method to model magnetic field configurations outside solar flux ropes has been applied to real data. Such configuration can be used for initialization of MHD simulations for space weather events.

Acknowledgements

This work was supported by INTAS grant 03-51-6206, AV ČR project S1003006, MŠMT ČR project ME501, and RFBR grant 03-02-16340.

References

Hoeksema, J. T. & Scherrer, P. H. 1986, *Sol. Phys.* 105, 205
Vandas, M., Romashets, E. P. & Watari, S. 2003, *Astron. Astrophys.* 412, 281
Romashets, E. P. & Vandas, M. 2004, in: A. V. Stepanov *et al.* (eds.), *Proc. IAU Symp. 223 'Multi-Wavelength Investigations of Solar Activity'*, in press

Coronal and Stellar Mass Ejections
Proceedings IAU Symposium No. 226, 2005
K. P. Dere, J. Wang & Y. Yan, eds.

© 2005 International Astronomical Union
doi:10.1017/S1743921305000980

Propagation of Magnetic Clouds – MHD Simulations of Real Events

M. Vandas[1], D. Odstrcil[2]†, S. Watari[3] and A. Geranios[4]

[1]Astronomical Institute, 14131 Praha 4, Czech Republic
email: vandas@ig.cas.cz

[2]University of Colorado/CIRES and NOAA/Space Environment Center,
Boulder, CO 80305, USA
email: Dusan.Odstrcil@noaa.gov

[3]National Institute of Information and Communications Technology,
4-2-1 Nukuikita, Koganei, Tokyo 184-8795, Japan
email: watari@nict.go.jp

[4]Physics Department, University of Athens,
Panepistimioupoli-Kouponia, Athens 15771, Greece
email: ageran@cc.uoa.gr

Abstract. The paper describes our approach to simulations of real interplanetary events, consisting of four steps: (i) determination of background solar wind, (ii) parameterization of a model flux rope, (iii) launching it into the solar wind, and (iv) calculating its propagation and evolution.

Keywords. Sun: coronal mass ejections (CMEs), solar wind, MHD.

1. Simulation procedure

Events considered are well-defined ICMEs containing an interplanetary flux rope (magnetic cloud) and with known solar sources. Our simulation programs solve time-dependent MHD equations in three dimensional inner heliosphere with a nearly full angular coverage (with exceptions around the solar poles) and a radial span starting from 0.14 AU (super-critical flow) and including 1 AU. The numerical codes are based on a TVD scheme in spherical coordinates (Odstrcil & Pizzo 1999) and, under development, on a finite volume method with a Riemann solver using an unstructured grid. Our simulation procedure consists of four steps.

1.1. Determination of the background solar wind for period of an event using photospheric magnetic field measurements

A quasi-steady state is determined by program runs, which are driven by conditions in the solar wind at the inner boundary. Two approaches are considered for specifying these conditions. A simple one uses a position of the heliospheric neutral sheet and an angular distance of a given point at the inner boundary to specify the radial velocity (Odstrčil, Dryer & Smith 1998). The second one is based on relationships between the radial velocity and divergence of the magnetic field (Arge & Pizzo 2000, Odstrcil 2003); see Figure 1, left part.

† On leave from the Astronomical Institute, Ondřejov Observatory, Czech Republic.

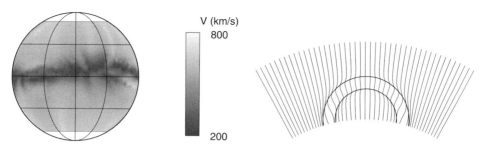

Figure 1. (left) A velocity distribution at the inner boundary based on photospheric magnetic field measurements (the method is described in Arge & Pizzo 2000 and Odstrcil 2003). (right) A toroidal flux rope merged into a radial field (the method is described in Romashets & Vandas 2004). Such a configuration can serve for initialization of MHD calculations (the method is described in Vandas, Odstrčil & Watari 2002).

1.2. *Estimation of flux rope parameters from observations*

It is necessary to determine orientation, size, velocity, magnetic field strength, and chirality of the flux rope. Probably vector magnetographs, coronal field measurements, and STEREO observations may help much in this difficult task.

1.3. *Launching of a model flux rope into the background solar wind*

For a simple field it has been done by Vandas, Odstrčil & Watari (2002) and by Vandas, Watari & Geranios (2003). For more realistic fields a method by Romashets & Vandas (2004) is considered (Figure 1, right part). A flux rope is introduced into the computational domain by time-dependent changes of quantities at the inner boundary.

1.4. *Running numerical simulations and comparing with observations*

Simulations of a flux rope propagation within our scheme in a simple background solar wind (unstructured and unipolar) have been done by Vandas, Odstrčil & Watari (2002) and by Vandas, Watari & Geranios (2003). Applications for real events have been and are performed with a model CME without a magnetic structure ("plasma clouds") (Odstrcil, Riley & Zhao 2004). We are developing procedures to include a magnetic flux rope topology within a more complex solar wind. Several examples of magnetic clouds observed near the Earth and their related CMEs near the Sun, which are well documented in the literature, will be considered in the simulations, the event of May 12, 1997 being the first.

Acknowledgements

This work is supported by INTAS grant 03-51-6206, AV ČR project S1003006, project ME501 from MŠMT ČR, GA ČR grant 205/03/0953, GA AV ČR grant A3003003, and DOD/AFOSR-MURI project. We also acknowledge support from the bilateral Czech-Greek agreement on collaboration in science and technology.

References

Arge, C. N. & Pizzo, V. J. 2000, *J. Geophys. Res.* 105, 10465
Odstrcil, D., Dryer, M. & Smith, Z. 1998, in: X. S. Feng, F. S. Wei & M. Dryer (eds.), *Advances in Solar Connection with Transient Interplanetary Phenomena*, Proceedings of the Third SOLTIP Symposium (Beijing: International Academic Publishers), 191
Odstrcil, D. & Pizzo, V. J. 1999, *J. Geophys. Res.* 104, 483
Odstrcil, D. 2003, *Adv. Space Res.* 32, 497
Odstrcil, D., Riley, P. & Zhao, X. P. 2004, *J. Geophys. Res.* 109, A02116

Romashets, E. P. & Vandas, M. 2004, this Proceedings

Vandas, M., Odstrčil, D. & Watari, S. 2002, *J. Geophys. Res.* 107, A9, SSH 2-1

Vandas, M., Watari, S. & Geranios, A. 2003, in: M. Velli, R. Bruno & F. Malara (eds.), *Solar Wind Ten,* AIP Conf. Proc. 679 (Melville, N. Y.: AIP), 691

Coronal and Stellar Mass Ejections
Proceedings IAU Symposium No. 226, 2005
K. P. Dere, J. Wang & Y. Yan, eds.

© 2005 International Astronomical Union
doi:10.1017/S1743921305000992

Initial Speeds of CMEs Estimated by Using Solar Wind Observations Near 1 AU

S. Watari[1], M. Vandas[2] and T. Watanabe[3]

[1]National Institute of Information and Communications Technology,
4-2-1 Nukuikita, Koganei, Tokyo 184-8795, Japan
email: watari@nict.go.jp

[2]Astronomical Institute, Academy of Sciences,
Bočni II 1401, 141 31 Praha 4, Czech Republic
email: vandas@ig.cas.cz

[3]Ibaraki University, 2-1-1 Bunkyo, Mito, Ibaraki 310-8512, Japan
email: watanabe@env.sci.ibaraki.ac.jp

Abstract. It is an important subject of space weather to forecast accurate arrival time of interplanetary coronal mass ejections (ICMEs) at the Earth. Determination of initial speeds of CMEs is an important factor for this. Here, we estimated the initial speeds of CMEs using solar wind observations near 1 AU and compared these speeds with CME speeds measured by the SOHO coronagraph.

Keywords. Solar wind, Sun:coronal mass ejections (CMEs), solar-terrestrial relations.

1. Introduction

Intense geomagnetic storms are often initiated by arrival of interplanetary coronal mass ejections (ICMEs). Therefore it is an important subject for space weather to predict arrival time of the ICMEs accurately. For the accurate prediction, we need to get accurate initial speeds of ICMEs as an input. The coronagraph on board the SOHO spacecraft enables to measure the CME speeds based on observations (Yashiro *et al.* 2004). The problem is that the measured CME speeds are apparent ones. Zhao, Plunkett & Liu (2002) tried to determine radial speeds of CMEs using the cone model. Dal Lago, Schwenn & Gonzalez (2003) presented a relationship between CME radial speeds and expansion speeds based on the "limb CME" observations. Here, we estimated the initial speeds of CME using solar wind speeds near 1 AU and compared them with the measured CME speeds. We used a simple model (Watari & Detman 1998) based on the shock time arrival (STOA) model developed by Dryer & Smart (1984) for this.

2. Comparison between measured and estimated speeds of CMEs

We assumed distance dependence of ICME speed to be

$$V = V_o + V_b \quad \text{for} \quad R = R_s \sim R_1 \tag{2.1}$$

$$V = V_o \left(\frac{R_1}{R} \right)^\alpha + V_b \quad \text{for} \quad R \geqslant R_1 \tag{2.2}$$

where V is speed of ICME at distance R from the Sun. R_s is the solar radius. V_o is initial speed of ICME. V_b is background solar wind speed. Deceleration of ICME starts at R_1.

433

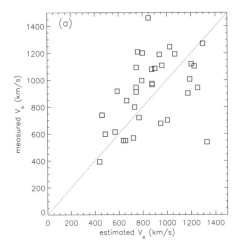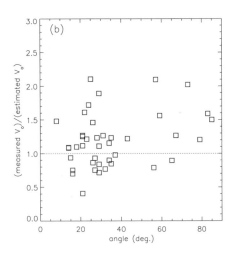

Figure 1. (a) A scatter plot of the measured and the estimated initial CME speeds. (b) A scatter plot of angles between Earth direction and main CME direction and ratios of the measured and estimated initial CME speeds.

We picked up 41 events which were clearly identified by the ACE observations between 2000 and 2003. For background solar wind speeds we choose solar wind speeds of quiet periods just before the shock arrival.

We assumed that α was 0.5 and R_1 was 0.15 AU (based on the result by Watari & Detman 1998) and estimated initial speeds of ICMEs using equation (2.2) and observed solar wind speeds near 1 AU. Figure 1a is a scatter plot of the estimated initial speeds of ICMEs and CME speeds measured by the SOHO observations. Figure 1b shows angles between Earth directions and main portions of CMEs and ratios of the measured and estimated speeds. The ratios tend to become larger as the angles increase. This shows the contribution of projection effect of CMEs in the measured CME speeds. The averaged value of the ratios for angles less than 45 degrees is approximately 1.1. This is consistent with the result of Dal Lago, Schwenn & Gonzalez (2003). This suggests the contribution of the expansion effect in the measured CME speeds.

Acknowledgements

The ACE solar wind data was provided from the ACE Science Center. The measured CME speeds are referred from the catalog generated and maintained by NASA and The Catholic University of America in cooperation with the Naval Research Laboratory. SOHO is a project of international cooperation between NASA and ESA. MV was supported by projects S1003006 from AV ČR and ME501 from MŠMT ČR.

References

Dal Lago, A. Schwenn, R. & Gonzalez, W. D. 2003, *Adv. Space Res.* 32, 2637
Dryer, M. & Smart, D. F. 1984, *Adv. Space Res.* 4, 291
Yashiro, S., Gopalswamy, N., Michalek, G., Cyr, O. C. St., Plunkett, S. P., Rich, N. B., & Howard, R. A. 2004, *J. Geophys. Res.* 109, A07105
Watari, S. & Detman, T. 1998, *Ann. Geophys.* 16, 370
Zhao, X. P., Plunkett, S. P., & Liu, W. 2002, *J. Geophys. Res.* 107, SSH13-1

Session 8

CMEs and Geomagnetic storms

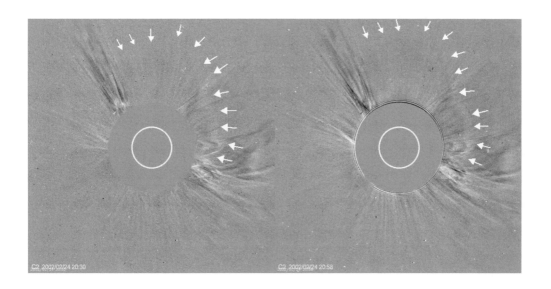

Coronal and Stellar Mass Ejections
Proceedings IAU Symposium No. 226, 2005
K. P. Dere, J. Wang & Y. Yan, eds.

© 2005 International Astronomical Union
doi:10.1017/S1743921305001006

Solar Sources of Geoeffective CMEs: a SOHO/EIT View

Andrei N. Zhukov[1,2]

[1]Royal Observatory of Belgium, Avenue Circulaire 3, B-1180 Brussels, Belgium
email: Andrei.Zhukov@oma.be

[2]Skobeltsyn Institute of Nuclear Physics, Moscow State University, 119992 Moscow, Russia

Abstract. Observations of the low solar corona, in particular in the EUV, are an effective means of identifying the solar sources of coronal mass ejections (CMEs). SOHO/EIT, with its continuous 24 hours per day coverage, is well suited to perform this task. Source regions and start times of frontside full and partial halo CMEs (that may be geoeffective) can thus be determined. The most frequent EUV signatures of CMEs are coronal dimmings. EIT waves, eruptive filaments and post-eruption arcades are also reliable signatures. Frontside halo CMEs with source regions close to the solar disc center have the strongest chance to hit the Earth. The inspection of the EIT data together with photospheric magnetograms may give an idea about the ejected interplanetary flux rope magnetic field and, in particular, about the presence or absence of southward (geoeffective) field. If a source region is situated close to the solar limb, the corresponding CME also may be geoeffective, as the CME-driven shocks have large angular extent. In this case the storm can be produced by the sheath plasma behind the shock, provided it contains strong enough southward interplanetary magnetic field. Some implications for the operational space weather forecast are discussed. EIT and LASCO are capable to identify the solar sources of the most of geomagnetic storms. In some cases, however, the identification is uncertain, so the observations by the future STEREO mission will be needed for the investigation of similar events.

Keywords. Sun: coronal mass ejections (CMEs), Sun: solar-terrestrial relations

1. Introduction

The study of coronal mass ejection (CME) phenomenon is very important for solar-terrestrial relations. It is now known that CMEs play a key role in producing geomagnetic storms (e.g. Gosling, Bame, McComas, *et al.* 1990; Kahler 1992). To be geoeffective, a CME or CME-associated disturbance (e.g. a post-shock sheath) should arrive to the Earth and contain suitable magnetic field orientation: the north – south interplanetary magnetic field (IMF) component B_z should be negative (southward), strong enough and long-lasting (Burton, McPherron & Russel 1975; Gonzalez & Tsurutani 1987).

Halo CMEs attract particular attention in the study of geoeffective solar eruptions. A full halo CME has a shape of a bright irregular ring completely surrounding the coronagraph occulter (Howard, Michels, Sheeley, *et al.* 1982), i.e. it is a CME with the angular width of 360°. Full halo CMEs are currently interpreted as an end-on view of CMEs propagating approximately along the Sun – Earth line (see e.g. discussion in Plunkett, Thompson, Howard, *et al.* 1998). A partial halo is a wide CME (angular width larger than e.g. 120°) which also can be directed towards the Earth.

CMEs are now routinely observed by the Large-Angle Spectroscopic Coronagraph (LASCO, see Brueckner, Howard, Koomen, *et al.* 1995) onboard the Solar and Heliospheric Observatory (SOHO). Coronagraph observations, however, cannot distinguish between frontside and backside halo CMEs, as the occulting disc obscures a direct view of the initiation site. This is why observations of the low corona are necessary. The Extreme-ultraviolet Imaging Telescope (EIT, see Delaboudinière, Artzner, Brunaud, *et al.* 1995) onboard SOHO observes the full disc of the Sun 24 hours per day in four extreme-ultraviolet (EUV) bandpasses, and its "CME Watch" data series (one image in the Fe XII (195 Å) bandpass every 12 minutes) is well suited for the detection of CME signatures in the low corona. The observations of the CME source region are crucial for the problem of CME initiation, and its solution may ultimately lead to the prediction of CME occurrences. Additionally, the observations of the low corona, combined with the photospheric magnetic field measurements may provide us with an important information on the magnetic field orientation in the resulting interplanetary CME (ICME).

2. Tracking ICMEs back to the Sun

The procedure of identification of the solar source of an ICME is now well established (Fox, Peredo & Thompson 1998; Webb, Cliver, Gopalswamy, *et al.* 1998; Brueckner, Delaboudiniere, Howard, *et al.* 1998; Bothmer & Schwenn 1998; Berdichevsky, Bougeret, Delaboudinière, *et al.* 1998; Webb, Cliver, Crooker, *et al.* 2000; Webb, Lepping, Burlaga, *et al.* 2000; Cane, Richardson & St. Cyr 2000; Gopalswamy, Lara, Yashiro, *et al.* 2001; Wang, Ye, Wang, *et al.* 2002; Zhang, Dere, Howard, *et al.* 2003; Cane & Richardson 2003; Zhao & Webb 2003; Zhukov, Veselovsky, Clette, *et al.* 2003). It can be summarized as follows. First, the ICME is identified using in situ plasma and magnetic field measurements. The average speed of the ICME is determined and the approximate start time from the Sun is calculated assuming the constant speed en route from the Sun to the Earth. All the full and partial halo CMEs that occurred close to the estimated start time are identified, using CME catalogs (e.g. Yashiro, Gopalswamy, Michalek, *et al.* 2004) or through the direct inspection of LASCO data. Their travel times to the Earth are estimated using the measured plane-of-the-sky velocities, and probable candidates are selected. The variation of the travel time depending on the CME speed in the plane of the sky has been investigated e.g. by Gopalswamy *et al.* (2001) and Cane & Richardson (2003).

The next step is to look at the low corona activity to determine the origin of the candidate halo CMEs. CME signatures observed by EIT (primarily in the Fe XII bandpass at 195 Å) are: coronal dimmings (including transient coronal holes, TCHs) – sudden local decreases in brightness; EIT waves – bright fronts often propagating from eruption sites; post-eruption arcades; erupting filaments (seen as prominences when observed above the limb); different limb signatures like loop opening, plasmoid rising, etc. Any of these features implies that a CME has occurred. Dimmings represent the most frequent CME signature in the low corona and are due to the evacuation of mass during CMEs. TCHs have been interpreted by Webb *et al.* (2000b) as footpoints of the ejected interplanetary flux rope (see, however, Kahler & Hudson 2001). Zhukov & Auchère (2004) showed that only a half of the CME mass observed by EIT during the event of May 12, 1997 is erupted from TCHs, the rest was ejected from weaker and larger dimming regions. EIT waves seem to be produced by compression during the opening of the field lines during the CME lifting. Arcades, prominences and loop opening are present in the "standard model" of a CME, see e.g. discussion by Hudson & Cliver (2001).

Often (especially during the years of low solar activity) there is only one halo CME with distinct signatures in EIT close to the estimated start time. In such cases the identification is relatively straightforward. Sometimes, however, it may happen that no halo CME is reported around the estimated start time (e.g. Cane & Richardson 2003). In principle, this may be due to the insufficient LASCO sensitivity. The Thomson scattering is most

APEV-259

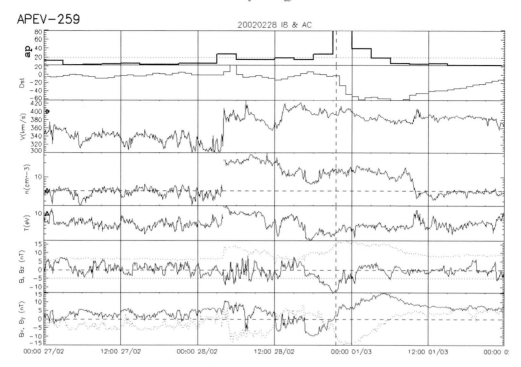

Figure 1. Solar wind (ACE) and geomagnetic data for the storm on February 28 – March 1, 2002. From top to bottom: 3-hour a_p index; 1-hour D_{st} index; solar wind speed; proton number density; proton temperature; IMF magnitude (dotted line) and its B_z component (solid line); IMF B_x (solid line) and B_y (dotted line) components. The plot is taken from the APEV database (http://observ.sinp.msu.ru/apev). All times are UT.

efficient close to the plane of the sky, so some Earth-directed events – which naturally have a lot of material out of the plane of the sky – may be missed by LASCO. Let us take a look at one of these events.

The geomagnetic storm occurred on February 28 – March 1, 2002 (peak $A_p = 80$, peak $D_{st} \sim -60$ nT). Figure 1 shows the Advanced Composition Explorer (ACE) observations of the solar wind structure that produced the storm. The shock arrived around 04:00 UT, followed by the ESW magnetic cloud (for the classification see e.g. Bothmer & Schwenn 1998; Mulligan, Russel & Luhmann 1998) starting approximately at 16:30 UT and ending about 9:30 UT on March 1. The ICME speed is around 400 km/s, so the disturbance has left the Sun around February 24 (assuming the constant propagation speed). LASCO CME catalog (Yashiro *et al.* 2004) and the LASCO operations scientist (http://lasco-www.nrl.navy.mil/cmelist.html) did not report any full or partial halos around this time (see also Cane & Richardson 2003); the last halo before the estimated start time occurred on February 20. CACTus software (Robbrecht & Berghmans 2004) did not detect any halos neither. All the reported CMEs were narrow and originated from the vicinity of the solar limb (so it is unlikely that they were directed towards the Earth, see below).

EIT data must now be inspected to identify all the CME signatures occurred during several days around the estimated start time. The eruptions associated with reported CMEs are rejected. Thus a coronal dimming not associated with any of the reported CMEs has been revealed (figure 2). It occurred next to the filament channel close to the disc center, with filament starting to rise slowly around 16:45 UT on February 24. Around

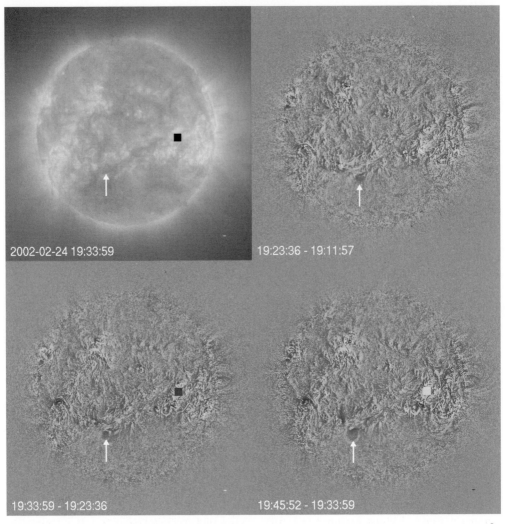

Figure 2. The eruption on February 24, 2002 observed by SOHO/EIT in the Fe XII (195 Å) bandpass. The top left panel is a plain image, other panels are running difference images. The arrows mark the place of the eruption in a filament channel (see top left panel) manifested as a dimming in running difference images. All times are UT.

19:23 UT a small, but clear dimming is visible to the south of the channel, indicating that an eruption indeed happened. Although no CMEs were reported, attentive inspection of the running difference LASCO C2 movie for the end of February 24 reveals an extremely weak partial halo CME starting around 18:30 UT. It spanned around 150° from NE to SW limb (figure 3). It has to be stressed that the identification of this CME is very difficult without a priori knowledge of its start time obtained using the EIT data. It seems that in this case it is easier to detect the dimming observed by EIT than the LASCO CME.

Therefore, presumable cases of ICMEs without any LASCO counterpart have to be double-checked. The described event shows that LASCO sensitivity allows us to detect even very weak CMEs, although sometimes EUV signatures are easier to find. A statistical study is needed to verify if ICMEs without corresponding LASCO CMEs indeed occurred.

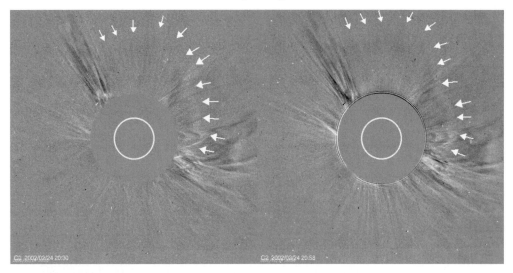

Figure 3. SOHO/LASCO C2 running difference images (20:30−19:54 UT, left, and 20:58−20:30 UT, right) showing a very weak partial halo CME marked by arrows.

3. Positions of source regions on the solar surface

Several studies addressed the distribution of the source regions of geoeffective CMEs on the solar surface (Lyons, Stalkton-Chalk & Lewis 1999; Cane *et al.* 2000; Wang *et al.* 2002; Zhang *et al.* 2003; Manoharan, Gopalswamy, Yashiro, *et al.* 2004; Srivastava & Venkatakrishnan 2004). There is a concentration of source regions near the solar disc center, approximately inside the circle with the radius of about 40°. In some studies (Wang *et al.* 2002; Zhang *et al.* 2003) it has been noted, however, that there is an east − west asymmetry in this distribution: geoeffective CMEs have a slight preference to originate from the western hemisphere. This finding is still controversial as e.g. Cane *et al.* (2000) and Srivastava & Venkatakrishnan (2004) did not find such an asymmetry. An explanation of this asymmetry has been proposed (Wang, Shen, Wang, *et al.* 2004): CMEs which are faster than the ambient solar wind are deflected to the east by the magnetic force of the ambient spiral IMF. The explanation seems to be plausible as the asymmetry seems to be more pronounced for fast events. A dynamic model of this interaction is still to be developed, and a statistical study including weaker events is needed to verify if the longitudinal asymmetry indeed exists.

Another interesting finding by Zhang *et al.* (2003) is that four major storms (with $D_{st} < -100$ nT) have been produced by the east-limb partial halo CMEs without any signatures on the solar disc. All these CMEs are very slow (around 200 km/s), and it seems possible that EUV dimmings in slow cases are continuously replenished with plasma and thus are not pronounced.

However, alternative sources for these storms can be proposed. For example, Zhang *et al.* (2003) identify the source of the storm on April 22, 1997 as a partial halo CME first seen in the LASCO C2 field of view on April 16, 07:35 UT. Indeed, EIT shows no signatures of this event, and this may indicate that it is a backside CME. On the other hand, an eruption close to the disc center was observed by EIT at 14:36 UT on April 16 (figure 4). A dimming is clearly seen, with a bright front ahead of it resembling the front of an EIT wave. Unfortunately, the eruptive signatures are seen only in one image as the data gap of 83 minutes followed. So, strictly speaking, we cannot state for sure if an EIT wave or a dimming indeed occurred. No CME has been observed by LASCO in

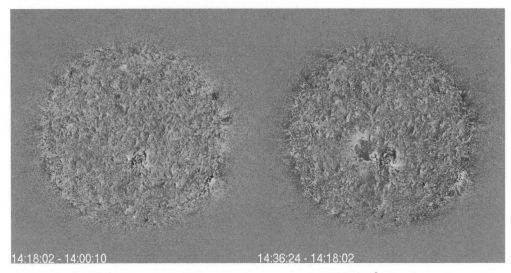

Figure 4. SOHO/EIT running difference images in the Fe XII (195 Å) bandpass showing the eruption on April 16, 1997 at 14:36 UT.

association with this event. It is unlikely that this eruption is the source of the CME identified by Zhang *et al.* (2003) as the CME started around 7 hours before the dimming.

So, in this case one has a choice between two alternative interpretations. On the one hand, the partial halo CME can be identified as the source of the storm, assuming that EIT did not observe the dimming because of a very slow CME speed. On the other hand, this CME can be classified as a backside one because of the absence of on-disc signatures, and a weak EIT event can represent a signature of a CME undetected by LASCO because of its insufficient sensitivity. Such a situation takes place in three out of four east-limb events identified by Zhang *et al.* (2003). In the fourth case it seems that EIT did not observe any alternative source. Thus, sources of slow CMEs seem to be the most difficult to identify.

Although there is a strong concentration of geoeffective CMEs' source regions close to the disc center, even CMEs originating at the limb can arrive to the Earth (provided they are wide enough) and produce geomagnetic storms. In most of such cases, however, only an interplanetary shock is observed (e.g. Manoharan *et al.* 2004) as the angular extent of the shock is larger than that of a corresponding CME. The CME thus misses the Earth and only the shock arrives.

An example of such an event is presented in figures 5–6. An EIT wave (figure 5) was observed above the east limb on October 21, 2003. It was propagating from the active region behind the limb (future NOAA AR 0486) as indicated by the rising post-eruption loops. However, the eruption was so powerful and wide that the corresponding CME was a full halo (figure 6): the south-western streamer was deflected by the CME-associated disturbance (probably the CME-driven shock). The geomagnetic storm ensued on October 24 as the interplanetary shock arrived around 15:00 UT. A short interval of cold plasma (22:00 UT, October 24 – 01:00 UT, October 25) may represent the CME matter following the hot post-shock sheath. This event illustrates that it can be misleading to consider CMEs as originating from a small source region – in this case such a region would be located on the back side of the Sun, right behind the east limb. Nevertheless, the corresponding disturbance arrived to the Earth due to the non-local nature of CME-associated structures.

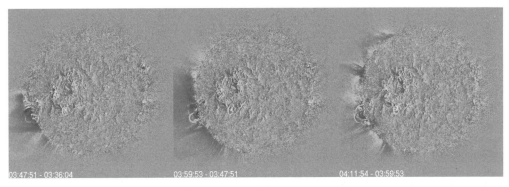

Figure 5. Running difference images illustrating the EIT wave observed by SOHO/EIT in the Fe XII (195 Å) bandpass above the eastern limb. All times are UT.

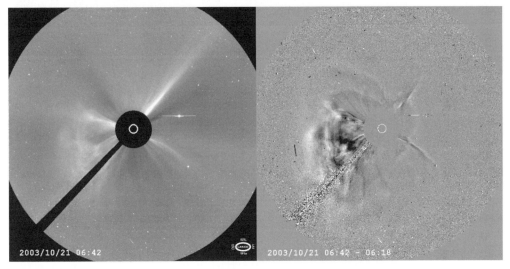

Figure 6. SOHO/LASCO C3 plain (left) and running difference (right) images showing the full halo CME on October 21, 2003. All times are UT.

4. Predicting the CME onset and the IMF direction?

Once the CME is observed, its arrival time to the Earth can be estimated on the base of the measured plane-of-the-sky speed (Brueckner *et al.* 1998; Gopalswamy *et al.* 2001; Cane & Richardson 2003). However, to assess the strength of a possible storm, not only the CME arrival, but also the IMF B_z component has to be predicted. Moreover, a big challenge is to predict the CME before it actually happened, so precursors of eruptions have to be identified.

Sigmoidal active regions (i.e. the ones displaying S-shaped structure in the soft X-rays) have been reported to have a higher probability to erupt than non-sigmoidal active regions (e.g. Canfield, Hudson & McKenzie 1999). Sigmoids can also be observed by EIT, especially in the Fe XV (284 Å) bandpass (compare the sigmoids observed by SXT and EIT in figure 4 of the paper by Sterling, Hudson, Thompson, *et al.* 2000). EIT has a narrower temperature response and thus is better suited for the investigation of the active regions fine structure. EIT observations demonstrate convincingly that a sigmoid is a collective feature (Glover, Ranns, Harra, *et al.* 2000; Zhukov *et al.* 2003) – a continuous S-shaped structure can almost never be traced. Pevtsov, Canfield & Zirin

(1996) and Glover, Ranns, Brown, et al. (2002) suggested that a single twisted unstable flux tube is formed of two sheared J-shaped loops right before the eruption. However, in such cases either the detector saturation effect is apparent in the middle of the sigmoid (Glover et al. 2002), or the S-shaped structure is inhomogeneous (Pevtsov et al. 1996), indicating that multiple flux systems are involved. EIT observations suggest that the sigmoid is better described by the separatrix surface model (Titov & Démoulin 1999) than by the kink-unstable twisted flux tube model (Rust & Kumar 1996).

This uncertainty of the observational definition of a sigmoid does not allow us to use sigmoids efficiently in the operational space weather forecast as different works give different statistics on the probability of eruption (Canfield et al. 1999; Glover et al. 2000). Moreover, the sigmoids can also appear right after the eruption (Glover, Harra, Matthews, et al. 2001; Zhukov et al. 2003) – a fact that does not seem to agree with the kink-unstable twisted flux tube eruption scenario. Finally, nothing indicates when the eruption will occur. An active region may have a sigmoidal shape during its whole passage from the east to the west limb, but produce only a couple of CMEs during this time (Gibson, Fletcher, Del Zanna, et al. 2002).

Although the eruption time cannot be predicted reliably now, one can obtain an indication on the resulting IMF orientation (in particular, on its B_z component which is crucial for the assessment of the strength of a possible geomagnetic disturbance). If the photospheric magnetic field of the CME source region has a bipolar configuration (it is often the case), one can determine the orientation of a neutral line and thus get an idea about the inclination of the axis of the ejected interplanetary flux rope (Marubashi 1997; Yurchyshyn, Wang, Goode, et al. 2001; McAllister, Martin, Crooker, et al. 2001; Bothmer 2003). If the shear of the magnetic field can be determined (looking e.g. at the post-eruption arcade in the EIT data), the direction of the magnetic field in the flux rope can be reasonably estimated.

Yurchyshyn et al. (2001) showed that for the full halo CME on February 17, 2000 the source region neutral line was oriented along the north – south direction, similarly to the resulting interplanetary flux rope orientation. As the axial field in the flux rope was northward, the magnetic cloud produced only a very weak geomagnetic disturbance. On the contrary, the halo CME on July 14, 2000 ("Bastille day") originated from an active region with the neutral line oriented along the east – west direction. This orientation will produce a negative B_z either in the leading or in the trailing part of the flux rope.

The inclinations of the flux rope axes close to the Sun and in the heliosphere do not always correspond to each other. The neutral line in the event of May 12, 1997 had the north – south orientation (Webb et al. 2000b), and, if its inclination is conserved, the flux rope had to have the ENW orientation, i.e. to be not geoeffective. However, the interplanetary flux rope was of the SEN type and produced a major geomagnetic storm. The reason for such a change of orientation seems to be that during the low activity years the CMEs are deflected by the fast flows from polar coronal holes (Cremades & Bothmer 2004) and thus have the tendency to have a small inclination with respect to the ecliptic plane. The CME on February 17, 2000 propagated without such an influence because of the absence of polar coronal holes during the activity maximum. Another explanation (Webb 2002) suggests that the interplanetary flux rope results rather from the large-scale dipole field than from the local bipolar field of the source region.

5. Conclusions

EIT and LASCO are capable to identify reliably the source regions of the most of geoeffective ICMEs. The identification works especially well if a CME is isolated and is not very slow. In some cases, however, the identification of the source region is not clear. These cases include very slow CMEs, when a partial halo CME observed by LASCO has no EIT source (Zhang *et al.* 2003), but an eruption seen by EIT close to the disc center has no LASCO counterpart or there are no EIT events at all. The identification can also be difficult in complicated cases of multiple (interacting) CMEs – a problem not discussed in this paper.

We have to note that these doubtful cases can correspond to major geomagnetic storms. To determine the sources of ICMEs more precisely, the propagation of CMEs has to be tracked from a vantage point out of the Sun – Earth line as well. STEREO mission (Solar – Terrestrial Relations Observatory) will for the first time provide such observations. So, although the combination of EIT and LASCO is sufficient in the most of the cases, the STEREO data will be necessary to identify unambiguously the solar sources of geomagnetic storms.

Acknowledgements

The author thanks the organizers of the Symposium for the invitation to give this talk and for financial support. EIT data have been used courtesy of SOHO/EIT consortium. The LASCO data used here are produced by a consortium of the Naval Research Laboratory (USA), Max-Planck-Institut für Aeronomie (Germany), Laboratoire d'Astronomie (France), and the University of Birmingham (UK). SOHO is a project of international cooperation between ESA and NASA. ACE SWEPAM and MAG instrument teams and the ACE Science Center are acknowledged for providing the ACE data, as well as the World Data Center for Geomagnetism, Kyoto, Japan, for the A_{p} and D_{st} indices. The author is grateful to V. Bothmer, A. V. Dmitriev, G. Lawrence, D. Berghmans, E. Robbrecht, R. Van der Linden, I. S. Veselovsky and N. Nitta for useful discussions. This work is partially supported by the INTAS Project 03-51-6206.

References

Berdichevsky, D., Bougeret, J.-L., Delaboudinière, J.-P., Fox, N., Kaiser, M., Lepping, R., Michels, D., Plunkett, S., Reames, D., Reiner, M., Richardson, I., Rostoker, G., Steinberg, J., Thompson, B., & von Rosenvinge, T. 1998, *Geophys. Res. Lett.* 25, 2473

Bothmer, V. & Schwenn, R. 1998, *Ann. Geophys.* 16, 1

Bothmer, V. 2003, in: A. Wilson (ed.), *Proc. ISCS 2003 Symposium, 'Solar Variability as an Input to the Earth's Environment'*, ESA SP-535, p. 419

Brueckner, G.E., Howard, R.A., Koomen, M.J., Korendyke, C.M., Michels, D.J., Moses, J.D., Socker, D.G., Dere, K.P., Lamy, P.L., Llebaria, A., Bout, M.V., Schwenn, R., Simnett, G.M., Bedford, D.K., & Eyles, C.J. 1995, *Sol. Phys.* 162, 357

Brueckner, G.E., Delaboudiniere, J.-P., Howard, R.A., Paswaters, S.E., St. Cyr, O.C., Schwenn, R., Lamy, P., Simnett, G.M., Thompson, B., & Wang, D. 1998, *Geophys. Res. Lett.* 25, 3019

Burton, R.K., McPherron, R.L., & Russell, C.T. 1975, *J. Geophys. Res.* 80, 4204

Cane, H.V., Richardson, I.G., & St. Cyr, O.C. 2000, *Geophys. Res. Lett.* 27, 3591

Cane, H.V. & Richardson, I.G. 2003, *J. Geophys. Res.* 108(A4), SSH 6-1

Canfield, R.C., Hudson, H.S., & McKenzie, D.E. 1999, *Geophys. Res. Lett.* 26, 627

Cremades, H. & Bothmer, V. 2004, *A&A* 422, 307

Delaboudinière, J.-P., Artzner, G.E., Brunaud, J., Gabriel, A.H., Hochedez, J.F., Millier, F., Song, X.Y., Au, B., Dere, K.P., Howard, R.A., Kreplin, R., Michels, D.J., Moses, J.D., Defise, J.M., Jamar, C., Rochus, P., Chauvineau, J.P., Marioge, J.P., Catura, R.C., Lemen, J.R., Shing, L., Stern, R.A., Gurman, J.B., Neupert, W.M., Maucherat, A., Clette, F., Cugnon, P., & van Dessel, E.L. 1995, *Sol. Phys.* 162, 291

Fox, N.J., Peredo, M., & Thompson, B.J. 1998, *Geophys. Res. Lett.* 25, 2461

Gibson, S.E., Fletcher, L., Del Zanna, G., Pike, C.D., Mason, H.E., Mandrini, C.H., Démoulin, P., Gilbert, H., Burkepile, J., Holzer, T., Alexander, D., Liu, Y., Nitta, N., Qiu, J., Schmieder, B., & Thompson, B.J. 2002, *ApJ* 574, 1021

Glover, A., Ranns, N.D.R., Harra, L.K., & Culhane, J.L. 2000, *Geophys. Res. Lett.* 27, 2161

Glover, A., Harra, L.K., Matthews, S.A., Hori, K., & Culhane, J.L. 2001, *A&A* 378, 239

Glover, A., Ranns, N.D.R., Brown, D.S., Harra, L.K., Matthews, S.A., & Culhane, J.L. 2002, *J. Atm. Sol.-Terr. Phys.* 64, 497

Gonzalez, W.D. & Tsurutani, B.T. 1987, *Planet. Space Sci.* 35, 1101

Gopalswamy, N., Lara, A., Yashiro, S., Kaiser, M.L., & Howard, R.A. 2001, *J. Geophys. Res.* 106, 29207

Gosling, J.T., Bame, S.J., McComas, D.J., & Phillips, J.L. 1990, *Geophys. Res. Lett.* 17, 901

Howard, R.A., Michels, D.J., Sheeley, N.R., & Jr., Koomen, M.J. 1982, *ApJ* 263, L101

Hudson, H.S. & Cliver E.W. 2001, *J. Geophys. Res.* 106, 25199

Kahler, S.W. 1992, *Ann. Rev. Astron. Astrophys.* 30, 113

Kahler, S.W. & Hudson, H.S. 2001, *J. Geophys. Res.* 106, 29239

Lyons, M., Stockton-Chalk, A.B., & Lewis, D.J. 1999, in: A. Wilson (ed.) *Proc. 9th European Meeting on Solar Physics, 'Magnetic Fields and Solar Processes'*, ESA SP-448, p. 943

Manoharan, P.K., Gopalswamy, N., Yashiro, S., Lara, A., Michalek, G., & Howard, R.A. 2004, *J. Geophys. Res.* 109, A06109

Marubashi, K. 1997, in: N. Crooker, J.A. Joselyn & J. Feynman (eds.), *Coronal Mass Ejections*, AGU Geophys. Monogr. Ser., vol. 99, p. 147

McAllister, A.H., Martin, S.F., Crooker, N.U., Lepping, R.P., & Fitzenreiter, R.J. 2001, *J. Geophys. Res.* 106, 29185

Mulligan, T., Russell, C.T., & Luhmann, J.G. 1998, *Geophys. Res. Lett.* 25, 2959

Pevtsov, A.A., Canfield, R.C., & Zirin, H. 1996, *ApJ* 473, 533

Plunkett, S.P., Thompson, B.J., Howard, R.A., Michels, D.J., St. Cyr, O.C., Tappin, S.J., Schwenn, R., & Lamy, P.L. 1998, *Geophys. Res. Lett.* 25, 2477

Robbrecht, E. & Berghmans, D. 2004, *A&A* 425, 1097

Rust, D.M. & Kumar, A. 1996, *ApJ* 464, L199

Srivastava, N. & Venkatakrishnan, P. 2004, *J. Geophys. Res.* 109, A10103

Sterling, A.C., Hudson, H.S., Thompson, B.J., & Zarro, D.M. 2000, *ApJ* 532, 628

Titov, V.S. & Démoulin, P. 1999, *A&A* 351, 707

Wang, Y.M., Ye, P.Z., Wang, S., Zhou, G.P., & Wang, J.X. 2002, *J. Geophys. Res.* 107(A11), SSH 2 - 1

Wang, Y., Shen, C., Wang, S., & Ye, P. 2004, *Sol. Phys.* 222, 329

Webb, D.F. 2002, in: A. Wilson (ed.), *Proc. SOHO 11 Symposium, 'From Solar Min to Max: Half a Solar Cycle with SOHO'*, ESA SP-508, p. 409

Webb, D.F., Cliver, E.W., Gopalswamy, N., Hudson, H.S., & St. Cyr, O.C. 1998, *Geophys. Res. Lett.* 25, 2469

Webb, D.F., Cliver, E.W., Crooker, N.U., St. Cyr, O.C., & Thompson, B.J. 2000, *J. Geophys. Res.* 105, 7491

Webb, D.F., Lepping, R.P., Burlaga, L.F., DeForest, C.E., Larson, D.E., Martin, S.F., Plunkett, S.P., & Rust, D.M. 2000, *J. Geophys. Res.* 105, 27251

Yashiro, S., Gopalswamy, N., Michalek, G., St. Cyr, O.C., Plunkett, S.P., Rich, N.B., & Howard, R.A. 2004, *J. Geophys. Res.* 109, A07105

Yurchyshyn, V.B., Wang, H., Goode, P.R., & Deng, Y. 2001, *ApJ* 563, 381

Zhao, X.P. & Webb, D.F. 2003, *J. Geophys. Res.* 108(A6), SSH 4 - 1

Zhang, J., Dere, K.P., Howard, R.A., & Bothmer, V. 2003, *ApJ* 582, 520

Zhukov, A.N., Veselovsky, I.S., Clette, F., Hochedez, J.-F., Dmitriev, A.V., Romashets, E.P., Bothmer, V., & Cargill, P. 2003, in: M. Velli, R. Bruno & F. Malara (eds.), *Proceedings of the Tenth International Solar Wind Conference*, AIP Conference Proceedings, vol. 679, p. 711

Zhukov, A. & Auchère, F. 2004, *A&A* 427, 705

Discussion

SCHWENN: "Predictions are always difficult, especially if they concern the future"—you give the impression of having control, but the situation is actually problematic for forward predictions. You need LASCO, and EIT can only help to discern front-side and backside event. Note: not even full or partial halo appearance always qualifies for good predictions.

ZHUKOV: I agree that the forward prediction is a very difficult task, and the identification method that I presented works well mostly for backward tracing, when an ICME has already arrived and the storm is over. The correspondence of LASCO CMEs and EIT eruptive signatures is good in both directions. Sometimes (e.g., magnetic storm on February 28-March 1, 2002) the eruptive signatures observed by EIT are much easier to identify than the corresponding LASCO CME, and in some cases (e.g. storm on April 21-22, 1997) the relevance of the partial halo CME seen by LASCO is questionable. But I agree that LASCO is needed to determine the direction and the extent of the CME.

DELABOUDINIERE: Does the mass in the dimmings compare to the mass of the CME?

ZHUKOV: The mass of the CME estimated on the base of LASCO observations was about three times larger than the mass that we found to be ejected from the low corona observed by EIT. The EIT DEM calculations give an order of magnitude estimate, so I am tempted to call this a good agreement.

JIE ZHANG: Comment regarding to J.-P. Delaboudiniere's comment: For the CME Andrei cited from my work, the slow partial halo CME that caused a large magnetic cloud and a major geomagnetic storm. The CME has no counterpart signature in EIT. We have also checked all other solar disk observations including SXT, Hα observations. We still can not find any eruptive signature.

ZHUKOV: I agree with this comment.

KOUTCHMY: You are calling our attention to the dimming phenomenon which could be a good proxy for predicting CMEs, etc. What is the interpretation of this phenomenon? You proposed to call them "transient coronal holes", but coronal holes are long lived and correspond to a magnetically open region. Here with CMEs we have a rising flux tube(s), which is the opposite of a coronall hole. I would suggest that the dimming effect is due to a low corona evaporation effect due to down flowing energetic particles along the field lines of the flux rope or a part of the flux rope.

ZHUKOV: Dimmings are not proxies for predicting CMEs, but rather the low corona manifestation of CME occurrence. The term "transient coronal holes" (TCHs) is often used in the literature and means that the TCH appearance resembles that of polar coronal holes. Indeed, the field lines coming out from TCHs are considered to be the footpoints of the interplanetary flux rope. However, as this flux rope has already erupted, the TCH field lines can be considered as "quasi-open" - the closure of the field lines occurs at very large distances from the Sun. The interpretation of dimmings as an evaporation of low corona is less probable than the removal of mass. The dimmings can be also observed by YOHKOH/SXT, which has the temperature response extended to quite high temperatures.

Coronal and Stellar Mass Ejections
Proceedings IAU Symposium No. 226, 2005
K. P. Dere, J. Wang & Y. Yan, eds.

Orientation and Geoeffectiveness of Magnetic Clouds as Consequences of Filament Eruptions

Yuming Wang[1], Guiping Zhou[2], Pinzhong Ye[1], S. Wang[1] and Jingxiu Wang[2]

[1] School of Earth & Space Sci., Univ. of Sci. & Tech. of China, Hefei, Anhui 230026, China.
email: ymwang@ustc.edu.cn

[2] National Astronomical Observatories of China.

Abstract. By investigating ten typical magnetic clouds (MCs) associated with large geomagnetic storms (Dst $\leqslant -100$ nT) from 2000 to 2003, the geoeffectiveness of MCs with various orientations is addressed. It is found that the Dst peak values during the geomagnetic storms are well estimated by applying flux rope model to these magnetic clouds. A high correlation between estimated and observed Dst values is obtained. Moreover, the effect of orientations of MCs on intensities of geomagnetic storms is studied. It is found that the favorable orientations of MCs are approximately at $\theta \sim 70°$ and $\phi \sim 40°$ in GSE coordinates to cause large geomagnetic storms. Further, by analyzing solar observations of four associated erupted filaments, the question who determine the orientations of MCs is studied. The likelihood of predicting the intensities of a geomagnetic storms several tens hours before their occurrences is also discussed.

Keywords. Sun: coronal mass ejections (CMEs), filament, magnetic fields, solar-terrestrial relations

---

1. Introduction

Geomagnetic storm is one of most important aspect in effecting the environment around the Earth. Large non-recurrent geomagnetic storms are usually caused by interplanetary ejecta, especially magnetic clouds (MCs), and shock sheaths preceding them (e.g., Sheeley, Jr. *et al.* 1985; Gosling *et al.* 1991). These notable perturbations change some properties of interplanetary medium greatly. The relationship between interplanetary parameters and intensities of geomagnetic storms are studied exhaustively in the past several decades (e.g., Burton *et al.* 1975; Gonzalez *et al.* 1989; Vassiliadis *et al.* 1999; Wang *et al.* 2003). Fast solar wind (V), strong southward component (B_s) of magnetic fields and long duration of B_s (Δt) are the most pivotal to create large geomagnetic storms (e.g., Gonzalez *et al.* 1994).

Intensity of a geomagnetic storm may be quantified by Dst indicator. Recent work suggested that there is a high correlation of VB_s, Δt (duration of B_s) with Dst peak values (Wang *et al.* 2003). Although a good estimation of Dst storm intensity can be excepted based on the observations of interplanetary medium, there is only rough an hour before arrival of peak of geomagnetic storm, i.e., only an almost real-time prediction can be made. Gonzalez *et al.* (2004) proposed recently that projected speeds of halo CMEs may be used to estimate Dst peak values. But the accuracy is bad (Kane 2004). How to predict intensity of a Dst storm several ten hours before its occurrence is an important and interesting topic in space weather research. This paper aims at it, and puts emphasis on filament-associated magnetic clouds, one of main source of large geomagnetic storms.

Table 1. List of typical magnetic clouds during 2000 – 2003

No.	Date	Observations			Fitted parameters									Dst	Associated filaments			$\Delta\alpha^l$
		V_0^a	B	B_s^b	B_0^c	H^d	θ^e	ϕ^f	R^g	D^h	χ^2/cc^i	Dst^j	Dst	Date	Location	Tilt^k	
1	2000.7.15-16	960	55	54	45.5	-1	0.2	70	7.9	0.051	0.065/0.92	-361	-301	7.14	N17W02	0°	0°
2	2000.8.10-11	440	14	13	11.5	-1	-44	3	19.2	0.417	0.057/0.97	-79	-106				
3	2000.8.12-13	630	34	30	35.9	-1	-33	115	11.5	0.174	0.069/0.94	-248	-235	8.9	N20E15	45°	9°
4	2000.10.28-29	400	20	18	18.3	-1	-57	187	20.8	0.168	0.017/0.99	-105	-127				
5	2000.11.6-7	550	25	16	24.5	-1	-7	118	10.6	0.217	0.037/0.97	-117	-159				
6	2001.3.19-21	400	22	21	26.0	-1	-72	286	20.5	0.459	0.032/0.96	-135	-149				
7	2001.4.21-23	370	16	14	14.0	-1	-45	289	11.5	0.261	0.036/0.96	-66	-102				
8	2002.4.17-19	500	15	13	12.0	1	-54	163	20.4	0.181	0.069/0.93	-89	-127				
9	2003.10.29-30	1500*	49	30	52.5	-1	-12	246	6.5	0.508	0.038/0.95	-156	-363	10.28	S16E15	-30°	17°
10	2003.11.20-21	640	56	54	48.0	1	-51	90	6.8	0.059	0.057/0.95	-421	-472	11.18	N03E18	0°	51°

[a] Center speed of magnetic cloud. [b] Maximum of southward component of magnetic field inside magnetic cloud. [c] Magnetic field magnitude at the axis of flux rope. [d] Sign of helicity of flux rope. [e] Elevation angle of axial field (i.e., axis) of flux rope in GSE coordinates. [f] Azimuthal angle of axial field (i.e., axis) of flux rope in GSE coordinates. [g] Radius of flux rope. [h] Distance of the closest approaching to flux rope. [i] Goodness of fit. [j] Estimated Dst peak value. [k] Tilt of filament to solar equator. [l] Angle projected in the plane perpendicular to the Sun-Earth line between filament and magnetic cloud. *The speed is not reliable.

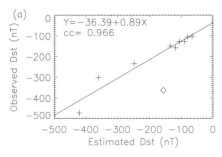
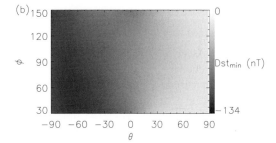

Figure 1. The left panel shows the correlation between the estimated and observed Dst peak values. The right panel exhibits Dst as a function of the elevation θ and azimuthal ϕ of the axis of a given cloud ($B_0 = 20$ nT, $R = 12$ hours $D = 0.2R$, and $V = 450$ km/s) in GSE coordinates.

2. Geoeffectiveness

Generally, MCs can be modeled by flux ropes (e.g., Burlaga 1988; Kumar and Rust 1996). It is obvious that B_s and Δt, the pivotal factors of causing geomagnetic storms, are dependent on the orientation of MC's axis for a magnetic cloud in consideration of flux rope model. Therefore, Zhao (2002) suggested that the orientation of MC is probably an important factor in geomagnetic storm. How good is the correlation between orientations of MCs and intensities of Dst storms? We check the MC-associated geomagnetic storms during 2000 to 2003. In selection of events, the following conditions are applied: (1) it was a large storm, i.e., Dst $\leqslant -100$ nT; (2) the associated MC was typical. Ten events are chose. Table 1 lists the observations of them as well as some fitted parameters by using force-free flux rope model. The goodness of fit of these clouds are all high as indicated by χ^2 and cc (correlation coefficient) listed in the 12th column. It is suggested that flux rope is a very close approximation to these typical MCs.

Wang *et al.* (2003) ever found an empirical formula to estimate Dst peak value during a geomagnetic storm. It is described as $Dst_{min} = -19.01 - 8.43(-\overline{VB_z})^{1.09}(\Delta t)^{0.30}$ nT. Based on the flux rope model, the values of B_s and Δt needed as input parameters by above formula may be derived from the fitted parameters of MCs, and therefore the Dst peak values can be estimated.

Figure 1(a) exhibits the result how consistent the estimated Dst values are with the observed values. Since the solar wind data was not reliable during the October 29, 2003 event and the speed of that MC is therefore uncertain, the October 29, 2003 event is excluded in our fitting procedure. It is obvious that except the October 29, 2003 event marked by the diamond in Fig. 1(a), all points are near the linear-fitting line. The correlation coefficient reaches 0.97. It implies that the flux rope model can be used to well predict the intensities of geomagnetic storms for typical magnetic clouds.

Further, for a magnetic cloud with given values of $B_0 = 20$ nT, $R = 12$ hours, $D = 0.2R$, and $V_0 = 450$ km/s, the estimated Dst peak values as a function of the orientations of this cloud is shown in Figure 1(b). The situation in the region of $180° \leqslant \phi < 360°$ is not represented, because it is the same as that in the region of $0° \leqslant \phi < 180°$. Moreover, if ϕ approaches to $0°$ or $180°$, i.e., the orientation is roughly parallel to the Sun-Earth line, Δt will become very long that is unreasonable. So the regions of $[0°, 30°]$ and $[150°, 180°]$ are also not plotted. From Fig 1(b), it is found that intense geomagnetic storms mainly concentrate in the negative θ, and the variation of storm's intensity is more sensitive in θ than in ϕ. The most favorable orientations of MCs to cause large geomagnetic storms are at $\theta \sim -70°$ and $\phi \sim 40°$.

3. Orientation

The last section suggests that the orientations of MCs do play an important role, and the prediction of intensities of geomagnetic storms by applying force-free flux rope model is feasible. Then, who determine orientations of MCs? Lots of previous work suggested that, for filament-associated magnetic clouds, their orientations are consistent with the directions of associated filaments (e.g., Bothmer and Schwenn 1994; Marubashi 1997; McAllister et al. 2001; Yurchyshyn et al. 2001). We address this problem again by analyzing the above events. There are only four MCs associated with obvious filament eruptions among all the ten events. The last four columns in Table 1 list them.

July 14, 2000 event. Figure 2(a) exhibits this event. The left upper panel is the EIT195Åimage showing the post-flare loops overlying the erupted filament. The right upper image is the photospheric magnetic field observed by MDI/SOHO. White denotes positive polarity and black denotes negative polarity. The thick line means the filament and the thin lines from positive polarity to negative polarity indicate the arcades overlying the filament. As a comparison, the sketch of the fitted interplanetary cloud is plotted at right lower corner, in which the thick arrow denotes the axial magnetic field, i.e., the orientation, and the thin arrow indicates the ring field, i.e., the rotation of magnetic field inside the cloud. The left lower picture shows a 3-D view of the MC and the associated filament. It is evident that the orientation of MC is very consistent with the direction of the filament ($\Delta \alpha \sim 0°$). The direction of the ring field of the cloud is also consistent with the direction of the arcades overlying the filament.

August 9, 2000 event. Figure 2(b) exhibits this event. Like the previous one, the orientation and the ring field of the cloud are both consistent with the solar observations of the erupted filament.

October 28, 2003 event. Figure 2(c) exhibits this event. Since this event is complicated, more observations are used. The upper middle image is obtained from Trace spacecraft, and the upper right image is the H_α observations overlapped by MDI observations. The orientation and ring field also primarily follow the erupted filament.

November 18, 2003 event. Figure 2(d) exhibits this event. The direction of the ring field of this magnetic cloud is roughly consistent with the direction of arcades overlying the erupted filament. Nevertheless, the orientation of the cloud deviates largely from the direction of the filament. The $\Delta \alpha$ reaches about $51°$ as shown in the left lower 3-D view.

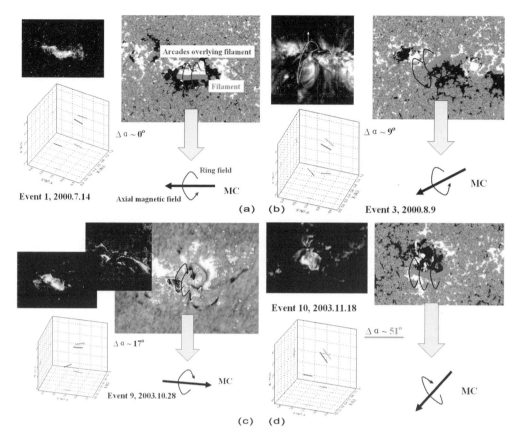

Figure 2. The solar observations of the four erupted filaments and their comparisons with the interplanetary magnetic clouds.

4. Discussions

According to the above analyses, the orientations of magnetic clouds are approximately along the directions of erupted filaments, but not always. This result indicates that using observations of erupted filaments to estimate orientations of magnetic clouds in the interplanetary medium takes a risk. Why is there an exception? We think that there are following three possible reasons: (1) The error in fitting interplanetary MCs by the flux rope model is too large; (2) the ambient solar wind deforms the configuration of magnetic fields inside MCs, and therefore makes their magnetic axes deviate from the initial directions; (3) the axes will *rotate* when MCs are propagating in the heliosphere. Fig. 1(b) suggests that the variation of intensities of geomagnetic storms is sensitive in θ. Thus, if the orientations of MCs really change when they are ejected out into the interplanetary medium, such change will significantly affects the level of possible geomagnetic storms.

For the four filament-associated MCs, the directions of the ring fields of them are all consistent with the directions of the arcades overlying the erupted filaments. It seems that the rotation of magnetic fields inside clouds can be derived from observations of arcades, i.e., from the polarities of photospheric magnetic fields beside erupted filaments. However, in theory, the ring fields are not necessary consistent with the arcades (refer to Low and Zhang (2002)). Basically, there are two categories of filament: normal type and inverse type. For inverse filaments, the ring fields are consistent with the arcades, whereas for normal filaments, it is reversed. According to this theory, above four events are all associated with inverse filaments. Observations suggest that the number of inverse filaments is much larger than that of normal filaments. This is a reason why there is no normal filament in our sample.

To predict intensities of geomagnetic storms several tens hours before their occurrences is difficult. Using observations of erupted filaments and flux rope model to do prediction needs to know not only the orientations (θ and ϕ) of magnetic clouds, but also B_0, R, and D. Thus, much more further work is required.

5. Conclusions

(1) For typical MCs, flux rope model can well estimate the intensities of geomagnetic storms. (2) To cause large geomagnetic storms, the favorable orientations of MCs are approximately at $\theta \sim 70°$ and $\phi \sim 40°$ in GSE coordinates. (3) The orientations of most MCs are roughly along the directions of associated erupted filaments, but not exactly. Occasionally, deviation is very large. (4) The ring fields of magnetic clouds are consistent with the arcades overlying the filaments. It is suggested that the erupted filaments are dominated by inverse type.

Acknowledgements

We acknowledge the use of the data from the SOHO, Trace and ACE spacecraft. This work is supported by the Chinese Academy of Sciences (KZCX2-SW-136), the National Natural Science Foundation of China (40404014, 40336052, 40336053), and the State Ministry of Science and Technology of China (G2000078405).

References

V. Bothmer and R. Schwenn 1994, *Space Sci. Rev.* **70**, 215.

L.F. Burlaga 1988, *J. Geophys. Res.* **93**, 7217.

R.K. Burton, R.L. McPherron, and C.T. Russell 1975, *J. Geophys. Res.* **80**, 4204.

W.D. Gonzalez, A. Dal Lago, A.L.C. de Gonzalez, L.E.A. Vieira, and B.T. Tsurutani 2004, *J. Atmos. Solar-Terres. Phys.* **66**, 161–165.

W.D. Gonzalez, J.A. Joselyn, Y. Kamide, H.W. Kroehl, G. Rostoker, B.T. Tsurutani, and V.M. Vasyliunas 1994, *J. Geophys. Res.* **99**, 5771.

W.D. Gonzalez, B.T. Tsurutani, A.L.C. Gonzalez, E.J. Smith, F. Tang, and S.I. Akasofu 1989, *J. Geophys. Res.* **94**, 8835.

J.T., Gosling, D.J. McComas, J.L. Phillips, and S.J. Bame 1991, *J. Geophys. Res.* **96**, 731.

R.P. Kane 2004, *J. Geophys. Res.* **submitted**.

A. Kumar and D.M. Rust 1996, *J. Geophys. Res.* **101**, 15667.

B.C. Low and M. Zhang 2002, *Astrophys. J* **564**, L53–L56.

K. Marubashi 1997, In: N. Crooker, J. A. Joselyn, and J. Feynman (eds.): *Coronal Mass Ejections.* pp. 147–156.

A.H. McAllister, S.F. Martin, N.U. Crooker, R.P. Lepping, and R.J. Fitzenreiter 2001, *J. Geophys. Res.* **106(A12)**, 29185–29194.

N.R. Sheeley, Jr., R.A. Howard, M.J. Koomen, D.J. Michels, R. Schwenn, K.H. Muhlhauser, and H. Rosenbauer 1985, *J. Geophys. Res.* **90(A1)**, 163.

D. Vassiliadis, A.J. Klimas, J.A. Valdivia, and D.N. Baker 1999, *J. Geophys. Res.* **104**, 24957.

Y. Wang, C.L. Shen, S. Wang, and P.Z. Ye 2003, *Geophys. Res. Lett.* **30(20)**, 2039.

V.B. Yurchyshyn, H. Wang, P.R. Goode, and Y. Deng 2001, *Astrophys. J.* **563**, 381–388.

X.P. Zhao 2002, In: H. Wang and R. Xu (eds.). *Proceedings of the COSPAR Colloquium held in the NAOC in Beijing, China.* p. 209.

Discussion

SCHMIEDER: 1. About the event of October 28, 2003, how do you identify the inversion line where the filament is located? Looking at Trace + MDI movies, the filament is over an inversion line more on the west part of your image between the main leading negative spots and the the following positive spots.
2. You are considering the direction of the arcades over the filaments and the filaments themselves.

WANG: 1. The identification of Oct. 28, 2003 event perhaps was not correct in our work. I will check it again.
2. My consideration is the correlation between the magnetic clouds and the filaments and the arcades overlying the filaments. Generally, magnetic clouds are formed by the ejected arcades overlying the filaments, so there should be a close correlation between them. My aim is to find them for predicting the properties of magnetic clouds, interplanetary space and potential geomagnetic storms intensity.

ZHUKOV: A comment concerning the discussion about what we see– filament or overlying arcade. In general, the filament material is extremely rarely observed in situ, so what we see as a magnetic cloud in the solar wind corresponds rather to the arcade overlying the filament.

WANG: Yes, maybe I did not clarify clearly in my talk that magnetic clouds are formed by the arcades (the blue curves in figures) overlying filaments (the red line in figure).

JINGXIU WANG: Comments: What he referred to about the direction of magnetic lines of force in the magnetic cloud (blue arrows in the figures) is the overall fields in magnetic arcades of the filaments, not the fields inside in the filaments. I agree with the comments made by Schmieder and Zhukov.

WANG: Yes, I mean that whether the rotation of magnetic field inside the magnetic cloud can be predicted by the arcades overlying the associated filaments. The results suggest the direction of the arcades overlying filaments are not always consistent with the rotation of magnetic field of magnetic clouds.

SCHWENN: Do "reverse" and "inverse" topologies both exist? Is there bservational evidence?

WANG: I believe the two types do both exist. In the events studied in my work, there is no observational evidence. It is just our supposition. I will try to find evidence in further work.

DELABOUDINIERE: What is the direction of the magnetic field in, in 1. the magnetic cloud?, 2. in the filament?, 3. in the arcade? The latter two are perpendicular. Which is parallel to which?

WANG: According to our result, the direction of the axial magnetic field in the magnetic cloud is (anti-) parallel to the direction of the filament, but not always. Perhaps the magnetic cloud will rotate when it is propagating in the heliosphere, so the orientation of the magnetic cloud is not always parallel to the direction of filament. The arcades overlying filaments form the magnetic cloud. But the rotation of the magnetic field in magentic cloud can not be determined by only the polarities of the photospheric field beside the filament, because there are two types of filament: normal and inverse.

Coronal and Stellar Mass Ejections
Proceedings IAU Symposium No. 226, 2005
K. P. Dere, J. Wang & Y. Yan, eds.

© 2005 International Astronomical Union
doi:10.1017/S174392130500102X

A Study of Intense Geomagnetic Storms and their Associated Solar and Interplanetary Causes

S. C. Kaushik

Government Autonomous P.G. College, Datiya, M.P., India

Abstract. Shocks driven by energetic coronal mass ejections and other interplanetary transients are mainly responsible for large disturbances in geomagnetic field of Earth and play a key role in producing a geomagnetic storm or substorm. A geomagnetic storm is a global disturbance in Earth's magnetic field usually occurred due to abnormal conditions in the IMF and solar wind plasma emissions caused by various solar phenomenon. Identifying intense geomagnetic storms with Dst decrease more than/or equal to 300 nT occurred during 1981-2001, a correlative study has been performed to analyze the associated solar and interplanetary causes of these 09 events using solar wind plasma, IMF and solar geophysical data. It is observed statistically that 55% storms have occurred during solar maximum and 45% occurred during minimum phase of solar cycles. Further, study reveals that 77% intense storms are associated with CMEs, which confirms earlier findings.

Keywords. Sun: coronal mass ejections (CMEs)

Discussion

JIE ZHANG: You said 77 % of super intense storms are associated with CMEs. I would expect 100% of them are with CMEs, e.g, for the period with SOHO/LASCO observation. My question is how you get this 77% number?

KAUSHIK: Yes,this will happen if all the events will be combined together and then study will performed. But because I want to study them separately from 1981 to 2002, the storms associated with CMEs are lesser in numbers. This is also that during solar minima there are coronal holes which are responsible for producing geomagnetic activities, rather than any other cause.

LAKHINA: Comments: I think this division of intense and super intense storm is rather artificial. If both the categories are combined together, the statistics of correlation with CME would go up from 77 % to nearly 100 percent.

KAUSHIK: Yes, that's true by doing so the storms associated with CME will be more in number, but because I wanted to study them separately, i.e., intense and super-intense, that's why I took this division of events, i.e.intense (up to $\sim$250 to 299nT) and super intense ($\sim$300nT and above). This is why the CME associated storms' number is low.

Coronal and Stellar Mass Ejections
Proceedings IAU Symposium No. 226, 2005
K. P. Dere, J. Wang & Y. Yan, eds.

Geoeffectiveness of CMEs in the Solar Wind

E. Huttunen

University of Helsinki, Finland

Abstract. The main drivers of strong geomagnetic activity at the Earth are interplanetary manifestations of coronal mass ejections. A magnetic storm can be caused by compressed sheath fields before the CME, by the CME ejecta or by the combination of these two structures. The most geoeffective subset of CMEs are magnetic clouds. When observed near 1 AU magnetic clouds are characterized by monotonous rotation of magnetic field direction through a large angle, high magnetic field magnitude, low temperature and low plasma beta. We have investigated the magnetic structure and the geomagnetic consequences of magnetic clouds identified from WIND and ACE data for the years 1997-2003. The geomagnetic response of a certain magnetic cloud depends greatly on its magnetic structure and orientation of sheath fields. We have investigated drivers of intense magnetic storms (Dst ¡ -100 nT) during the interval of 1997-2002, i.e. rising, maximum and early declining phases of solar cycle 23. Sheath regions and post-shock streams caused nearly half of all intense storms. Importance of sheath regions as storm drivers even increased as the level of the storm increased. In 2003 two most intense geomagnetic storms of the solar cycle 23 took place. Both of these were driven by southward fields embedded in a magnetic cloud that had axis highly inclined to the ecliptic plane. Though sheath regions alone efficiently drive intense Dst storms (¡ -100 nT) the largest storms (Dst ¡ -300 nT) require exceptionally long-time and intense southward magnetic fields that presumably only magnetic clouds can provide. High solar wind dynamic pressure seems to be important in generating extremely intense Dst storms. As an example we show solar wind condition during Nov 19-20, 2003 magnetic cloud that caused the largest storm of the solar cycle 23.

Magnetic clouds have smoothly changing magnetic field direction combined with low solar wind dynamic pressure. Sheath regions typically have rapidly varying magnetic field direction and high dynamic pressure. Thus, these two solar wind drivers put magnetosphere under different type of driving. We also studied the responses of the Dst index that aims to measure the strength of the equatorial ring current and the Kp index that records more global and higher latitude activity than Dst to different storm drivers. We found that in general sheath regions generate higher Kp activity when compared to the level of the the Dst disturbance than magnetic clouds. In some cases rapidly fluctuating magnetic field in the sheath region caused very strong high-latitude activity (Kp 8-9) though the Dst index was significantly less enhanced. This suggest that magnetospheric current systems have different responses to different solar wind drivers.

Keywords. Sun: coronal mass ejections (CMEs)

Discussion

DERE: Did you also consider geomagnetic storms that were not associated with MCs.

HUTTUNEN: The results were shown from two different studies: 1)magnetic clouds from 1997 to 2003; 2)drivers of all storms with Dst <-100 nT from 1997-2002. In 2) we considered all drivers of storms, 25% of 53 storms were not associated to MCs or sheath regions/shocks. They were driven by ICMEs without flux rope structure or solar wind structures not related to CMEs (e,g., compression regions of slow and high (solar wind streams) speed).

Jie Zhang: You attribute storms either to shock sheath or to MC. But in some cases, it must be a combinedl effect. For example, an ICME has both shock and sheath. How do you distinguish when this occurs?

Huttunen: The storm period was determined according to Kamide etal.1998 (whether one storm with two steps or two different storm). The driver of the storm was defined as a feature causing 85% of the Dst minimum for that storm. We approximated the time when the front edge of a magnetic cloud reached the magnetopause and added a hour to that time to define the time when the effect of the magnetic cloud started. Only in small subset of cases it was difficult to separate the effect of sheath and magnetic cloud (for SN type magnetic clouds)

Coronal and Stellar Mass Ejections
Proceedings IAU Symposium No. 226, 2005
K. P. Dere, J. Wang & Y. Yan, eds.

© 2005 International Astronomical Union
doi:10.1017/S1743921305001043

Double Star Program in China

Z. X. Liu

Center for Space Science and Applied Research, Chinese Academy of Sciences, Beijing, China

Abstract. The Geospace Double Star Project (DSP) contains two satellites operating in the near-earth equatorial and polar regions respectively. The tasks of DSP are: (i) to provide high-resolution field, particle and wave measurements in several important near-earth magnetosphere active regions which have not been covered by existing ISTP missions in the geospace, such as the near-earth plasma sheet and its boundary layer, the ring current, the radiation belts, the dayside magnetopause boundary layer, and the polar region; (ii) to investigate the trigger mechanisms of magnetic storms, magnetospheric substorms, and magnetospheric particle storms, as well as the responses of geospace storms to solar activities and interplanetary disturbances; (iii) to set up the models describing the spatial and temporal variations of the near-earth space environment. To complete the mission, there are eight instruments on board the equatorial satellite and the polar satellite, respectively. The orbit of the equatorial satellite with a perigee at 565.5km and an apogee at 78959.9km, and the inclination is 28.17 ; while the orbit of the polar satellite is proposed with a perigee at 700km and an apogee at 40000km, as well as an inclination about 90 . The equatorial satellite has been launched successfully in December 2003. Now the equatorial satellite (TC-1) and instruments operate normally. Payloads have provided good quality of data of fields and particles. Already very good conjunction in the dayside magnetopause and magnetotail with Cluster, CME effects could be investigated. The first results of data analysis have already shown great interesting. The polar (TC-2) satellite has been launched successfully in July 2004. Now the satellite operates normally. The commissioning of the payload has started since the end of July 2004 and will be finished in the middle of Sep.2004. The instrument has been normally operation and downlink the data.

Keywords. Sun: coronal mass ejections (CMEs), Sun: solar-terrestrial relations

Coronal and Stellar Mass Ejections
Proceedings IAU Symposium No. 226, 2005
K. P. Dere, J. Wang & Y. Yan, eds.

© 2005 International Astronomical Union
doi:10.1017/S1743921305001055

The Possible Sources of the Relativistic Electrons in the Magnetosphere

L. Xie†, Z.Y.Pu and Y.Lu

Department of Geophysics, Peking University, Beijing, 100871, China
email:xielun@pku.edu.cn

Abstract. Using data from ACE and low-altitude polar orbit satellite of NOAA, we investigate the possible sources of the enhancements of relativistic electrons in the magnetosphere. The observations from NOAA for the different geomagnetic activity periods show that substorms injections provide seed electrons for MeV electron enhancement associated with geomagnetic storms and the energetic electrons in the solar wind provide an alternative source for the relativistic electrons in the magnetosphere during the SEP events.

Keywords. Sun: coronal mass ejections (CMEs), particle emission, solar-terrestrial relations, solar wind

1. Introduction

The magnetosphere is an efficient accelerator and effective trapping device for energetic electrons. The relativistic electron population has long been observed in the Earth's outer radiation belt. The fluxes of these electrons are highly variable and dynamics, showing enhancements of several orders of magnitude occurring on timescales of about one day in many storm periods (Baker *et al.* 1994). The trigger for the flux enhancement of high-energy electron associated with geomagnetic storms is known mostly to be the enhanced solar wind velocity and north-south changes of the interplanetary magnetic field (IMF). Relativistic electron flux enhancements were further often also shown strong recurrent tendencies with the 27-day rotation period of the sun, particularly in the years approaching solar minimum (e.g., Baker *et al.* 1979). These have led to the view that the MeV electron flux enhancements are linked to solar energetic particle (SEP), coronal mass ejections (CMEs), and corotating interaction regions (CIRs) associated with high-speed solar wind stream.

The sources, acceleration mechanisms and transport processes of relativistic electron are primary issues in the research field of magnetospheric physics and remain unsolved problems for further studies. It has been proposed that relativistic electrons in the solar wind, either from the sun or from Jupiter, may be a source of the relativistic electron in the magnetosphere (e.g., Baker *et al.* 1979; Baker *et al.* 1986). Although the acceleration mechanisms of relativistic electron have not fully understood, in the meantime a few mechanisms responsible for acceleration of relativistic electron population in the Earth's magnetosphere have been proposed: radial diffusion (Li, 1997), re-circulation (Fujimoto, 1990) and wave-particle interaction (Liu, 1999), etc. These acceleration mechanisms all need a source that provides seed electrons (approximately hundreds of keV to one or two MeV) to be accelerated to the range of a few to ten MeV. The origin of the MeV electrons in the magnetosphere remains an unsolved problem. Two notable possible sources have been noticed in the literature: the energetic electrons near the outer boundary of the inner magnetosphere and in the solar wind. It has been suggested that substorms may

† Present address: Peking University, Beijing, China.

effectively produce the seed population of energetic electron that can be accelerated by a not yet defined process to be relativistic electrons (Baker *et al.* 1998). High-energy particles within in the Earth's radiation belt can be accelerated in the magnetosphere (Schulz and Lanzerotti, 1974). The seed population can also originate from source beyond the Earth's immediate influence (Scholer, 1979). It has been established that energetic protons and other ions from the sun can be penetrated into the terrestrial magnetosphere (Scholer, 1979; Fennel, 1973). In this paper we aim to identify the evidences for the sources of the relativistic electrons in the magnetosphere mainly based on the observations from low-altitude polar orbit NOAA satellite.

2. Data

NOAA/Polar Orbiting Environmental Satellites (POES) were launched into a nearly polar orbit with an altitude about 850km, an inclination of 98°. The spacecraft carry a suite of instruments that detect and monitor the flux of energetic ions and electrons into the atmosphere and the particle radiation environment at the altitude of the satellite. The present study examines the observations of electrons from the (Medium Energy Proton and Electron Detector) MEPED on board the NOAA-15 and NOAA-16. The MEPED instrument measures electrons at angle of 10° and 80° to the local vertical measurements of trapped electron in three differential energy channels E1(>30keV), E2(>100keV), E3(>300keV) at 80° angle were used in this study. It worthwhile to point that we have try to exclude the electron contamination by the proton based on the suggestion by the principle scientist of MEPED/NOAA. The ACE electron flux in the energy range of 175–315keV, proton flux with the energy greater than 10MeV, and the solar wind velocity are provided by the energy particle and solar wind instrument on board ACE spacecraft.

3. Energetic electron source in the solar wind

On Oct. 28, 2003 the sun produced an extreme large flares X17/4B peaking at 1110 UT. A very fast (near 2000km/s) earthward coronal mass ejection (CME) was observed on SOHO/LASCO. Successively at 1150UT after X-ray burst, beginning of a strong SEP event was, detected by GOES spacecraft at the geosynchronous orbit. Two fast-moving magnetic clouds from the sun swept past Earth and produced extreme geomagnetic storms. The storm suddenly commenced at 0611UT on Oct. 29 and at 1637UT on Oct. 30, respectively. Figure 1 shows the counts rate of >0.3MeV electrons of Oct. 27–30 from MEPED instrument on NOAA-16. This figure is color-coded spectrogram showing the logarithm of counts rate, the scale being indicated by the color bar on right. The data are binned in 0.1 L-values. The total L range 1<L<15 is shown in the figure. The time between adjacent columns of pixels is 2 hours. The electron counts rate had been relatively low until an enhancement across a wide range from high L-values to low L-value occurred around 1200 UT on Oct. 28. This enhancement stared before the magnetic storm on Oct. 29 and was directly correlated with the SEP event in the upstream solar wind. The observation by the NOAA satellite shows that the abrupt electron enhancement extends across all high L-value (L>6) without notable time difference. Figure 2 shows the variation of energetic electron (175–315 keV) and proton (>10MeV) flux of Oct. 28–29 measured by the ACE. Measurements by ACE indicated that the massive solar energetic electrons with energy being a few hundreds keV could be produced during the period of the SEP event. The increases of the solar energetic electron beginning around 1200

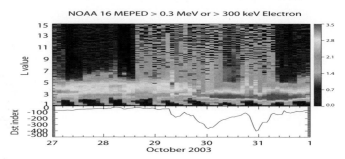

Figure 1. The counts rate of >0.3MeV electrons as a function of time and L-shell. The Dst index is shown below.

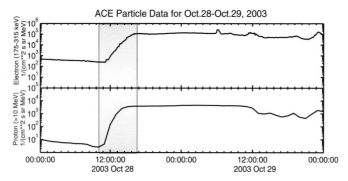

Figure 2. The variation of energetic electron (175–315keV) and proton (>10MeV) flux of Oct. 28–29 measured by the ACE.

UT are coincident with the enhancement of relativistic electrons in the magnetosphere observed by NOAA.

Following up on the above observations, a question arises as to whether solar energetic electron population during the SEPs is a significant source of MeV electron enhancement in the magnetosphere? Figure 3 plot the NOAA data of Oct. 28–29, 2003. The measurements were made in the north polar cap region with a selection criterion applied such that magnetic latitude was greater than 70°. In the open magnetosphere model, this criterion would imply that at such high latitudes NOAA was sampling essentially interplanetary-connected field lines (Baker *et al.* 1997) As is seen in Figure 3(a), the >0.3MeV electrons measured by MEPED /NOAA enhanced rapidly above background levels around 0600UT on 28 October. Simultaneously an increase of energetic electrons was detected by ACE. The intensities of electrons remained high level for more than one day. This observation suggests that massive solar energetic electrons can be seen in the polar cap region. Figure 3(b) plots data of the electron counts rate in the energy range Of >0.3 MeV at L = 6.6 for the same time period as figure 3(a). In Figure 3(b) an enhancement of electrons in the inner magnetosphere observed while strong access of solar electrons to the polar cap was seen. In Figure 4, we show the measurement of spin average integral electron rates with energy >0.3MeV of measured by POLAR for the same time period. Similar enhancement can be seen in this figure. All these electrons require a channel to penetrate into the magnetosphere. It has been suggested that solar particles would have rather directed access to polar cap region following the open magnetic filed lines (Fennel, 1973). Sheldon *et al.* (1998) also presented the POLAR observation of MeV electrons population in the cusp region. The following solar wind electron entry scenario

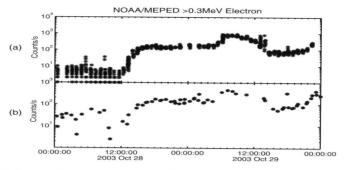

Figure 3. (a) Polar cap electron counts rates (>0.3MeV) measured by NOAA during October 28-29 2003. (b) Same as (a) but showing electron counts rate at low latitudes for L=6.6

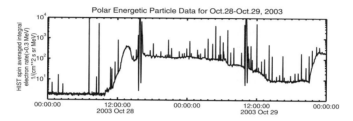

Figure 4. Spin average integral electron rates with energy >0.3MeV of POLAR for the same period of time as in Figure 4.

is then suggested in the present paper: When the solar wind energetic electrons in arrived at the Earth, they penetrate into the terrestrial magnetosphere along the open magnetic filed lines through the cusp and polar cap regions. With this scenario and based on the observations in Figures 3 and 4, we postulate that the solar wind energetic electrons in the solar wind may provide a source for the relativistic electrons in the magnetosphere during the period of the SEP events.

4. Energetic electron source near the Outer boundary of the magnetosphere

In the late 2000, two successive geomagnetic storms began on Nov 6 and 10. Figure 5(a) and (b) show the NOAA observations of >0.3MeV and 30-300keV electrons, respectively, for the interval from November 4 to 13. There was strong substorm activities during the storm time. Figure 5(b) shows a number of substorm particle injections during the storm time. In Figure 5(a)we see a numbers reduction of relativistic electrons at 3<L<6 during the main phase of the storm and an enhancement during the recover phase.

What is the source population of MeV electron filling the outer belt during the storm recover phase? In Figure 5(b) we see substorm produced intense injections of energetic electrons in the storm main phase (as well as in the recovery phase) before the increase of relativistic electrons in the inner magnetosphere. The substorm energetic electrons provided by substorm can be accelerated up to MeV by local energization mechanisms, such as wave-particle interaction, etc. Observations in Figure 5(a) and (b) thus strongly support the view that substorms can provide seed electrons for MeV electron enhancement associated with geomagnetic storms.

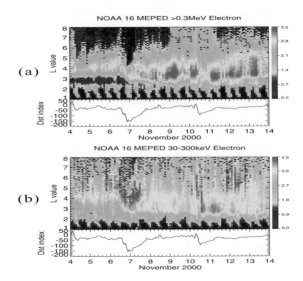

Figure 5. The counts rate of >0.3MeV electrons as a function of time and L-shell. The Dst index is shown below.

5. Conclusions

In the present paper we investigate the source of enhancement of relativistic electrons in the magnetosphere based on NOAA measurements. Upon all the observations we address the following views: (1) Substorm injections provide seed electrons for MeV electron enhancement associated with geomagnetic storms. (2) The Energetic electrons in the solar wind provide an alternative source for the relativistic electrons in the magnetosphere during the SEP events.

Acknowledgements

This work is supported by the Chinese Research Project G20000784 and the CNSF Grant 40390150.We would like to acknowledge the Daniel Wilkinson and David Evans at NOAA for the NOAA/POES data.

References

Baker, D.N., Higbie, P.R., Belian, R.D., & Hones Jr., E.W. 1979, *Geophys. Res. Lett.* 6,531
Baker, D.N., Blake, L.B., S., & Higbie, P.R. 1986, *J. Geophys. Res.* 91,4256
Baker, D.N., Blake, L.B., Callis, L.B., Cummings, J.R., Hovestadt, D., Kanekal, S., Klecker, B., Mewaldt, R.A., & Zwickl R.D. 1994, *Geophys. Res. Lett.* 409
Baker, D.N., Pulkkinen, T.I., Li, X., Kanekal, J.B., Blake, L.B., Selesnick, R.S., Henderson, M.G., Reeves, G.D., Spence, H.E., & Rostoker G. 1998, *J. Geophys. Res.* 17,279
Fennell, J.F. 1973, *J. Geophys. Res.* 78,1036
Fujimoto, M. & Nishida, A. 1990, *J. Geophys. Res.* 95, 4265
Liu, W.W., Rostoker, G., & Baker, D.N. 1999, *J. Geophys. Res.* 1,7391
Li, X.L., Baker, D.N., Temerin, M., Cayton, T., Reeves, G.D., Araki, T., Singer, H., Larson, D., Lin, R.P., & Kanekal, S.G. 1997, *Geophys. Res. Lett.* 104,4467
Liu, W.W., Rostoker, G., & Baker, D.N. 1999, *J. Geophys. Res.* 1, 7391
Scholer, M. 1979, in: R.F.Donnelly (eds.), *Solar-terrestrial Prediction Proceedings* 2,446
Schulz, M. & Lanzerotti, L.J. 1974, *Particle Diffusion in the Radiation Belts* (New York: Springer)

Sheldon, R.B., Spence, H.E., Sullivan, J.D., Fritz, T.A., Chen, J. 1998, *Geophys. Res. Lett.* 25,1825

Discussion

YOUSEF: It is very interesting that you found those relativistic electron. I have been looking for such events in order to test my model of acceleration found in this volume. Have you found other relativistic nuclei(ions) also during such events?

XIE: I didn't focus on the other relativistic ions during the SEP events. I only research the relativistic electron enhancement during the SEPs.

(NAME NOT WRITTEN): Do the relativistic electrons in the solar wind follow Parker's spiral?

XIE: The spacecraft used to measure relativistic electrons inside the Earth are NOAA, Polar, Sampex etc. We do not think electron detector onboard these satellites can tell us whether or not these electrons are along Parker's spiral.

KOUTCHMY: Can you say of your relativistic electron follow or not the Archimedes spiral (Parker) of the magnetic field coming from the sun? What is the angle of arrival at the Earth?

XIE: I think the relativistic electrons perhaps coming from the sun following the Parker orbit, but we have not directly observations. The angle of MeV electron measured by NOAA satellite is isotropic.

Coronal and Stellar Mass Ejections
Proceedings IAU Symposium No. 226, 2005
K. P. Dere, J. Wang & Y. Yan, eds.

© 2005 International Astronomical Union
doi:10.1017/S1743921305001067

CMEs Associated with Eruptive Prominences: How to Predict?

B.P. Filippov, O.G. Den and A.M. Zagnetko

Institute of Terrestrial Magnetism, Ionosphere and Radio Wave Propagation, Russian
Academy of Sciences, Troitsk Moscow Region, 142190, Russia
email: bfilip@izmiran.ru

Abstract. Solar prominences can be viewed as pre-eruptive states of coronal mass ejections (CMEs). Eruptive prominences are the phenomena most related to CMEs observed in the lower layers of the solar atmosphere. The most probable initial magnetic configuration of a CME is a flux rope consisting of twisted field lines which fills the whole volume of the dark cavity stretched in the corona along the photospheric polarity inversion line. Cold dense prominence matter accumulates in the lower parts of helical flux tubes, which serve as magnetic traps in the gravitation field. Coronal cavity is rather inconvenient feature for observation owing to reduced emission, so prominences and filaments are the best tracers of the flux ropes in the corona long before the beginning of eruption. Thus, the problem of the CME prediction can be reduced to the analysis of the filament equilibrium and estimation of the stability store. The height of a prominence (or a filament when observed against the disk) increases with its age and the death of a filament is usually an eruption which is followed by a CME. The filament height, then, can be a measure of its age and its readiness for eruption. In inverse-polarity models the equilibrium height of a filament is related to the value of the filament electric current. The stronger the electric current, the greater the height of the filament. However, the equilibrium and stability of a filament depend not only on its current but also on the characteristics of the external magnetic field. In order to estimate the probability of eruption, we should therefore compare the observed prominence height with a value characterizing the photospheric magnetic field. This value is the critical height, which can be found in the distribution of the magnetic field vertical gradient above the polarity inversion line. We had analyzed three dozens of filaments and found that eruptive prominences were near the limit of stability a few days before eruptions. We believe that the comparison of the real heights of prominences with the calculated critical heights could be a basis for predicting filament eruptions and following CMEs.

Keywords. Sun: corona, coronal mass ejections (CMEs), filaments

1. Introduction

The "classical" structure of a coronal mass ejection (CME) consists of three parts and one of these parts, a bright core, is the remnant of an eruptive prominence (Crifo, Picat & Cailloux 1983; Sime, MacQueen & Hundhausen 1984; Hundhausen 1999). Sometimes it is possible to follow the whole process of eruption beginning from filament activation and up to the CME formation. In figure 1 one can see the polar crown filament eruption on 14 June 1999 visible in SOHO EIT He II line and the following CME observed by SOHO LASCO. The projection of the erupting filament was favorable to recognize the twisted structure of the filament loop in the core of the CME up to the distances of 3 solar radii. Of course, not every CME could be associated with observed filament eruption. The reason for it may be that large-scale phenomena high in the corona are registered now more effectively than near-surface phenomena. At least half of eruptions could originate on the backside of the Sun. Nevertheless, eruptive prominences are the phenomena most

SOHO EIT, He II line, 304 A

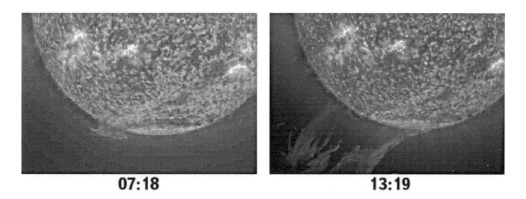

07:18   13:19

SOHO LASCO, C2

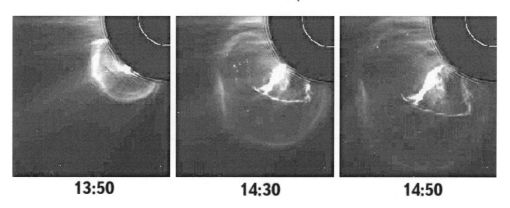

13:50   14:30   14:50

Figure 1. Polar crown filament eruption on 14 June 1999 visible in SOHO EIT He II line (top row) and the following CME observed with LASCO C2 (bottom row). (Courtesy of SOHO/EIT and SOHO/LASCO consortia. SOHO is a joint ESA-NASA project.)

related to CMEs observed in the lower layers of the solar atmosphere (Munro *et al.* 1979; Webb & Hundhausen 1987; St. Cyr & Webb 1991). So, solar prominences can be considered pre-eruptive states of coronal mass ejections (CMEs).

2. Filament stability and the critical height

A lot of filaments and prominences are seen on the Sun at epoch of rather high activity. Each of them could be assumed as the place where a CME could originate from. Thus, the problem of the CME prediction could be reduced to the analysis of the filament equilibrium and estimation of the store of stability. The height of a prominence increases with its age and the death of a filament is usually an eruption (Rompolt 1990), which is followed by a CME. The filament height, then, can be the measure of its age and its readiness for eruption. However, for the quantitative description we need a scale, which can be used to estimate whether a filament is sufficiently high or low. It is evident that as far as a filament is supported by the magnetic force, the scaling factor should depend on the scale of the background magnetic field.

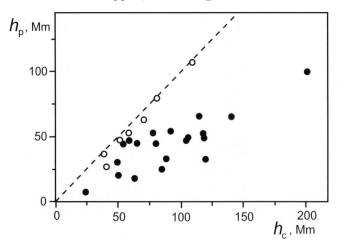

Figure 2. Critical height of stable filament equilibrium h_c versus observed prominence height above the limb h_p. The dotted line corresponding to equality of these quantities is the stability boundary. The solid circles correspond to the filaments which safely passed the west limb. The open circles correspond to the filaments which disappeared from the disk.

We analyzed the situation from the point of view of a model of inverse polarity filament or a flux rope model (Kuperus & Raadu 1974; Van Tend & Kuperus 1978; Molodensky & Filippov 1987; Priest, Hood & Anzer 1989; Forbes & Isenberg 1991; Low & Hundhausen 1995; Aulanier & Demoulin 1998) . In this model, the background magnetic field acts on the filament current with the Lorenz force directed downward and the only supporting force is filament current's repulsion from the currents induced in the photosphere. The filament current should be great enough to create the magnetic field dominating inside the filament and its vicinity. Cold dense prominence plasma could be accumulated at the bottoms of helical field lines. Sometimes helical threads easily observed in the fine structure of prominences and filaments though generally their structure is tangled.

In the simplest model, the flux rope could be assumed as a straight linear current. If the photospheric background field falls off with height faster than $1/h$, the equilibrium of the linear current could not be stable. So for a given photospheric magnetic field distribution B, the critical height h_c for stable equilibrium exists defined by equation (Filippov & Den 2000; Filippov & Den (2001))

$$h_c = \frac{B}{dB/dh|_{h_c}}.$$

Figure 2 shows the relationship between the prominence height measured on limb and the critical height calculated at the time when the filament crosses the central meridian. The filaments, which safely pass the west limb, are marked with the solid circles while the filaments, which disappear on the disc, are marked with the open circles. It is seen that the solid circles more or less evenly fill the angle between the bisector showing the limit of stability and the horizontal axis while the open circles tend to cluster about the bisector. This shows that eruptive prominences were near the limit of stability a few days before an eruption.

3. The problem of the filament height measurements

Figure 3 shows an example of a filament eruption observed on the disk. The prominence appeared on the limb on August 16. Then it was seen as well developed quiescent filament

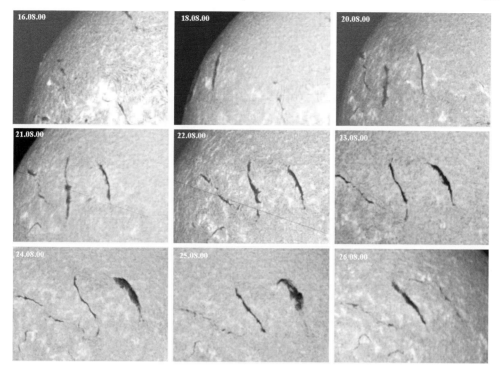

Figure 3. Filament pass through the solar disk in August 2000. Filament eruption happened on 25 August after which filament disappeared completely. It was reconstructed in part the next day. (Courtesy of Meudon Observatory.)

during eight days of its transit through the disk. The width of the filament increase from day to day revealing the increase of its size and height. On August 24 there were some indications of activation and internal motions but the filament was still stable. On August 25 at 10:40 UT the filament began to ascend rapidly and disappeared from the disk. The eruption is visible quite clearly also in the SOHO/EIT movie in Fe XII line.

We calculated the critical height h_c for this filament. It varied slightly during the days before the eruption but on average it was about 60 Mm. A small bipolar region appeared near the northern end of the filament on August 24. It could be assumed as the destabilizing factor for the filament equilibrium. However, the bipolar region seems to be well developed several hours before the eruption. On the same time, we see that rapid ascend of the filament began only after it reached high altitude. It's much harder to estimate the height of a prominence when it is projected against the disk as a filament. There were attempts to estimate the height of a filament above the photosphere based on the observed changes of its projection due to the rotation of the Sun (Vrsnak *et al.* 1999). We tried to measure the height of the filament before the eruption using different technique. At first, we found the tilt of the filament body to the vertical direction using the technique proposed by d'Azambuja & d'Azambuja (1948). It was about 20°. Then taking into account the filament heliocoordinates and its visible width we found the height of the filament h_p to be 40 Mm on August 23, about 50 Mm on August 24, and about 70 Mm just before the eruption.

4. Conclusions

The most probable magnetic configuration of a CME source region is a flux rope. This configuration allows a catastrophic process. So the onset of eruption does not need a powerful trigger. It is not so easy to observe a flux rope on the Sun because most of its volume is filled with rear plasma and has only weak emissions. Prominences and filaments are the best tracers of the flux ropes in the corona long before the beginning of an eruption. The filament height can be a measure of its age and readiness for eruption. The comparison of the real heights of prominences with the calculated critical heights could be a basis for predicting filament eruptions and following CMEs.

Acknowledgements

B.P.F. is grateful to the Organizing Committee of the IAU 226 Symposium for the financial support. This work was supported in part by RFBR (grant 03-02-16093).

References

Aulanier, G. & Demoulin, P. 1998, *Astron. Astrophys.* 329, 1125
d'Azambuja, M. & d'Azambuja, L. 1948, *Ann. Obs. Paris, Meudon* 6, Fasc. VII
Crifo, F., Picat, J.P., & Cailloux, M. 1983, *Solar Phys.* 83, 143
Filippov, B.P. & Den, O.G. 2000, *Astron. Lett.* 26, 322
Filippov, B.P. & Den, O.G. 2001, *J. Geophys. Res.* 106, 25177
Forbes, T.G. & Isenberg, P.A. 1991, *Astrophys. J.* 373, 294
Hundhausen, A.J. 1999, in: K.T. Strong *et al.* (eds.), *The Many Faces of the Sun: A Summary of the Results From NASA's Solar Maximum Mission,* (New York: Springer-Verlag), p. 143
Kuperus, M. & Raadu, M.A. 1974, *Astron. Astrophys.* 31, 189
Low, B.C. & Hundhausen, J.R. 1995, *Astrophys. J.* 443, 818
Molodensky, M.M. & Filippov, B.P. 1987, *Soviet Astron.* 31, 564
Munro, R.H., Gosling, J.T., Hildner, E., MacQueen, R.M., Poland, A.E., & Ross, C.L. 1979, *Solar Phys.* 61, 201
Priest, E.R., Hood, A.W., & Anzer, U. 1989, *Astrophys. J.* 344, 1010
Rompolt, B. 1990, *Hvar Obs. Bull.* 14, 37
Sime, D.G., MacQueen, R.M., & Hundhausen, A.J. 1984, *J. Geophys. Res.* 89, 2113
St. Cyr, O.C. & Webb, D.F. 1991, *Solar Phys.* 136, 379
Van Tend, W. & Kuperus, M. 1978, *Solar Phys.* 59, 115
Vrsnak, B., Rosa, D., Bozic, H. *et al.* 1999, *Solar Phys.* 185, 207
Webb, D.F. & Hundhausen, A.J. 1987, *Solar Phys.* 108, 383
Zagnetko, A.M., Filippov, B.P., & Den, O.G. 2005, *Astron. Reports* (in press)

Discussion

FORBES: Could you say more about the physical significance of your stability criterion?

FILIPPOV: The criterion arises from the height dependence of the magnetic field of the so called "mirror" current that simulate the magnetic field of the currents induced by the changing of the coronal current in the photosphere. This field falls with height as $1/h$ in our model. So in the power law expansion of the background magnetic field the power degree should be less than 1.

YOUSEF: 1) Was this filament eruption accompanied by a CME? 2) Filament do migrate as they get older towards the pole. That means they have passed the critical height you are talking about. Have you ever seen a polar filament eruption? 3) If this filament occurred in an AR free region, would you comment on this?

FILIPPOV: 1) The filament was accompanied by only a small CME that could be recognized only in differential images. Maybe it was masked by other CME that came from the opposite solar limb. The source region of this second CME was not observed on the disk. Maybe it was on the backside of the sun. 2) Yes we observed prominence eruptions above the poles. These happened just nearly before the polar magnetic field reversal. However in my talk I was speaking about critical height of filaments, not the critical latitude. 3) Yes, this was a quiescent filament, observed rather far from active regions. Maybe it was the reason that the eruption was not very energetic and flare ribbons were not observed after the eruption.

KOUTCHMY: In the case of the last event you showed, did you find evidence of: 1) spiraling motion(twist)? 2) brightening near the feet of the flux rope?

FILIPPOV: 1)The spiraling motion was not visible very clearly in this event but some hints of its presence could be recognized.
2) One bright spot was visible near the northern end of the filament. But it appeared a day before and was related to a emerging small bipolar region. Brightenings which lasted only during the eruption were not observed.

Coronal and Stellar Mass Ejections
Proceedings IAU Symposium No. 226, 2005
K. P. Dere, J. Wang & Y. Yan, eds.

© 2005 International Astronomical Union
doi:10.1017/S1743921305001079

Helicity of Magnetic Clouds and Solar Cycle Variations of their Geoeffectiveness

Katya Georgieva[1] and Boian Kirov[2]

[1]Solar-Terrestrial Influences Laboratory at the Bulgarian Academy of Sciences, Bl.3
Acad.G.Bonchev Str, 1113 Sofia, Bulgaria, email: kgeorg@bas.bg

[2]Solar-Terrestrial Influences Laboratory at the Bulgarian Academy of Sciences, Bl.3
Acad.G.Bonchev Str, 1113 Sofia, Bulgaria, email: bkirov@space.bas.bg

Abstract. Coronal mass ejections (CMEs) are sources of the strongest geomagnetic disturbances. From sunspot minimum to sunspot maximum, the intensity of storms associated with CMEs increases but the degree of association decreases. We divide the CMEs in the last solar cycle (1996–2002) into magnetic clouds (MCs)and CMEs which are not magnetic clouds. MCs are much more geoeffective than non-MC CMEs, and the portion of CMEs which are MCs is maximum in sunspot minimum and minimum at sunspot maximum, corresponding to the net helicity transferred from the solar interior into the corona. The smaller portion of the more geoeffective MCs is the explanation of the smaller degree of association of CMEs with geomagnetic disturbances in sunspot maximum.

Keywords. Sun: coronal mass ejections (CMEs), magnetic fields, solar-terrestrial relations.

1. Introduction

Coronal mass ejections (CMEs) are large-scale bubbles of plasma and magnetic fields expelled from the Sun. When they hit the Earth's magnetosphere, CMEs often produce intense geomagnetic disturbances. From sunspot minimum to sunspot maximum, the intensity of storms associated with CMEs increases, however the degree of association between CMEs and storms decreases. Several possible explanations have been proposed for this solar cycle dependence (see for example Webb (2000), and the references therein), but the question remains open. A special class of CMEs are magnetic clouds (MCs) which are distinguished by the smooth rotation of the magnetic field (Lepping, Jones & Burlaga 1990). Here we study a list of 202 CMEs in the last solar cycle (1996–2002), which we divide into MCs, i.e. CMEs with magnetic field rotation (74 cases) compiled from several sources: Fenrich & Luhmann (1998); Leamon, Canfield & Pevtsov (2002); Vilmer, *et al.* (2003); SOHO LASCO CME catalog; WIND MFI magnetic cloud list, and 124 cases of CMEs without magnetic field rotation which hence are not MCs - what is left from the Richardson & Cane 2003 list of CMEs after removing all cases identified as MCs.

2. Geoeffective parameters and geoeffectiveness of CMEs and MCs

The main factor for the geoeffectiveness of an interplanetary structure is the prolonged period of southward magnetic field ($B_z < 0$) providing coupling with the Earth's magnetic field (Gonsalez, Tsurutani & Clua de Gonsalez, 1999). Additional factors are the total magnetic field magnitude B and the velocity V. MCs, because of the magnetic field rotation, do have prolonged periods of $B_z < 0$. Our data show that MCs also have significantly higher B than CMEs.

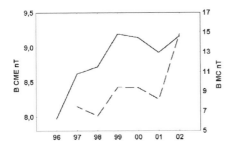

(a) Solar cycle variation of the magnetic field intensity B in MCs (solid line) and CMEs (broken line).

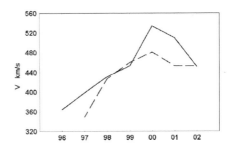

(b) Solar cycle variations of the velocity V_{of} MCs (solid line) and CMEs (broken line).

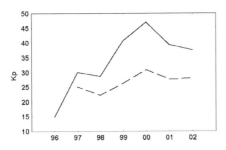

(c) Solar cycle variations of MC-associated (solid line) and CME-associated (broken line) K_p index.

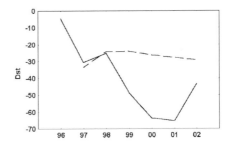

(d) Solar cycle variations of MC-associated (solid line) and CME-associated (broken line) D_{st} index.

Figure 1. Solar cycle variations of MC associated properties

B in both MCs and CMEs follows the sunspot cycle, and B in MCs is persistently higher than in CMEs in all phases of the solar cycle (figure 1a), the difference between them also being solar cycle dependent. V of both MCs and CMEs is higher in solar maximum than in solar minimum, and in all phases of the sunspot cycle it is higher for MCs than for CMEs, or equal (figure 1b). The geomagnetic disturbances caused by MCs as measured by K_p index are greater than the ones caused by CMEs (figure 1c), and for both are solar cycle dependent. MC-associated D_{st} index is much greater than the CME-associated, and shows strong solar cycle variations for MCs but no solar cycle variations for CMEs (figure 1d). This means that solar cycle variations of CME-related D_{st} index reported by previous studies (Webb, 2002) in which CMEs haven't been divided into MCs and non-MCs, are due to the solar cycle variations of the MC-related D_{st}.

3. Occurrence frequency of MCs and CMEs

It has been noted that the occurrence frequency of CMEs follows the sunspot cycle (Gopalswamy *et al.* 2003) while the occurrence frequency of MCs follows neither the sunspot cycle nor the occurrence frequency of CMEs (Wu *et al.*, 2003). Different estimations have been made about what portion of CMEs are MCs: 30% (Gosling 1990), 50% (Bothmer 1996), 60–70% (Webb 2002), until it was finally realized that this ratio

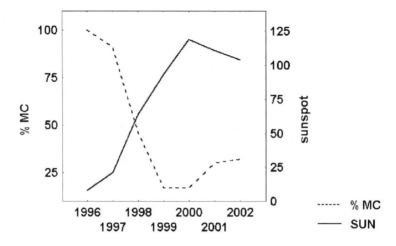

Figure 2. Solar cycle variations of the percentage of CMEs which are MCs.

varies with the sunspot cycle - from 100% at solar minimum though with poor statistics, to 15% at solar maximum (Richardson & Cane 2003). Our data confirm this conclusion (figure 1). A possible explanation of this solar cycle dependence is the solar cycle variation in the net helicity transferred from the solar interior to the surface (Berger & Ruzmaikin 2000): it is maximum in sunspot minimum and minimum in sunspot maximum. It could be speculated that the greater amount of helicity is contained in the solar corona, the greater part of CMEs expelled from it will contain helical, or twisted, magnetic fields and will be registered at the Earth's orbit as MCs.

4. Conclusion

MCs are significantly more geoeffective than CMEs which are not MCs. In sunspot maximum the portion of MCs among CMEs is small which is the reason for the lower degree of association of CMEs with geomagnetic storms than in sunspot minimum when practically all CMEs are MCs. The reason for this solar cycle variation of the ratio between MCs and CMEs is the solar cycle variation of the helicity transferred from the solar dynamo region into the corona.

References

Berger, M.A. & Ruzmaikin A. 2000, *J. Geophys. Res.* 105, A5, 10481
Bothmer, V. & Schwenn, R. 1996, *Adv. Space Res.* 17, 319
Gopalswamy, Lara, A., Yashiro, S., Nunes, S., & Howard, R. A. 2003, in: Solar variability as an input to the Earth's environment, *ISCS Symposium*, 403, 2003
Gosling, J. 1990, *AGU Geophys.Monogr.* 58, 343
Fenrich, F.R. & Luhmann, J.G. 1998 *Geophys. Res. Lett*, 25, 15, 2999
Koch, W. 1983, *J. Sound Vib.* 88, 233
Gonzalez, W.D., Tsurutani, B.T. & Clua de Gonzalez, A.L. 1999, *Space Sci.Rev.*,88, 259
Leamon, R. J. Canfield, R. C. & Pevtsov, A. A. 2002, *J. Geophys. Res.* 95, A8, 11957
Lepping, R. P. Jones, J. A. & Burlaga, L. F. 1990, *J. Geophys. Res.* 107, A9, SSH 1-1
Richardson, I.G. & Cane, H.V. 2003, *AGU - EUG Joint Assembly* 6603
Vilmer, N., Pick, M., Schwenn, R., Ballatore, P., & Villain, J. P. 2003, *Annales Geophys.* 847.
Webb 2000, *SOHO 11 Symp.on From Solar Min to Max: Half a Solar Cycle with SOHO*,Noordwijk: ESA Publications Division, p. 409

Coronal and Stellar Mass Ejections
Proceedings IAU Symposium No. 226, 2005
K. P. Dere, J. Wang & Y. Yan, eds.

Energetic Electrons in Magnetosphere during Gradual Solar Energetic Particle Event Observations by Cluster

C.J. Xiao[1], Z.Y. Pu[2], H.F. Chen[2], L. Xie[2], Q.G. Zong[3], T.A. Fritz[3] and P.W. Daly[4]

[1]National Astronomical Observatories, Chinese Academy of Sciences, Beijing 100012, China
email: cjxiao@pku.edu.cn

[2]Institute of Space Physics and Applied Technology, Peking University,Beijing 100871, China

[3]Center for Space Physics, Boston University, U.S.A.

[4]Max-Planck-Institut for Aeronomie, Katlenburg-Lindau, Germany

Abstract. More than 40 gradual SEPs have been observed by GOES and SOHO from 2001 to 2003. During 12 SEPs of all these events, energetic electron flux enhancements with energies from 38 keV to 337 keV were observed by RAPID onboard Cluster spacecraft. During these 12 events the variation of the energetic electron flux measured by RAPID/Cluster was closely associated to the variation of the solar energetic proton flux. The observed energetic electron flux was independent on the location of Cluster in the magnetosphere and the background level of magnetospheric electrons (even when the Cluster spacecraft was crossing the magnetopause, the plasma sheet and the low latitude boundary layer). The similar variation of the enhancement of energetic electron flux has also been observed by POLAR and Geotail in some of these events. In some of these events, the electron measurement of RAPID/Cluster will not be contaminated by the SEP protons.

Keywords. Sun: coronal mass ejections (CMEs), particle emission

1. Introduction

The gradual solar energetic particle events (SEPs) in association with coronal mass ejections (CME) have important consequences for space weather. The geoeffectiveness of SEPs is one of key issues in space science research (Reames (1999) and references herein).

The Cluster mission is designed to study the small-scale structures and multi-scale dynamics in the magnetosphere(Escoubet Schmidt & Goldstein (1997)). The RAPID spectrometer (Research with Adaptive Particle Imaging Detectors) for the Cluster mission is an advanced particle detector for the measurement of energetic particles in the energy range from 20–400 keV for electrons, 40 keV–4000 keV for proton and other heavier ions (Wilken, *et al.* (1997)).

More than 40 gradual SEPs have been observed by GOES and SOHO from 2001 to 2003. During 12 SEPs of all these events, energetic electron flux enhancement were observed by RAPID/Cluster. During these 12 events the variation of the energetic electron flux measured by RAPID/Cluster was closely associated to the solar energetic proton flux variation. The observed energetic electron flux was not affected by the location of Cluster in the magnetosphere and the background level of magnetospheric electrons (even when the Cluster spacecraft were crossing the magnetopause, the plasma sheet and the low latitude boundary layer). The similar variation of the enhancement of energetic electron flux has also been observed by POLAR and Geotail in some of these events.

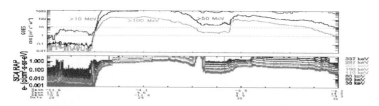

Figure 1. Observations of the energetic particles on Oct. 28–31, 2003 event.*top*:The integral flux of energetic ions measured by GOES; *bottom*: the differential flux of energetic electrons measured by Cluster and the position of Cluster in GSM coordinate system.

2. Oct. 28–31, 2003 event

From Oct. 28 11:00 to Oct. 31, 20:00 UT, the Cluster spacecraft traversed the magnetosheath, magnetopause, plasmasphere and radiation belts. The flux of energetic electrons observed by RAPID/Cluster was unusually high. The enhancement lasted during the SEPs (Figure 1), while the flux of energetic ions observed by RAPID-IIMS did not show any special change. The similar results were measured by the Geotail, ACE and WIND spacecraft. The variations of the energetic electron flux measured by those satellites were all closely associated to the variations of solar energetic proton flux measured by the GOES spacecraft in the geosynchronous orbit. The observed energetic electron flux was independent on the locations of Cluster or Geotail in the magnetosphere and the background level of the magnetospheric electrons.

3. Discussions and Conclusions

Electrons, even with the energy as high as 1MeV, are very difficult to entry the magnetosphere by directly penetrating across the magnetopause. Furthermore, the hard x-ray or γ-ray burst could not keep to contaminate the energetic electron measurement for several days. Therefore the question arises: whether or not the electron flux enhancements observed by RAPID/Cluster were contaminated by the solar energetic protons?

It is known that the solid state detectors of the RAPID/Cluster can be shielded from the solar energetic protons with the energy lower than 100 MeV (Wilken, *et al.* (1997)), while during some events the fluxes of solar energetic protons with energy more than 50 MeV (observed by GOES) were two-order smaller than the fluxes of energetic electrons. During the Mar. 23, 2002 event, while the flux of energetic protons higher than 10 MeV maintained on the background level, the energetic electrons flux enhancements observed by RAPID/Cluster. So the solar energetic protons can not be the contamination source in these events.

In a summary, at lease in some of these 12 SEPs, neither solar energetic protons nor the x-ray or γ-ray bursts can be the main contamination source of the electron measurement by RAPID/Cluster. The other physical reasons need to be discussion further.

Acknowledgements

This work is supported by the Chinese DSP-Cluster Science Team and by the CNSF Key Project 40390150 and Chinese Fundamental Research Project G200000784.

References

Escoubet, C.P., Schmidt, R., and Goldstein M.L. 1997 *Space Sci. Rev.* 79, 11
Reames, D.V. 1999, *Space Sci. Rev.* 90, 413
Wilken, B., Axford, W.I., and Daglis, I. *et al.* 1997, *Space Sci. Rev.* 79, 399

Coronal and Stellar Mass Ejections
Proceedings IAU Symposium No. 226, 2005
K. P. Dere, J. Wang & Y. Yan, eds.

CMEs and Long-Lived Geomagnetic Storms: A Case Study

H. Xie[1,2], N. Gopalswamy[2], P.K. Manoharan[3], S. Yashiro[1,2], A. Lara[4], and S. Lepri[5]

[1]The Catholic University of America, Washington DC, USA

[2]NASA Goddard Space Flight Center, Greenbelt, Maryland, USA
[3]National Center for Radio Astronomy, Ooty, INDIA
[4]Instituto de Geofisica, UNAM, Mexico
[5]University of Michigan, Michigan, USA

Abstract. We studied the relationship between successive coronal mass ejections (CMEs) and a long-lived geomagnetic storm (LLGMS) by examining the 1998 May 4 event. Five successive CMEs from the same active region and four interplanetary shocks were found to be associated with this LLGMS. We investigated the effect of successive and interacting CMEs on the LLGMS.

Keywords. Sun: coronal mass ejections (CMEs), solar-terrestrial relations.

1. Introduction

It is now well established that front-side halo coronal mass ejections (CMEs) are the major cause for large geomagnetic storms (e.g., Burlaga *et al.* 2002; Cane *et al.* 2000; Gopalswamy *et al.* 2000; Webb *et al.* 2000; Zhang *et al.* 2003;). Isolated geomagnetic storms typically have a recovery phase less than ~ 1 day. Some storms have main and recovery phases exceeding ~ 3 days. We call them long-lived geomagnetic storms (LL-GMS). LLGMSs occur mostly when successive CMEs ejected from the Sun and impact Earth. In this paper, we present a case study of a LLGMS involving successive CMEs and CME interaction.

2. Data and models

The LLGMS of interest as defined by the Dst (disturbance storm time) index, and reported by the World Data Center in Kyoto (http://swdcwww.kugi.kyoto-u.ac.jp/dstdir/) was observed from 1998 May 2 - 7. We use Fe charge state data and the component of the interplanetary magnetic field B_z to help identify the interplanetary CMEs (ICMEs). The associated CMEs were from the catalog of CMEs observed by the Solar and Heliospheric Observatory (SOHO) mission's coronagraphs (http://cdaw.gsfc.nasa.gov/CME_list). The height-time profiles of CMEs were plotted to examine if there was any possible interaction between successive CMEs.

Figure 1(a) shows the Dst, Fe charge state (Q_{Fe}), B_z, and the height-time profiles of the associated CMEs for the 1998 May 4 event. The LLGMS lasted for 5 days from May 2 to May 7 (main phase ~ 2 days and recovery phase ~ 3 days). The Dst_{min} was $\sim$ -205 nT.

Five CMEs were found to be associated with the LLGMS. The CMEs originated from AR8210 when it was at S17E23, S18W05, S20W07, S15W15, and S13W34. From CME height-time profiles, we can see that CME 1 and CME 5 were well separated, but CME 2, CME 3, and CME 4 were ejected in quick succession. The speeds of CME 2 and CME 3

476　　　　　　　　　　　　　　　　　　Xie *et al.*

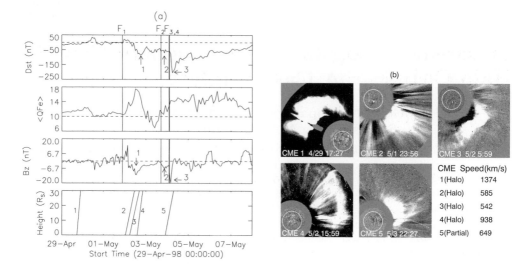

Figure 1. (a) The Dst, Q_{Fe}, B_z, and CME height-time profiles. The vertical solid lines (F_1, F_2, F_3, F_4) indicate the ICME shock fronts. Note that F_3 and F_4 are very close to each other, and the drop in $Q_{Fe} \sim 6$ before F_2 is due to the instrumental noise. The numbers on the height-time plots indicate the associated CMEs. (b) The five CMEs associated with the LLGMS: C2 images superposed with EIT images. The times and speeds of the CMEs are indicated.

were 585 and 542 (km/s), respectively, and their trajectories were nearly parallel. CME 4 was faster (938 km/s) than CME 2 and CME 3. It is likely that CME 4 would catch up with CME 3, causing CME 3 to speed up and CME 4 to slow down, thus producing two successive strong shocks F_3 and F_4. Possible interaction occurred between CME 3 and CME 4, but they were not totally merged at 1 AU. CME 3 and CME 4 were separated by ~ 8.5 hrs near the Sun. But they were separated by only ~ 0.5 hrs at 1 AU. Four shocks shown in the figure are related to CME 1, CME 2, and possible interaction of CME 3 and CME 4, respectively.

　　Three dips in the main phase of the storm and in the B_z profile can be seen in Fig. 1 (pointed by arrows): the first related to the interplanetary MC associated with CME 1, the second related to CME 2, and the last DST_{min} caused by the possible interaction of CME 3 and CME 4. Successive CMEs increase the duration of LLGMS. CME interaction enhances the intensity of LLGMS. However, our statistical study indicates that successive interacting CMEs can also lead to modest LLGMS (details will be presented elsewhere).

Acknowledgements

　　This research was supported by NASA LWS and NSF SHINE (ATM 0204558) programs.

References

Burlaga, L.F., Plunkett, S.P., St., & Cyr, O.C. 2002, *J. Geophys. Res.* 107, SSH 1-1
Cane, H.V., Richardson, I.G., St., & Cyr, O.C. 2000, *Geophys. Res. Lett.* 27, 3591
Gopalswamy, N., Lara, A., Lepping, R.P., Kaiser, M.L., Berdichevsky, D., & St. Cyr, O.C. 2000, *Geophys. Res. Lett.* 27, 145
Webb, D.F., Cliver, E.W., Crooker, N.U., St., Cry, O.C., & Thompson, B. 2002, *J. Geophys. Res.* 105, 7491
Zhang, J., Dere, K.P., Howard, R.A., & Bothmer, V. 2003, *Astrophys. J.* 582, 520

Coronal and Stellar Mass Ejections
Proceedings IAU Symposium No. 226, 2005
K. P. Dere, J. Wang & Y. Yan, eds.

© 2005 International Astronomical Union
doi:10.1017/S1743921305001109

Successive Impacts Of The Earth by Several Halo CMEs From Active Region NOAA 652

Shahinaz Yousef[1], M.S. El Nawawy[1], M. El-Nazer[1] and Mohamed Yousef[2]

[1]Astronomy& Meteorology Dept, Faculty of Science, Cairo University,
email: shahinazyousef@yahoo.com

[2]National Research Institute of Astronomy and Geophysics, Cairo, Egypt

Abstract. Several Halo CMEs hit the Earth in the second half of July 2004. They were produced by the very large complex active region NOAA 652 (Yousef *et al.* 2005). For CME details consult the web (ftp://lasco6.nascom.nasa.gov/pub/lasco/status/LASCO_CME_List_004).

We focus on the 26th -27th of July CME hit. This CME was associated with the long-duration M1 flare at 25/15:14. It made a very fast Sun to Earth transit - just over 31 hours (SGAS 27 July 2004). A greater than 10 MeV proton event began at 25/18:55. Solar wind speed remained elevated from 500 to over 700 km/s. A Severe Geomagnetic storm was observed and the aurora was seen as far as California.

A strong shock impacted the ACE spacecraft at 26/22:28. A sudden impulse (SI) of 96 nT was observed on the Boulder magnetometer at 22:51. The IMF Bz component was turned negative (-18 nT). Generally speaking, according to de Pater and Lissauer (2001), since a strong CME disturbance in the solar wind is usually preceded by an interplanetary shock followed by an enhanced density and velocity, the field strength first increases when the disturbance hits the magnetosphere, inducing an increase in the ring current. Several hours(up to over 25 hrs) the field strength Dst decreases dramatically during the storm main phase which typically lasts for a day The main phase is caused by an increase in the ring current, resulting from an enhanced particle flow towards the Earth. It is well known that geomagnetic storms tend to occur when IMF is directed southward. Magnetic reconnection occurs between the negative IMF and the magnetosphere thus opens the field lines with one end connected to the Earth (Dungey 1963). This magnetic reconnection allowed the protons and electrons to leak in. The proton and electron flux maximums occurred around the time of geomagnetic storm commencement which lasted for about 27 h (fig. 1). This is in agreement with the statement of Robinson (2003) that large numbers of energetic protons are constrained to occupy the region around the IP shock. The IMF Bz component dropped to -20 nT on 27 of July at 12:00 UT as measured by ACE satellite while Kp reached a maximum of 9 around 15:00 UT at the storm maximum as seen in fig. 2.

Keywords. Sun: coronal mass ejections (CMEs), solar-terrestrial relations

References

De Pater, I. and Lissauer, J.J. 2001. Planetary Sciences. Cambridge University Press.

Dungey, J.W. 1963. The structure of the exosphere or adventures in velocity space. in Geophysics: The Earth Environment, edited by C.Dewitte, J. Hiebolt, and A. Lebeau, pp. 505–550. Gordon and Breach, New York.

Robinson, I.M. 2003. Proton Signatures of Halo CMEs at L1. Proceedings of ISCS. Solar Variability as an input to the Earth's environment. Tatranska Lomnica,Slovak Republic, 23–28 June, pp. 593–602.

SGAS Number 209 Issued at 0245Z on 27 July 2004: Joint USAF/NOAA Solar and Geophysical Activity Summary.

Yousef, S., El Nazer, M., and Bebars, A. 2005, This proceeding pp. 145–146.

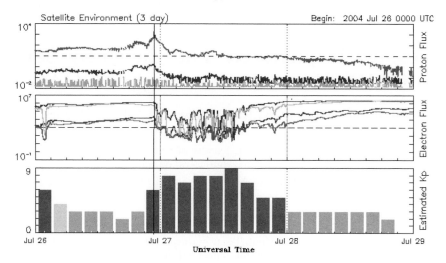

Figure 1. Three day plots of the 26-27 of July 2004 halo CME hit of the Earth. Top is the proton flare (Goes 11 proton flux 5 minutes data in three energy domains 10, 50 and 100 MeV from top to bottom). Middle is the Goes 10 and 12 electron flux (>2, > 0.6 MeV) . Lower curve is the Kp index showing the magnetic storm that started around 26d 23h ,maximum Kp was 9 on the 27th of July around 15:00. There is a coincidence of the maximums of proton and electron fluxes with the start of the magnetic storm. Plots are from the web site http://sec.noaa.gov/ftpmenu/plots/2004_plots/satenv.htm

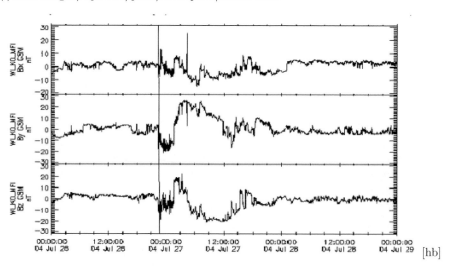

Figure 2. The three IMF components. Note that Bz dropped to negative value of -18nT. The CME shock arrived ACE satellite at 26/22:28. The Bz component stayed negative for few hours before it was switched to positive also for few hours and then back to negative. Data is provided by R.Lepping at NASA/GSFC and CDAweb

Session 9

Stellar ejections

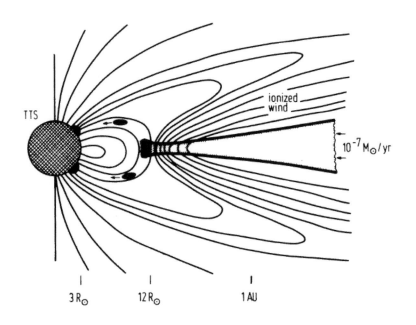

Coronal and Stellar Mass Ejections
Proceedings IAU Symposium No. 226, 2005
K. P. Dere, J. Wang & Y. Yan, eds.

© 2005 International Astronomical Union
doi:10.1017/S1743921305001110

Winds and Ejecta from Cool Stars

Moira Jardine[1]†

[1]School of Physics and Astronomy, University of St. Andrews, North Haugh, St Andrews,
KY16 9SS, Scotland
email: mmj@st-and.ac.uk

Abstract. As young stars evolve from the time of their formation onto the main sequence, they lose mass in a variety of ways. At the very earliest stages, mass loss may be in the form of jets associated with accretion from a surrounding disk. These cool jets carve out the surrounding gas and their changes over time may indicate changes in the star-formation process. At later stages, mass loss is predominantly in the form of a hot, magnetically channelled wind that carries mass, but more importantly angular momentum, away from the star. This wind determines the rotational evolution of cool stars and is intimately connected to the process of field generation deep inside the star. Mass loss also occurs in a sporadic way in the form of the ejection of cool clouds of coronal gas. The coronal distribution and evolution of these clouds (or prominences) gives us vital clues about the structure and short-timescale evolution of stellar coronae. In this review I will discuss recent advances in our understanding of the coronae, winds and accretion processes in cool stars and show how these processes may be related at different stages of evolution.

Keywords. stars: magnetic fields, stars: coronae, stars: imaging.

1. Introduction

In this review I would like to take a step back from the detailed studies of the solar corona that we have been hearing about to consider the types of mass loss that occur on other, solar-like stars. Mass loss also occurs on stars much more massive than the Sun, and on more evolved stars, but I will focus only on those stars that are of mass similar to (or less than) one solar mass. In this review I will concentrate on the so-called "main sequence" stars - those that have ignited Hydrogen in their cores - and the next review will focus on the earlier phase of evolution, when stars are still contracting out of their parent molecular clouds.

There are two ways in which mass loss can occur: in a continuous form through winds or jets, and in an intermittent form, as ejected prominences. In contrast to solar studies, however, we are not so interested in the amount of mass or energy that is removed in these mass ejections, but more in the amount of angular momentum that is removed. We can see the effect of this loss of angular momentum if we look at distributions of stellar rotation speeds in clusters of different ages. If we compare these distributions for α Persei (aged 50 Myr), the Pleiades (aged 70 Myr), and the Hyades (aged 600 Myr) we find that as we go towards the older clusters, the number of rapid rotators diminishes, until all the stars are rotating slowly (Soderblom *et al.* 1993). It appears that some mechanism causes stars to spin down as they age. This is only part of the story however. When stars are very young - when they are still surrounded by a dusty disk - the presence of the disk appears to act as a "governor", preventing the spin up that would naturally occur as the star continues to contract. Once the disk has dissipated however, the star

† Present address: School of Physics and Astronomy, University of St. Andrews, North Haugh, St Andrews, KY16 9SS, Scotland

is free to spin up. This increase in rotation rate is accompanied by enhanced dynamo activity and hence - since these stars possess a magnetically-controlled wind - by enhanced angular momentum loss. Once the loss of angular momentum in the wind is sufficient to counteract the spin-up due to the star's own contraction, the star will begin the slow spin-down that continues until well beyond the age of the Sun. This theory of stellar spin down is well established and reproduces the observed scaling of rotation rate as a function of time ($\Omega \propto t^{-1/2}$) (Weber & Davies 1967; Skumanich 1972). In order for the theory to reproduce the cluster rotation distributions, however, it was necessary to force the angular momentum loss of the most rapid rotators to be inhibited. This was achieved by assuming that at the highest rotation rates, the dynamo saturates so that the field strength no longer increases with increasing rotation rate (Endal & Sofia 1981; Charbonneau & MacGregor 1993; Collier Cameron & Li 1994).

The need for this assumption has been questioned recently however. Solanki *et al.* (1997) showed that angular momentum loss can be reduced if the star's magnetic flux is concentrated at the rotation pole, rather than being distributed uniformly over the stellar surface as had previously been assumed. A wind emerging from close to the rotation pole has a much smaller lever arm than one that emerges from the equator. The net result is that this can mimic the effect of dynamo saturation on the stellar spin down.

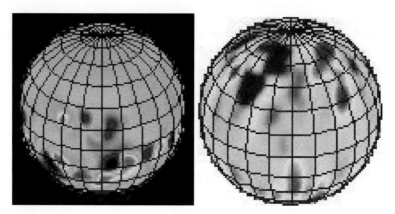

Figure 1. Shown (left) is a solar magnetogram taken at around cycle maximum. Red denotes positive polarity, blue denotes negative polarity. The appearance of bipoles at low to mid latitudes is clearly seen. Shown (right) is a magnetogram produced by Zeeman-Doppler imaging of AB Dor, a young, solar type star which rotates 50 times faster than the Sun (P=0.514 days). Mixed polarity flux is present all all latitudes.

2. Where does flux appear on stellar surfaces?

There are good reasons for expecting that on rapid rotators flux should emerge through the surface at high latitudes. Schuessler & Solanki (1992) showed that flux tubes formed at the base of the convective region in rapid rotators do not have time to exchange angular momentum with their surroundings as they rise buoyantly towards the surface and so they are forced to rise parallel to the rotation axis to emerge at high latitudes.

In addition to these theoretical reasons for expecting high-latitude spots, Doppler images of these stars do indeed typically show dark polar caps, although they also tend to show spots at low latitudes too (Strassmeier 1996). One of the best-studied examples is AB Dor, a marginally pre-main sequence, rapidly-rotating K0 dwarf (period 0.514 days) aged 20-30 million years (Cameron & Foing 1997; Innis *et al.* 1988). The surface of this star shows spots at all latitudes and in particular all the way up to the pole. The inclination of its rotation axis at about 60° does however mean that little information can be gleaned about the hemisphere that is tilted away from the observer.

Curiously, although AB Dor is such a rapid rotator, its latitudinal differential rotation is close to that of the Sun. It takes 120 days for its equator to pull ahead of the polar regions by one complete rotation, compared to 110 days for the Sun (Donati & Collier Cameron 1997). Similar results have now been found for other such stars, and indeed it appears that the differential rotation is a function of stellar mass, falling to zero towards the low-mass end of the main sequence (Barnes *et al.* 2004).

Brightness images only tell part of the story however. If we are to understand the structure of the coronal magnetic fields of these stars properly, we need information about the polarity of the magnetic elements. This is now possible with Zeeman-Doppler imaging (Donati & Collier Cameron 1997; Donati *et al.* 1999). As shown in Fig. 1 the pattern of surface flux on AB Dor is quite different to that of the Sun, with mixed-polarity regions extending up to the rotation pole. There are also indications that AB Dor (and indeed the Sun) possess so-called active longitudes (Jetsu *et al.* 1993; Korhonen *et al.* 1999; Berdyugina *et al.* 2002; Berdyugina & Usoskin 2003; Benevolenskaya 2002; Bigazzi & Ruzmaikin 2004). These are longitudes of enhanced flux, which in the case of the Sun can only be detected by studying a very long time series of data.

In common with other rapid rotators, AB Dor has a high X-ray emission measure (10^{53} cm^{-3}), but this shows very little rotational modulation (Kürster *et al.* 1997). This suggests that the emitting gas is rarely eclipsed by the star - something that can be achieved by having a very extended corona, or a fairly compact one, but one where the emitting gas is confined at high latitudes where it is always in view.

The question of the extent of stellar coronae has been raised also by the recent results from the X-ray satellites XMM-Newton and Chandra, as well as from FUSE. Densities derived for stellar coronae can be very large, as high as 10^{13} cm^{-3} (Dupree *et al.* 1993; Schrijver *et al.* 1995; Brickhouse & Dupree 1998) and for AB Dor values range from $10^9 - 10^{12}$ cm^{-3} (Maggio *et al.* 2000; Güdel *et al.* 2001; Sanz-Forcada *et al.* 2003). Differential emission measures also show that much of the emission comes from temperatures as high as 10^7K, something only seen on the Sun during flares. These very high densities in particular suggest that the coronae of these stars must be compact, since large loops could not be stable at such high densities. Additional support for the suggestion that these coronae are compact comes from Beppo-SAX observations of two flares on AB Dor (Maggio *et al.* 2000). The flare decay phase lasted for more than one complete stellar rotation, but showed no rotational self-eclipse. Modelling of the flare decay phase suggested that the loops were small (0.3 R$_\star$) and so the only solution is to have the flaring loop(s) at high latitude where they are never self-eclipsed.

3. Modelling stellar coronae

Clearly, trying to learn about the structure of stellar coronae from the X-ray data alone is an extremely challenging problem, and one that is unlikely to have a unique solution. A more useful approach is to tackle the forward problem: to specify the form of the magnetic field and then to predict the observable signatures and compare them with the real observations. This was originally done for the Sun over 30 years ago by taking surface magnetograms and extrapolating a potential coronal field assuming that at some radius (the "source surface") the field is forced open by the effects of the stellar wind to become purely radial (Altschuler & Newkirk, Jr. 1969). Fig. 2 shows a field extrapolation for AB Dor based on data acquired in December 2002 (Donati *et al.* 2003). Since one half

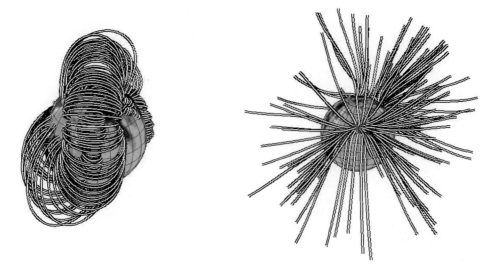

Figure 2. Shown is a potential field extrapolation for a surface magnetogram of AB Dor from December 2002. The source surface has been set to 3.4 R$_\star$. Shown (left) are the closed and (right) open field lines.

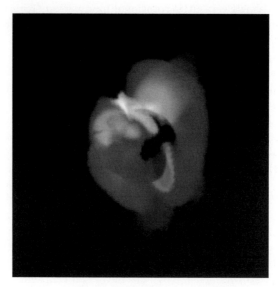

Figure 3. Shown is an X-ray image at a temperature of 10^7 K for the field extrapolations shown in Fig. 2.

of the star is of predominantly positive polarity and the other half is of predominantly negative polarity, on the largest scales the field lines simply connect these polarities to produce large East-West loops than form an arcade that runs from the equator up over the pole (Jardine *et al.* 2002a). Of course, on smaller scales the field is much more complex with many low-lying loop structures. Indeed, we find loops on all scales in the corona. While this figure shows a potential field extrapolation, it is worth commenting that it is possible to do this for non-potential fields. Hussain *et al.* (2002) have developed a method for fitting non-potential fields directly to the Stokes profiles. On the large scales, however, the global field structure is very similar to that for a potential field.

The behaviour of the open field is of course the most interesting from the point of view of stellar winds. As Fig. 2 shows, the open field regions for this magnetogram lie in two discrete mid-latitude regions separated by 180° of longitude. This suggests that much of the coronal volume is filled with open field, and that the stellar wind may indeed leave not from the polar regions, but from mid-latidues. This has important consequences for angular momentum loss in stellar winds (Solanki *et al.* 1997; Holzwarth 2004).

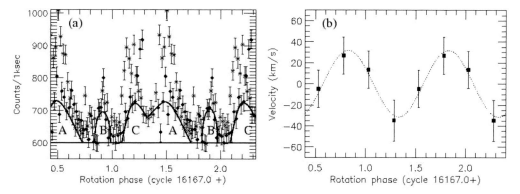

Figure 4. (a) Chandra/LETG lightcurve folded with AB Dor's rotation period. Asterisks and diamonds represent consecutive rotation cycles. The horizontal solid line represents the flat, unmodulated emission level while the curves trace the quiescent modulated emission from the star. (b) The phase-folded mean velocity shifts in the line centroids of the OVIII 18.97 Angstrom profile (+ is red-shifted while - is blue-shifted). The dotted line is the best-fit sine-curve. Figure taken from Hussain *et al.* (2004).

If we are to test these field extrapolations, however, we need to calculate the predicted observable signatures and compare them with what is observed. We can do this by calculating the X-ray emission from the field structure (Jardine *et al.* 2002b).We do this by assuming that the coronal plasma is isothermal and in hydrostatic balance along each field line. We can therefore determine the pressure everywhere with only one free parameter, which is the plasma pressure at the base of the corona. We set this proportional to the magnetic pressure so that $p_0 = RB_0^2$. In addition, we set the plasma pressure to zero if a field line is open, or if anywhere along a field line the plasma pressure exceeds the local magnetic pressure and so should have opened up that field line. As the constant of proportionality R is increased, so is the coronal density and hence the emissivity which is proportional to n_e^2. We increase R until the emission measure matches the observed value for AB Dor which is around 10^{53} cm^{-3} Maggio *et al.* (2000). This then naturally produces an emission measure-weighted density that is in the observed range of about 10^{11} cm^{-3} and a low rotational modulation of the X-ray emission, since most of the emitting regions are at high latitude and so never pass behind the star (see Fig. 3).

We can then compare the variation of the X-ray emission as the star rotates with what was observed by Chandra at the time when the ground-based optical observations were being undertaken (Hussain *et al.* 2004). Fig. 4 shows the Chandra counts phased on the rotation period for AB Dor. Almost 2 successive stellar rotations were observed, allowing a determination of the degree of rotational modulation in the X-ray emission. This Figure shows a) a base level of emission that is always present, b) rotational modulation that is outlined by the thick line and c) transient emission superimposed on this. The feature marked "B" appears only very briefly as the star rotates and so is likely to have been located close to the pole of the star that points away from the observer. Features "A" and "C" are reproduced by the forward modelling procedure. Shown on the right panel is the velocity shift of the centroid of the OVIII line due to the rotation of the star. The value of 30 kms^{-1} is also consistent with the forward modelling.

While a coordinated observing campaign that combines ground-based observations (to provide surface brightness and magnetic maps) with space-based observations (to determine the X-ray coronal spectrum) is capable of determining the structure of a stellar corona, it can only provide a snapshot. Stellar coronae, just like that of the Sun, are not static and can show significant variations from year to year. An example is shown in Fig. 5 which shows the open field lines for LQ Hya which has a rotation period of 1.6 days. In December 2000, the open field structure was similar to that of AB Dor, with much of the open field emerging from two mid-latitude regions, 180° apart in longitude. One year later however, the open field emerged mainly from the pole (Donati *et al.* 2003; McIvor *et al.* 2004).

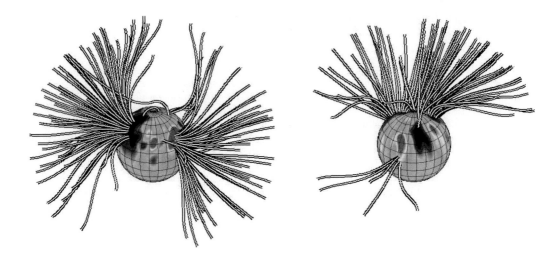

Figure 5. Shown are the open field lines from a potential field extrapolation for surface magnetograms of LQ Hya from December 2000 (left) and December 2001 (right) (Donati *et al.* 2003). The source surface has been set to 3.4R$_\star$.

To what extent do the details of the field structure really affect the behaviour of the stellar wind over time? Solanki *et al.* (1997) showed the effect of a poleward concentration of flux, but what about other field structures, such as we see in these field extrapolations? To answer this question, we can consider the evolution of a solar mass star from the point where it is released from its disk (Holzwarth & Jardine 2004). We assume that the star has one of four types of flux distribution, shown in Fig. 6. Based on these, we calculate the angular momentum loss at each timestep as the star evolves and so follow the star's rotational history. We can then compare the stellar rotation rate as a function of time with what would have been predicted if the star had possessed a field that was uniform at the surface (as assumed by the classic model of Weber & Davies (1967)). Fig. 7 shows the deviation from the Weber-Davis prediction as a function of time for the different types of flux distribution. Shown are stars that were released from their disks with three different rotation rates. It is clear from this that the detailed structure of the stellar field can be as important as the original rotation rate in determining the rotational evolution.

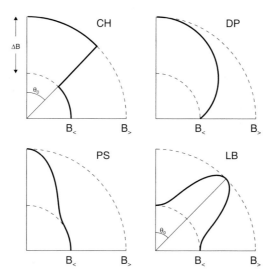

Figure 6. Field distributions close to the stellar surface: coronal hole (CH); dipolar field (DP); polar spot (PS) and latitudinal belt (LB). Dashed lines indicate the upper and lower field strengths. Figure taken from Holzwarth & Jardine (2004).

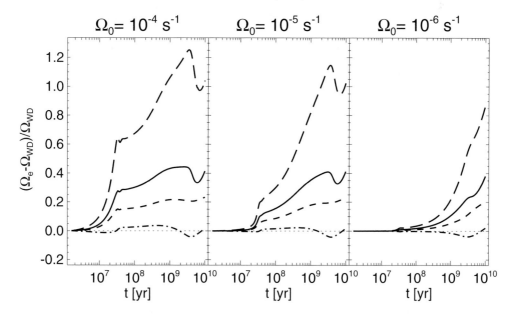

Figure 7. Deviations of the stellar rotation rate from the value obtained using the Weber-Davis formalism. Shown are curves for the four different field geometries of Fig. 6: coronal hole (solid); dipolar field (short dashed); polar spot (long dashed) and latitudinal belt (dashed dotted). Curves are shown for stars with three different initial rotation rates. Figure taken from Holzwarth & Jardine (2004).

4. Stellar prominences

The other way in which these stars lose mass is by the intermittent ejection of stellar prominences. These are observed as transient absorption features that move through the $H\alpha$ line as the prominence passes between the star and the observer. They form on a

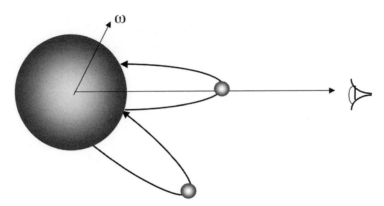

Figure 8. Schematic illustration of the effect of the stellar inclination on the visibility of prominences. Both prominences are at the same distance from the rotation axis, but only the high-latitude one occults the disk as seen by the observer.

timescale of about one day, and typically live for one or two days. At any one time, there are typically some 5 or 6 present in the observable part of the corona of AB Dor. Their masses have been estimated to be about 10^{17}g, about 100 times that of a large quiescent solar prominence (Collier Cameron & Robinson 1989a,b). These prominences typically form at or just beyond the Keplerian co-rotation radius, which is the point at which the outward pull of centrifugal forces just balances the inward pull of gravity. For the Sun, this point is at some 40 solar radii and has little impact on the physics of the corona, but for these rapidly rotating stars, the point of centrifugal balance is inside the corona. These prominences are detected both in binary stars and in single stars, and have been found in 90% of young (pre-) main sequence stars with rotation periods less than one day.

In addition to AB Dor and four G dwarfs in the α Per cluster (Collier Cameron & Woods 1992), other stars for which prominences have been detected include HD197890 (Jeffries 1993); HK Aqr (Byrne *et al.* 1996); RE J1816+541 (Eibe 1998); PZ Tel (Barnes *et al.* 2000) and RXJ1508.6-4423 (Donati *et al.* 2000). This last star is particularly interesting as the prominences were seen in emission (and hence by solar terminology are indeed correctly called prominences, not filaments). This star is viewed at a very low inclination and so the Hα-emitting clouds are never eclipsed by the star.

The ejection of these prominences does not remove significant amounts of angular momentum or mass; indeed the mass loss rate in prominences is probably only comparable to that in the solar wind, and hence is several orders of magnitude less than in a stellar wind (Wood *et al.* 2004). The reason for such interest in these prominences is more to do with what they can tell us about the structure and dynamics of stellar coronae. If providing enough mass to fuel a solar prominence is a challenge for the theorists, the same problem for stellar prominences with their greater mass and greater heights is much more acute. The very number of prominences also gives us a great deal of information about the structure of the corona. Fig. 8 shows two prominences both located at the same distance from the stellar rotation axis. The prominence in the equatorial plane would, however, be unobservable, since it would never cross in front of the observer. Similarly, any prominences in the lower hemisphere would also never be observed. The observed number of prominences is therefore a lower limit to the number that may be present in the stellar corona. Their presence shows that the coronae of these stars still retain a high degree of complexity even out to 3-5 $R_{\star}$ where the prominences are held in co-rotation.

If these stars had simple dipole fields after all, the prominences would by symmetry all lie in the equatorial plane and would never be observed.

Some progress has been made in modelling the mechanical and thermal equilibria of these prominences (Collier Cameron 1988; van den Oord 1988; Jardine & Collier Cameron 1991; Ferreira & Jardine 1995; Ferreira & Mendoza-Briceño 1997; Ferreira 2000; Jardine *et al.* 2001). Progress so far has however been limited to simple field configurations and static equilibria. Studies of the dynamics of prominence formation and ejection could yield important insights into the fundamental timescales governing the evolution of stellar coronae, which may in turn shed some light onto the problem of coronal mass ejection on the Sun.

References

Altschuler, M. D. & Newkirk, Jr., G. 1969, Solar Phys., 9, 131

Barnes, J., Collier Cameron, A., Donati, J.-F., *et al.* 2004, MNRAS, in press

Barnes, J., Collier Cameron, A., James, D. J., & Donati, J.-F. 2000, MNRAS, 314, 162

Benevolenskaya, E. E. 2002, Advances in Space Research, 29, 1941

Berdyugina, S. V., Pelt, J., & Tuominen, I. 2002, A&A, 394, 505

Berdyugina, S. V. & Usoskin, I. G. 2003, A&A, 405, 1121

Bigazzi, A. & Ruzmaikin, A. 2004, ApJ, 604, 944

Brickhouse, N. & Dupree, A. 1998, ApJ, 502, 918

Byrne, P., Eibe, M., & Rolleston, W. 1996, A&A, 311, 651

Cameron, A. & Foing, B. 1997, The Observatory, 117, 218

Charbonneau, P. & MacGregor, K. B. 1993, ApJ, 417, 762

Collier Cameron, A. 1988, MNRAS, 233, 235

Collier Cameron, A. & Li, J. 1994, MNRAS, 269, 1099

Collier Cameron, A. & Robinson, R. D. 1989a, MNRAS, 238, 657

Collier Cameron, A. & Robinson, R. D. 1989b, MNRAS, 236, 57

Collier Cameron, A. & Woods, J. A. 1992, MNRAS, 258, 360

Donati, J.-F., Cameron, A. C., Semel, M., *et al.* 2003, MNRAS, 345, 1145

Donati, J.-F. & Collier Cameron, A. 1997, MNRAS, 291, 1

Donati, J.-F., Collier Cameron, A., Hussain, G., & Semel, M. 1999, MNRAS, 302, 437

Donati, J.-F., Mengel, M., Carter, B., Cameron, A., & Wichmann, R. 2000, MNRAS, 316, 699

Dupree, A., Brickhouse, N., Doschek, G., Green, J., & Raymond, J. 1993, ApJ, 418, L41

Eibe, M. T. 1998, A&A, 337, 757

Endal, A. S. & Sofia, S. 1981, ApJ, 243, 625

Ferreira, J. 2000, MNRAS, 316, 647

Ferreira, J. & Jardine, M. 1995, A&A, 298, 172

Ferreira, J. & Mendoza-Briceño, C. 1997, A&A, 327, 252

Güdel, M., Audard, M., den Boggende, A., *et al.* 2001, in Proceedings of "X-ray astronomy 2000", ed. S. S. R. Giaconni, L. Stella (ASP conference series)

Holzwarth, V. 2004, MNRAS, in press

Holzwarth, V. R. & Jardine, M. 2004, in Proceedings 13th Cool Stars Workshop

Hussain, G., Brickhouse, N., Dupree, A., *et al.* 2004, ApJ, in press

Hussain, G. A. J., van Ballegooijen, A. A., Jardine, M., & Collier Cameron, A. 2002, ApJ, 575, 1078

Innis, J. L., Thompson, K., Coates, D. W., & Evans, T. L. 1988, MNRAS, 235, 1411

Jardine, M. & Collier Cameron, A. 1991, Solar Phys., 131, 269

Jardine, M., Collier Cameron, A., & Donati, J.-F. 2002a, MNRAS, 333, 339

Jardine, M., Collier Cameron, A., Donati, J.-F., & Pointer, G. 2001, MNRAS, 324, 201

Jardine, M., Wood, K., Collier Cameron, A., Donati, J.-F., & Mackay, D. H. 2002b, MNRAS, 336, 1364

Jeffries, R. 1993, MNRAS, 262, 369

Jetsu, L., Pelt, J., & Tuominen, I. 1993, A&A, 278, 449

Korhonen, H., Berdyugina, S. V., Hackman, T., *et al.* 1999, A&A, 346, 101

Kürster, M., Schmitt, J., Cutispoto, G., & Dennerl, K. 1997, A&A, 320, 831

Maggio, A., Pallavicini, R., Reale, F., & Tagliaferri, G. 2000, A&A, 356, 627

McIvor, T., Jardine, M., Collier Cameron, A., Wood, K., & Donati, J.-F. 2004, MNRAS, in press

Sanz-Forcada, J., Maggio, A., & Micela, G. 2003, A&A, 408, 1087

Schrijver, C., Mewe, R., van den Oord, G., & Kaastra, J. 1995, A&A, 302, 438

Schuessler, M. & Solanki, S. K. 1992, A&A, 264, L13

Skumanich, A. 1972, ApJ, 171, 565

Soderblom, D. R., Stauffer, J. R., Hudon, J. D., & Jones, B. F. 1993, ApJS, 85, 315

Solanki, S., Motamen, S., & Keppens, R. 1997, A&A, 324, 943

Strassmeier, K. 1996, in IAU Symposium 176: Stellar Surface Structure, ed. Strassmeier, K.G. & Linsky, J.L. (Kluwer), 289–298

van den Oord, G. 1988, A&A, 205, 167

Weber, E. & Davies, L. 1967, ApJ, 148, 217

Wood, B. E., Müller, H.-R., Zank, G. P., Izmodenov, V. V., & Linsky, J. L. 2004, Advances in Space Research, 34, 66

Discussion

GOPALSWAMY: What is the origin of mass in the prominences? Is prominence material observed between the surface and the critical height where usually prominences reside?

JARDINE: The mass must come from the surface, but we never see it "in transit" (i.e. between the surface and the co-rotation radius). This means that the material must flow from the surface out to $\sim 3R_*$ in less than 12 hours (the rotation period of AB Dor).

YOUSEF: 1) What is the size of a starspot?
2) Have you detected star flares?

JARDINE: 1) The smallest starspots we can detect are about the same size as the largest sunspot.
2) Yes, these stars flare often - so often that you can not directly associate any particular flare with any particular prominence.

SCHMIEDER: 1) Have you a Doppler signal in your Hα observation? 2) Do you observe magnetic field lines (Zeeman effect) at the same time than your Hα lines? Do you cross correlate the result? 3) With same Hα modeling, you could distinguish if the prominence is on the surface or a cloud.

JARDINE: 1) No, in the time when the prominence system is visible, there is no significant evidence of radial velocity.
2) Yes, Hα is observed at the same time as the photospheric lines used for Zeeman-Doppler imaging, but I'm not aware that they have been cross-correlated (perhaps because there are so many magnetic signatures).
3) Yes, the time taken for the features to cross the line profile show that they are not a surface feature.

Coronal and Stellar Mass Ejections
Proceedings IAU Symposium No. 226, 2005
K. P. Dere, J. Wang & Y. Yan, eds.

© 2005 International Astronomical Union
doi:10.1017/S1743921305001122

Disks and winds in Young Solar-Type Stars: the Magnetic Connection

C. Dougados[1], J. Bouvier[1], J. Ferreira[1] and S. Cabrit[2]

[1]Laboratoire d'Astrophysique de Grenoble, BP53, 38041 Grenoble Cédex 9 France

[2] Observatoire de Paris, LERMA, 61 Avenue de l'Observatoire, 75014 Paris, France
email: Catherine.Dougados@obs.ujf-grenoble.fr

Abstract. I discuss in this contribution the accretion and ejection processes occurring in solar-type young stars. Understanding these two important processes, and their link, is one of the major issues in star formation. The magnetic field is thought to play a central role in both extracting the angular momentum from the disk and directing the accretion flow onto the star. I will focus on the well studied T Tauri stars, optically-revealed pre-main sequence stars with ages 1-10 Myrs and mass $\simeq 0.5$ M$_\odot$. In the first part of this contribution, I present the current paradigm for magnetically channeled accretion, where the stellar magnetic field truncates the disk and directs the accretion flow and discuss recent observations, which indicate that this process is non-axisymmetric and time-dependent. I then turn to the study of the supersonic collimated jets observed in young stars. Magneto-hydrodynamic processes are the most likely driving mechanism. I present the main steady and non-steady outflow models, as well as constraints brought by recent high-resolution studies. I finally discuss the origin of time variability in jets.

Keywords. stars:pre-main sequence, stars:mass loss, stars:magnetic fields, accretion disks

1. Introduction

I discuss in this contribution the accretion and ejection processes occurring in solar-type young stars and their possible connection to the magnetic field. I will focus on the well studied T Tauri stars, pre-main sequence sources with typical mass 0.5 M$_\odot$, effective temperature 3500-4000 K (spectral types K-M) and radius 2 R$_\odot$. These young stars, with an age of a few million years, are still contracting towards the main sequence. A sub-class of T Tauri stars, the so-called Classical T Tauri stars or CTTs, are actively accreting matter from a circumstellar disk at a typical rate of $10^{-8} M_\odot yr^{-1}$ (see e.g. Ménard & Bertout 1999 for a review) and are driving supersonic jets. Accretion and ejection signatures are strongly correlated in young stars (Cabrit *et al.* 1990, Hartigan *et al.* 1995). Understanding these two important processes, and their link, is one of the major issue in star formation. The magnetic field is thought to play a central role in both extracting the angular momentum from the disk and directing the accretion flow onto the star. I will focus in this contribution on two main critical, yet unsolved, issues: 1) the nature of the star-disk interaction, 2) the driving mechanism for the supersonic jets.

The exact nature of the star-disk interaction in T Tauri stars holds important clues on: the mechanism responsible for launching the jets, the regulation of stellar angular momentum as well as the physical parameters in the inner disc which control its evolution and the models of planetary formation. I will first describe in section §2 our current knowledge of surface magnetic fields in young stars, then review the arguments that have led to establish that dipolar stellar fields truncate the accretion disk and direct the accretion flow onto the star, in the so-called *magnetospheric accretion* scenario. I will

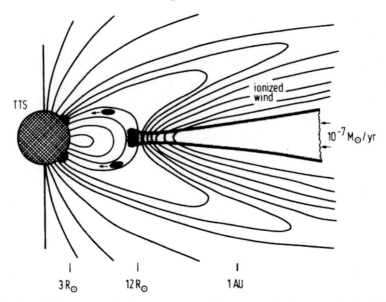

Figure 1. A sketch illustrating the basic concept of magnetospheric accretion in T Tauri stars (from Camenzind 1990).

then show results from recent monitoring studies that challenge the standard steady and axisymmetric picture of magnetically channeled accretion.

I turn in §3 to the origin of the supersonic atomic flows. I first describe their main observational characteristics, review the current magneto-hydrodynamic ejection models, both steady and non-steady, then present recent constraints obtained from high-angular resolution observations of the small scale jets from T Tauri stars. I also briefly discuss the origin of structures (knots) in jets and their relation to time variability. I finally summarize and discuss open questions in section §4.

2. Accretion flow onto the star: the magnetospheric paradigm

2.1. *Stellar magnetic field*

The first evidence for magnetic activity at the surface of young pre-main sequence stars came from their powerful X-ray and centimetric radio emissions (Montmerle 2002, André 1987). T Tauri stars show a strong correlation between X-ray and bolometric luminosities with an average L_X/L_{bol} ratio of 10^{-4}, about 2 orders of magnitude higher than observed in the active Sun. X-ray flares, with properties similar to the ones observed in the Sun, are also detected in T Tauri stars. The light-curves show rapid rise in luminosity followed by an exponential decline over a characteristic time of $\simeq 10$ hours. The spectral properties of these flares indicate thermal bremsstrahlung emission from a hot cooling plasma with *coronal* electronic densities and temperatures. By analogy with the Sun, it is assumed that these flares arise in confined plasma loops. Using purely radiative cooling flare models and assuming equipartion field, loop sizes of a few stellar radii and surface magnetic fields on the order $\simeq 1$ kG are inferred. Indeed, mean field strengths of 2-3 kG have been recently detected in a dozen T Tauri stars from Zeeman broadening measurements of photospheric lines (Guenther *et al.* 1999, Johns-Krull *et al.* 1999). This method does not provide information on the global geometry of the magnetic field. Circular polarization has been detected in the He I 5876 Åemission line (Johns-Krull *et al.* 1999), which is thought to form at the footpoint of the accretion funnel. Magnetic polarity therefore appears uniform in localized regions at the surface of TTs, suggestive of magnetic poles of a dipolar component. In the following, it will be assumed that such a component dominates at large distances from the star. In principle, the detailed geometry of the magnetic field can be reconstructed with Zeeman-Doppler imaging technics, which have been already extensively applied to the study of nearby cool stars (see the contribution by Moira Jardine, these proceedings, for more details). The high resolution spectro-polarimeter ESPADONS being currently installed at the CFH telescope in Hawaii will soon allow to apply such mapping techniques to T Tauri stars.

2.2. *Magnetically channeled accretion*

The strong magnetic fields present at the surface of T Tauri stars are believed to significantly alter the accretion flow from the circumstellar disk to the stellar surface. Uchida (1983) and Uchida & Shibata (1984) first suggested that CTTs magnetic fields disrupt the inner accretion disk, lifting material out of the disk plane towards the stellar magnetic pole. Camenzind (1990) and Königl (1991) showed that the disk is expected to be truncated by the stellar magnetosphere at a few stellar radii, where the stellar magnetic field torque balances the viscous accretion disk torque, for typical mass accretion rates of 10^{-8} M$_\odot$ yr^{-1} and dipolar magnetic fields of 1-3 kG. The basic concept of magnetospheric accretion is illustrated in Figure 1. Disk material is channeled from the disk inner edge to the stellar surface along magnetic field lines. The accretion flow then hits the stellar surface with close to free-falling velocities, giving rise to accretion shocks near the magnetic poles. The successes and limits of the current magnetospheric accretion models to account for the observational properties of CTTs are reviewed for example in Bouvier *et al.* (2004). Magnetospheric accretion models have been in particular successful at reproducing the shape and strength of the main hydrogen emission lines (Hartmann *et al.* 1994; Muzerolle *et al.* 2001), where all previous wind-based or chromospheric models had failed.

One of the main strength of the magnetospheric scenario is that it allows, in principle, to solve for the angular momentum problem in T Tauri stars. Young stars with accretion disks are observed to rotate slowly, at $\simeq 1/10$ of their break-up velocity, although they are expected to accrete large angular momentum material from the disk. For the magnetospheric accretion configuration to be stable on long timescales, the disk truncation radius must lie at or close to the co-rotation radius, ie the location in the disk where the keplerian velocity matches the equatorial velocity of the star (on the order of a few stellar radii in T Tauri stars). Stellar field lines threading the disk outside the co-rotation radius allow to transfer angular momentum outwards with a net braking effect on the star (Collier-Cameron & Campbell 1993; Armitage & Clarke 1996; Shu *et al.* 1994), providing the basis of what is referred to as the disk-locking mechanism.

2.3. *Challenges to the standard model*

2.3.1. *Departure from axisymmetry*

Young stars with active accretion disks show variations in luminosity that can be interpreted with bright (T $\simeq 8000$ K) surface spots covering of order one percent of the stellar surface. These hot spots are identified as emission from shocked regions formed close to the magnetic poles at the footpoint of the accretion funnel (Bouvier & Bertout 1989; Vrba *et al.* 1993). Rotational modulation is also detected in the fluxes and line-of-sight velocities of the main emission lines, formed in the hot plasma filling the accretion columns. These variabilities are interpreted as signatures of a magnetosphere *tilted* with respect to the rotation axis of the star (see e.g. Johns & Basri 1995).

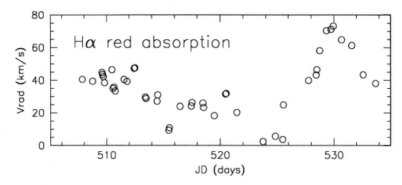

Figure 2. The observed variation over $\simeq 1$ month of the radial velocity of the redshifted absorption component in the H_α line (from Bouvier *et al.* 2003). This variation is interpreted as a signature of magnetospheric inflation.

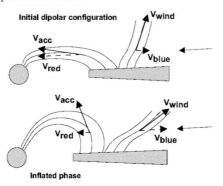

Figure 3. A sketch of the magnetospheric inflation scenario. The arrow on the right side indicates the line of sight to the AA Tau system (from Bouvier *et al.* 2003).

Further insight into the geometry of the star-disk interaction and its temporal evolution can be derived from spectro-photometric monitoring studies. The star AA Tau has been recently followed twice during $\simeq 4$ stellar rotation periods (Bouvier *et al.* 1999, 2003). The disk in this system is seen close to edge-on (inclination to the line-of-sight of i $\simeq$ 75°). This peculiar geometry allows to study in details the critical region where the stellar magnetosphere interacts with the inner accretion disk. Recurrent eclipses of the photospheric light, with a period of 8.2 days, are reported in this star (Bouvier *et al.* 1999). These periodic occultations are interpreted as signatures of a non-axisymmetric warp of the inner disk located close to the co-rotation radius. The puffed inner disk rim periodically obscures the central star as it rotates at the keplerian rate. Such a warp of the inner disk is indeed expected to develop as a result of the interaction of the disk material with an *inclined* dipolar magnetosphere (Terquem & Papaloizou 2000; Lai 1999; Romanova *et al.* 2003).

2.3.2. *A dynamical process?*

Magnetospheric accretion models assume that the stellar magnetosphere truncates the disk close to the co-rotation radius. However, the stellar magnetosphere interacts with the inner disk over a finite radial distance. The footpoints of the field lines anchored into the star and the ones anchored into the disk will therefore not rotate at the same rate.

Recent numerical simulations indicate that such differential rotation induces a substantial distortion of the field lines over a timescale of a few keplerian periods at the inner disk. The response of the magnetic configuration to the differential rotation critically depends upon the magnitude of the magnetic diffusivity in the disk, a parameter which is still poorly constrained. If magnetic diffusivity is large, field lines can diffuse outwards in the disk, leading to magnetic flux expulsion (Bardou & Heyvaerts 1996). In the opposite case, field lines are predicted to expand, open and eventually reconnect, leading to a restoration of the initial configuration (Hayashi *et al.* 1996, Goodson *et al.* 1997, Romanova *et al.* 2003). In such models, magnetospheric inflation cycles are expected to occur, accompanied by violent, episodic outflows and accretion events onto the star as the field lines open and reconnect.

A few observations seem to indicate that magnetospheric accretion in young stars is indeed a dynamical process. Time delays in the appearance of velocity components in the emission line profiles have been interpreted for possible evidence for magnetic field lines being twisted by differential rotation in SU Aur by Oliveira *et al.* (2000). Quasi-periodic X-ray flarings have been also reported for the embedded proto-stellar source YLW 15 (Tsuboi *et al.* 2000). They have been interpreted as recurrent reconnection events in a twisted magnetosphere. Our observations of the AA Tau system also indicate a strongly variable accretion process (Bouvier *et al.* 2003). In AA Tau the line-of-sight velocity of the redshifted absorption component in the H_{alpha} line is seen to slowly vary over typically 3 stellar rotation periods. The magnitude of this velocity component is thought to measure the curvature of the magnetospheric field lines which intersect the disk towards the observer (see Figure 3). In addition, a simultaneous blueshifted absorption feature is observed, strongly correlated in velocity with the redshifted absorption component. We interpret these velocity variations as signatures of *magnetospheric inflation* in AA Tau. As magnetic field lines expand due to differential rotation, the line-of-sight velocity of the accretion (resp. wind) flow decreases (resp. increases) due to projection effects. At the maximum inflation of the magnetosphere (ie when the radial velocity of the redshifted absorption component is minimum), all accretion signatures onto the star disappear, as would be expected due to the unfavorable configuration of field lines to launch disk material. Shortly thereafter, strong accretion resumes and the radial velocity of the redshifted component strongly increases, indicating that the magnetosphere has recovered its initial configuration. Obviously, monitoring campaigns on longer timescales would be required to fully confirm this scenario.

3. Origin of the supersonic jets

3.1. *Properties of the atomic jets*

At all stages where young stars are actively accreting matter (*i.e* for ages $\leqslant$ a few 10^6 yrs), strong mass-loss signatures are also observed. The most spectacular ejection events occur during the earliest stages of star formation for ages $\leqslant 10^5$ yrs, when the central protostar is still embedded in its natal cocoon. Strongly collimated, supersonic ($v_{flow} =$ 300-400 km s^{-1}) atomic flows, the so-called Herbig-Haro (hereafter HH) jets, are detected out to distances of a few parsecs from their driving sources. These flows emit mostly in low excitation lines of weakly ionized atoms (the most prominent being [O I] 6300 Å, [S II] 6716,6731 Å and [N II] 6584 Å), characteristic of excitation temperatures in the range of 8000-10^4 K. At later ages ($\geqslant 10^6$ yrs), when the surrounding envelope dissipates and the central source becomes an optically visible T Tauri star, both accretion and ejection signatures strongly decrease in strength but their correlation persists. With the

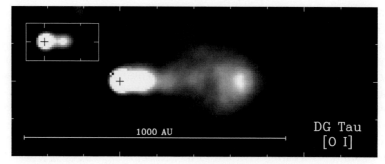

Figure 4. Map in the [O I] 6300 Å line of the small scale jet around the T Tauri star DG Tau. The cross locates the centroid of the continuum emission (star position). The insert at the upper left corner shows the central arc-second of the jet emission. Adapted from Dougados *et al.* (2000).

recent advance of high-angular resolution technics, small scale jets extending out to a few hundred AUs have been clearly identified in the vicinity of T Tauri stars (see Figure 4). These microjets are the smaller scale version of the more massive HH jets emerging from younger sources, clear confirmation that the same physical mechanism is at play through all phases of star formation and that it is intimately connected with the accretion process. Although less powerful, these small scale jets are precious in that they give access to the innermost regions of the flow (central 100 AU) where models predict that most of the collimation and acceleration processes occur.

3.2. *Magneto-centrifugal ejection*

The physical mechanism by which mass is ejected from young stars and collimated into jets remains a fundamental open issue in star formation theory. The strong correlation between ejection and accretion found in PMS stars, with a mass flux ratio as high as 0.1 (Cabrit *et al.* 1990; Hartigan *et al.* 1995) has favored accretion-driven magneto-hydrodynamic (MHD) wind models. A large scale magnetic field, either originating in the disk or in the star, provides a mechanism to extract angular momentum from the star-disc system and collimate the flow on small scales. However, it is not yet established whether the jet originates from the stellar surface, the magnetosphere/disk interface, or a wide range in disk radii; whether it is launched mostly by magneto-centrifugal forces or by a strong thermal pressure gradient in an accretion-heated corona; and whether it is a steady process or not. Answering these questions is crucial not only for the jet phenomenon in itself, but also for models of exoplanet formation, as they have distinct implications on the internal structure, angular momentum transfer, and heating/irradiation processes in the inner regions of protoplanetary disks.

3.3. *Steady wind models*

Three classes of stationary magnetic wind models have been developed: pure stellar winds (e.g. Sauty & Tsinganos 1994), extended disk winds (Blandford & Payne 1982; Ferreira 1997) and winds originating from the star-disk interaction (X-wind: Shu *et al.* 1995, reconnection X-winds: Ferreira *et al.* 2000). These models are distinguished by whether they are powered by accretion (disk winds and X-winds) or by the protostellar rotational energy (stellar winds, Reconnection X-winds) and whether the disk is threaded by a significant magnetic field or not (extended disc winds vs X-wind and stellar winds). In the case of the stellar and disk winds, self-similar solutions have been found. In the X-wind solution of F. Shu & collaborators, mass-loss originates from a single point in the disk (the inner disk radius), which leads to an intrinsically non self-similar behavior. Self-similar steady disk wind solutions require a large-scale dipolar magnetic field threading the disk and that the magnetic torque dominates over the viscous torque. Recent numerical simulations have confirmed that indeed, under such conditions, steady disk ejection can occur (Casse & ferreira 2004).

Observational predictions have been recently computed for steady disk and X-wind models (Garcia *et al.* 2001, Shang *et al.* 2002), which can be confronted to observations of the inner regions of jets from T Tauri stars. Both these models quantitatively reproduce some general global properties of jets such as: ejection to accretion rates, terminal velocities, collimation scales and jet widths. The most stringent constraint to date on the jet launching radius comes from the detection of rotation signatures in a few T Tauri jets (Bacciotti *et al.* 2002; Coffey *et al.* 2004). The ratio of poloidal velocity to specific angular momentum (r×v$_\phi$) measured in the jet, at a given altitude above the disk, provides a model-independent measure of the launching radius (Bacciotti *et al.* 2002; Anderson *et al.* 2003). Observations suggest launching radii in the range 0.1-3 AU, which indicate that a substantial disk wind component is present at the T Tauri phase. Observational predictions strongly depends on the assumed heating mechanism. The properties of the line emission from these jets requires that temperatures on the order of 8000 K be sustained on scales $\geqslant$ 1000 au. This is still an unsolved problem. Detailed studies of line ratios in 2 jet systems have shown that shock heating best reproduce the observations (Lavalley-Fouquet *et al.* 2000, Dougados *et al.* 2000). Time variability may therefore play an important role in structuring the jets. We will come back to this issue later.

3.4. *Non-steady ejection models*

Another class of models have considered the possibility that jets in young stars originate from intrinsically non-steady processes. In the cyclic magnetospheric inflation scenario of Hayashi *et al.* (1996) and Goodson *et al.* (1997), episodic outflows occur as a consequence of reconnection events in the magnetosphere. These CME-like ejections could in particular account for some of the variable outflow signatures observed in young stars. Indeed, strong variability in the high-velocity wings of the line emission profiles are observed in young stars (Solf 1994). An expanding shocked bubble has also been reported in the case of XZ Tau (Coffey *et al.* 2004). These episodic ejection events will however require an external pressure, in excess of the one provided by the interstellar medium, to collimate them.

Time dependent numerical simulations of wind generation from the disk have been performed by numerous authors. In these simulations, the disk is usually treated as a boundary condition (Ustyugova *et al.* 1999, 2000; Ouyed & Pudritz 1997, 1999; Krasnopolsky *et al.* 1999). Simulations taking into account the accretion flow, but still in the ideal MHD case, have been computed by Uchida & Shibata (1985), Shibata & Uchida (1986, 1990), Stone & Norman (1994), Kudoh *et al.* (1998), Kato *et al.* (2002). However, realistic computations with both the accretion flow and diffusivity effects in the disk are still scarce. Kuwabara *et al.* (2000) investigated the effect of magnetic diffusivity in the disk and found that when this parameter is low, mass accretion and jet formation takes place intermittently.

3.5. *Origin of jet structures*

One important morphological property of jets is the fact that the emission is dominated by knots, usually spatially compact close to the central source, but resolved into extended bow-shaped structures at larger distances. Several competing models have been proposed to explain the ubiquitous presence of knots in stellar jets: (1) variability in the jet ejection velocity and/or direction (Raga & Kofman 1992), (2) Kelvin-Helmholtz instabilities at the jet/cloud interface (e.g. Micono *et al.* 2000), (3) internal MHD or current-driven instabilities developing in the jet (e.g. Appl *et al.* 2000; Baty & Keppens 2002). Knot kinematical properties, in particular proper motions, seem to favor the first alternative. Very different variability timescales are involved: from $\Delta T \simeq 10^4$ yrs, corresponding to the bow-shocks observed on distances of typically 0.1 pc in HH jets from embedded sources, to $\Delta T \simeq$ 1-10 yrs for the small scale knots located at the base (d < 1000 AU) of T Tauri jets. The extended bow-shocks seem related to the strong thermal instability events massive disks undergo during the earliest stages of their evolution, leading to powerful outbursts in luminosity of the central system (the so-called Fu Ori events). A significant fraction of the mass of the central protostar could be accreted during this stage. Small scale knots prevalent at the wind base, on the other hand, correspond to moderate shock velocities ($\leqslant$ 50-70 km s^{-1}), *i.e.* moderate perturbations of the underlying average flow. One possible origin for the short timescale variations could be instabilities at the interface region between the stellar magnetosphere and the disk leading to episodic mass ejection events, such as the cyclic inflation scenario presented earlier. Numerical simulations currently fail however to reproduce the typical 1-10 yrs timescales (cf *e.g.* Matt *et al.* 2002): the natural timescale coming out of these simulations correspond to the orbital period at the inner disk edge $\simeq$ 1 week. One alternative possible could be that the short timescale variations are linked to a magnetic cycle similar to the one observed in the Sun, with recurrent polarity reversals induced by the stellar dynamo. We note that although the origin of these small-scale perturbations is still unknown, they should not significantly affect the conclusions reached by comparing the average flow properties with a stationary wind model.

4. Summary and Open questions

There is now strong observational evidence that the accretion flow in T Tauri stars is channeled by the stellar magnetosphere and that the star-disk interaction is a non-axisymmetric and time dependent process. Important questions are still raised: 1) what is the geometry of the stellar magnetosphere ? Does the dipolar component dominate at the distance of $\simeq$ 0.1 au, where interaction with the inner accretion disk is predicted to occur ? We can hope to get some answers to this question in the near-future with the application of Zeeman-Doppler imaging technics to the study of T Tauri stars. Another theoretical question that is raised is: can the magnetosphere sustain the accretion flow at the rates deduced from the observations and how ? What is the stability of this structure and how is the helicity evacuated ? In particular do reconnection events and episodic outflows occur in young stars as a consequence of differential rotation ? Long-term spectro-photometric monitoring studies of a large sample of T Tauri stars are required to tackle this issue.

Large scale magnetic fields seem required to launch and collimate the supersonic atomic jets. Recent detection of rotation rates in jets indicate that an extended disk wind component is present at the T Tauri phase. This implies that the inner disk (out to $\simeq$ 3 AUs typically) is magnetized, which may have strong implications for planetary formation and migration models (Terquem 2003). The origin of this large scale magnetic field, whether advected from the interstellar medium or dynamo, is still an open question. The existence of an additional (unsteady?) wind component originating from the star-disk interface is also still under debate. A related unsolved issue is the origin of the outflow variability observed on short timescales (1-10 yrs) and its connection to magnetic processes. Finally, how does the transition occur between these massive young winds and the solar-type winds observed on the main sequence ?

Acknowledgements

I would like to thank the organizers for this very pleasant and instructive meeting, that allowed me to get introduced to the world of coronal mass ejections as well as get a glimpse of Beijing. I warmly thank them for their hospitality and the financial support provided.

References

Anderson, J.M., Li, Z.-Y., Krasnopolsky, R., & Blandford, R.D. 2003 ApJ 590, L107

André, P. 1987, in: *Protostars and Molecular Clouds*, eds. T. Montmerle & C. Bertout, p. 143

Appl, S., Lery, T., & Baty, H. 2000, A&A, 355, 818

Armitage, P.J. & Clarke, C.J. 1996, MNRAS, 280, 458

Bacciotti, F., Ray, T., Mundt, R., Eislöffel, J., & Solf, J. 2002, ApJ 576, 222

Bardou, A. & Heyvaerts, J. 1996, A&A, 307, 1009

Baty, H. & Keppens, R. 2002, ApJ 580, 800

Bouvier, J. & Bertout, C. 1989, A&A, 211, 99

Blandford, R.D. & Payne D.G. 1982, MNRAS 199, 883

Bouvier, J., Chelli, A., Allain, S., *et al.* 1999, A&A, 349, 619

Bouvier, J., Grankin, K., Alencar, S., *et al.* 2003, A&A, 409, 16

Bouvier, J., Dougados, C., & Alencar, S. 2004 Ap&SS, 292, 6599

Cabrit, S., Edwards, S., Strom, S.E., & Strom, K.M. 1990, ApJ, 354, 687

Camenzind, M. 1990, Reviews of Modern Astronomy, 3, 234

Coffey, D., Bacciotti, F., Woitas, J., Ray, T.P., & Eislffel, J. 2004, ApJ 604, 758

Collier Cameron, A. & Campbell, C.G. 1993, A&A, 274, 309

Casse, F. & Ferreira, J. 2004, ApJ 601, L139

Dougados, C., Cabrit, S., Lavalley, C., & Ménard, F. 2000, A&A 357, L61

Ferreira, J. 1997, A&A 319, 340

Ferreira, J., Pelletier, G., & Appl, S. 2000, MNRAS 312, 387

Garcia, P.J.V., Cabrit, S., Ferreira, J., & Binette, L. 2001, A&A 377, 609

Goodson, A.P., Winglee, R.M., & Boehm, K. 1997, ApJ, 489, 199

Guenther, E.W., Lehmann, H., Emerson, J.P., & Staude, J. 1999, A&A, 341, 768

Hartigan, P., Edwards, S., & Ghandour, L. 1995, ApJ, 452, 736

Hartmann, L., Hewett, R., & Calvet, N. 1994, ApJ, 426, 669

Hayashi, M.R., Shibata, K., & Matsumoto, R. 1996, ApJ, 468, L37

Johns, C.M. & Basri, G. 1995a, ApJ, 449, 341

Johns-Krull, C.M., Valenti, J.A., & Koresko, C. 1999, ApJ, 516, 900

Kato, S.X., Kudoh, T., & Shibata, K. 2002 ApJ, 565, 1035

Königl, A. 1991, ApJ, 370, L39

Krasnopolsky R., Li, Z.Y., & Blandford, R. 1999, ApJ 526, 631

Kudoh, T., Matsumoto, R., & Shibata, K. 1998, ApJ 508, 186

Kuwabara, T., Shibata, K., Kudoh, T., & Matsumoto, R. 2000, PASJ 52, 1109

Lavalley-Fouquet, C., Cabrit, S., & Dougados, C. 2000, A&A 356, L41

Matt, S., Goodson, A.P., Winglee, R.M., *et al.* 2002, Ap&J, 574, 232

Ménard, F. & Bertout, C. 1999, NATO ASIC Proc. 540: The Origin of Stars and Planetary Systems, 341

Micono, M., Bodo, G., Massaglia, S., Rossi, P., Ferrari, S., & Rosner, R. 2000, A&A 360, 795

Montmerle, T. 2002, EAS Publications Series, Volume 3, Proceedings of *Star Formation and the Physics of Young Stars*, held 18-22 September, 2000 in Aussois France. Edited by J. Bouvier and J.-P. Zahn. EDP Sciences, p 85.

Muzerolle, J., Hartmann, L., & Calvet, N. 2001, ApJ, 550, 944

Oliveira, J.M., Foing, B.H., van Loon, J.T., & Unruh, Y.C. 2000, A&A, 362, 615

Ouyed, R. & Pudritz, R.E. 1997, ApJ 484, 794

Ouyed, R. & Pudritz, R.E. 1999, MNRAS 309, 233

Raga, A. & Kofman, L. 1992, ApJ 386, 222

Sauty, C. & Tsinganos, K. 1994, A&A 287, 893

Shang, H., Glassgold, A.E., Shu, F.H., & Lizano, S. 2002, ApJ 564, 853

Shibata, K. & Uchida, Y. 1986, PASJ 38, 631

Shibata, K. & Uchida, Y. 1990, PASJ 42, 39

Stone, J.M. & Norman, M.L. 1994 ApJ 433, 746

Shu, F., Najita, J., Ostriker, E., Wilkin, F., Ruden, S., & Lizano, S. 1994, ApJ, 429, 781

Shu, F.H., Najita, J., Ostriker, E.C., & Shang, H. 1995, ApJ 455, L155

Solf, J. 1994 in ASP Conf. Ser., Vol 57, *Stellar and Circumstellar Astrophysics*, eds. G. Wallerstein & A. Noriega-Crespo, p 22

Terquem, C. 2003, MNRAS 341, 1157

Terquem, C. & Papaloizou, J.C.B. 2000, A&A, 360, 1031

Tsuboi, Y., Imanishi, K., Koyama, K., *et al.* 2000, ApJ 532, 1089

Romanova, M.M., Ustyugova, G.V., Koldoba, A.V., Wick J.W., & Lovelace, R.V.E. 2003, ApJ, 595, 1009

Uchida, Y. 1983, in *Activity in Red-Dwarf Stars*, Reidel, Dordrecht, P.B Byrne & M. Rodono Eds., p. 625.

Uchida, Y. & Shibata, K. 1984, PASJ 36, 105

Uchida, Y. & Shibata, K. 1985, PASJ 37, 515

Ustyugova, G.V., Koldoba, A.V., Romanova, M.M., Chechetkin, V.M., & Lovelace, R.V.E. 1999, ApJ 516, 221

Ustyugova G.V., Lovelace, R.V.E., Romanova, M.M., Li, H., & Colgate, S.A. 2000, ApJ 541, L21

Vrba, F.J., Chugainov, P.F., Weaver, W.B., & Stauffer, J.S. 1993, AJ, 106, 1608

Discussion

SHIBATA: − You mentioned that the first resistive MHD simulations of jets from a disk has been done by Casse, but this is not correct. Actually, Kawabara *et al.* (2002 PASJ, v52, 1109) was the first. In relation to this, you mentioned that the steady solution was obtained by a resistive model, while the ideal MHD model does not lead to steady state if a disk is included, and gave an impression that the resistive steady model is better. But we should remember that a resistive steady model is very idealized and probably very unrealistic. This is because magnetorotational instability occurs in the disk to produce time variability, turbulence, and effective viscosity. Actual disk and associated ejection of jets would be very dynamic and time dependent like solar corona. You showed interesting Hα observations of star-disk interaction by Bouvier *et al.* (2003). Does it show only accretion? Is there a possibility that mass ejections are associated?

DOUGADOS: The radial velocity component I discussed is associated to the accretion process. However there is also a blueshifted radial velocity component anti-correlated with the redshifted radial velocity and which we believe traces the base of the wind. See Bouvier *et al.* (2003) for more details.

Coronal and Stellar Mass Ejections
Proceedings IAU Symposium No. 226, 2005
K. P. Dere, J. Wang & Y. Yan, eds.

© 2005 International Astronomical Union
doi:10.1017/S1743921305001134

Plage and flare Activity of the RS CVn-type Star UX Arietis during 2001-2002

Sheng-hong Gu †

National Astronomical Observatories/Yunnan Observatory, Chinese Academy of Sciences,
Kunming, China
e-mail: shenghonggu@ynao.ac.cn

Abstract. The very active RS CVn-type star UX Ari was observed using high-resolution echelle spectrograph attached to the 2.16m telescope of Xinglong station in Nov.-Dec. 2001 and Dec. 2002. By means of synthetic spectral subtraction method, the information about chromospheric activity of the system was obtained through several chromospheric activity indicators HeI D_3, NaI D_1D_2, H_α, and CaII IRT lines. Based on the analysis for these activity indicators, we found that the chromospheric activity of UX Ari showed obvious orbital modulation phenomenon, and the favorite active longitudes were around the quadratures of the binary system. During the two observing runs, hot plage and very strong optical flare events were detected, which were always happened around the favorite active longitudes of the system. Moreover, they were linked with the photospheric starspots in spatial structure, and appeared just above the main starspots.

Keywords. binaries: spectroscopic, stars: chromospheres, stars: activity, stars: flare

1. Introduction

UX Ari is one of the most active RS CVn-type stars, which shows significant photometric variability, long-term starspot activity (Aarum & Henry 2003; Aarum *et al.* 1999; Gu *et al.* 2003; Gu *et al.* 2004); strong H_α emission above continuum spectrum, CaII H&K, IRT core emission, and optical flares (Aarum & Engvold 2003; Gu *et al.* 2002; Montes *et al.* 1996, 2000). It is a triple-lined spectroscopic triple system with spectral type G5V+K0IV+K5V and orbital period $6^d.4378553$ for the binary (primary and secondary) (Duemmler & Aarum 2001), and its observational spectrum is very complicated. The previous research demonstrates that the chromospheric activity of this system is mainly associated with the cooler secondary component. In order to investigate the scale and evolution of the solar-like activity on UX Ari, we began a long-term monitoring project for it by means of high-resolution spectroscopy in 2000. Here we present the preliminary result about its chromospheric activity based on several chromospheric activity indicators HeI D_3, NaI D_1D_2, H_α, and CaII IRT lines in 2001-2002.

2. Observation and data reduction

The new observations were made with Coude echelle spectrograph (Jiang 1996; Zhao & Li 2001) of the 2.16m telescope at Xinglong station of NAOC during two observing runs: Nov. 23-Dec. 1 2001 and Dec. 13-17 2002. The spectral resolution is about 36000. A 1k×1k Tektronix CCD detector was used in the observations. The reciprocal dispersions were 0.081 Å/pixel for HeI and NaI, 0.091 Å/pixel for H_α, and 0.120 Å/pixel for CaII line spectral regions, respectively. The signal to noise ratio (S/N) was more than 100 for most of the observations. All observed spectral images were reduced by using IRAF

† Present address: P.O. Box 110, Kunming, Yunnan Province, China.

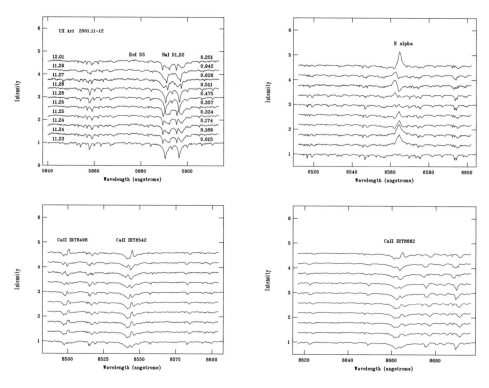

Figure 1. The chromospheric activity indicators in Nov.-Dec. 2001.

package according to the standard fashion. Several chromospheric activity indicators in our observations are displayed in Figure 1 and 2.

3. Spectral synthesis

Because UX Ari is a triple system, the spectra of its components are mixed together in the observed spectra. Therefore, it is hard to measure chromospheric emission of the active component directly. In order to derive the pure chromospheric emission information, we use synthesized spectral subtraction technique (Barden 1985; Montes *et al.* 1995) to do data analysis. In this method, the synthesized spectrum for the system is constructed from artificially rotationally broadened, radial-velocity shifted, and weighted spectra of three inactive stars with the same (similar) spectral-type and luminosity class as three components of the system. Then the excess chromospheric emission can be derived by removing the synthesized spectrum from the observed one. In our observing runs, we observed three inactive template stars HR3309 (G5V), HR3351 (K0IV) and HR753 (K3V) for the primary, secondary and tertiary, respectively. The examples for spectral subtraction technique are shown in Figure 3.

4. Orbital modulation, plage and flare

From Figure 1 and 2, it can be found that the chromospheric activity indicators vary following orbital phase, and the behavior of each indicator is consistent to the others. This suggests that there is an orbital modulation phenomenon for the chromospheric activity of UX Ari. The measured equivalent width EWs of H_α excess emission in our two

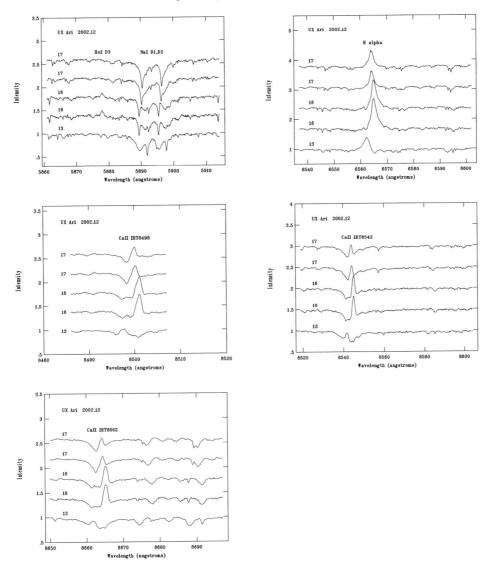

Figure 2. The chromospheric activity indicators in Dec. 2002.

observing runs have been plotted vs the orbital phase in Figure 4. In the first observing run, the spectra at orbital phases 0.169, 0.174, and 0.251 show stronger excess emission for all indicators. In these phases, the HeI D_3 line shows weaker emission, this means that we detected a hot plage-like event in that run (Berdyugina *et al.* 1999). In the second observing run, the spectra at orbital phases 0.280 and 0.292 exhibit very strong excess emission in all indicators. In these phases, the H_α and CaII IRT emission lines show a dramatic increase, the HeI D_3 line changes to emission, and the NaI D_1D_2 lines show a stronger fill-in. This means that a flare-like event was detected in the second run. By means of the calibration of Hall (1996) as the relation of color index (V-R), we convert the excess emission EW of H_α line into luminosity, and derive the energy released in the flare, 4.7×10^{31} erg/s. Compared with the energy release in previous flare detected

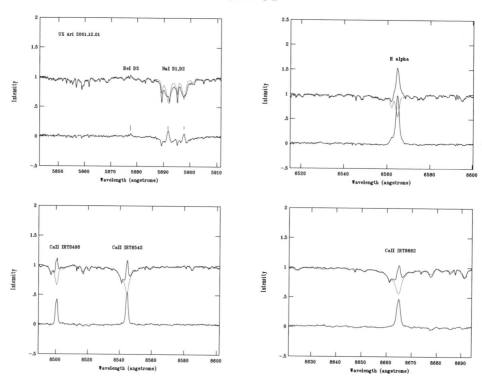

Figure 3. The examples for spectral subtraction technique.

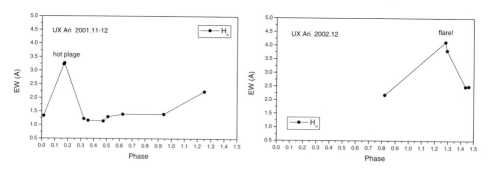

Figure 4. The EW of the excess emission vs orbital phase for H$_\alpha$ line.

(1.7×10^{31} erg/s (Montes *et al.* 1996) and 2.1×10^{31} erg/s (Gu *et al.* 2002)), it can be found that this is the strongest flare detected for UX Ari in the optical wavelength.

5. Correlation between chromospheric and photospheric activity

In our first observing run, the orbital phase coverage is good so that we can reconstruct the starspot pattern of the photospheric surface (Gu *et al.* 2004). This permits us to make a comparison between chromospheric and photospheric activity for UX Ari. In this run, the chromospheric active region is around phase 0.2. Based on the Doppler image of UX Ari (Gu *et al.* 2004) in this observing run, two main starspot active regions locate around phase 0.25. Thus, it is obvious that there is a spatial connection between the main starspots and the hot plage. This demonstrates that the hot plage is associated with the starspot active regions in spatial structure.

6. Summary

From our new observations and the previous observations for UX Ari, we can see that UX Ari is a very active system, and the plage-like and flare-like events are frequent. There is a obvious orbital modulation phenomenon for the chromospheric activity of the system, and the favorite active longitudes are around two quadratures of the system (i.e. orbital phase 0.25 and 0.75). The flare just appears around the position where the hot plage appeared before. The chromospheric active region is associated with the main starspots in spatial structure. In order to understand the behavior of the magnetic activity of UX Ari, it is necessary to do further more observations for the system.

Acknowledgements

I would like to thank Dr. Jian-yan Wei and Dr. Xiao-jun Jiang for supporting our research project in 2.16m telescope of Xinglong station. This work is supported by NSFC under grant No. 10373023.

References

Aarum, V. *et al.* 1999, *Proceedings of "Astrophysics with the Nordic Optical Telescope"*, Karttunen, H. & Piirola, V. eds., University of Turku, 222

Aarum, V. & Engvold, O. 2003, *A&A* 402, 1043

Aarum, V. & Henry, G.W. 2003, *A&A* 402, 1033

Barden, S.C. 1985, *ApJ* 295, 162

Berdyugina, S. *et al.* 1999, *A&A* 349, 863

Duemmler, R. & Aarum, V. 2001, *A&A* 370, 974

Gu, S.-h. *et al.* 2002, *A&A* 388, 889

Gu, S.-h. *et al.* 2003, *IAU Symp. 219*, in press

Gu, S.-h. *et al.* 2004, *Proceeding of 13th Cambridge Workshop "Cool Stars, Stellar Systems, and the Sun"*, in press

Hall, J.C. 1996, *PASP* 108, 313

Jiang, S.-y. 1996, in *Ground-based Astronomy in Asia*, ed. Kaifu, N., 335

Montes, D. *et al.* 1995, *A&A* 294, 165

Montes, D. *et al.* 1996, *A&A* 310, L29

Montes, D. *et al.* 2000, *A&AS* 146, 103

Zhao, G. & Li, H.-b. 2001, *CJAA* 1, 555

Discussion

CHENG FANG: What is the duration of the plage and sunspots? Did you find several periodical behavior of the emission of the plage?

GU: 1) Because the time resolution is lower in our observation, we can not estimate the duration of the plages and starspots properly. 2) Due to the above reason, we also can not determine the periodical behavior of the plage.

DERE: Does this star show X-ray signatures of flare activity?

GU: I don't notice the behavior of this star in X-ray wavelength, but for most RS-CVn stars, there is very strong X-ray emission when the flare event is occurring. So, I guess this star should show stronger X-ray emission during the flare event.

Coronal and Stellar Mass Ejections
Proceedings IAU Symposium No. 226, 2005
K. P. Dere, J. Wang & Y. Yan, eds.

A Two-Temperature Model for LBVs

J. H. Guo[1,2], Y. Li[1], and H. G. Shan[1]

[1]National Astronomical Observatories/Yunnan Observatory, Chinese Academy of Sciences,
P.O. Box 110, Kunming 650011, China
email: guojh@ynao.ac.cn

[2]Graduate School of Chinese Academy of Sciences

Abstract. The continuum energy distributions of R127 and R110 in the outburst phase are fitted by use of a optically envelope model. Both stars show two peaks in the continuum energy distributions in which one lies in the short-wavelength range (near 1250Å) and the other in the optical band. We suggest that the fluxes in the UV and optical bands may have different origins: the UV flux comes from the central star and the optical flux comes from the expanded optically envelope. We construct such a model for LBVs with the use of two LTE atmosphere models with different temperatures, and find it to be in satisfactory agreement with the observed spectral energy distributions of R127 and R110.

Keywords. stars: mass loss, variables: other

1. Introduction

Luminous blue variables (LBVs) are a separate class of massive stars. They are very evolved post main sequence stars or post red super giants. Their most distinct character is irregular variations of brightness on different timescales. The amplitudes of such variations vary from very small to a few magnitudes, when the timescales vary from the order of days to a few hundred years (van Genderen 2001). A general discussion of the LBV phenomenon can be found in Nota & Lamers (1997). Recent investigations focus on the instability of the moderate variations, which is characterized by visual magnitude changes of 1 to 2 mag and timescales from years to decades. Whereas the normal LBVs at the quiescent phase occupy a wide temperature range, the eruptive LBVs are almost in a vertical strip on the HR diagram (Wolf 1989). Their continuum radiation shows the spectra of hot super giants at the minimum or quiescent phase. During the visual maximum period, the atmosphere of an LBV resembles a much cooler supergiant of spectral type A or F (Humphreys & Davidson 1994).

Investigations on the continuum spectra of the LBVs are rare but desirable, because they may indicate the basic configuration and physical conditions of the LBVs. Stahl *et al.* (1983, 1990), Shore *et al.* (1996), and Szeifert *et al.* (1993) have studied R127, R110, R40, and AG Car. Their results hint a general feature, namely the brightness variations in the UV and optical bands are anti-correlated. However, an important feature on the continuum spectra of R127 and R110, namely the over-luminous component in the short-wavelength range (SWP, 1150-1980Å) at the maximum phase, has not provoked enough interest of investigators up to now. Figure 1 shows the continuum energy distributions of R127 and R110 during the maximum phase. Evidently, the continuum show two peaks. On the other hand, the non-spherically symmetric wind(mass loss) of some LBVs have been identified by many astronomers(Nota *et al.*, 1992; Leitherer *et al.*, 1994; Schulte-Ladbeck *et al.*, 1993). It is usually assumed that the deposition of the matter can form optically thick regions (Davidson 1987), but optically thin regions can also be formed simultaneously due to the anisotropy of the stellar wind or mass ejection. This shows

Table 1. The photometric data of R127

Data	System	U	B	V	J	H	K	L	State	Ref.
Dec. 1969	BVRI			11.16					min	Mendoza (1970)
Feb. 16, 1983	UBV	9.21	10.23	10.13	9.76	9.58	9.4	8.850	min*	Stahl *et al.* (1983)
		u	v	b	y					
Nov. 30, 1988	uvby	9.82	9.36	9.10	8.84				max	LTPV

* a relative minimum during the interval of 1983-1992

Table 2. The photometric data of R110

Data	System	U	B	V	R	I	J	H	K	State	Ref.
Jan. 1989	UBV	10.26	10.34	9.99		9.77	9.53	9.28	9.19	9.07	max	Stahl *et al.* (1990)
		u	v	b	y							
Sep. 26, 1990	uvby	12.31	10.90	10.27	9.91						max	LTPV

that the asymmetric mass loss can result in a coexistence of optically thin and thick regions around the star, which allows the radiation from the star to be seen by observers. We suggest, that the over-luminous continuum in the SWP range could be a part of the radiation from the central star. The radiation from the optically thick regions induced by the asymmetric mass loss concentrates on the optical band, because these regions are far from the central star and therefore have lower temperatures. The combination of both radiation components may reproduce the observed continuum energy distributions. So, we decided to fit the observed continuum spectra of the LBVs during the maximum phase with a model of two stellar atmospheres, in which one denotes the central star and the other represents the optically thick regions formed around the central star by the anisotropic mass loss.

2. Observations

The photometry of R127 (=HDE 269858) with the UBV photometer has already been published and summarized by Stahl *et al.* (1983) and Spoon *et al.* (1994). In addition, a long-term project to monitor slow variations (LTPV) originated by Sterken (1983) also included R127 and R110, which was done in the Strömgren uvby-system. In Table 1 we summarize these observational data for R127. The photometric history of R110 is summarized in Table 2.

All spectra discussed in this paper have been obtained through the database of the IUE satellite. These data are consisted of short-wavelength part (SWP, 1150-1980Å) and long-wavelength part (LWP and LWR, 1850-3350Å) in low (R=300) and high (R=10 000) resolution modes. All raw data have been reduced by the NEW SPECTRAL IMAGE PROCESSING SYSTEM (ref. IUE NEWSIPS manual; Nichols & Linsky 1996). In Table 3 we collect the observe time, data format, exposure time, and so on.

3. Continuum energy distributions

The ground based photometric data combined with the IUE data were used to constitute the continuum energy distributions. We corrected the observed IUE spectra for the interstellar reddening using the interstellar law of Savage and Mathis (1979) and Nandy *et al.* (1981) for the galactic foreground extinction and the absorption within the LMC. Considering both stars in the LMC, the same value of the interstellar extinction

Table 3. IUE observations of R127 and R110.

Object	Date	Camera	Aperture[a]	Dispersion[b]	Exposure time[min]	Image number	State
R127	1989 Feb.11	LWP	L	L	3	15013	max
	1989 Feb.11	SWP	L	L	15	35533	max
	1983 Mar.02	SWP	L	L	10	19372	min
	1983 Mar.02	LWR	L	L	6	15407	min
R110	1989 Apr.25	LWP	L	L	30	15404	max
	1989 Apr.25	SWP	L	L	75	36088	max

[a] Aperture: L=Large [b] Dispersion: L=Low

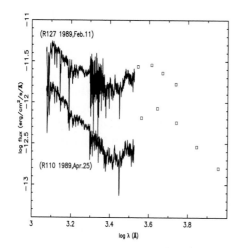

Figure 1. The de-reddened continuum energy distributions of R127 and R110 in the eruptive phase.

Figure 2. The spectral energy distributions of R127 at the minimum and maximum states. The crosses denote the photometric data at the maximum state, and the diamonds are at the minimum state.

was used. The standard value E(B-V)=0.05 was used for the galactic foreground extinction, and the value E(B-V)=0.10 was used for the LMC. The photometric data were de-reddened using the reddening law of Code *et al.* (1976) and Leitherer and Wolf (1984) for the visual and infrared bands, respectively. For the Strömgren uvby measurements the absolute calibration was obtained from Gray (1998).

3.1. *R127*

A minimum state spectral energy distribution of R127 on March 2, 1983 is shown in Fig. 2. We fitted the continuum energy distribution with an LTE atmosphere model of $T_{eff} = 17000K$, $R = 135R_\odot$, and $\log g = 2.5$, which is almost identical with the result of Stahl *et al.* (1983). Subsequently, R127 entered a period of outburst till 1989 during which the visual brightness increased slowly with the decrease of the UV flux. The continuum energy distributions of the two states (cf. Fig. 2) are similar in the short wavelength part, which could reflect the correlation between the quiescent and outburst states. Evidently, any atmosphere model cannot fit the essential profile of the continuum in the outburst phase due to the existence of two peaks. The first peak, which is located at about 1250Å, reflects the feature of a B-type star (R127 itself). The second peak emerges around u band in the Strömgren system as a normal A-type star (the optically thick regions in the wind). We have already known that the flux in the SWP at the

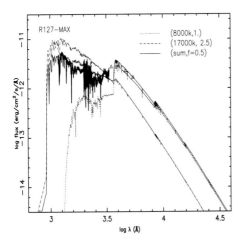

Figure 3. The model of the circumstellar envelope for R127 at the outburst phase. The continuum energy distribution is fitted by two atmosphere models of $T_{eff} = 8000$ K (dotted line) and $T_{eff} = 17000$K (dashed-line), and $f = 0.5$. The absolute fluxes at the effective wavelengths of the Strömgren uvby filters are given by triangles.

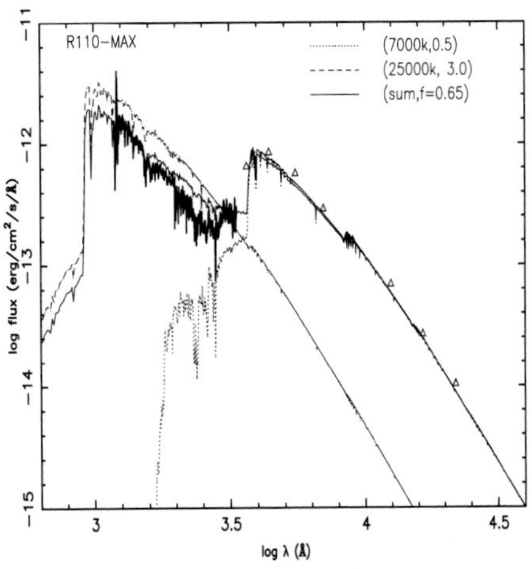

Figure 4. The model of R110 at the maximum phase. The continuum energy distribution is fitted by two atmosphere models of $T_{eff} = 7000$K (dotted line) and $T_{eff} = 25000$K (dashed line), and $f = 0.65$.

minimum state is higher than that at the maximum state, thus the overall continuum energy distribution at the maximum state can be constructed with a permeating factor f, which measures how much light from the central star can permeate the optically thin regions in the wind and finally arrive at observers. Simply, the model in the quiescent state will be multiplied by the factor f (less than 1) to account for the observed SWP continuum in the outburst phase. Finally, the model at the outburst phase is constructed by adding the model of an A-type star to the modified minimum model of the central star (multiplied by factor f). We fitted the profile of the optical spectrum at the outburst phase with a model of $T_{eff} = 8000$K, $R = 485R_{\odot}$ and $\log g = 1$, which represented the average physical parameters of the optically thick regions. In our numerical experiments, we found that the permeating factor $f = 0.5$ was appropriate to recover most characters of the observations of R127. The fitted result is shown in Fig. 3, which shows satisfactory agreement with the observations.

3.2. *R110*

R110 increased its visual brightness by 0.5 mag from 1980 to 1989 (Stahl *et al.* 1990). However, the physical parameters in the minimum phase are not known clearly due to the rare observations. Around 1989 R110 attained the maximum phase during which the continuum energy distribution appeared like R127 (cf. Fig. 4). Therefore, we fitted its continuum spectrum with the same technique for R127. Over the visual spectral rang a model of $T_{eff} = 7000$K, $R = 350R_{\odot}$ and $\log g = 0.5$ was applied, which was comparable with the parameters derived by Stahl *et al.* (1990). For the whole continuum including the UV and visual bands the second component of $T_{eff} = 25000$K, $R = 27R_{\odot}$, and $\log g = 3.0$ was adopted, and the permeating factor f was chosen to be 0.65. The result is shown in Fig. 4, which is satisfactory consistent with the observations.

4. Conclusions

In this paper we discuss the continuum energy distributions of R127 and R110. We point out that the over-luminous continuum in the SWP range is the contribution from the central star itself due to the asymmetry of the external matter of the star. The light emitted from the central star leaks out through the optically thin regions of the non-homogeneous envelope. Based on this detailed research we conclude that the continuum spectra of both stars at the outburst phase are composed of the radiation fluxes of the central star and the optically thick matter around the star. They make up the observed UV and optical continuum spectra, respectively.

Acknowledgements

We are grateful to P.S. Chen and S.H. Gu for the data process, and T. Szeifert for kind advice to transform the uvby magnitudes into absolute fluxes.

References

Code, A.D., Davis, J., Bless, R.C., & Brown., R.H. 1976, *ApJ* 203, 417

Davidson, K. 1987, *ApJ* 317, 760

van Genderen, A.M. 2001, *A&A* 366, 508

Gray, R.O. 1998, *AJ* 116, 482

Humphreys, R.M. & Davidson, K. 1994, *PASP* 106, 704

Leitherer, C. & Wolf, B. 1984, *A&A* 132, 151

Leitherer, C., Allen, R., Altner, B., Damineli, A., Drissen, L., Idiart, T., Lupie, O., Nota, A., Robert, C., Schmutz, W., & Shore, S.N. 1994, *ApJ* 428, 292

Nandy, K., Morgan, D.H., Willis, A.J., Wilson, R., & Gondhalekar, P.M. 1981, *MNRAS*, 196, 955

Nichols, J.S. & Linsky, J.L. 1996, *AJ* 111, 517

Nota, A., & Lamers, H.J.G.L.M., eds. 1997, Luminous Blue Variables: Massive Stars in Transition, ASP Conference Series, 120

Nota, A., Leitherer, C., Clampin, M., Greenfield, P., & Golinowski, D.A. 1992, *ApJ* 398, 621

Savage, B.D. & Mathis, J.S. 1979, *Ann. Rev. Astron. Astrophys.* 17, 73

Schulte-Ladbeck, R.E., Leitherer, C., Clayton, G.C., Robert, C., Meade, M.R., Drissen, L., Nota, A., & Schmutz, W. 1993, *ApJ* 407, 723

Shore, S.N., Altner, B., & Waxin, I. 1996, *AJ* 112, 2744

Spoon, H.W.W., de Koter, A., Sterken, C., Lamers, H.J.G.L.M., & Stahl, O. 1994, *A&AS* 106, 141

Stahl, O., Wolf, B., Klare, G., Cassatella, A., Krautter, J., Persi, P., & Ferrari-Toniolo, M. 1983, *A&A* 127, 49

Stahl, O., Wolf, B., Klare, G., Jüttner, A., & Cassatella, A. 1990, *A&A* 228, 379

Sterken, C. 1983, ESO *The Messenger*, No. 33, 10

Szeifert, T., Stahl, O., Wolf, B., Zickgraf, F.J., Bouchet, P., & Klare, G. 1993, *A&A* 280, 508

Wolf, B. 1989, *A&A* 217, 87

Coronal and Stellar Mass Ejections
Proceedings IAU Symposium No. 226, 2005
K. P. Dere, J. Wang & Y. Yan, eds.

Prominence Mapping of the RS CVn system HR 1099

P. Petit[1], J.-F Donati[2], M. Jardine[3] and A. Collier Cameron[3]

[1]Max-Planck-Institut für Sonnensystemforschung, Max-Planck-Str. 2,
37191 Katlenburg-Lindau, Germany, email: petit@linmpi.mpg.de

[2]Observatoire Midi-Pyrénées, 14 avenue Edouard Belin, 31400 Toulouse, France, email:
donati@ast.obs-mip.fr

[3]School of Physics and Astronomy, University of Saint Andrews, Saint Andrews KY16 9SS,
UK, email: mmj@st-and.ac.uk, acc4@st-and.ac.uk

Abstract. We investigate temporal fluctuations in the Hα emission profiles of the RS CVn system HR 1099 from a monitoring using the MuSiCoS spectropolarimeter (Observatoire du Pic du Midi, France) in 2001, between December 01 and December 18. Part of the observed emission fluctuations is consistent with rotational modulation, which we interpret as the spectral signature of a dense and complex prominence system trapped in the magnetosphere of HR 1099 and forced to co-rotate with the binary system. The distribution of emitting material is mapped by means of Doppler tomography. We discuss the evolution of prominences over the observing window.

Keywords. stars: flare, stars: activity

1. Observations

A total of 164 high-resolution spectra of the RS CVn system HR 1099 was collected in 2001, between Dec. 01 and Dec. 18, with the MuSiCoS spectropolarimeter (Donati *et al.* 1999) at Observatoire du Pic du Midi (France). The Hα line of the system is always in emission with a peaked, asymmetric, fast-changing profile revealing frequent flaring events (Foing *et al.* 1996). On sufficiently small timescales, and as long as the system is observed during a quiescent state, most of the structures seen in Hα repeat over several rotation cycles. It is therefore possible to reconstruct the spatial distribution of the emitting material by Doppler tomography. We use here the maximum entropy code developed by Marsh & Horne (1988) and Donati *et al.* (2000).

2. Doppler tomography of Hα emission

Fig. 1 shows the distribution of Hα emitting material in the velocity space. Two close-by epochs are considered, each one providing a good rotational coverage of the system. In the case of co-rotating structures, velocity maps represent a straightforward image of spatial distribution. Most of the co-rotating clouds are reconstructed around the extremely active primary sub-giant. Prominences are mostly concentrated as a single high-contrast structure in the first epoch, while their distribution is more diffuse in the second map.

A comparison between observed and synthetic line profiles shows that a large part of the profile fluctuations are correctly reproduced by the imaging procedure. Fast, non-rotational changes in the profile shape (as observed in particular at epoch 2) have only a marginal impact on the final map.

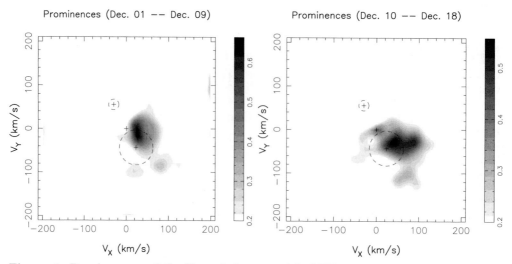

Figure 1. Doppler maps of the Hα-emitting material of HR 1099 in the velocity space, as reconstructed from data sets collected between Dec. 01 and Dec. 09 (left) and between Dec. 10 and Dec. 18 (right). The photosphere of both components of the binary system are represented by dashed circles of radius v.sini. The K1 primary subgiant shows up on bottom-right. The G5 secondary dwarf is on top-left. The center of mass of the system is shown as a cross at null velocity.

3. Future work

The spatial distribution of prominences provides valuable information about the magnetic geometry of the corona of HR 1099. A photospheric magnetic map of the primary star was reconstructed from the same data set as that employed for Hα tomography (Petit *et al.*2004) and can be used as boundary conditions to extrapolate the three-dimensional structure of the coronal field (Jardine *et al.* 2002). Simple assumptions on the field geometry of the secondary component can also be derived from the measurements of Donati (1999). The resulting coronal topology will then be tested by comparison with prominence maps.

Nearly simultaneous observations of HR 1099 obtained at the Anglo-Australian Telescope are available and should allow us to enlarge our observing window up to early January of 2002. Preliminary work demonstrates that intrinsic variability is much more important in this data set, to a point where the basic assumption of co-rotation of circumstellar material does not hold, even over short timescales. Further work is needed to extract an adequate sub-set and possibly merge it with MuSiCoS spectra.

References

Donati, J.-F. 1999, MNRAS 302, p. 457

Donati, J.-F., Mengel, M., Carter, B.D., Marsden, S., Collier Cameron, A., & Wichmann, R. 2000, MNRAS 316, p. 699

Foing, B. H., *et al.* 1996, IAU Colloq. 153: Magnetodynamic Phenomena in the Solar Atmosphere - Prototypes of Stellar Magnetic Activity, p. 283

Jardine, M., Collier Cameron, A., & Donati, J.-F. 2002, MNRAS 333, p. 339

Marsh, T. R. & Horne, K. 1988, MNRAS 235, p. 269

Petit, P., *et al.* 2004, MNRAS 348, p. 1175

Author Index